The **GUINNESS** book of

ASTRONOMY

Patrick Moore

Fifth Edition

GUINNESS PUBLISHING

Acknowledgements

My grateful thanks are due to those who have read through various sections of the book, and made many helpful comments. Any errors or omissions remaining are entirely my responsibility! So my thanks to:

Dr Peter Cattermole
Dr Dale Cruikshank
Dr Gilbert Fielder
Bruce Hardie
Dr Eleanor Helin
Michael Hendrie
Professor Garry Hunt
John Isles
Dr John Mason
Dr Paul Murdin
Iain Nicolson
Dr John Rogers
Dr Richard Stephenson
Dr David Whitehouse
Professor Iwan Williams
Professor Sir Arnold Wolfendale.

I am also most grateful to Paul Doherty, for his splendid illustrations, and to Béatrice Frei.

All illustrations by Patrick Moore except where specifically credited.

PATRICK MOORE

Author's Note

In this book, I have retained references to the USSR with respect to past results. Now that the USSR has broken up, future developments come under the headings of the new Commonwealth of Independent States.

Metric Conversion

The current practice of giving lengths in metric units rather than the imperial ones has been followed. To help in avoiding confusion, the following table may be found useful.

centimetres	inches		kilometres		miles
2·54	1	0·39	1·61	1	0·62
5·08	2	0·79	3·22	2	1·24
7·62	3	1·18	4·83	3	1·86
10·16	4	1·58	6·44	4	2·49
12·70	5	1·97	8·05	5	3·11
15·24	6	2·36	9·66	6	3·73
17·78	7	2·76	11·27	7	4·35
20·32	8	3·15	12·88	8	4·97
22·86	9	3·54	14·48	9	5·59
25·40	10	3·94	16·09	10	6·21
50·80	20	7·87	32·19	20	12·43
76·20	30	11·81	48·28	30	18·64
101·6	40	15·75	64·37	40	24·86
127·0	50	19·69	80·47	50	31·07
152·4	60	23·62	96·56	60	37·28
177·8	70	27·56	112·7	70	43·50
203·2	80	31·50	128·7	80	49·71
228·6	90	35·43	144·8	90	55·92
254·0	100	39·37	160·9	100	62·14

Editor: Béatrice Frei
Design: Amanda Sedge
Picture Editing and Research: Image Select International

The Work Copyright © Patrick Moore, 1995

The right of Patrick Moore to be identified as Author of this Work has been asserted in accordance with the Copyright, Design & Patents Act 1988.

First published in 1979 by Guinness Superlatives Ltd
   Second edition 1983
   Third edition 1988 by Guinness Publishing Ltd
   Fourth edition 1992
   Fifth edition 1995
   Reprint 10 9 8 7 6 5 4 3 2 1 0

This publication Copyright © Guinness Publishing Ltd 1995, 33 London Road, Enfield, Middlesex, Great Britain

'GUINNESS' is a registered trade mark of Guinness Publishing Ltd

Typeset in Great Britain by Ace Filmsetting Ltd, Frome, Somerset
Printed and bound in Italy by New Interlitto Italia SpA, Milan

A catalogue record for this book is available from the British Library.

ISBN 0–85112–643–X

CONTENTS

THE SOLAR SYSTEM

THE SUN

DATA ....

Distance from the Earth:
mean: 149 597 893 km (1 astronomical unit)
max: 152 103 000 km
min: 147 104 000 km
Mean parallax: 8 sec·794
Distance from centre of Galaxy: 25 000 light-years
Velocity round centre of Galaxy: 220 km/s
Period of revolution round centre of Galaxy: about 225 000 000 years (= 1 'cosmic year')
Velocity toward solar apex: 19·5 km/s
Apparent diameter:
mean: 32 min 01 sec
max: 32 min 35 sec
min: 31 min 31 sec
Equatorial diameter: 1 392 000 km
Density, water = 1: 1·409
Mass: 2×10^{27} tonnes (99 per cent of the mass of the entire Solar System)

Mass, Earth = 1: 332 946
Volume, Earth = 1: 1 303 600
Surface gravity, Earth = 1: 27·90
Escape velocity: 617·5 km/s
Luminosity: $3·86 \times 10^{26}$ watts
Solar constant: 1370 w.m² (solar radiation per second vertically incident at unit area at 1 a.u. from the Sun)
Mean apparent magnitude: −26·8 (=600 000 times as brilliant as the full moon)
Absolute magnitude: +4·83
Spectrum: G2
Surface temperature: 5500 °C
Core temperature: 15 000 000 °C
Rotation period:
sidereal, mean: 25·380 days
synodic, mean: 27·275 days
Time taken for light from the Sun to reach the Earth: 499·012 sec = 8·3 min

The Sun is by far the nearest star; it is 270 000 times closer than α Centauri. It is therefore the only star which may be studied in detail. It is a normal Main Sequence star. From observations of the movements of the tiny asteroid Icarus, W. Landgraf of Göttingen has calculated that the difference between the polar and equatorial radii of the Sun is 25 km.

✻ *The first known estimate of the distance of the Sun* was made by the Greek philosopher Anaxagoras (500–428 BC). He assumed the Earth to be flat, and gave the Sun's distance as 6500 km (using modern units) with a diameter of over 50 km. A much better estimate was made by Aristarchus of Samos, about 270 BC. His value, derived from observations of the angle between the Sun and the exact half-moon, was approximately 4 800 000 km. Ptolemy (circa AD 150) increased this to 8 000 000 km, but about

AD 1543 Copernicus reverted to only 3 200 000 km. Kepler, in 1618, gave a value of 22 500 000 km.

✻ *The first reasonably accurate estimate of the mean distance of the Sun* was made by G. D. Cassini in 1672. He gave a value of 138 370 000 km. Successive determinations have been as table below:

✻ *The first suggestion of measuring the Sun's distance (astronomical unit) by using transits of Venus* was made by J. Gregory in 1663, and was extended by Edmond Halley in 1678. The method was sound in theory, but was affected by the 'Black Drop' – the apparent effect of Venus drawing a strip of blackness after it has passed on to the Sun's disk, thus making exact timings difficult. (Captain Cook's famous voyage, during which he discovered Australia, was made in order to take the

astronomer Green to a suitable position from which to observe the transit of 1769.) Parallax measurements of the planets and asteroids were more accurate, but Spencer Jones' value as derived from the close approach of Eros in 1931 was too high, and was later revised by Rabe to 149 493 000 km. The modern method – radar – was introduced in the early 1960s in the United States. The present accepted value for the astronomical unit is accurate to a tiny fraction of one per cent.

✻ *The first comments upon the Sun's rotation* were made by Galileo, following his observations of sunspots from 1610. He gave a value of rather less than one month.

✻ *The discovery of the Sun's differential rotation*, i.e. that the Sun does not rotate as a solid body would do, the equatorial rotation being shorter than the polar, was made by Richard Carrington in 1863. In order to help in identifying specific rotations of the Sun, Carrington had introduced a numbering system, beginning with Rotation No 1 on 9 November 1853. Rotation No 1500 began on 19 October 1965. *The first observations of Doppler shifts at opposite limbs due to the solar rotation* were made by H. C. Vogel in 1871.

Synodic rotation periods for features at various heliographic latitudes are as follows:

Lat	Average synodic rotation period, days	Lat	Average synodic rotation period, days
0	24·6	50	29·2
10	24·9	60	30·9
20	25·2	70	32·4
30	25·8	80	33·7
40	27·5	90	34·0

✻ *The first serious attempt to measure the solar constant* was made by Sir John Herschel in 1837–8, using an actinometer (basically a bowl of water; the estimate was made by seeing the rate at which the bowl was heated). He gave a value which is about half the actual figure. The solar constant may be defined as the amount of energy in the form of solar radiation which is normally received on unit area at the top of the Earth's atmosphere; it is roughly equal to the amount of energy reaching ground level on a clear day. The modern value is 1.95 calories per square cm per minute (1370 watts per square metre).

✻ *The first photograph of the Sun* (a Daguerreotype) seems to have been taken by Lerebours, in France, in 1842. However, the first good Daguerreotype was taken by Fizeau and Foucault, also in France, on 2 April 1845, at the request of F. Arago. In 1854 J. B. Reade used a dry collodion plate to

Year	Authority	Method	Distance, km
1672	Cassini	Parallax of Mars	138 370 000
1770	Euler	Transit of Venus, 1769	151 225 000
1771	Lalande	Transit of Venus, 1769	154 198 000
1814	Delambre	Transit of Venus, 1769	153 841 000
1823	Encke	Transits of Venus, 1761 and 1769	153 303 000
1862	Foucault	Velocity of light	147 459 000
1867	Newcomb	Parallax of Mars	148 626 000
1872	Le Verrier	Masses of the planets	148 459 000
1875	Galle	Parallax of asteroid Flora	148 290 000
1877	Airy	Transit of Venus, 1874	150 152 000
1878	Stone	Transit of Venus, 1874	148 125 000
1881	Puiseux	Transit of Venus, 1874	146 475 000
1931	Spencer Jones	Parallax of Eros	149 645 000
1976	various	Radar to Venus	149 597 000
1992	various	Radar	149 597 893

show mottling on the Sun. *The first systematic series of solar photographs* was taken at Kew (London) from 1858 to 1872, using equipment designed by the British astronomer Warren de la Rue. Nowadays the Sun is photographed daily at many observatories throughout the world.

✳ *The world's largest solar telescope* is the McMath telescope at Kitt Peak, Arizona. It looks like a large, white, inclined tunnel. At the top is an upper mirror or heliostat, 203 cm in diameter; it can be rotated, and sends the sunlight down the tunnel in a fixed direction. At the bottom of the 183 m tunnel is a 152 cm mirror, which reflects the rays back up the tunnel on to the half-way stage, where a flat mirror sends the rays down through a hole into the solar laboratory, where the analyses are carried out. This means that the heavy equipment in the solar laboratory does not have to be moved at all.

SUNSPOTS AND ASSOCIATED PHENOMENA

*T*he discovery of sunspots was made in 1610–11. Naked-eye spots had been previously recorded, but had not been explained; one given in a Chinese record of 28 BC is described as 'a black vapour as large as a coin', and there is a Chinese record of an 'obscuration' in the Sun, which may well have been a spot, as early as 800 BC. However, the first observer to publish telescopic observations of them was J. Fabricius, from Holland, in 1611, and though his drawings are undated he probably saw the spots toward the end of 1610. C. Scheiner at Ingolstädt recorded spots in March 1611, with his pupil C. B. Cysat. Scheiner wrote a tract, which came to the notice of Galileo, who claimed to have been observing sunspots since November 1610. No doubt all these observers recorded spots telescopically at about the same time (the period was close to solar maximum). However, interpretations differed. Galileo's explanation was basically correct; Scheiner regarded the spots as dark bodies moving round the Sun at a distance close to the solar surface; Cassini, later, regarded them as mountains protruding through the bright surface!

✳ *The first to describe the projection method of observing sunspots* may have been Galileo's pupil B. Castelli. Galileo himself certainly used the method, and said (correctly) that it was 'the method that any sensible person will use'. (This seems to dispose of the legend that he ruined his eyesight by looking directly at the Sun through a telescope.)

✳ *The Wilson effect* was announced by A. Wilson, of Glasgow, in 1774. He observed that with a regular spot, the penumbra to the limbward side seemed to become broadened, as against the opposite side, as the spot neared the edge of the disk. From this, Wilson deduced that the spots must be hollows. The original observations were made in 1769.

✳ *The largest spot-group on record* was that of April 1947; it covered an area of 18 130 000 000 km², reaching its maximum on 8 April. To be visible with the naked eye, a spot-group must cover 500 millionths of the visible hemisphere. (One millionth of the hemisphere is equal to 3 000 000 km².)

✳ *The longest-lived spot group* lasted for 200 days, between June and December 1943. Very small spots (pores) may have lifetimes of less than an hour.

✳ *The first suggestion of a solar cycle* seems to have come from the Danish astronomer Horrebow in 1775–6, but his work was not published until 1859, by which time the cycle had been definitely identified.

✳ *The 11-year solar cycle was discovered* by H. Schwabe, a Dessau pharmacist, who began observing the Sun regularly in 1826 – mainly to see whether he could observe the transit of an intra-Mercurian planet. In 1851 his findings were popularized by Humboldt. A connection between solar activity and terrestrial phenomena was found by E. Sabine in 1852, and in 1870 E. Loomis, at Yale, established the link between the solar cycle and the frequency of aurorae.

✳ *The mean value of the length of the solar cycle since 1715* is 11·04 years. *The longest interval between successive maxima* has been 17·1 years (1788 to 1805); *the shortest interval* has been 7·3 years (1829·9 to 1837). Since 1715, when reasonably accurate records began, the *most energetic maximum* has been that of 1957·9. *The least energetic maximum* was that of 1816.

✳ *The 'Maunder Minimum'* was discovered, from examination of old records, by the British astronomer E. W. Maunder in 1890. (It had been noted independently by F. G. W. Spörer.) He found that between 1645 and 1715 there were virtually no spots at all, so that the solar cycle was suspended; more recent research indicates that the corona may also have been virtually absent. It may be significant that there was freak weather in England during the 1680s, when the Thames froze regularly and 'frost fairs' were held on it. Aurorae, too were lacking; Halley noted that he saw his first aurora

only in 1716, after forty years of watching. There may have been an earlier spotless period of the same type from 1400 to 1510, though the records are very incomplete.

Further evidence of these prolonged minima comes from tree-ring studies. High-energy cosmic rays which pervade the Galaxy transmute a small amount of atmospheric nitrogen to the radioactive isotope of carbon-14. When trees assimilate carbon dioxide, each growth ring contains a small percentage of carbon-14, which decays exponentially with a half-life of 5730 years. At sunspot maximum, the magnetic field ejected by the Sun deflects some of the cosmic rays away from the Earth, and reduces the level of carbon-14 in the atmosphere, so that the tree-rings formed at sunspot maximum have a lower amount of the carbon-14 isotope. Careful studies were carried out by F. Vercelli, who examined a tree which lived between about 275 BC and AD 1914. Then, in 1976, J. Eddy compared the carbon-14 record of solar activity with historical records of sunspots, aurorae and climatic data, and confirmed Maunder's suggestion of a dearth of spots from 1645 to 1715. Yet strangely, though there were hardly any records of telescopic sunspots during this period, naked-eye spots were recorded from China in 1647, 1650, 1655, 1656, 1665 and 1694; whether or not these observations are reliable must be a matter for debate. There is strong evidence for a longer cycle superimposed on the 11-year one.

✳ *The law relating to the latitudes of sunspots (Spörer's Law)* was discovered by the German amateur F. G. W. Spörer in 1861. At the start of a new cycle after minimum, the first spots appear at latitudes between 30 and 45 degrees north or south. As the cycle progresses, spots appear closer to the equator, until at maximum the average latitude of the groups is only about 15 degrees north or south. After maximum the spots become less common, but the approach to the equator continues, reaching only about 7 degrees north or south. The spots of the old cycle then die out (before reaching the equator), but even before they have completely disappeared the first spots of the new cycle are seen at the higher latitudes.

✳ *The first 'butterfly diagram'*, showing the effects of Spörer's Law, was drawn by Maunder in 1904.

✳ *The Wolf or Zürich sunspot number for any given day*, indicating the state of the Sun at that time, was worked out by R. Wolf of Zürich in 1852. The formula is: $R = k(10g + f)$, where R is the Wolf number, g is the number of groups seen, f is the total number of individual spots seen, and k is a constant depending on the equipment and

Outburst on the Sun. Clouds of particles are sent out during a solar flare, together with various radiations. (NASA)

SUNSPOT MAXIMA AND MINIMA, 1718–1992		
Maxima		
1718·2	1816·4	1917·6
1727·5	1829·9	1928·4
1738·7	1837·2	1937·4
1750·5	1848·1	1947·5
1761·5	1860·1	1957·9
1769·7	1870·6	1968·9
1778·4	1883·9	1979·9
1788·1	1894·1	1990·8
1805·2	1907·0	
Minima		
1723·5	1823·3	1923·6
1734·0	1833·9	1933·8
1745·0	1843·5	1944·2
1755·2	1856·0	1954·3
1766·5	1867·2	1964·7
1775·5	1878·9	1976·5
1784·7	1889·6	1986·8
1798·3	1901·7	
1810·6	1913·6	

site of the observer. (k is not usually far from unity.) The Wolf number may range from 0 for a clear disk up to over 200. A spot less than about 2500 km in diameter is officially known as a **pore**.

✳ *The magnetic fields associated with sunspots* were discovered by G. E. Hale, in the USA, in 1908. This resulted from the Zeeman effect (discovered in 1896 by the Dutch physicist P. Zeeman), according to which the spectral lines of a light source are split into two or three components if the source is associated with a magnetic field. Hale also found that in a spot-group the leader and the follower are of opposite polarity – and that the conditions are the same over a complete hemisphere of the Sun, though reversed in the opposite hemisphere. At the end of each cycle the whole situation is reversed, and there are grounds for supposing that the true cycle is 22 years in length rather than 11. Thus in the present cycle (beginning in 1986) the leader of a northern-hemisphere group is a south-polarity spot, the follower a north-polarity; in the southern hemisphere of the Sun, it is the leader which has north polarity. The magnetic fields of spots are very powerful, and may exceed 4000 gauss. *The strongest field observed* was with a group seen in 1967; the field was 5000 gauss, according to Steshenlio.

✳ *Plages* are bright, active regions in the Sun's atmosphere, usually seen around sunspot groups; when observed in monochromatic light (usually hydrogen or calcium) they are termed *flocculi*. The brightest features of this type are seen in integrated light as *faculæ*.

The discovery of faculæ was made by C. Scheiner, probably about 1611. Faculæ (Latin, 'torches') are clouds of incandescent gases lying above the brilliant surface; they are composed largely of hydrogen, and are best seen near the limb, where the photosphere is less bright than at the centre of the disk (in fact, the limb has only two-thirds of the brilliance of the centre, because at the centre we are looking down more directly into the hotter material). Faculæ may last for over two months, though their average lifetime is about 15 days. Also, faculæ often appear in areas where a spot-group is about to appear, and persist after the group has disappeared.

Polar faculæ are different from those of the more central regions, and are much less easy to observe; they are most common near sunspot minimum, and have latitudes higher than 65 degrees, with lifetimes ranging from a few days to no more than 12 minutes. They may well be associated with coronal plumes.

✷ *Flares* are violent, short-lived outbursts, usually occurring above active spot-groups. They emit charged particles as well as radiation, ranging from very short gamma-rays up to long-wavelength radio waves. They are most energetic in the X-ray and EUV (extreme ultra-violet) regions of the electromagnetic spectrum. They produce shock-waves in the corona and chromosphere, and considerable quantities of plasma may be expelled from the Sun. A typical flare lasts for around 20 minutes, though extreme cases may extend over a few hours, and they are phenomena of the chromosphere and lower corona.

Flares studied in the light of H (hydrogen) are classified according to their area. The classification is as follows:

Area square degrees	Classification
Over 24·7	4
12·5–24·7	3
5·2–12·4	2
2·0–5·1	1
Less than 2	s

F = faint, N = normal, B = bright. Thus the most important flares are classified as 4B.

Flares are thought to be due to the sudden release of magnetic energy which has been built up in the magnetic fields of complicated active regions, but it is true to say that their mechanism is not yet fully understood. The *first flare* to be observed was that of 1 September 1859, by R. Carrington, in white light, but very few have since been seen except with spectroscopic equipment or the equivalent.

✷ *The greatest solar-flare* was that of 16 August 1989; it lasted for 13 hours.

✷ *The modern sunspot theory* was proposed by H. Babcock in 1961. His magnetic dynamo model uses two basic properties of the Sun: its convective envelope, and its differential rotation. It may be assumed that the solar magnetic lines of force run from one magnetic pole to the other below the bright surface. An initial polar magnetic field is located in the convection zone. The differential rotation means that the field is 'stretched' more at the equator than at the poles. After many rotations, the field has become concentrated as toroids either side of the equator. Convection causes the field to become tangled, and kinks break through at the surface, steadying it and cooling it and producing pairs of sunspots – predictably, with opposite polarities. At the end of one 11-year cycle the toroid fields have diffused polewards and formed a polar field with reversed polarity. After 22 years the magnetic field properties revert to their original state.

Every spot-group has its own characteristics, but in general an 'average' two-spot group begins as two tiny pores at the limit of visibility. The pores develop into proper spots, growing and separating in longitude. Within two weeks the group has reached its maximum length, with a fairly regular leading spot together with a less regular follower of opposite polarity – and, of course, various minor spots and clusters. The darkest part of a spot – the umbra – still has a temperature of about 4500 °C, while the surrounding penumbra is at over 5000 °C; this means that a spot is by no means black, and if it could be seen shining on its own the surface brilliancy would be greater than that of an arc-lamp. After the group has

Gigantic sunspot group photographed on 28 Feb. 1967. (W. M. Baxter)

reached its peak a slower decline sets in. The leader is generally the last survivor. Roughly 75 per cent of groups fit into this general pattern, but of the remainder some do not conform, and there are also frequent single spots.

Even in non-spot zones, the solar surface is not calm. The brilliant surface or photosphere has a granular structure; each granule is about 1000 km in diameter (1 sec ·3) and has a life of about 8 minutes. It is estimated that the whole surface includes about 4 000 000 granules at any one time; they represent upcurrents, and the general situation has been compared with 'boiling' of a liquid, though the photosphere is of course entirely gaseous. The granular structure is easy to observe, though the first really good pictures of it were obtained from a balloon (*Stratoscope II*).

Supergranulation involves large organized cells, usually polygonal and of the order of 30 000 km in diameter, each of which contains several hundreds of individual granules; matter wells up in the centre of the cell, spreading out to the edges before sinking again.

Results from the Solar Maximum Mission spacecraft showed that the Sun's energy output varies by 1 part in 500 on a time-scale of days to weeks. These fluctuations seem to be due to projected areas of dark spots and bright faculæ on the disk.

SPECTRUM AND COMPOSITION OF THE SUN

*T*he first solar spectrum was obtained by Isaac Newton in 1666, but he never took his investigations much further, though he did of course prove the complex nature of sunlight.

✳ *The discovery of dark lines in the solar spectrum* was made in England by W. H. Wollaston, in 1802. However, Wollaston merely took the lines for the boundaries between different colours of the rainbow spectrum.

✳ *The first systematic studies of the dark lines* were carried out in Germany by J. von Fraunhofer, from 1814. Fraunhofer realized that the lines were permanent; he recorded 574 of them, and mapped 324.

✳ *The first explanation of the dark 'Fraunhofer lines'* was given by G. Kirchhoff

in 1859 (initially working with R. Bunsen). Kirchhoff found that the photosphere yields a rainbow or continuous spectrum; the overlying gases produce a line spectrum, but since these lines are seen against the rainbow background they appear dark instead of bright. Since their positions and intensities are not affected, each line may be tracked down to a particular element or group of elements. In 1861–2 Kirchhoff produced the *first detailed map of the solar spectrum*. (His eyesight was affected, and the work was actually finished by his assistant, K. Hofmann.) In 1869 A. Ångström, the Swedish physicist, studied the solar spectrum by using a grating instead of a prism, and in 1889 Rowland produced a detailed photographic map of the spectrum.

The most prominent Fraunhofer lines in the visible spectrum are:

Letter	Wavelength Ångströms	Identification	Letter	Wavelength Ångströms	Identification
A	7593	O_2			
a	7183	H_2O			
B	6867	O_2			

(These three are telluric lines – due to the Earth's intervening atmosphere.)

C(Hα)	6563	H	b₄	5167	Mg
D₁	5896		F(Hβ)	4861	H
D₂	5890	Na	f(Hγ)	4340	H
E	5270	Ca, Fe	G	4308	Fe, Ti
	5269	Fe	g	4227	Ca
b₁	5183	Mg	h(Hδ)	4102	H
b₂	5173	Mg	H	3967	Ca^II
b₃	5169	Fe	K	3933	

(One Ångström is equal to one hundred-millionth part of a centimetre; it is named in honour of Anders Ångström. The diameter of a human hair is roughly 500 000 Å.
To convert Ångströms into nanometres, divide all wavelengths by 10, so that, for instance, Hα becomes 656·3 nm.)

By now many of the chemical elements have been identified in the Sun. The list of elements which have and have not been identified is as follows:

THE CHEMICAL ELEMENTS, AND THEIR OCCURRENCE IN THE SUN

The following is a list of elements 1 to 103
* = detected in the Sun.
R = included in H. A. Rowland's list published in 1891.

Atomic No.	Name	Atomic Weight	Occurrence in the Sun
1 H	Hydrogen	1·008	* R
2 He	Helium	4·003	*
3 Li	Lithium	6·939	* (in sunspots)
4 Be	Beryllium	9·013	* R
5 B	Boron	10·812	* (in compound)
6 C	Carbon	12·012	* R
7 N	Nitrogen	14·007	*
8 O	Oxygen	16·000	*
9 F	Fluorine	18·999	* (in compound)
10 Ne	Neon	20·184	*
11 Na	Sodium	22·991	* R
12 Mg	Magnesium	24·313	* R
13 Al	Aluminium	26·982	* R
14 Si	Silicon	28·090	* R
15 P	Phosphorus	30·975	*
16 S	Sulphur	32·066	*
17 Cl	Chlorine	35·434	
18 A	Argon	39·949	* (in corona)
19 K	Potassium	39·103	* R
20 Ca	Calcium	40·080	* R
21 Sc	Scandium	44·958	* R
22 Ti	Titanium	47·900	* R
23 V	Vanadium	50·944	* R
24 Cr	Chromium	52·00	* R
25 Mn	Manganese	52·94	* R
26 Fe	Iron	55·85	* R
27 Co	Cobalt	58·94	* R
28 Ni	Nickel	58·71	* R
29 Cu	Copper	63·55	* R
30 Zn	Zinc	65·37	* R
31 Ga	Gallium	69·72	*
32 Ge	Germanium	72·60	* R
33 As	Arsenic	74·92	
34 Se	Selenium	78·96	
35 Br	Bromine	79·91	
36 Kr	Krypton	83·80	
37 Rb	Rubidium	85·48	* (in spots)
38 Sr	Strontium	87·63	* R
39 Y	Yttrium	88·91	* R
40 Zr	Zirconium	91·22	* R
41 Nb	Niobium	92·91	* R
42 Mo	Molybdenum	95·95	* R
43 Tc	Technetium	99	
44 Ru	Ruthenium	101·07	*
45 Rh	Rhodium	102·91	* R
46 Pd	Palladium	106·5	* R
47 Ag	Silver	107·87	* R
48 Cd	Cadmium	112·41	* R
49 In	Indium	114·82	* (in spots)
50 Sn	Tin	118·70	* R
51 Sb	Antimony	121·76	*
52 Te	Tellurium	127·61	
53 I	Iodine	126·91	
54 Xe	Xenon	131·30	
55 Cs	Cæsium	132·91	
56 Ba	Barium	137·35	* R
57 La	Lanthanum	138·92	* R
58 Ce	Cerium	140·13	* R
59 Pr	Praseodymium	140·91	*
60 Nd	Neodymium	144·25	* R
61 Pm	Promethium	147	
62 Sm	Samarium	150·36	*
63 Eu	Europium	151·96	*
64 Gd	Gadolinium	157·25	*
65 Tb	Terbium	158·93	*
66 Dy	Dysprosium	162·50	*
67 Ho	Holmium	164·94	
68 Er	Erbium	167·27	* R
69 Tm	Thulium	168·94	*
70 Yb	Ytterbium	173·04	*
71 Lu	Lutecium	174·98	*
72 Hf	Hafnium	178·50	*
73 Ta	Tantalum	180·96	*
74 W	Tungsten	183·86	*
75 Re	Rhenium	186·3	
76 Os	Osmium	190·2	*
77 Ir	Iridium	192·2	*
78 Pt	Platinum	195·1	*
79 Au	Gold	197·0	*
80 Hg	Mercury	200·6	

Atomic No.	Name	Atomic Weight	Occurrence in the Sun
81 Tl	Thallium	204·4	
82 Pb	Lead	207·2	* R
83 Bi	Bismuth	209·0	
84 Po	Polonium	210	
85 At	Astatine	211	
86 Rn	Radon	222	
87 Fr	Francium	223	
88 Ra	Radium	226	
89 Ac	Actinium	227	
90 Th	Thorium	232	*
91 Pa	Protoactinium	231	
92 U	Uranium	238	

The remaining elements are 'transuranic' and radioactive, and have not been detected in the Sun. They are:

Atomic No.	Name	Atomic Weight	Occurrence in the Sun
93 Np	Neptunium	237	
94 Pu	Plutonium	239	
95 Am	Americium	241	
96 Cm	Curium	242	
97 Bk	Berkelium	243	
98 Cf	Californium	244	
99 Es	Einsteinium	253	
100 Fm	Fermium	254	
101 Md	Mendelevium	254	
102 No	Nobelium	254	
103 Lw	Lawrencium	257	
104 Rf	Rutherfordium	–	
105 Ha	Hahnium	–	

For elements 43, 61, 85–89, 91, 93–103 the mass number is that of the most stable isotope.

The fact that the remaining elements have not been identified in the Sun does not necessarily indicate that they are completely absent. They may be present, though no doubt in very small amounts.

So far as relative mass is concerned, the most abundant element by far is hydrogen (71 per cent). It is followed by helium (27 per cent). All the others combined make up only 2 per cent.

For every million atoms of hydrogen, the numbers of other atoms are as follows:

Helium	63 000
Oxygen	690
Carbon	420
Nitrogen	87
Silicon	45
Magnesium	40
Neon	37
Iron	32
Sulphur	16
Aluminium	3
Carbon	2
Sodium	2
Nickel	2
Argon	1

Helium was identified in the Sun (by Sir Norman Lockyer, in 1868) before being found on Earth. Lockyer named it after the Greek ἥλιος, the Sun. It was detected on Earth in 1894, by Sir William Ramsay, as a gas occluded in cleveite. For a time it was believed that the corona contained another element unknown on Earth, and it was even given a name – Coronium – but the lines, described initially by Harkness and Young at the eclipse of 1869, proved to be due to elements already known. The brilliant photosphere is surprisingly thin – only about 300 km thick. *The limb darkening was first explained* by K. Schwarzschild in 1906; as already noted, it is due to the fact that when we look at the centre of the disk we are seeing into deeper and hotter layers. Old ideas about the Sun sound strange today. Sir William Herschel believed that below the bright surface there was a cool layer, which could well be inhabited; up to the time of his death, in 1822, he maintained this view. As recently as 1952 a German lawyer, Godfried Büren, stated that the Sun had a vegetation-covered inner globe, and offered a prize of 25 000 marks to anyone who could prove him wrong. The leader German astronomical society took up the challenge, and won a court case. (Whether the prize was paid does not seem to be on record!)

✳ *The spectroheliograph, enabling the Sun to be photographed in the light of one element only*, was invented by G. E. Hale in 1892. The visual equivalent, the *spectrohelioscope*, was invented in 1923, also by Hale. In 1933 B. Lyot, of France, developed the *Lyot filter*, which is less versatile but more convenient, and also allows the Sun to be studied in the light of one element only.

✳ *The meteoritic theory of solar energy* was discussed by J. R. Mayer in 1848. Mayer found that a globe of hot gas the size of the sun would cool down in 5000 years or so if there were no other energy source, while a Sun made of coal, and burning furiously enough to produce as much heat as the Sun actually does, would last for only 4600 years. He therefore assumed that the energy was produced by meteorites striking the solar surface.

✳ *The contraction theory* was proposed in 1834 by H. von Helmholtz. He calculated that if the Sun contracted by 60 m per year, the energy produced would suffice for 15 000 000 years. This theory was supported later by the great British physicist Lord Kelvin. However, it had to be abandoned when astronomers concluded that the Sun is certainly at least as old as the Earth (about 4600 million years) and probably older.

✳ *The nuclear transformation theory* was worked out by H. Bethe in 1938, during a train journey from Washington to Cornell University. Hydrogen is being converted to helium, so that energy is released and mass is lost: the decrease in mass amounts to 4 000 000 tonnes per second. Bethe assumed that carbon and nitrogen were used as catalysts, but C. Critchfield, also in America, subsequently showed that in solar-type stars the proton-proton reaction is dominant.

Slight variations in output occur, and some authorities maintain that it is these minor changes which have produced the Ice Ages which have affected the Earth now and then throughout its history, but obviously the Sun cannot continue radiating for ever.

Recent calculations indicate that the luminosity is gradually increasing, so that in 1·1 thousand million years from now the Sun will be about 10 per cent more powerful

The Kitt Peak solar telescope, 1980. Patrick Moore's figure, in the foreground, brings out the size of the telescope.

The Sun in X-rays. Image taken on 25 Nov. 1990 from the Japanese satellite Yohkoh.

than it is at present. This will probably make the Earth uninhabitable. In approximately 6·5 thousand million years hence the Sun will leave the Main Sequence and become a red giant star; by 7·8 thousand million years from now it will have swelled out to a diameter 170 times that at the present time, engulfing Mercury and increasing to over 2000 times its present luminosity. The Earth will become molten, though the Sun's loss of mass, and the weakening of its gravitational pull, may mean that the Earth will move outward to a larger orbit and will escape total destruction. Subsequently, the Sun will throw off its outer layers and go through the planetary nebula stage; the remaining core will be a white dwarf, small in size but of very high density. Finally, after a much longer period, the Sun will lose all its light and heat and will become a cold, dead black dwarf. If this scenario is correct, the Earth will remain habitable for only a little more than 1 thousand million years in the future – a rather shorter time than has been previously believed.

It now seems that the solar core, in which thermonuclear reactions are taking place, extends out to 0·25 of the solar radius from the centre of the globe. Out to 0·83 of the radius, energy is transported outward by radiative diffusion, while in the outer layers it is convection which is the transporting agency.

✳ *The first indications of a solar oscillation* were obtained in 1960. The period was found to be 5 minutes, and was thought to be a surface ripple affecting not more than the outermost 10 000 km of the Sun's globe. *Solar vibrations were discovered in 1973* by R. H. Dicke, who was attempting to make measurements of the polar and equatorial diameters of the Sun to see whether there was any observable flattening. Dicke found that the Sun is 'quivering like a jelly', so that the equator bulges as the poles are flattened, but the maximum amplitude is only about 5 km and the velocity about 10 m/s, so that the observations are immensely difficult and delicate. There are also some indications that the diameter of the Sun is not absolutely constant, though any changes at the present epoch must be very minor and probably periodical.

There are other, longer-term oscillations; one of 50 minutes, one of 2 hours 40 minutes discovered in 1976 by a Russian team together with a team from Birmingham, and so on. These may represent global pulsations of the Sun as a whole, and by now the

study of what has been called 'solar seismology' has become of immense importance.

✳ *Solar neutrinos* are presenting astronomers with many problems. Neutrinos are particles with no mass and no electric charge, so that they are extremely difficult to detect. Theoretical considerations indicate that the Sun should emit quantities of them, and in 1966 efforts to detect them were begun by a team from the Brookhaven National Laboratory in the USA, led by R. Davis. The 'telescope' is located in the Homestake Gold Mine in South Dakota, inside a deep mineshaft, and consists of a tank of 454 600 litres of cleaning fluid (tetrachloroethylene). Only neutrinos can penetrate so far below ground level, and on the rare occasions when a chlorine atom happened to be struck by a neutrino, radioactive argon would be produced; this could therefore provide a key to the number of neutrinos. However, the observed flux was much smaller than had been expected, and the detector recorded only about one-third the anticipated number of neutrinos. The same was found by a team in Russia, using 100 tonnes of liquid scintillator and 144 photo-detectors in a mine in the Donetsk Basin. Further confirmation came from Kamiokande in Japan, using light-sensitive detectors on the walls of a tank holding 3000 tonnes of water. When a neutrino hits an electron it produces a spark of light, and the direction of this, as the electron moves, tells the direction from which the neutrino

has come – which the Homestake detector cannot do. Yet another sort of detector in Russia uses gallium-71; if hit by a neutrino this gallium would be converted to germanium-71. Another gallium experiment is at Gran Sasso, deep in the Apennines. Again the flux is unexpectedly low, and to make matters even more curious the Homestake results seem to indicate that neutrinos are least plentiful around the time of sunspot maximum.

Various theories have been proposed to explain the paucity of solar neutrinos. It is known that neutrinos are of several different kinds, and the Homestake detector can record only those with lower energies; even so, the number of events recorded each month should have been around 25, whereas actually it was on average 9. If the Sun's core temperature is no more than about 14 000 000°C, as against the usually-assumed 15 000 000°, the neutrino flux would be easier to explain, but this raises other difficulties. Another suggestion is that the core temperature is reduced by the presence of WIMPs (Weakly Interaction Massive Particles). A WIMP is quite different from ordinary matter, and is said to have a mass from 5 to 10 times that of a proton; but the existence of WIMPs has not been proved, and many authorities are decidedly sceptical. At the moment it is fair to say that the neutrino problem remains unsolved.

✳ *The chromosphere* of the Sun lies above the photosphere. The temperature increases from 4200°C in the low chromosphere to 8000°C at an altitude of 1500 km, and then increases rapidly until the chromosphere merges with the corona. The dark lines in the solar spectrum are produced in the upper photosphere and lower chromosphere, often called the *reversing layer*. The chromosphere is 2000 to 10000 km deep.

✴ *The corona*, visible with the naked eye only during a total solar eclipse, lies above the chromosphere. The mean temperature is nearly 2000000°C, but the density is so low – less than one million millionth of the Earth's air at sea-level – that there is very little 'heat'. (The density of the Earth's air at sea-level is about 10^{19} particles per cubic centimetre; of the Sun's photosphere, 10^{17}; of the corona, only 10^5 particles per cubic centimetre.) The corona sends out only one-millionth as much light as the photosphere. It was once believed that the high temperature was due to sound waves, produced by turbulence in the photosphere, but it now seems more likely that magnetic phenomena are responsible. Areas in the corona where the temperatures and pressures are much lower than average are known as *coronal holes*. The corona has no definite boundary, but simply thins out until its density has become no greater than that of the inter-planetary medium.

✴ *The first coronagraph* was built by B. Lyot in 1930, and tested at the Pic du Midi Observatory (altitude 2870 m). It depends upon producing an 'artificial eclipse' instrumentally. With it, Lyot was able to study the inner corona and its spectrum.

✴ *Solar Wind* is the radial, continuous outflow of charged particles from the Sun. It was first predicted in 1951 by Biermann, following a study of comets' tails, though indications go back to a 1900 paper by Sir Oliver Lodge. The solar wind, consisting of 'plasma', is made up of protons, helium nuclei (alpha-particles) and electrons; the velocity away from the Sun ranges between 200 and 900 km/s, with the average velocity past the Earth given as 200–600 km/s. Low-velocity streams come from the loops in the corona, and high-velocity streams from coronal holes, particularly evident at the solar poles, where the magnetic field lines are open. The solar wind affects the Earth's magnetosphere, and during periods of great activity on the Sun, when the wind is enhanced particles cascade down into the lower part of the atmosphere, producing auroræ. The region in space where the solar wind ceases to be appreciable is known as the *heliopause*. The heliopause may lie at between 100 and 150 astronomical units from the Sun. Apparently a dense, high-speed shell of plasma ejected by the Sun in

mid-1991 reached the heliopause in late 1992, and triggered a massive release of radio energy which was detected by both the Voyager probes. It is hoped to track some current space-vehicles (Pioneers 10 and 11, and Voyagers 1 and 2) until they reach the heliopause.

✴ *The first X-ray observations of the Sun* were made in 1949 by workers in the United States Naval Research Laboratory. Because the X-rays cannot penetrate the Earth's atmosphere, all observations have to be undertaken by space research methods. From X-ray studies it has been found that there are relatively dark, cool regions in the solar corona ('coronal holes') and that there are bright, active, short-lived points.

✴ *Cosmic rays from the Sun* were detected by Forbush in 1942. In 1954 Forbush established that cosmic-ray intensity decreases when solar activity increases (Forbush effect).

✴ *Prominences* were first described in detail by the Swedish observer Vassenius at the total eclipse of 1733, though he believed that they belonged to the Moon rather than the Sun. (They may have been seen earlier – by Stannyan in 1706, from Berne.) It was only after the eclipse of 1842 that astronomers became certain that they were solar rather than lunar. They were formerly termed 'red flames', but are in fact regions of hot hydrogen gas, reddish in colour. *Quiescent* prominences may persist for months; *eruptive* prominences show violent motions, and may attain heights of several thousand km.

Prominences are visible with the naked eye only during a total eclipse. However, following the eclipse of 19 August 1868, J. Janssen (France) and Sir Norman Lockyer (England) developed the method of observing them spectroscopically without an eclipse. By observing at hydrogen wavelengths, prominences may be seen against the bright disk of the Sun as dark filaments, sometimes termed flocculi. (Bright flocculi are due to calcium.)

✴ *The discovery of ultra-violet radiation from the Sun* was made in 1801 by J. Ritter, using a prism to produce a solar spectrum and noting that paper soaked in NaCl was darkened if held in a region beyond the violet end of the visible spectrum.

✴ *The discovery of radio emission from the Sun* was due to J. S. Hey and his team in 1942 (27–8 February). Initially, the effect was thought to be due to German jamming of the radar transmitters! Various types of emission are now known, and there are 'bursts', some of which are associated with flares. In June every year the 'radio sun'

occults the Crab Nebula, itself a radio source, and the phenomenon yields valuable information about the corona. The first observations of this kind were carried out in June 1952.

✴ *The first radar contact with the Sun* was made in 1959, by Eshleman and his colleagues at the Stanford Research Institute USA.

✴ *Sunrise* is defined as the moment when the Sun's upper limb appears above the horizon. As refraction reduces the apparent zenith distance of the Sun by 34 minutes of arc at the horizon, and as the Sun's semidiameter is about 16 min, the moment of sunrise is defined as being the instant when the Sun has a zenith distance of 90 degrees 50 minutes. *Sunset* is the moment when the upper limb of the Sun disappears below the horizon: the zenith distance is again 90 degrees 50 minutes. Because of refraction effects, it has sometimes been possible to see the Sun and the full Moon visible simultaneously above opposite horizons.

The apex of the Sun's way (that is to say, the point in the sky toward which the Sun is at present moving) is in Hercules, at R.A. 18 h, declination 34N. The *antapex* is in Columba (R.A. 6 h, declination 34S).

ECLIPSES OF THE SUN

An eclipse of the Sun occurs when the Moon passes in front of the Sun; strictly speaking, the phenomenon is an occultation of the Sun by the Moon. Solar eclipses may be total (when the whole of the photosphere is hidden), partial, or annular (when the Moon's apparent diameter is smaller than that of the Sun, so that a ring of light is left round the lunar disk; Latin *annulus*, a ring).

✴ *The greatest number of eclipses possible in one year* is 7; thus in 1935 there were 5 solar and 2 lunar eclipses, and in 1982 there were 4 solar and 3 lunar.

✴ *The least number of eclipses possible in one year* is 2, both of which must be solar, as in 1984.

The length of the Moon's shadow varies between 381000 km and 365000 km, with a mean of 372000 km. As the mean distance of the Moon from the Earth is 384400 km, the shadow is on average too short to reach the Earth's surface, so that annular eclipses are more frequent than total eclipses in the ratio of 5 to 4. The Moon's shadow cone diameter ranges between 269 and 313 km. On average there are 238 total eclipses per century.

✷ *The longest possible duration of totality for a solar eclipse* is 7 min 31 sec. This has never been observed, but at the eclipse of 20 June 1955 totality over the Philippine Islands lasted for 7 min 8 sec.

✷ *The shortest possible duration of totality* may be a fraction of a second. This happened at the eclipse of 3 October 1986, which was annular along most of the central track, but was total for about a tenth of a second over a restricted area in the North Atlantic Ocean.

✷ *The longest possible duration of the annular phase of an eclipse* is 12 min 24 sec.

✷ *The longest totality ever observed* was during the eclipse of 30 June 1973. A Concorde aircraft, specially equipped for the purpose, flew underneath the Moon's shadow and kept pace with it, so that the scientists on board (including the British astronomer John Beckman) saw a totality lasting for 72 minutes! They were carrying out observations at millimetre wavelengths, and at their height of 55 000 feet were above most of the water vapour in our atmosphere, which normally hampers such observations. They were also able to see definite changes in the corona and prominences over the full period. The Moon's shadow moves over the Earth at over 3000 km/h.

✷ *The widest track for totality across the Earth's surface* is 272 km. In most total eclipses, of course, the width is considerably less than this; and a partial eclipse is seen to either side of the track of totality.

✷ *The first eclipse predictions* were made by studies of the Saros period. This is the period after which the Sun, Moon and node arrive back at almost the same relative positions; it amounts to 6585·321 solar days, or approximately 18 years 11 days. Therefore, an eclipse tends to be followed by another eclipse in the same Saros series 18 years 11 days later, though conditions are not identical, and the Saros is at best a reasonable guide. *The first known prediction* was made by the Greek philosopher Thales, who forecast the eclipse of 25 May 585 BC. This occurred near sunset in the Mediterranean area, and is said to have put an end to a battle between the forces of King Alyattes of the Lydians and King Cyaxares of the Medes; the combatants were so alarmed by the sudden darkness that they concluded a hasty peace. One Saros series lasts for 1150 years; it includes 64 eclipses, of which 43 or 44 are total, while the rest are partial eclipses seen from the polar zones of the Earth.

✷ *The least successful political eclipse manœuvre* was used by General William Harrison (later President of the United States), when he was Governor of Indiana Territory. He was opposed by the Shawnee prophet Tenskwatawa, and decided to ridicule him by asking whether he could cause 'the Sun to stand still and the Moon to alter its course'. Unluckily for him, Tenskwatawa knew that an eclipse was due on 16 July 1806, so he said that he would show his power by blotting out the Sun. A crowd gathered at the camp, and the prophet timed the eclipse with full dramatic effect, so that the gallant General was nonplussed (though in 1811 he did destroy the Shawnees at the Battle of Tippecanoe).

✷ *The first recorded solar eclipse* seems to have been that of 2136 BC (22 October), seen in China during the reign of the Emperor Chung K'ang. The next eclipse which may be dated with any certainty is that of 1375 BC, described on a clay tablet found at Ugarit (Syria).

✷ *The first solar eclipse recorded in Britain* was that of 15 February 538, described in the Anglo-Saxon Chronicles; it occurred four years after the death of Cerdic, first king of the West Saxons. The Sun was two-thirds eclipsed in London. English total eclipses over the past thousand years have been those of 1140, 1715, 1724 and 1927. The eclipse of 3 May 1715 was observed by Halley from London and by Flamsteed from Greenwich; that of 22 May 1724 was observed only by Dr Stukeley from Haradon Hill, near Salisbury, because of generally cloudy conditions. Totality was, however, seen from France. The eclipse of 29 June 1927 was total over part of north England; the conditions were generally poor, and the best photographs were obtained from Giggleswick – totality lasted for less than half a minute. The eclipse of 30 June 1954 was total over parts of Norway and Sweden; the extreme southern edge of the track just brushed the Shetland Isles. The next English total eclipse will be that of 11 August 1999, when the track will cross Cornwall. No total eclipse was seen from London between 875 and 1715, though those of 968 and 1140 were nearly total there. The 1927 eclipse was 95 per cent total in London. There were total eclipses over parts of Scotland in 1133, and on 17 June 1433, when the length of totality reached 4·5 minutes at Inverness and caused what became known as 'Black Hour'. The eclipses of 1598 and 1652 were also total over parts of Scotland, and that of 1699 nearly so.

The maximum theoretical length for a British total eclipse is 5·5 minutes. That of 15 June 885 lasted for almost 5 minutes, and the same will be true for the Scottish total eclipse of 20 July 2381.

✷ *The first total eclipse recorded in the United States* was that of 24 June 1778, when the track passed from Lower California to New England. *The first official American total eclipse expedition* was that of 21 October 1780, when a party went to Penobscot, Maine; it was led by S. Williams of Harvard, and it was given 'free passage' by the British forces, but unfortunately a mistake in the calculations meant that the party remained outside the track of totality! *The first American eclipse expedition to Europe* was that of 28 July 1851, when G. P. Bond led a team to Scandinavia.

✷ *The only emperor to have died of fright because of an eclipse* was Louis of Bavaria, in 840 (his three sons then proceeded to indulge in a ruinous war over the succession).

✷ *The only astronomer to have escaped from a besieged city in a balloon to study a total eclipse* was Jules Janssen. The eclipse was that of 22 December 1870, and Janssen flew out from Paris, which was surrounded by the German forces. He arrived safely at Oran, but clouds prevented him from making any observations.

✷ *The first mention of the corona* may have been due to Plutarch, who lived from about AD 46 to 120. Plutarch's book 'On the Face in the Orb of the Moon' contains a reference to 'a certain splendour' round the eclipsed Sun which could well have been the corona. The corona was definitely recorded from Corfu during the eclipse of 22 December 968. The astronomer Clavius saw it at the eclipse of 9 April 1567, but regarded it as merely the uncovered edge of the Sun; Kepler showed that this could not be so, and attributed it to a lunar atmosphere. After observing the eclipse of 16 June 1806 from Kindehook, New York, the Spanish astronomer Don José Joaquin de Ferrer pointed out that if the corona were due to a lunar atmosphere, then the height of this atmosphere would have to be 50 times greater than that of the Earth, which was clearly unreasonable. However, it was only after careful studies of the eclipses of 1842 and 1851 that the corona and the prominences were shown unmistakably to belong to the Sun rather than to the Moon.

There is some evidence that during eclipses which occurred during the 'Maunder Minimum' (1645–1715) the corona was absent, though it is impossible to be sure. Certainly the shape of the corona at spot-maximum is more symmetrical than at spot-minimum, when there are 'polar streamers'. This was first recognized after studies of the eclipses of 1871 and 1872.

✷ *The first observation of Baily's Beads* seems to have been made by Halley during the eclipse of 1715. They were also seen by Maclaurin, from Edinburgh, at the annular eclipse of 1 March 1737. They were described in detail by the English astronomer Francis Baily (after whom they are named) in 1836 during the annular eclipse of 15 May. The brilliant bead-like effect is caused by the Sun's rays shining through valleys on the lunar limb immediately before and immediately after totality. They were *first photographed* at the eclipse of 7 August 1869 by C. F. Hines and members of the Philadelphia Photographic Corps, observing from Ottumwa, Iowa.

✳ *The first observation of the shadow-bands* was made by H. Goldschmidt in 1820. These are wavy lines seen across terrestrial features before and after totality; they are, of course, produced in the Earth's atmosphere.

✳ *The first attempt to photograph a total eclipse* was made by the Austrian astronomer Majocci on 8 July 1842. He failed to record totality, though he did succeed in photographing the partial phase.

✳ *The first successful photograph of a total eclipse*, showing the corona and prominences, was taken by Berkoswki on 28 July 1851, using the 6·25 in Königsberg heliometer and giving an exposure time of 24 sec.

✳ *The first photograph of the flash spectrum* was taken by the American astronomer C. Young on 22 December 1870. (The flash spectrum is the sudden, brief change in the Fraunhofer lines from dark to bright, when the Moon blots out the photosphere in the background, and the chromosphere is left shining 'on its own'). The flash spectrum was first observed during an annular eclipse by Pogson in 1872.

✳ *The only cricket match to have been interrupted by an eclipse of the Sun* was the Jubilee Test between India and England in February 1980. A solar eclipse was due that afternoon, and the Indian Board, in agreement with the English team, did not want the responsibility of a crowd of 50 000 damaging their eyes by looking at the Sun when the eclipse began. So that day was declared a 'rest day' – and the Test Match continued on the next morning!

✳ *The first attempt to show a total eclipse on television from several stations along the track* was made by the BBC at the eclipse of 15 February 1961. The attempt was successful, and totality was shown successively from stations in France, Italy and Yugoslavia.

✳ *The last total eclipse* will probably occur in about 1 250 000 years from now (exact calculations are not yet made). By then the Moon will have receded to a distance of about 28 780 km further from the Earth, and the disk will no longer appear large enough to cover the Sun.

The Sun from Skylab 4, 19 Dec. 1973. To the upper left is one of the most spectacular flares ever recorded, spanning 588 000 km across the solar surface. The photograph was taken in the light of ionized helium by the extreme ultra-violet spectroheliograph. (NASA)

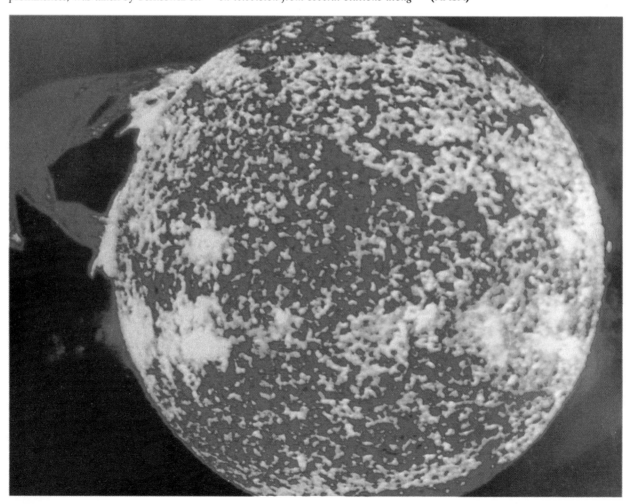

SOLAR ECLIPSES, 1923–1990

T = total, P = partial, A = annular

Date	Type	Area
1923 March 16	A	S Africa
1923 Sept. 10	T	California, Mexico
1924 March 5	P	S Africa
1924 July 31	P	Antarctic
1924 Aug. 29	P	Iceland, N Russia, Japan
1925 Jan. 24	T	North-eastern USA
1925 July 20/1	A	New Zealand, Australia
1926 Jan. 14	T	E Africa, Indian Ocean, Borneo
1926 July 9/10	A	Pacific
1927 Jan. 3	A	New Zealand, S America
1927 June 29	T	England, Scandinavia
1927 Dec. 24	P	Polar zone
1928 May 19	T	S Atlantic
1928 June 17	P	N Siberia
1928 Nov. 12	P	England to India
1929 May 9	T	Indian Ocean, Philippines
1929 Nov. 1	A	Newfoundland, C Africa, Indian Ocean
1930 April 28	T	Pacific
1930 Oct. 21/2	T	S Pacific to S America
1931 April 17/18	P	Arctic
1931 Sept. 12	P	Alaska, N Pacific
1931 Oct. 11	P	S America, S Pacific, Antarctic
1932 March 7	A	Antarctic
1932 Aug. 31	T	USA
1933 Feb. 24	A	S America, C Africa
1933 Aug. 21	A	Iran, India, N Australia
1934 Feb. 13/14	T	Pacific
1934 Aug. 10	A	S Africa
1935 Jan. 5	P	No land surface
1935 Feb. 3	P	N America
1935 June 30	P	Britain
1935 July 30	P	No land surface
1935 Dec. 25	A	New Zealand, south S America
1936 June 19	T	Greece, Turkey, Siberia, Japan
1936 Dec. 13/14	A	Australia, New Zealand
1937 June 8	T	Pacific, Chile
1937 Dec. 2/3	A	Pacific
1938 May 29	T	S Atlantic
1938 Nov. 21/2	P	E Asia, Pacific coast of N America
1938 April 19	A	Alaska, Arctic
1939 Oct. 12	T	Antarctic
1940 April 7	A	USA, Pacific
1940 Oct. 1	T	Brazil, S Atlantic, S Africa
1941 March 27	A	S Pacific, S America
1941 Sept. 21	T	China, Pacific
1942 Mar. 16/17	P	S Pacific, Antarctic
1942 Aug. 12	P	Invisible in Britain
1942 Sept. 10	P	Britain
1943 Feb. 4/5	T	Japan, Alaska
1943 Aug. 1	A	Pacific
1944 Jan. 25	T	Brazil, Atlantic, Sudan
1944 July 20	A	India, New Guinea
1945 Jan. 14	A	Australia, New Zealand
1945 July 9	T	Canada, Greenland, N Europe
1946 Jan. 3	P	Invisible in Britain
1946 May 30	P	S Pacific
1946 June 29	P	Arctic
1946 Nov. 23	P	N America
1947 May 20	T	Pacific, Equatorial Africa, Kenya
1947 Nov. 12	A	Pacific

Date	Type	Area
1948 May 8/9	A	E Asia
1948 Nov. 1	T	Kenya, Pacific
1949 April 28	P	Britain
1949 Oct. 21	P	New Zealand, Australia
1950 March 18	A	S Atlantic
1950 Sept. 12	T	Siberia, N Pacific
1951 March 7	A	Pacific
1951 Sept. 1	A	Eastern USA, C and S Africa
1952 Feb. 25	T	Africa, Arabia, Russia
1952 Aug. 20	A	S America
1953 Feb. 13/14	P	E Asia
1953 July 11	P	Arctic
1953 Aug. 9	P	Pacific
1954 Jan. 5	A	Antarctic
1954 June 30	T	Iceland, Norway, Sweden, Russia, India
1954 Dec. 25	A	S Africa, S Indian Ocean
1955 June 20	T	S Asia, Pacific, Philippines
1955 Dec. 14	A	Sudan, Indian Ocean, China
1956 June 8	T	S Pacific
1956 Dec. 2	T	Europe, Asia
1957 April 29/30	A	Arctic
1957 Oct. 23	T	Antarctica
1958 April 19	A	Indian Ocean, Pacific
1958 Oct. 12	T	Pacific
1959 April 8	A	S Indian Ocean, Pacific
1959 Oct. 2	T	N Atlantic, N Africa
1960 March 27	P	Australia, Antarctica
1960 Sept. 20/1	P	N America, E Siberia
1961 Feb. 15	T	France, Italy, Greece, Yugoslavia, Russia
1961 Aug. 11	A	S Atlantic, Antarctica
1962 Feb. 4/5	T	Pacific
1962 July 31	A	S America, C Africa
1963 Jan. 25	A	Pacific, S Africa
1963 July 20	T	Japan, north N America, Pacific
1964 Jan. 14	P	Tasmania, Antarctica
1964 July 9	P	N Canada, Arctic
1964 Dec. 3/4	P	NE Asia, Alaska, Pacific
1965 May 30	T	Pacific, New Zealand, Peruvian coast
1965 Nov. 23	A	Russia, Tibet, E Indies
1966 May 20	A	Greece, Russia
1966 Nov. 12	T	S America, Atlantic
1967 May 9	P	N America, Iceland, Scandinavia
1967 Nov. 2	T	S Atlantic
1968 March 28/9	P	Pacific, Antarctica
1968 Sept. 22	T	Arctic, Mongolia, Siberia
1969 March 18	A	Indian Ocean, Pacific
1969 Sept. 11	A	Peru, Bolivia
1970 March 7	T	Mexico, USA, Canada
1970 Aug. 31/ Sept. 1	T	East Indies, Pacific
1971 Feb. 25	P	Europe, NW Africa
1971 July 22	P	Alaska, Arctic
1971 Aug. 20	P	Australasia, S Pacific
1972 Jan 16	A	Antarctica
1972 July 10	T	Alaska, Canada
1973 Jan. 4	A	Pacific, S Atlantic
1973 June 30	T	Atlantic, N Africa, Kenya, Indian Ocean
1973 Dec. 24	A	Brazil, Atlantic, N Africa
1974 June 20	T	Indian Ocean
1974 Dec. 13	P	N and C America
1975 May 11	P	Europe, N Asia, Arctic
1975 Nov. 3	P	Antarctic

Date	Type	Area
1976 April 29	A	NW Africa, Turkey, China
1976 Oct. 23	T	Tanzania, Indian Ocean, Australia
1977 April 18	A	Atlantic, SW Africa, Indian Ocean
1977 Oct. 12	T	Pacific, Peru, Brazil
1978 April 7	P	Antarctic
1978 Oct. 2	P	Arctic
1979 Feb. 26	T	Pacific, USA, Canada, Greenland
1980 Aug. 10	A	S Pacific, Brazil
1981 Feb. 4	A	Pacific, S Australia, New Zealand
1981 July 31	T	Russia, N Pacific
1982 Jan. 25	P	Antarctic
1982 June 21	P	Antarctic
1982 July 20	P	Arctic
1982 Dec. 15	P	Arctic
1983 June 11	T	Indian Ocean, E Indies, Pacific
1983 Dec. 4	A	Atlantic, Equatorial Africa
1984 May 30	A	Pacific, Mexico, USA, N Africa
1984 Nov. 22/3	T	E Indies, S Pacific
1985 May 19	P	Arctic
1985 Nov. 12	T	S Pacific, Antarctica
1986 April 9	P	Antarctic
1986 Oct. 3	T	N Atlantic
1987 Mar. 29	T	Argentina, C Africa, Indian Ocean
1987 Sept. 23	A	Russia, China, Pacific
1988 Mar. 18	T	Indian Ocean, E Indies, Pacific
1989 Mar. 7	P	Arctic
1989 Aug. 31	P	Antarctic
1990 Jan. 26	A	Antarctic
1990 July 22	T	Finland, Russia, Pacific

SOLAR PROBES

*T*he Sun emits not only visible light, but also radiations in other parts of the electromagnetic spectrum – from very long radio waves through to ultra-short X-rays. *The first suggestion of a 'solar wind'*, made up of streams of low-energy atomic particles sent out constantly in all directions, was made on theoretical grounds by S. Chapman and V. C. A. Ferraro in 1931. In 1958 E. Parker showed that the corona must be in a stage of constant expansion, so that particles will escape all the time and by-pass the Earth at 400 to 800 km/s. The existence of the solar wind was proved soon after the start of the Space Age, by the early Russian vehicles of 1959–61 and by the US Pioneer and Explorer probes.

✳ *The first attempt at solar observations from high altitude* was made in 1914 by Charles Abbott, using an automated pyrheliometer launched from Omaha by hydrogen-filled rubber balloons. The altitude reached was 24·4 km. In 1935 a balloon, Explorer II, took a two-man crew to the same height.

SOLAR ECLIPSES 1991–2006

Date	UT (h)	Type	Max. duration (if total or annular) m s	% eclipsed (if partial)	Area
1991 Jan. 15	24	A	7 55	–	Pacific, New Zealand, Tasmania, SW Australia
1991 July 11	19	T	6 54	–	Pacific, Mexico, Hawaii
1992 Jan. 4	23	A	11 42	–	Pacific
1992 June 30	12	T	5 20	–	Atlantic
1992 Dec. 24	01	P	–	84	Arctic
1993 May 21	14	P	–	73	Arctic
1993 Nov. 13	22	P	–	93	Antarctic
1994 May 10	17	A	6 14	–	Pacific, Mexico, USA, Canada
1994 Nov. 3	14	T	4 23	–	Peru, Brazil, S Atlantic
1995 Apr. 29	18	A	6 38	–	S Pacific, Peru, S Atlantic
1995 Oct. 24	05	T	2 10	–	Iran, India, E Indies, Borneo, Pacific
1996 Apr. 17	23	P	–	88	Antarctic
1996 Oct. 12	14	P	–	76	Arctic
1997 Mar. 9	01	T	2 50	–	Siberia, Arctic
1997 Sept. 2	00	P	–	90	Antarctic
1998 Feb. 26	17	T	4 08	–	Pacific, Atlantic
1998 Aug. 22	02	A	3 14	–	Indian Ocean, E Indies, Pacific
1999 Feb. 16	07	A	1 19	–	Indian Ocean, Australia, Pacific
1999 Aug. 11	11	T	2 23	–	Atlantic, England (Cornwall), France, Turkey, India
2000 Feb. 5	13	P	–	56	Antarctic
2000 July 31	02	P	–	60	Arctic
2000 Dec. 25	18	P	–	72	Arctic
2001 June 21	12	T	4 56	–	Atlantic, South Africa
2001 Dec. 14	21	A	3 54	–	Central America, Pacific
2002 June 10	24	A	1 13	–	Pacific
2002 Dec. 4	08	T	2 04	–	S Africa, Indian Ocean, Australia
2003 May 31	04	A	3 37	–	N Scotland, Iceland
2003 Nov. 23	23	T	1 57	–	Antarctic
2004 Apr. 19	14	P	–	74	Antarctic
2004 Oct. 14	03	P	–	93	Arctic
2005 Apr. 8	21	T	0 42	–	Pacific, America, N of S America
2005 Oct. 3	11	A	4 32	–	Atlantic, Spain, Africa, Indian Ocean
2006 Mar. 29	10	T	4 07	–	Atlantic, Africa, Turkey, Russia
2006 Sept. 22	12	A	7 09	–	NE of S America, Atlantic, S Indian Ocean

There will be total eclipses on 1 August 2008, 22 July 2009, 11 July 2010, 13 November 2012, 20 March 2015, 9 March 2016, 21 August 2017, 2 July 2019, 14 December 2020, 4 December 2021, 8 April 2024, 12 August 2026, 2 August 2027, 22 July 2028, 25 November 2030, 14 November 2031, 30 March 2033, 20 March 2034 and 2 September 2035.

Following 11 August 1999 (Cornwall, Channel Islands), there will be British total eclipses on:

2015 Mar. 20	(narrowly misses North Scotland)
2081 Sept. 3	(Channel Islands)
2090 Sept. 23	(South Ireland, Isles of Scilly, Cornwall, Channel Islands)
2133 June 3	(South Scotland, Shetland Islands)
2135 Oct. 7	(South Scotland)
2142 May 25	(brushes Southern England)
2151 June 14	(Belfast, Isle of Man, Liverpool, Birmingham)
2160 June 4	(South-West Ireland, Isles of Scilly, Cornwall)

Annular eclipses visible from Britain will fall on 31 May 2003 and 23 July 2093. In fact, the track in 2003 will brush N Scotland and annularity will be well seen from Iceland – clouds permitting!

The following British total eclipses have occurred since AD 1 (*indicates a near miss):

London	158, 228, 319, 393, 878, 1715
Edinburgh	183, *303, 594, 865, 1133, 1433, 1598, 1652
Belfast	183, 228, 393, 594, 664, 865, 878, *885, 1652
Cardiff	*129, 228, 319, 458, 1023, 1140, 1715, 1724
Isle of Scilly	19, 28, *118, *458, *758, 968, 1023, 1598, *1630, 1715. As these islands will again see totality in *2081, 2090, 2160 and 2189, they seem to be particularly fortunate!

London will see totality in the year 2600.

Many solar probes have now been launched. Most of them carry out general studies of the Sun as well as specialized programmes. The following list includes some of the more important vehicles, with specialized investigations given.

✳ **The first space vehicle devoted entirely to studies of the Sun** was OSO 1 (Orbiting Space Observatory 1), launched on 4 September 1964. It carried 13 experiments, including those designed to obtain solar data at ultra-violet, X-ray and gamma-ray wavelengths.

✳ **The first detailed solar observations from a manned space-craft** were made from the US station Skylab, launched on 14 May 1973 (though some solar work had been carried out by the Apollo lunar astronauts, and solar wind particles had been collected by equipment deployed on the Moon's surface). The equipment on Skylab could monitor the Sun at wavelengths from visible light through to X-rays. The last of the three crews left on 8 February 1974, and Skylab itself decayed in the atmosphere in 1979.

✳ **The most elaborate solar vehicle to date** has been the Solar Maximum Mission, SMM ('Solar Max'), launched on 14 February 1980 into a circular, 574-km orbit. It was designed to study the Sun during the peak of a cycle. Following a fault, the vehicle was repaired in April 1984 by a crew from the Space Shuttle, and finally decayed on 2 December 1989.

✳ **The first vehicle designed to survey the poles of the Sun** is the European spacecraft Ulysses, launched on 6 October 1990 from the Shuttle. It swung past Jupiter on 8 February 1992, at 430000 km; Jupiter's gravity sent the probe south of the ecliptic to survey the Sun's south pole (June to October 1994). It recrossed the ecliptic in February 1995, and surveyed the Sun's north pole from June to September 1995. The results have been decidedly unexpected, and in particular the Sun's south magnetic pole appears to be very ill-defined. It may be that many of our ideas about the Sun's magnetic structure will have to be revised.

✳ **The first attempt at solar research by a modern-type rocket** was made in 1946, when a captured and converted German V2 was launched from White Sands, New Mexico. It reached 55 km, and recorded the solar spectrum down to 2400 Ångströms. **The first X-ray solar flares** were recorded in 1956, from balloon-launched rockets, though solar X-rays had been detected as early as 1948.

✳ **The closest deliberate approach to the Sun** was that of the German-built Helios 2, in 1976. The minimum distance from the Sun was 45000000 km.

SOLAR MISSIONS

Name	Launch data	Nationality	Experiments
Pioneer 4	3 Mar. 1959	American	Lunar probe, but in solar orbit: solar flares.
Vanguard 3	18 Sept. 1959	American	Solar X-rays.
Pioneer 5	11 Mar. 1960	American	Solar orbit, 0·806 × 0·995 a.u. Flares, solar wind. Transmitted until 26 June 1960, at 37 000 000 km from Earth.
OSO 1	7 Mar. 1962	American	Orbiting Solar Observatory 1. Earth orbit, 553 × 595 km. Lost on 6 Aug. 1963.
Cosmos 3	24 Apr. 1962	Russian	Earth orbit, 228 × 719 km; decayed after 176 days. Solar and cosmic radiation.
Cosmos 7	28 July 1962	Russian	Earth orbit, 209 × 368 km. Monitoring solar flares during Vostok 3 and 4 missions. Decayed after 4 days.
Explorer 18–IMP	26 Nov. 1963	American	Interplanetary Monitoring Platform 1. Earth orbit, 125 000 × 202 000 km. Provision for flare warning during manned missions.
OGO 1	4 Sept. 1964	American	Orbiting Geophysical Observatory 1. Earth–Sun relationships.
OSO 2	3 Feb. 1965	American	Earth–Sun relationships. Flares.
Explorer 30–Solrad	18 Nov. 1965	American	Solar radiation and X-rays; part of the IQSY programme (International Year of the Quiet Sun).
Pioneer 6	16 Dec. 1965	American	Solar orbit, 0·814 × 0·985 a.u. First detailed space analysis of solar atmosphere.
Pioneer 7	17 Aug. 1966	American	Solar orbit, 1·010 × 1·125 a.u. Solar atmosphere. Flares.
OSO 3	8 Mar. 1967	American	Earth–Sun relationships. Flares.
Cosmos 166	16 June 1967	Russian	Earth orbit, 260 × 577 km. Solar radiation. Decayed after 130 days.
OSO 4	18 Oct. 1967	American	Earth–Sun relationships. Flares.
Pioneer 8	13 Dec. 1967	American	Solar orbit, 1·00 × 1·10 a.u. Solar wind; programme with Pioneers 6 and 7.
Cosmos 215	19 Apr. 1968	Russian	Solar orbit, 260 × 577 km. Solar radiation. Decayed after 72 days.
Pioneer 9	8 Nov. 1968	American	Solar orbit, 0·75 × 1·0 a.u. Solar wind, flares etc.
HEOS 1	5 Dec. 1968	American	High-Energy Orbiting Satellite. Earth orbit, 418 × 112 440 km. With HEOS 2, monitored 7 of the 11-year solar cycle.
Cosmos 262	26 Dec. 1968	Russian	Earth orbit, 262 × 965 km. Solar X-rays and ultra-violet.
OSO 5	22 Jan. 1969	American	Earth orbit, 550 km, inclination 32°·8. General solar studies.
OSO 6	9 Aug. 1969	American	Earth orbit, 550 km, inclination 32°·8. General solar studies, as with OSO 5.
Shinsei SS1	28 Sept. 1971	Japanese	Earth orbit, 870 × 1870 km. Operated for 4 months.
OSO 7	29 Sept. 1971	American	Earth orbit, 329 × 575 km. General studies: solar X-ray, ultra-violet, EUV. Operated until 9 July 1974, despite having been put into the wrong orbit.
HEOS 2	31 Jan. 1972	American	Earth orbit. High-energy particles, in conjunction with HEOS 1.
Prognoz 1	14 Apr. 1972	Russian	First of a series of Russian solar wind and X-ray satellites. (Prognoz = Forecast.) Earth orbit, 965 × 200 000 km.
Prognoz 2	29 June 1972	Russian	Earth orbit, 550 × 200 000 km. Solar wind and X-ray studies.
Prognoz 3	15 Feb. 1973	Russian	Earth orbit, 590 × 200 000 km. General solar studies, including X-ray and gamma-rays.
Intercosmos 9	19 Apr. 1973	Russian–Polish	Earth orbit, 202 × 1551 km, inclination 48°. Solar radio emissions.
Skylab	14 May 1973	American	Manned mission. 3 successive crews. Decayed 11 July 1979.
Intercosmos 11	17 May 1974	Russian	Earth orbit, 484 × 526 km. Solar ultra-violet and X-rays.
Explorer 52–Injun	3 June 1974	American	Solar wind.
Helios 1	1 Dec. 1974	German	American-launched. Close-range studies; went to 48 000 000 km from the Sun.
Aryabhāta	19 Apr. 1975	Indian	Russian-launched. Solar neutrons and gamma-radiation.
OSO 8	18 June 1975	American	General studies, including solar X-rays.
Prognoz 4	22 Dec. 1975	Russian	Earth orbit, 634 × 199 000 km. Continuation of Prognoz programmes.
Helios 2	15 June 1976	German	American-launched. Close-range studies; went to 45 000 000 km from the Sun.
Prognoz 5	25 Nov. 1976	Russian	Earth orbit, 510 × 199 000 km. Carried Czech and French experiments.
Prognoz 6	22 Sept. 1977	Russian	Earth orbit. Effects of solar X-rays and gamma-rays on Earth's magnetic field.
Solar Maximum Mission	14 Feb. 1980	American	Long-term vehicle. Repaired in orbit April 1984; decayed December 1989.
Ulysses	6 Oct. 1990	European	American-launched. Solar Polar probe.
Yohkoh	30 Aug 1991	Japanese	X-ray studies of the Sun.
Koronos–1	2 May 1994	Russian	Longterm studies; carries coronagraph and X-ray telescope.

In addition, some temporary solar satellites have been released from Shuttles – for example Spartan-201, released from the Shuttle *Discovery* on 11 April 1993 and retrieved two days later; it carried out dynamical studies of the Sun's outer atmosphere.

THE MOON

DATA

Distance from the Earth, centre to centre:	**Orbital eccentricity:** 0·0549
mean: 384 400 km	**Diameter:** 3475·6 km
closest (perigee): 356 410 km	**Apparent diameter from Earth:**
furthest (apogee): 406 697 km	*mean:* 31 min 5 sec
Distance from Earth, surface to surface:	*max:* 33 min 31 sec
mean: 376 284 km	*min:* 29 min 22 sec
closest (perigee): 348 294 km	**Reciprocal mass, Earth = 1:** 81·3
furthest (apogee): 398 581 km	**Density, water = 1:** 3·342
Revolution period: 27·321 661 days	**Mass** = $7·35 \times 10^{25}$ g
Axial rotation period: 27·321 661 days	**Mass, Earth = 1:** 0·0123
Synodic period: 29 d 12 h 44 m 2·9 s	**Volume, Earth = 1:** 0·0203
Mean orbital velocity: 3680 km/h	**Escape velocity:** 2·38 km/s
Axial inclination of equator, referred to	**Surface gravity, Earth = 1:** 0·1653
ecliptic: 1 deg 32 min	**Albedo:** 0·07
Orbital inclination: 5 deg 09 min	**Mean magnitude, at full:** −12·7

*T*he Moon is officially ranked as the Earth's satellite. Relative to its primary, it is however extremely large and massive, and it might well be better to regard the Earth-Moon system as a double planet.

The earliest sighting of the 'new moon' seems to have been made on 15 March 1972 by R. Moran of California, using 10 × 50 binoculars. The Moon was then 14 h 53 m past conjunction.

✳ *The first British political party to time its meetings according to the phases of the Moon* was probably the Newcastle Green Party, in 1992. It announced that it would meet at new moon to discuss ideas and full moon to act on them!

✳ *The first suggestion that the Moon is mountainous* was made by the Greek philosopher Democritus (460–370 BC). Earlier, Xenophanes (*c.* 450 BC) had supposed that there were many suns and moons according to the regions, divisions and zones of the Earth!

✳ *The first map of the Moon* was drawn by W. Gilbert, probably around 1600 (Gilbert died in 1603) though not published until 1651. This was compiled with the naked eye, as telescopes had not then come into use.

✳ *The first telescopic map of the Moon* was drawn by Thomas Harriot in July 1609. It shows a number of identifiable features,

and was much more accurate than the drawings made by Galileo from 1610.

✳ *The first British telescopic observations of the Moon following those of Harriot* seem to have been made by Sir William Lower, from Wales, from about 1611. Lower's drawings have not survived, but he compared the appearance of the Moon with a tart that his cook had made!

✳ *The first serious attempts to measure the heights of lunar mountains* were made by Galileo, from 1610. He measured the heights of the Lunar Apennines, and though he over-estimated the altitudes his results were of the right order. Much better results were obtained by J. H. Schröter, from 1778.

✳ *The first explanation of the faint visibility of the 'night' side of the Moon*, when the Moon is in the crescent stage, was given by Leonardo da Vinci (1452–1519). Leonardo correctly stated that the phenomenon is due to light reflected on the Moon from the Earth.

✳ *The first systems of lunar nomenclature were introduced* in 1645 by Langrenus, and in 1647 by Hevelius; but these have not survived. (For instance, the crater now called Plato was called by Hevelius 'the Greater Black Lake'.) The modern system was introduced in 1651 by G. Riccioli, who named the features in honour of scientists – plus a few

The Lunar crater Linné, from Apollo 15 (1971). The diameter is 2150 m.

others. He was not impartial; for instance he allotted a major formation to himself and another to his pupil Grimaldi, but he did not believe in the 'Copernican' theory that the Earth moves round the Sun, and so he 'flung Copernicus into the Ocean of Storms'. Riccioli's principle has been followed since, though clearly all the major craters were 'used up' and later distinguished scientists have had to be given formations of lesser importance.

✳ *The first system of lunar co-ordinates* was introduced by T. Mayer in 1775. The first accurate measures with a heliometer were made by F. W. Bessel in 1839.

✳ *The first detailed statement that the Moon has a synchronous rotation* was made by G. D. Cassini in 1693, though Galileo may also have realized it. The Moon's rotation is equal to its revolution period (27·3 Earth-days), and so the same face is turned Earthward all the time, though the eccentricity of the lunar orbit means that there are 'libration zones' which are brought alternately into and out of view. From the Earth, 59 per cent of the Moon's surface can be studied at one time or another. The remaining 41 per cent is inaccessible. There is no mystery about the behaviour; tidal forces over the ages have been responsible. Most other planetary satellites also have synchronous rotation with respect to their primaries.

✳ *The first detailed descriptions of lunar rills (also: rilles or clefts)* were given by J. H. Schröter, from 1778, though rills had been observed earlier by Huygens.

✳ *The first really good map of the Moon* was published in 1837–8 by W. Beer and J. H. Mädler, of Berlin. Though they used a small telescope (Beer's 3·75 in refractor) their map was amazingly good, and remained the best for several decades. They also published a book, *Der Mond*, the first detailed description of the lunar surface.

✳ *The first detailed English book about the Moon* was published in 1876 by E. Neison (real name, Nevill). Neison included a map, which was basically a revision of that by Beer and Mädler.

✳ *The largest lunar map compiled by a visual observer* was that by the Welsh selenographer H. P. Wilkins. It was 300 in in diameter. The final revision, to one-third this scale, appeared in 1959.

✳ *The first photograph of the Moon* (a Daguerreotype) was taken on 23 March 1840 by J. W. Draper, using a 5 in reflector. The image was 1 in across, and the exposure time was 20 minutes. *The first good photographic atlas* was published in 1899 by the French astronomers Loewy and Puiseux.

Crescent Moon. (Henry Brinton)

✳ *The first report of a 'lunar city'* was made in 1822 by the German astronomer F. von P. Gruithuisen, who described a structure with 'dark gigantic ramparts' – which, however, proved to be low and quite haphazard ridges! In 1790 Sir William Herschel had stated his belief in an inhabited Moon, and in 1835 there came the famous 'lunar hoax', when a paper, the New York *Sun*, published some quite imaginary reports of discoveries made by Sir John Herschel at the Cape of Good Hope. The reports were written by a reporter, R. A. Locke, and included descriptions of bat-men and quartz mountains. The first article appeared on 25 August, and the *Sun* admitted the hoax on 16 September.

✳ *The first serious suggestion of change on the Moon* was made in 1866 by J. Schmidt, from Athens. Schmidt believed that a crater, Linné on the Mare Serenitatis, had been transformed into a pit surrounded by a white patch. The controversy continued for many years, but it is not now believed that any change occurred there.

✳ *The first reports of active volcanoes on the Moon* were made by Sir William Herschel, in 1783 (May 4) and again in 1787 (April 19 and 20). However, it does not seem

that Herschel saw anything more significant than bright areas shining by earthlight. In modern times, lunar events or Transient Lunar Phenomena (TLP) have been reported on many occasions. *The first photographic confirmation* was obtained on 3 November 1958 by N. A. Kozyrev, at the Crimean Astrophysical Observatory, of an event in Alphonsus. On 30 October 1963 an event in the Aristarchus area was observed at the Lowell Observatory by J. Greenacre and J. Barr. *The first comprehensive catalogue* of TLP reports was published by NASA, in 1967, by B. Middlehurst and Patrick Moore; Moore published a supplement in 1971. The total number of events reported in these catalogues is 713: though there is no doubt that many of them are due to observational error or misinterpretation, it is generally agreed that some are genuine, and that the Moon is not totally inert.

✳ *The most famous 'blue moon'* of recent times was that of 26 September 1950. The phenomenon was due to dust in the Earth's atmosphere, raised by vast forest fires in

Canada. There was also a blue moon on 27 August 1883, caused by material sent up from the Krakatoa outburst; and green moons were seen in Sweden in 1884 – at Kalmar on 14 February, for 3 minutes, and at Stockholm on 17 January, also for 3 minutes.

✳ *The temperatures on the lunar surface* were first measured with reasonable accuracy by the fourth Earl of Rosse, from Birr Castle in Ireland. Lord Rosse's papers, from 1869, indicated that at noon the surface temperature approached 100°C, though later Langley erroneously concluded that the temperature never rose above 0°C. It is known that the equatorial noon temperature is about 117°C, falling to –163°C at minimum.

✳ *The first radar echoes from the Moon* were obtained in 1946 by Z. Bay, from Hungary. One recent determination of the Moon's distance, by a laser measure (J. Faller, Lick) gave 353 911·218 km ±45 m from which the mean distance can be derived. At the Moon, the laser beam has a diameter of 4 km. The telescope used was the 1·4 m reflector at Catalina (Arizona).

✳ *The first good X-ray picture of the Moon* was obtained on 29 June 1990, from the Rosat satellite.

✳ *The largest regular 'sea'* is the Mare Imbrium, diameter 1300 km. It is 7 km deeper in its centre than at its periphery. Of the other regular maria, Orientale has a diameter of 965 km; Serenitatis, 690 km × 885 km; Crisium, 590 km × 460 km; Humorum, 423 km × 460 km; Humboldtianum, 305 km × 410 km. Mare Crisium is longest in an East–West direction, though from Earth foreshortening makes it appear longest North–South. Of the irregular seas, the Oceanus Procellarum is much larger than the Mare Imbrium. On average the Oceanus Procellarum is only 1 km lower than the mean sphere, whereas the deeper seas, such as Mare Crisium, are 4 km lower. The Mare Orientale (discovered by Wilkins and Moore in 1938) is the only major sea which extends on to the Moon's far side, and is very badly placed for observation from Earth.

✳ *The largest basin on the Moon* may be the South Pole-Aitken Basin, which is 2500 km across and 13 km deep. It was surveyed by the Clementine spacecraft in 1994, and covers almost a quarter of the Moon's circumference. If it is due to meteoritic impact, as most astronomers believe, its age may be from 3·8 to 4·3 thousand million years, and to have been produced by the impact of an object 200 km across.

✳ *The largest crater* is Bailly, diameter 295 km (depth 3·96 km). Next, insofar as the Moon's near side is concerned, is Clavius

The Lunar surface: in the centre is the Mare Serenitatis, 100 in photograph, Mount Wilson.

(232 km diameter, 4·91 km deep). Many of the large structures are much shallower – such as Ptolemæus (diameter 148 km, depth 2·4 km). Craters with central structures include Theophilus (101 km diameter, 4·4 km deep); Copernicus (97 km, 3·8 km) and Tycho (84 km, 4·8 km). The great dark formation Plato has a depth of 2·4 km.

✳ *The deepest crater* is thought to be Newton, with a floor 8·85 km below the crest of its wall; the wall rises 2·25 km above the outer surface.

✳ *The brightest crater* on the Moon is Aristarchus, which often appears prominently when earthlit. *The darkest area* is the floor of Grimaldi.

✳ *The greatest ray-crater* is Tycho, whose rays dominate the scene at or near full moon. Second to it is Copernicus, on the Oceanus Procellarum. Other important ray-centres include Kepler, Olbers, Anaxagoras, and Thales.

✳ *The most famous plateau* on the Moon is Wargentin, which is lava-filled. All other known plateaux are much smaller and less regular.

✳ *The most famous rills* on the Moon are those in the Mare Vaporum area (Hyginus, Ariadæus) and the Hadley Rill in the Apennine area, visited by the Apollo 15 astronauts; the Hadley Rill is 135 km long, 1 to 2 km wide, and 370 m deep. Other famous rills or rill-systems include those associated with Sirsalis, Bürg, Hesiodus, Triesnecker, Ramsden and Hippalus. Extending from Herodotus, near Aristarchus, is the imposing valley known as Schröter's Valley in honour of its discoverer, J. H. Schröter (in a way rather misleading, since the crater

named after Schröter is in a completely different area).

✳ *Domes*, up to 80 km in diameter, occur in various parts of the Moon – for instance near Arago in the Mare Tranquillitatis, Prinz in the general area of Aristarchus, and on the floor of Capuanus. Many of them have summit pits; their slopes are gentle. Domes were first described in detail by the British selenographer R. Barker in 1932.

✳ *The first description of the 'grid system'* (families of linear features, aligned in definite directions) was given by the American geologist J. E. Spurr in 1945; Spurr also invented the term. Most astronomers believe it to be of considerable significance in studies of the Moon's past history.

✳ *The origin of the lunar formations* has caused considerable controversy. There have been many eccentric theories; for example P. Fauth, who died in 1943, supported the idea that the Moon is covered with ice; Weisberger, who died in 1942, denied the existence of any lunar mountains or craters, and attributed the effects to storms and cyclones in a dense atmosphere; Ocampo, in 1951, claimed that the craters were the result of an atomic war between two races of Moon-men (the fact that some craters have central peaks while others have not proves, of course, that the two sides used different kinds of bombs). However, in recent times the only serious question has been as to whether the craters were produced by internal action – that is to say, vulcanism – or whether they are due to impact.

There are still a few supporters of the theory that the main craters are of internal origin, and are related to calderas; these include Jack Green in the United States, Gilbert Fielder and the author of this book in Britain. It is pointed out that the craters are not distributed at random, as they would be if produced by bombardment, and that the samples brought home are volcanic, with meteoritic material in short supply. Some small features are unquestionably of internal origin. However, almost all authorities now support the impact theory, which was proposed by F. von Paula Gruithuisen in 1824, revived by G. K. Gilbert in 1892, and put into its modern form by Ralph Baldwin in 1949.

According to the impact theory, the sequence of events was as follows. The Moon was formed at about the same time as the Earth (about 4600 million years ago). The heat generated during the formation made the outer layers molten down to a depth of several hundred kilometres; less dense minerals then separated out to the surface and, in the course of time, produced a crust.

Tremendous meteoritic impacts then produced the maria and the major craters.

The oldest maria, such as Mare Tranquillitatis and Mare Fœcunditatis, are at least 4500 million years old; Mare Imbrium is 3900 million years, and the youngest mare, Mare Orientale, around 3850 million years (of course, each mare covers a span of ages). Then, for 700 million years – between 3900 and 3200 million years ago – there was widespread volcanic activity, with lava pouring out from below the crust and flooding the mare floors. The lava-flows ended rather suddenly, and since then the Moon has seen little activity apart from the formation of occasional impact craters. For example, it has been claimed that Copernicus is no more than 1000 million years old, and that some small craters are even younger.

✳ **The first reports of mascons, or mass concentrations,** were made by P. Muller and W. L. Sjögren in 1968, from studies of the movements of the probe Orbiter 5. Exceptionally dense areas of material were found to lie below various regular features – Imbrium, Crisium, Smythii, Serenitatis, Humorum, Nectaris, Humboldtianum, Orientale, Æstuum and Grimaldi. On the other hand, the material below some of the other features, such as Copernicus, was of much lower density. (In 1968 B. Middlehurst and Patrick Moore had shown that TLP are often associated with features near the boundaries of the regular seas.)

✳ **The first attempted lunar probe** was Able 1, launched by the Americans on 17 August 1958. It was not successful.

✳ **The first successful probe** to the Moon was Luna 1 (2 January 1959). It by-passed the Moon, sending back valuable data.

✳ **The first probe to land on the Moon** was Luna 2 (13 September 1959).

✳ **The first photographs of the Moon's far side** were obtained from Luna 3, in October 1959.

✳ **The first good close-range pictures of the Moon** were obtained from Ranger 7 (31 July 1964). The region, not far from Guericke, has been named the Mare Cognitum, or Known Sea, though the name has not come into general use.

✳ **The first successful soft landing on the Moon** was made by Luna 9, launched on 31 January 1966.

✳ **The first lunar satellite** was Luna 10, launched on 31 March 1966.

✳ **The first lunar orbiter** to be photographed from Earth while in a path round the Moon was Orbiter 5, imaged on 21 January 1968 by J. Fountain, S. Larson and G. Kuiper, using the 100 in reflector at Catalina. The magnitude ranged between 12 and 15.

✳ **The first manned flight round the Moon** was that of Apollo 8, in December 1968.

✳ **The first landing on the Moon** was that of Apollo 11, on 20 July 1969. Early on 21 July Neil Armstrong stepped out on to the lunar surface, followed shortly afterwards by Edwin Aldrin.

✳ **The first unmanned probe to return samples from the Moon** was Luna 16, launched on 12 September 1970.

✳ **The first 'crawler' on the Moon** was Lunokhod 1. It was taken to the Moon by Luna 17, and landed there on 17 November 1970.

On 11 December 1993 all rights to the Russian Lunokhods were auctioned at Sotheby's in London; the price was $68 500. Presumably the buyer will have to collect them from the Moon!

✳ **The first 'Moon Car' or LRV** (Lunar Roving Vehicle) was taken to the Moon in Apollo 15 (30 July 1971). Astronauts Scott and Irwin used it to drive around in the Apennines area.

✳ **The first professional geologist to go to the Moon** was Harrison ('Jack') Schmitt, in Apollo 17 (11 December 1972). It was Schmitt who found the famous 'orange soil' near the crater which was known unofficially as Shorty. The colour proved to be due to small glassy particles which were very ancient; the age is 3·7 to 3·8 æons. (One æon: a thousand million years.)

✳ **The last man on the Moon** (so far!) was Eugene Cernan, in Apollo 17. When he re-entered the lunar module, following Dr Schmitt, the first stage in lunar manned exploration was over.

✳ **The first Japanese lunar probe,** Hagomoro, was launched on 24 January 1990. On 15 February 1992 the satellite Hiten (Space Flyer) was ejected and put into lunar orbit; on 10 April 1993 it crash-landed near the crater Furnerius.

✳ **The last lunar probe** (to 1995) was the US Clementine, named after the character in the old mining song who was 'lost and gone forever'! It was a joint NASA–USAF venture, and was launched not from Canaveral but from Vandenberg Air Force Base, on 25 January 1994. On 21 February it entered lunar orbit, and continued mapping until 23 April; by the time that the probe left lunar orbit, on 4 May, the whole of the Moon's surface had been mapped. Unfortunately, a fault developed, making it impossible to go on to an encounter with an asteroid (Geographos) as had been hoped. The last lunar flyby was on 20 July, the 25th anniversary of the Apollo 11 landing, after which Clementine entered a solar orbit. The mini-

mum distance from the Moon had been 425 km.

Clementine revealed various new features, such as a large basin (Coulomb-Sarton) in the Orientale area. The far-side crater Aitken was found to have record depth (12 km). There were reports of signs of ice inside some of the polar craters, which are always shadowed and are intensely cold. This seems very unlikely – the samples brought home have given no indication of hydrated materials – and the existence of ice has not been confirmed; further investigations are needed.

Clementine was among the cheapest of all lunar probes. The cost was $55 000 000.

The Apollo results have changed many of our ideas about the Moon. There is a loose upper layer or regolith, from 1 to 20 metres deep (on average 4 to 5 metres in the maria, 10 in the highlands); this layer contains many different ingredients. On the highlands, the rock fragments are chiefly anorthosites, with the minerals plagioclase, pyroxene, and/or olivine; on the maria, the rock fragments are basalt, with pyroxene, plagioclase and ilmenite. Below the regolith is a kilometre-thick layer of shattered bedrock. Under this is a layer of more solid rock going down to about 25 km in the Mare regions. Beneath this is a layer of material 30 km thick, with properties similar to the anorthosites and the feldspar-rich gabbros of the highlands. The bottom of this layer is 50 km from the surface. At 50 km another kind of rock appears; there is a sharp boundary. These new rocks are pyroxene and olivine rich, and are dense; they go down to at least 150 km, possibly much more. Beneath this is the core region, which is metallic, and may be 1000 to 1500 km in diameter. A 1-tonne meteoroid which hit the Moon on 17 July 1972 indicated, from the moonquake waves, that there is a zone 1000 to 1200 km deep inside the Moon where the rocks are hot enough to be molten. The crust on the far side of the Moon may be thicker than that on the Earth-turned side (75 to 100 km deep).

All the rocks brought home for analysis are igneous or breccias produced by impact processes; the Apollo missions recovered 2196 samples, with a total weight of 381·69 kg, now divided into 35 600 samples. The youngest rocks seem to be about 3·1 thousand million years old; the oldest, 4·6 thousand million – about the same as the age of the Moon itself. There are no sedimentary or metamorphic rocks. The main epoch of lava-flowing ended about 3 æons ago.

In the lavas, basalts are dominant. They contain more titanium than terrestrial lavas;

over 10 per cent in the Apollo 11 samples (3 to 5 per cent in the Apollo 12 samples, as against 1 to 3 per cent in terrestrial basalts). Small amounts of metallic iron were found. Generally, lunar rocks have one-tenth as much sodium and potassium as terrestrial rocks. A new mineral – an opaque oxide of iron, titanium and magnesium, not unlike ilmenite – has been named armalcolite (in honour of *Arm*strong, *Al*drin and *Col*lins).

All the returned highland rocks are of igneous origin, with twice as much calcium and aluminium as the Mare lavas, but less titanium, magnesium and iron. Plagioclase makes up at least half the highland rocks; there is little sodium or potassium, so that the volatiles must have been lost from the Moon at an early stage. The average age of highland rocks is from 4 to 4·2 æons. The so-called 'Genesis Rock' (Apollo 15) is anorthosite, with an age of 4 æons.

Breccias have been collected from every landing site on the Moon. These are complex rocks made up of shattered, crushed and sometimes once-melted pieces of rock. The fragments may be igneous rocks of various types, glasses, or bits of other breccias.

The Apollo 12 sample 12013 (collected by Conrad from the Oceanus Procellarum) is unique. It is about the size of a lemon, and contains 61 per cent of SiO_2, whereas the associated lavas have only 35 to 40 per cent of SiO_2. It also contains 40 times as much potassium, uranium and thorium, making it one of the most radioactive rocks found anywhere on the Moon. It is composed of a dark grey breccia, a light grey breccia and a vein of solidified lava.

(*Pyroxene*, a Ca-Mg-Fe silicate, is the most common mineral in lunar lavas, making up about half of most specimens; it forms yellowish-brown crystals up to a few centimetres in size. *Plagioclase* or feldspar, a Na or Ca-Al silicate, forms elongated white crystals. *Anorthosite* is a rock type containing the minerals plagioclase, pyroxene and/or olivine in various proportions. *Basalt* is a rock type containing the minerals plagioclase, pyroxene and ilmenite in varying proportions. *Olivine*, a Mg-Fe silicate, is made up of pale green crystals a few millimetres in size, and is not uncommon in the anorthosites. *Ilmenite*, present in the basalts, is an Fe-Ti oxide. It is notable that there are no hydrated materials on the Moon, so that evidently water has never existed there.)

✳ *Moonquakes* have been recorded regularly; there are about 3000 per year, centred mainly from 600 to 800 km below the surface, though very shallow moonquakes also occur. The impacts from the crashed lunar modules of the Apollo missions showed that the outermost few kilometres of the Moon are made up of cracked and shattered rock, so that signals can echo back and forth for hours; it was even said that after an impact of a module the Moon 'rang like a bell'! Researches by J. Green, B. Middlehurst and the author of this book have confirmed that moonquakes are most common at the time of lunar perigee, when the Moon's crust is under maximum strain.

There is virtually no overall magnetic field, though it seems that one did exist in the remote past. But, there are areas of magnetized material, particularly in and near the far-side crater Van de Graaff. All recording Apollo stations on the Moon were shut down on 30 September 1977, partly on the grounds of expense and partly because some of the equipment was showing signs of failure.

There is no Earth-type atmosphere, and the so-called lunar atmosphere is a collisionless gas in which hydrogen, helium, neon and argon have been identified; the density is about 2×10^5 molecules per cubic centimetre, which corresponds to what we normally call a very high laboratory vacuum. There is no trace of life, either past or present, and the quarantining of Apollo astronauts was dropped after the first three successful missions.

There have been a great deal of discussion about the *origin of the Moon*. The attractive tidal theory due to G. H. Darwin in the last century – that the Earth rotated so rapidly that it threw off a large piece of material, which became the Moon – is mathematically untenable, but in 1974 a new 'catastrophe' theory was proposed by W. Hartmann and D. R. Davis, involving a collision between the Earth and a large impacting body. According to this theory, the cores of the Earth and the impactor merged, and mantle débris ejected during the collision formed a transient cloud round the Earth which subsequently aggregated to form the Moon. Certainly the Earth and the Moon are of about the same age. The Earthward-turned axis of the Moon is 2·2 km longer than the rotational axis, to that there is a slight Earthward bulge; there are good reasons for believing that in its early history the Moon was much closer to the Earth than it is now. It is also important to note that the Moon is the only large satellite to move round a small planet (unless we include the Pluto–Charon pair), and in many ways it seems better to regard the Earth–Moon system as a double planet. The Moon's orbit is always concave to the Sun.

Because of tidal effects, the Moon is receding from the Earth at the rate of 3.8 cm per year; also, the same effects mean that the length of the Earth's rotation is lengthening, on average, by 0·0000002 s/day.

No minor Earth satellites of natural origin seem to exist. Careful searches for them have been made, as in 1957 by Tombaugh (discoverer of Pluto), but without result. Clouds of loose material in the Moon's orbit, at the Langrangian points, were reported by the Polish astronomer K. Kordylewski in 1961, but remain unconfirmed and efforts made on various occasions to photograph them have been unsuccessful.

NAMED MARE AREAS ON THE NEAR SIDE OF THE MOON

MARE		Lat	Long	
Mare Anguis	The Serpent Sea	22 N	67 E	Length 130 km, area 10 000 km². Narrow darkish area NE of M. Crisium.
Mare Australe	The Southern Sea	46 S	91 E	Irregular, patchy area in the SE. Area about 149 000 km². Extends to far side.
Mare Cognitum	The Known Sea	10 S	23 W	Part of M. Nubium, E of Riphæans. Landing site of Ranger 7 in 1964.
Mare Crisium	The Sea of Crises	17 N	59 E	Well-defined Mare, separate from main system. Diameter 450 × 560 km. Area 200 000 km² (similar to Great Britain).
Mare Fœcunditatis	The Sea of Fertility	SE quadrant		Irregular; confluent with M. Tranquillitatis. Area 344 000 km².
Mare Frigoris	The Sea of Cold	Far northern		Irregular; elongated. Area about 441 000 km² (including Lacus Mortis).
Mare Humboldtianum	Humboldt's Sea	57 N	80 E	Limb sea, beyond Endymion; fairly regular.
Mare Humorum	The Sea of Humours	24 S	39 W	Circular; leads off M. Nubium. Area 118 000 km². Diameter 380 km.
Mare Imbrium	The Sea of Showers	NW quadrant		Largest regular sea; 863 000 km² (equal to Great Britain and France combined). (Contains Palus Nebularum, Palus Putredinis.)
Mare Insularum	The Sea of Isles	7 N	22 W	Part of M. Nubium; Kepler/Encke region. Diameter 900 km. Contains Copernicus. Ill-defined.

MARE

		Lat	Long	
Mare Marginis	The Marginal Sea	12 N	88 E	Limb sea, beyond M. Crisium. Fairly well-defined. Area 62 000 km². Diameter 360 km.
Mare Nectaris	The Sea of Nectar	15 S	35 E	Leads off M. Tranquillitatis. Circular, diameter 350 km, area 100 000 km². Central part of a very ancient basin, whose border is marked by the Altai Scarp.
Mare Nubium	The Sea of Clouds	20 S	15 W	Ill-defined N border. Area 265 000 km², with M. Cognitum.
Mare Orientale	The Eastern Sea	20 S	95 W	Limb sea. Visible only under favourable libration, and then only in part. Vast ringed structure. Diameter 300 km.
Oceanus Procellarum	The Ocean of Storms	NW, SW quadrants		Area 2 290 000 km². Rather irregular.
Mare Serenitatis	The Sea of Serenity	NE quadrant		More or less circular. Area 314 000 km². Contains few large craterlets.
Mare Smythii	Smyth's Sea	2 S	87 E	Well-defined limb sea; area 104 000 km².
Mare Spumans	The Foaming Sea	1 N	65 E	Darkish area S of M. Crisium. Area 16 000 km².
Mare Tranquillitatis	The Sea of Tranquillity	NE quadrant		Confluent with M. Serenitatis and M. Nectaris; lighter and patchier than Serenitatis. Area 440 000 km². Rather irregular.
Oceanus Procellarum	Ocean of Storms	NW, SW quadrants		Area 2 290 000 km². Rather irregular.
Mare Undarum	The Sea of Waves	7 N	69 E	Darkish irregular area E of Firmicus. Area 21 000 km².
Mare Vaporum	The Sea of Vapours	13 N	3 E	Area 55 000 km². Diameter 230 km. SE of the Apennines. Contains some very dark patches.

LACUS

		Lat	Long	
Lacus Æstatis	Summer Lake	15 S	69 W	Two dark areas N of Crüger. Combined area about 1000 km².
Lacus Autumni	Autumn Lake	14 S	82 W	Dark patches in the Cordilleras. Extend for between 200 and 250 km.
Lacus Bonitatis	Lake of Goodness	23 N	44 E	Small darkish area near Macrobius. Diameter 130 km.
Lacus Doloris	Lake of Grief	17 N	9 E	Darkish area N of Manilius. Diameter 110 km.
Lacus Excellentiæ	Lake of Excellence	36 S	43 W	Vague Mare area near Clausius. Length about 150 km.
Lacus Felicitatis	Lake of Happiness	19 N	5 E	Small darkish area N of M. Vaporum. Diameter about 90 km.
Lacus Gaudii	Lake of Joy	17 N	13 E	Darkish area between Manilius and the Haemus Mountains. Diameter 100 km.
Lacus Hiemis	Winter Lake	15 N	14 E	Small darkish area SW of Menelaus. Diameter 50 km.
Lacus Lenitatis	Lake of Tenderness	14 N	12 E	80 km darkish area E of Manilius.
Lacus Mortis	The Lake of Death	45 N	27 E	Adjoins Lacus Somniorum. Area 21 000 km². Diameter 150 km. Dark; contains the Bürg rills.
Lacus Odii	Lake of Hate	19 N	7 E	70 km darkish patch W of the Haemus Mountains.
Lacus Perseverantiæ	Lake of Perseverance	8 N	62 E	70 km darkish patch adjoining Firmicus to the W.
Lacus Somniorum	Lake of the Dreamers	40 N	30 E	Irregular darkish area leading off M. Serenitatis. Area 70 000 km².
Lacus Spei	Lake of Hope	43 N	65 E	80 km dark strip between Messala and Zeno.
Lacus Temporis	Lake of Time	46 N	57 E	Irregular area between Mercurius and Atlas. Diameter about 250 km.
Lacus Timoris	Lake of Fear	39 S	28 W	Narrow darkish patch E of Hainzel. Length about 130 km.
Lacus Veris	Spring Lake	13 S	87 W	Narrow irregular area in Rook Mountains. Total diameter 540 km, area 12 000 km².

PALUS

		Lat	Long	
Palus Epidemiarum	Marsh of Epidemics	32 S	27 W	Darkish area adjoining Mercator and Campanus. Diameter 300 km. Area 27 000 km².
Palus Nebularum	Marsh of Clouds	40 N	6 W	Part of M. Imbrium, W of Aristillus and Autolycus. Name sometimes deleted.
Palus Putredinis	Marsh of Decay	27 N	0	Part of M. Imbrium, near Archimedes. Diameter about 180 km.
Palus Somnii	Marsh of Sleep	12 N	42 E	Curiously coloured area bounded by bright rays from Proclus.

SINUS

		Lat	Long	
Sinus Æstuum	Bay of Heats	12 N	8 W	Fairly regular dark area SE of Eratosthenes. Area 40 000 km². Diameter 230 km.
Sinus Amoris	Bay of Love	19 N	38 E	Part of M. Tranquillitatis, E of Maraldi.
Sinus Asperitatis	Bay of Asperity	6 S	25 E	Rough area between Theophilus and Hypatia; part of M. Nectaris. Diameter 180 km.
Sinus Concordiæ	Bay of Concord	11 N	43 E	160 km bay S of Palus Somnii.
Sinus Fidei	Bay of Faith	18 N	2 E	Outlet of M. Vaporum, to the N. Length 70 km.
Sinus Honoris	Bay of Honour	12 N	18 E	Edge of M. Tranquillitatis, NW of Maclear. Length about 100 km.
Sinus Iridum	Bay of Rainbows	45 N	32 W	Beautiful bay extending from M. Imbrium.
Sinus Lunicus	Bay of the Lunas	32 N	1 W	In M. Imbrium, between Aristillus and Archimedes; probable landing site of Luna 2 in 1959.
Sinus Medii	Central Bay	0	0	Small bay in the apparent centre of the disk. Area 52 000 km². Diameter 350 km.
Sinus Roris	Bay of Dew	52 N	58 W	Area joining M. Frigoris to Oceanis Procellarum.
Sinus Successus	Bay of Success	1 N	58 E	Ill-defined darkish area W of M. Spumans. Diameter 100 km.

CAPES (PROMONOTORIA)

	Lat	Long	
Ænarium (or Tænarium)	19 S	8 W	E edge of M. Nubium, N of the Straight Wall.
Agarum	14 N	66 E	E side of M. Crisium.
Agassiz	42 N	2 E	E edge of M. Imbrium; in the Alps.
Archerusia (or Acherusia)	17 N	22 E	N of M. Serenitatis; between Plinius and Tacquet.
Deville	43 N	1 E	E of M. Imbrium; in the Alps, S of Mont Blanc.
Fresnel	29 N	5 E	Apennines, E of Palus Putredinis.
Heraclides	41 N	34 W	N cape of Sinus Iridum.
Kelvin	27 S	33 W	M. Humorum, near Hippalus.
Laplace	46 N	26 W	S cape of Sinus Iridum.

These 'seas' are not alike. Some, such as Mare Imbrium, are regular and largely mountain-bordered, whereas others, such as Mare Undarum, are discontinuous, and consist only of patches of dark mare material. Most of the seas are connected; for instance Palus Nebularum (a name omitted from some lists) is part of the Mare Imbrium, while Sinus Iridum is a bay leading out from the Mare Imbrium. Isolated seas are rare; the Mare Crisium is a good example. The Mare Orientale spreads on to the far side of the Moon. The name was given to it by H. P. Wilkins and the present writer, but owing to a policy decision by the International Astronomical Union, reversing lunar east and west, the *Eastern Sea* is now on the *western* limb of the Moon.

MOUNTAIN RANGES

Name	Lat	Long	Notes
Agricola	29 N	54 W	In Aristarchus highlands; length 160 km.
Alps	45 N	1 E	N border of M. Imbrium; length 250 km; peaks ·1800–2400 m.
Altai	24 S	23 E	On M. Nectaris Basin; better termed the Altai Scarp; length 480 km.
Apennines	20 N	3 W	Border of M. Imbrium; length 600 km; some peaks 5000 m.
Archimedes	26 N	5 W	S of Archimedes; covers an area 140 km across.
Carpathians	15 N	25 W	400 km range along the S border of M. Imbrium.
Caucasus	39 N	9 E	Continuation of Apennines; length 520 km; peaks up to 6000 m; separated from the Apennines by a 50 km strait.
Cordilleras	20 S	80 W	Outer wall of M. Orientale; length 1500 km.
Haemus	17 N	13 E	Border of M. Serenitatis; length 400 km.
Harbingers	27 N	41 W	Mountain clumps near Aristarchus; covering almost 100 km².
Juras	47 N	37 W	Border of Sinus Iridum; 'Jewelled Handle' effect!
Pyrenees	14 S	41 E	Range S of Gutenberg; length 250 km.
Riphæans	7 S	28 W	In M. Nubium; near Euclides.
Rook	20 S	83 W	Length 900 km; one of the ranges surrounding M. Orientale.
Secchi	3 N	43 E	Low peaks NW of Secchi.
Spitzbergen	35 N	5 W	Chain N of Archimedes; length 60 km; peaks up to 1500 m.
Straight Range	48 N	20 W	Very regular range in N of M. Imbrium; W of Plato.
Taurus	26 N	36 E	Mountain clumps near Rømer.
Teneriffe	48 N	13 W	Mountain clumps E of the Straight Range.
Urals	5 S	26 W	North extension of the Riphæans.

MOUNTAINS

Name	Lat	Long	Notes
Ampère	19 N	4 W	Massif in the central Apennines; length 30 km.
Argæus	19 N	29 E	Border of M. Serenitatis; S of Littrow; 50 km massif.
Bradley	22 N	1 E	In Apennines, near Conon; 30 km massif.
De l'Isle	29 N	38 W	Mountain SW of De l'Isle; base diameter 30 km.
Hadley	27 N	5 E	25 km massif in the northern Apennines.
Hadley Delta	26 N	4 E	Near Hadley; near site of Apollo 15 landing.
Hansteen	12 S	52 W	Bright triangular massif; diameter 30 km.
Herodotus	23 N	50 W	5 km base massif N of Schröter's Valley.
Huygens	20 N	3 W	5400 m massif in the central Apennines; length 40 km.
Lahire	28 N	25 W	Isolated peak-massif in M. Imbrium; NW of Lambert; 10 × 20 km.
Maraldi	20 N	35 E	15 km long massif NE of Maraldi, on Sinus Amoris.
Mont Blanc	45 N	1 E	In Alps; base diameter 25 km; height 3600 m.
Moro	12 S	20 W	Highland SW of Bonpland; length 10 km; lies on a ridge.
Penck	10 S	22 E	Massif 4000 m high; length 30 km; NW of Theophilus.
Pico	46 N	9 W	Prominent triple-peaked mountain in M. Imbrium; 2400 m high; base 15 × 25 km; lies on the S edge of a ghost ring, whose N edge touches Plato (this ring was once known as Newton).
Piton	41 N	1 W	Bright mountain with a summit pit; height 2250 m; base diameter 25 km; on M. Imbrium; SE of Pico.
Rümker	41 N	58 W	Dome complex; very unusual; diameter of the formation 70 km; in many catalogues it is listed as a crater.
Usov	12 N	63 E	Just in M. Crisium; S of Agarum and near the E border; base diameter 15 km.
Vinogradov	22 N	43 W	Mountain group between Euler and Brayley; base diameter 25 km.
Vitruvius	19 N	31 E	Mountain; base diameter 15 km; between Vitruvius and Littrow.
Wolff	17 N	7 W	Massif in the SW cape of the Apennines; base diameter 35 km.

LUNAR CRATERS ON THE VISIBLE FACE OF THE MOON

Name	Lat	Long	Diam km	Notes
Abbot	6 N	55 E	10	Uplands; E of Taruntius; (Apollonius K).
Abel	35 S	86 E	114	Flooded wall plain; limb; E of Furnerius.
Abenezra	21 S	12 E	43	W of Nectaris; well-formed; pair with Azophi.
Abetti	20 N	28 E	7	Obscure; dark floor; in Serenitatis; NW of Argæus.
Abulfeda	14 S	14 E	64	Abenezra area; pair with Almanon.
Acosta	6 S	60 E	13	N of Langrenus. (Langrenus C).
Adams	32 S	69 E	64	E of Vendelinus; irregular walls.
Agatharchides	20 S	31 W	48	Humorum area; remains of c.p.; irregular walls.
Agrippa	4 N	11 E	48	Vaporum area; regular; c.p.; pair with Godin.
Airy	18 S	6 E	35	Pair with Argelander; irregular walls.
Al-Bakri	14 N	20 E	12	NW of Plinius. (Tacquet A).
Al-Biruni	18 N	93 E	77	Libration zone; Marginis area.
Al-Marrakushi	10 S	46 E	8	W of Langrenus. (Langrenus D).
Albategnius	12 S	4 E	129	Companion to Hipparchus; c.p.; irregular, terraced walls.
Aldrin	1·4 N	22 E	3·4	On Tranquillitatis; E of Sabine. (Sabine B).
Alexander	40 N	14 E	105	N end of Caucasus; darkish floor; low walls.
Alfraganus	6 S	19 E	19	NW of Theophilus; very bright; minor ray-centre.
Alhazen	18 N	70 E	32	Near border of Crisium. (Not Schröter's Alhazen.)
Aliacensis	31 S	5 E	84	Pair with Werner; Walter area; high walls; c.p.
Almanon	17 S	15 E	48	Regular; pair with Abulfeda.
Alpetragius	16 S	4 W	43	Outside Alphonsus; high, terraced walls; vast c.p. with summit pit.
Alphonsus	13 S	3 W	129	Ptolemæus group; low c.p.; rills on floor.
Ameghino	3 N	57 E	9	Uplands SW of Apollonius.
Ammonius	9 S	1 W	9	Prominent; in Ptolemæus. (Ptolemæus A).
Amontons	5 S	47 E	3	Craterlet on Fœcunditatis; S of Messier.

Name	Lat	Long	Diam km	Notes
Amundsen	84 S	83 E	105	Well-formed; pair with Scott.
Anaxagoras	75 N	10 W	52	N polar area; c.p.; distorts Goldschmidt; ray-centre.
Anaximander	66 N	48 W	87	Pythagoras area; pair with Carpenter; no c.p.
Anaximenes	75 N	45 W	72	Near Philolaus; rather low walls.
Ândël	10 S	13 E	40	Highlands W of Theophilus; low, irregular walls.
Andersson	50 S	95 W	13	Libration zone; S of Guthnick and Rydberg.
Ångström	30 N	42 W	11	N of Harbingers; in Imbrium.
Ansgarius	14 S	82 E	80	Distinct; E of Fœcunditatis; pair with La Peyrouse.
Anuchin	49 S	101 E	57	Libration zone; beyond Australe.
Anville	2 N	50 E	11	Fœcunditatis; E of Secchi. (Taruntius G).
Apianus	27 S	8 E	63	Aliacensis area; high walls.
Apollonius	5 N	61 E	48	Uplands; S of Crisium; well-formed.
Arago	6 N	21 E	29	On Tranquillitatis; domes nearby.
Aratus	24 N	5 E	13	In Apennines; bright; not regular in outline.
Archimedes	30 N	4 W	97	On Imbrium; very regular; darkish floor; no c.p.
Archytas	59 N	5 E	34	On Frigoris; bright; distinct; c.p.
Argelander	17 S	6 E	42	Albategnius area; pair with Airy; c.p.
Ariadaeus	5 N	17 E	15	Vaporum area; associated with great rill.
Aristarchus	24 N	48 W	37	Brightest crater on Moon; terraced walls; c.p.; inner bands.
Aristillus	34 N	1 E	56	Archimedes group; fine c.p.
Aristoteles	50 N	18 E	97	High walls; pair with Eudoxus.
Armstrong	1·4 N	25 E	3·4	M. Tranquillitatis; E of Sabine. (Sabine E).
Arnold	67 N	38 E	81	Low walls; NW of Democritus; N of Frigoris.
Arrhenius	56 S	91 W	40	Libration zone; beyond Inghirami.
Artemis	25 N	25 W	2	Pair with Verne; between Euler and Lambert.
Artsimovich	28 N	37 W	9	On Imbrium. (Diophantus A).
Aryabhata	6 N	35 E	22	Between Maskelyne and Cauchy; flooded; irregular. (Maskelyne E).
Arzachel	18 S	2 W	97	Ptolemæus group; high walls; c.p.
Asada	7 N	50 E	12	Distinct; NW of Taruntius; edge of Fœcunditatis. (Taruntius A).
Asclepi	55 S	26 E	32	S uplands; W of Hommel; distinct.
Aston	33 N	88 W	43	Limb; beyond Ulugh Beigh.
Atlas	47 N	44 E	89 ·	Pair with Hercules; high walls; much interior detail.
Atwood	6 S	58 E	29	(Langrenus K). Trio with Biharz and Naonobu; NW of Langrenus.
Autolycus	31 N	1 E	36	Archimedes group; regular; distinct.
Auwers	15 N	17 E	16	Foothills of Haemus; not very bright.
Auzout	10 N	64 E	31	Outside Crisium; low c.p.
Avery	1 S	81 E	9	Limb; N of Smythii. (Gilbert U).
Avicenna	40 N	97 W	74	Libration zone; N of Lorentz.
Azophi	22 S	13 E	43	W of Nectaris; well-formed. Pair with Abenezra.
Baade	45 S	82 W	55	Limb; beyond Schickard and Inghirami.
Babbage	58 N	52 W	145	Irregular enclosure near Pythagoras.
Babcock	4 W	94 E	99	Libration zone; Smythii area; beyond Neper.
Back	1 N	81 E	35	Limb; S of Schubert; (Schubert B).
Baco	51 S	19 E	64	Licetus area; high walls; low c.p.
Baillaud	75 N	40 E	80	Uplands; NE of Meton; rather low walls.
Bailly	66 S	65 E	294	'Field of ruins'; uplands on the S limb.
Baily	50 N	31 E	19	Frigoris area; N of Bürg.
Balboa	19 N	83 W	70	SE of Otto Struve.
Ball	36 S	8 W	40	On edge of Deslandres; high walls.
Balmer	20 S	71 E	112	Ruined walled plain W of Hekataeus.
Banachiewicz	5 N	80 E	92	Very foreshortened; N of Schubert.
Bancroft	28 N	6 W	13	Deep crater NW of Archimedes. (Archimedes A).
Banting	27 N	16 E	5	Craterlet in Serenitatis; E of Linné. (Linné E).
Barkla	11 S	67 E	43	Complex; between Langrenus and Kapteyn. (Langrenus A).
Barnard	30 S	86 E	100	Limb; beyond Ansgarius and Legendre.
Barocius	45 S	17 E	80	Outside Maurolycus; high but broken walls.
Barrow	73 N	10 E	87	NW of W.C. Bond; N polar area; low, broken walls.
Bartels	25 N	90 W	55	Limb; beyond Otto Struve.
Bayer	51 S	35 W	52	Outside Schiller; high, terraced walls.
Beals	37 N	87 E	48	Limb; beyond Gauss.
Beaumont	18 S	29 E	48	Bay on Nectaris.
Beer	27 N	9 W	11	On Imbrium; twin with Feuille.
Behaim	16 S	71 E	56	E of Vendelinus; high walls; c. craters.
Beketov	16 N	29 E	8	Craterlet N of Jansen; SW of Vitruvius. (Jansen C).
Bela	25 N	2 E	11 × 2	Elongated; origin of the Hadley Rill.
Belkovich	62 N	90 E	198	Humboldtianum area; high walls with 2 craters; c.peaks.
Bell	22 N	96 W	86	Libration zone; beyond Einstein.
Bellot	12 S	48 E	19	Edge of Fœcunditatis; bright floor.
Bernoulli	34 N	60 E	40	E of Geminus; fairly regular.
Berosus	33 N	70 E	61	E of Cleomedes; pair with Hahn; terraced walls.
Berzelius	37 N	51 E	24	Taurus area; darkish floor; c.p.
Bessarion	15 N	37 W	10	Deep craterlet on Imbrium; S of Brayley.

Name	Lat	Long	Diam km	Notes
Bessel	22 N	18 E	19	On Serenitatis; associated with a long ray.
Bettinus	63 S	45 W	66	Bailly area; one of a line; high walls.
Bianchini	49 N	34 W	40	In Jura Mtns; rather irregular; c.p.
Biela	55 S	52 E	74	Vlacq area; high walls; c.p.
Bilharz	6 S	56 E	43	Trio with Naonobu and Atwood. (Langrenus F).
Billy	14 S	50 W	42	Pair with Hansteen; S edge of Procellarum; very dark floor.
Biot	23 S	51 E	16	In Fœcunditatis; very bright.
Birmingham	64 S	10 W	106	N of Frigoris; low-walled; irregular.
Birt	22 S	9 W	18	On Nubium; rill to the W; W of St. Wall; profile irregular.
Black	9 S	80 E	18	NE of La Peyrouse. (Kästner F).
Blagg	1 N	2 E	6	In Sinus Medii; quite distinct.
Blancanus	64 S	21 W	92	Near Clavius; pair with Scheiner; high walls.
Blanchard	58 S	94 W	40	Libration zone; N of Hausen, beyond Pingré.
Blanchinus	25 S	3 E	53	N of Werner; uneven walls; rough floor.
Bobillier	20 N	16 E	7	In Serenitatis; E of Sulpicius Gallus. (Bessel E).
Bode	7 N	2 W	18	Outside S Æstuum; bright; minor ray-centre.
Boethius	6 N	72 E	10	Well-formed; (Dubiago U).
Boguslawsky	75 S	45 E	97	Southern uplands; high walls.
Bohnenberger	16 S	40 E	35	Edge of Nectaris; low walls.
Bohr	13 N	86 W	71	Beyond Vasco da Gama.
Boltzmann	75 S	91 W	76	Libration zone; closely N of Drygalski.
Bombelli	5 N	56 E	10	E of Apollonius. (Apollonius T).
Bond, G.P.	32 N	36 E	23	E of Posidonius; fairly regular.
Bond, W.C.	64 N	3 E	160	N of Frigoris; old and broken.
Bonpland	8 S	17 W	58	In Nubium; Fra Mauro group; fairly regular.
Boole	64 N	87 W	63	Limb; beyond Pythagoras.
Borda	25 S	47 E	42	W of Petavius; low walls.
Borel	22 N	26 E	5	In Serenitatis, SW of Le Monnier. (Le Monnier C).
Born	6 S	67 E	15	Distinct; NE of Langrenus. (Maclaurin Y).
Boscovich	10 N	11 E	43	On edge of Vaporum; irregular; low walls; very dark floor.
Boss	46 N	39 E	47	Limb, beyond Mercurius.
Bouguer	52 N	36 W	24	In Jura uplands; very distinct.
Boussingault	70 S	50 E	78	S uplands; made up of 3 rings.
Bowen	18 N	9 E	9	N of Manilius; flattish floor. (Manilius A).
Brackett	18 N	24 E	9	In Serenitatis; N of Plinius; inconspicuous.
Bragg	42 N	103 W	84	Libration zone; beyond Gerard.
Brayley	21 N	37 W	16	On Procellarum; low c.p.
Breislak	48 S	18 E	32	SE of Maurolycus; fairly regular; no c.p.
Brenner	39 S	39 E	90	Broken; adjoins Janssen (S uplands).
Brewster	23 N	36 E	11	Between Rømer and Littrow. (Rømer L).
Brianchon	75 S	86 W	145	Large limb enclosure; limb; beyond Carpenter.
Briggs	26 N	69 W	38	On Procellarum (Otto Struve area); similar to Seleucus.
Brisbane	50 S	65 E	47	Australe area; fairly regular.
Brown	46 S	18 W	40	NE of Longomontanus; irregular.
Bruce	1 N	0	9	In Sinus Medii; distinct.
Brunner	10 S	91 E	53	Libration zone; beyond Kästner; Hirayama area.
Buch	39 S	18 E	48	Maurolycus area; adjoins Büsching.
Bullialdus	21 S	22 W	50	On Nubium; massive, terraced walls; c.p.
Bunsen	41 N	85 W	52	Between Gerard and La Voisier.
Burckhardt	31 N	57 E	48	N of Cleomedes; member of a complex group.
Burnham	14 S	7 E	29	Albategnius area; low walls.
Bürg	45 N	28 E	48	N of L. Mortis; large c.p.; with pit; major rill system nearby.
Büsching	38 S	20 E	58	Adjoins Buch, but is less regular.
Byrd	83 N	10 E	94	N polar area; walled plain; adjoining Gioja.
Byrgius	25 S	65 W	64	W of Humorum; Byrgius A; on its E crest is a ray-centre.
Cabaeus	85 S	20 W	140	S polar uplands; fairly high walls.
Cajal	13 N	31 E	9	On Tranquillitatis; E of Jansen. (Jansen F).
Calippus	39 N	11 E	31	N end of Caucasus; irregular.
Cameron	6 N	46 E	11	Intrudes into NW wall of Taruntius. (Taruntius C).
Campanus	28 S	28 W	38	W edge of Nubium; pair with Mercator; with a lighter floor.
Cannizzare	56 N	100 W	56	Libration zone; in Poczobut.
Cannon	20 N	81 E	57	Limb; N of Marginis; light floor.
Capella	8 S	36 E	48	Uplands N of Nectaris; large c.p.; cut by crater-valley.
Capuanus	34 S	26 W	56	Edge of Epidemiarum; domes on floor.
Cardanus	13 N	73 W	52	On Procellarum; c.p.; pair with Krafft.
Carlini	34 N	24 W	8	Bright craterlet on Imbrium.
Carmichael	20 N	40 E	20	Well-formed; on Mare; W of Macrobius.
Carpenter	70 N	50 W	65	N of Frigoris; adjoins Anaximander.
Carrel	11 N	27 E	16	Central crater; on Tranquillitatis; SW of Jansen. (Jansen B).
Carrillo	2 S	81 E	16	N of Kästner.
Carrington	44 N	62 E	25	Messala group.
Cartan	4 N	59 E	16	Regular; closely W of Apollonius. (Apollonius D).
Casatus	75 S	35 W	104	High walls; intrudes into Klaproth; S of Clavius.
Cassini	40 N	5 E	58	Edge of Nebularum; low walls; contains a deep crater, A.
Cassini, J.J.	68 N	16 W	50	Near Philolaus; irregular ridge-bounded area.
Catalan	46 S	87 W	25	Limb crater beyond Schickard and Baade.

Name	Lat	Long	Diam km	Notes
Catharina	18 S	24 E	89	Theophilus group; rough floor; no c.p.
Cauchy	10 N	39 E	13	Bright crater on Tranquillitatis.
Cavalerius	5 N	67 W	64	In Hevel group; central ridge.
Cavendish	25 S	54 W	52	W of Humorum; fairly high walls.
Caventou	30 N	29 W	3	On Imbrium; fairly prominent. (Lahire D).
Cayley	4 N	15 E	13	Very bright; uplands; W of Tranquillitatis.
Celsius	34 S	20 E	45	Rabbi Levi group; rather elliptical.
Censorinus	0	32 E	5	Brilliant craterlet; uplands; SE of Tranquillitatis.
Cepheus	41 N	46 E	48	Somnerium area; pair with Franklin.
Chacornac	30 N	32 E	48	Edge of Serenitatis; adjoins Posidonius.
Chadwick	53 S	101 W	30	Libration zone; beyond Pingré; near de Roy.
Challis	78 N	9 E	56	N polar area; contact pair with Main.
Chamberlin	59 S	96 E	58	Libration zone; beyond Hanno.
Chapman	50 N	101 W	71	Libration zone; beyond Galvani.
Chappe	61 S	91 W	59	Libration zone.
Chevallier	45 N	52 E	47	Atlas area; low walls.
Ching-te	20 N	30 E	4	On Tranquillitatis; SW of Littrow; craterlet.
Chladni	4 N	1 E	8	Bright crater; abuts on Murchison; Sinus Medii area.
Cichus	33 S	21 W	32	Just S of Nubium; well-formed.
Clairaut	48 S	14 E	64	Maurolycus area; broken walls.
Clausius	37 S	44 W	24	Schickard area; bright and distinct.
Clavius	56 S	14 W	232	S uplands; massive walls; no c.p.; curved line of craters on floor.
Cleomedes	27 N	55 E	126	Crisium area; distorted by Tralles.
Cleostratus	60 N	74 W	70	Pythagoras area; distinct.
Clerke	22 N	29 E	7	On Tranquillitatis; W of Littrow. (Littrow B).
Collins	1·3 N	24 E	2·4	On Tranquillitatis; E of Sabine. (Sabine D).
Colombo	15 S	46 E	81	In Fœcunditatis; irregular; broken by large crater, A.
Compton	56 N	105 E	162	Libration zone; beyond Humboldtianum; crossed by rill.
Condamine	53 N	28 W	48	Edge of Frigoris; fairly regular.
Condon	2 N	60 E	35	Dark floor; low walls; N of Webb. (Webb R).
Condorcet	12 N	70 E	72	Outside Crisium; regular; no c.p.
Conon	22 N	2 E	21	In Apennine uplands; fairly distinct.
Cook	18 S	49 E	42	Edge of Fœcunditatis; darkish floor.
Copernicus	10 N	20 W	97	Great ray centre; massive walls; c.mountain group; terraced.
Couder	5 S	92 W	21	Libration zone; N of Orientale.
Cremona	68 N	91 W	85	Libration zone; limb; beyond Pythagoras.
Crile	14 N	46 E	9	Distinct; S of Proclus. (Proclus F).
Crozier	14 S	51 E	24	Fœcunditatis area; SE of Colombo; c.p.
Crüger	17 S	67 W	48	SW of Procellarum; very dark floor; regular; no c.p.
Curie	23 S	92 E	138	Libration zone; beyond W. Humboldt; pair with Sklodowska.
Curtis	15 N	57 E	3	On Crisium; E of Picard. (Picard Z).
Curtius	77 S	5 E	81	Moretus area; massive, terraced walls.
Cusanus	72 N	71 E	63	Limb; beyond Democritus; well-formed; no c.p.
Cuvier	50 S	10 E	66	Licetus group; high walls; c.p.
Cyrillus	13 S	24 E	97	Theophilus group; low c.p.; rather irregular in form.
Cysatus	66 S	7 W	47	Moretus area; high walls.
D'Arrest	2 N	15 E	30	E of Godin; low, broken wall.
Da Vinci	9 N	45 E	38	N of Taruntius; low, broken walls; irregular outline.
Daguerre	12 S	34 E	32	In Nectaris; very low walls.
Dale	10 S	83 E	22	NE of La Peyrouse; trio with Black and Kreiken.
Dalton	17 N	84 W	61	Limb crater; E of Einstein and W of Krafft.
Daly	6 N	60 E	17	Well-formed; twin with Apollonius F. (Apollonius P).
Damoiseau	5 S	61 W	35	E of Grimaldi; very irregular.
Daniell	35 N	31 E	24	Adjoins Posidonius; no c.p.
Darney	15 S	24 W	17	Bright crater in Nubium; Fra Mauro area.
Darwin	20 S	69 W	130	Grimaldi area; low walls; contains a large dome.
Daubrée	16 N	15 E	14	SW of Menelaus; darkish floor. (Menelaus S).
Davy	12 S	8 W	32	Edge of Nubium; irregular walls.
Dawes	17 N	26 E	23	Distinct; between Serenitatis and Tranquillitatis.
De Gasparis	26 S	50 W	48	W of Humorum; S of Mersenius; fairly regular.
De la Rue	67 N	56 E	160	Very low; broken walls; adjoins Endymion to the NW.
De Morgan	3 N	15 E	8	Bright craterlet; uplands W of Tranquillitatis.
De Roy	55 S	99 W	43	Libration zone; beyond Pingré; pair with Chadwick.
De Sitter	80 N	40 E	65	Limb crater; N of Euctemon.
De Vico	20 S	60 W	24	Deep crater; W of Gassendi.
Debes	29 N	52 E	48	Outside Cleomedes; fusion of two rings.
Dechen	48 N	67 W	12	In NW of Procellarum; not bright.
Delambre	2 S	18 E	52	Tranquillitatis area; high walls.
Delaunay	22 S	3 E	45	Albategnius area, near Faye; irregular.
De L'Isle	30 N	36 W	22	On Procellarum; pair with Diophantus; c.p.
Delmotte	27 N	60 E	31	N of Crisium; not prominent.
Deluc	55 S	3 W	45	Clavius area; walls of moderate height.
Dembowski	3 N	7 E	23	Low walls; E of Sinus Medii.
Democritus	62 N	35 E	37	Highlands; N of Frigoris; very deep.
Demonax	85 S	35 E	121	Boguslawsky area; fairly regular.
Desargues	70 N	73 W	85	Limb formation; beyond Anaximander.

Name	Lat	Long	Diam km	Notes
Descartes	12 S	16 E	48	NE of Abulfeda; low, broken walls.
Deseilligny	21 N	21 E	9	Distinct craterlet on Serenitatis.
Deslandres	32 S	61 W	186	Ruined enclosure W of Walter.
Dionysius	3 N	17 E	19	Brilliant crater on edge of Tranquillitatis.
Diophantus	28 N	34 W	18	On Procellarum; pair with De L'Isle; c.p.
Doerfel	69 S	108 W	68	Libration zone; beyond Hausen.
Dollond	10 S	14 E	10	W of Theophilus; borders a large 'ghost'.
Donati	21 S	5 E	35	S of Albategnius; pair with Faye; irregular; c.p.
Donner	31 S	98 E	58	Libration zone; beyond W. Humboldt.
Doppelmayer	28 S	41 W	68	Bay in Humorum; remnant of c.p.
Dove	47 S	32 E	18	Janssen area; low walls.
Draper	18 N	22 W	10	In Imbrium; S of Pytheas; one of a pair.
Drebbel	41 S	49 W	29	Schickard area; well-formed.
Dreyer	10 N	97 E	61	Libration zone; beyond Marginis.
Drude	39 S	92 W	24	Libration zone; in Cordilleras; beyond Orientale.
Drygalski	80 S	87 W	163	Cabaeus area; irregular.
Dubiago	4 N	70 E	48	Smythii area; regular.
Dugan	64 N	103 E	50	Libration zone; Humboldtianum area; beyond Belkovich.
Dunthorne	30 S	32 W	18	Edge of Epidemiarum; broad walls.
Dziewulski	21 N	99 E	63	Libration zone; beyond Marginis.
Eckert	17 N	58 E	3	Craterlet on Crisium; NE of Picard.
Eddington	22 N	72 W	125	Flooded plain between Seleucus and Otto Struve.
Edison	25 N	99 E	62	Libration zone; Marginis area; adjoins Lomonosov.
Egede	49 N	11 E	49	Near Alpine Valley; low walls; lozenge-shaped.
Eichstädt	23 S	80 W	52	Orientale area; regular.
Eimmart	24 N	65 E	42	Near edge of Crisium; regular.
Einstein	18 N	86 W	160	W of Procellarum; Otto Struve area; central crater.
Elger	35 S	30 W	24	Edge of Epidemiarum; low-walled; imperfect.
Ellison	55 N	108 W	36	Libration zone; beyond Xenophanes and Poczobut.
Elmer	10 S	84 E	17	Limb crater; beyond La Peyrouse.
Encke	5 N	37 W	32	On Procellarum; Kepler area.
Endymion	55 N	55 E	117	Humboldtianum area; darkish floor; no c.p.
Epigenes	73 N	4 W	52	N of Frigoris; broad walls.
Epimenides	41 S	30 W	24	One of a pair E of Hainzel.
Eppinger	8 S	26 W	6	Deep craterlet; NE of Riphæans. (Euclides D).
Eratosthenes	15 N	11 W	61	End of Apennines; terraced; very deep; c.p.
Erro	6 N	99 E	61	Libration zone; Smythii area; beyond Babcock.
Esclangon	22 N	42 E	16	On Tranquillitatis; W of Macrobius; low walls. (Mercurius L).
Euclides	7 S	29 W	12	Near Riphæans; lies on bright nimbus.
Euctemon	76 N	31 E	62	Walled plain beyond Meton.
Eudoxus	44 N	16 E	64	S of Frigoris; pair with Aristoteles.
Euler	23 N	29 W	25	On Imbrium; minor ray-centre.
Fabbroni	19 N	29 E	11	Edge of Mare; NW of Vitruvius. (Vitruvius E).
Fabricius	43 S	42 E	89	Intrudes into Janssen; rough floor; c.p.
Fabry	43 N	101 E	17	Libration zone; in Harkhebi.
Fahrenheit	13 N	62 E	6	On Crisium; W of Agarum. (Picard X).
Faraday	42 S	8 E	64	Intrudes into Stöfler; irregular.
Faustini	87 S	77 E	39	Libration zone.
Fauth	6 N	20 W	13	On Procellarum; S of Copernicus; double crater, with A.
Faye	21 S	4 E	36	S of Albategnius; irregular; c.p.; pair with Donati.
Fedorov	28 N	37 W	7	On Imbrium; W of Diophantus.
Fenyi	45 S	105 W	38	Libration zone; beyond Rydberg and Guthnick.
Fermat	23 S	20 E	40	Altai area; distinct.
Fernelius	38 S	5 E	64	N of Stöfler; rather irregular.
Feuillé	27 N	10 W	13	On Imbrium; twin with Beer.
Finsch	24 N	21 E	4	In Serenitatis; NE of Bessel; darkish floor.
Firmicus	7 N	64 E	56	S of Crisium; dark floor; no c.p.
Flammarion	3 S	4 W	72	NW of Ptolemæus; irregular enclosure.
Flamsteed	5 S	44 W	19	On Procellarum; associated with 100 km 'ghost'.
Focas	34 S	94 W	22	Libration zone; beyond Rook Mountains; well-formed.
Fontana	10 S	57 W	48	Between Billy and Crüger; well-formed; c.p.
Fontenelle	63 N	19 W	45	N edge of Frigoris; deep, distinct.
Foucault	50 N	40 W	24	Jura Mtns area; bright and deep.
Fourier	30 S	53 W	58	W of Humorum; terraced; with central crater.
Fox	0·5 N	98 E	24	Libration zone; regular; Smythii area.
Fra Mauro	6 S	17 W	81	On Nubium; low walls; trio with Bonpland and Parry.
Francastorius	21 S	33 E	97	Bay at S of Nectaris.
Franck	23 N	36 E	12	On Mare; S of Rømer. (Rømer K).
Franklin	39 N	48 E	55	Regular; SE of Atlas.
Franz	16 N	40 E	34	Edge of Somnii; low walls.
Fraunhofer	39 S	59 E	48	S of Furnerius; NW wall broken by craters.
Fredholm	18 N	47 E	15	Highlands; S of Macrobius. (Macrobius D).
Freud	26 N	52 W	3	Craterlet adjoining Schröter's Valley.
Froelich	80 N	110 W	58	Libration zone; pair with Lovelace; beyond Mouchez.
Fryxell	21 S	101 W	18	Libration zone; beyond Orientale.
Furnerius	36 S	60 E	129	In Petavius chain; rather broken walls.

Name	Lat	Long	Diam km	Notes
Galén	22 N	5 E	10	Crater S of Aratus; E of Conon. (Aratus A).
Galilaei	10 N	63 W	15	On Procellarum; inconspicuous.
Galle	56 N	22 E	24	On Frigoris; distinct.
Galvani	50 N	85 W	80	Limb; beyond Repsold.
Gambart	1 N	15 W	26	On Procellarum; SSE of Copernicus; regular; low walls.
Ganswindt	80 S	110 E	74	Libration zone; beyond Demonax; adjoins Schrödinger.
Gardner	18 N	34 E	18	Distinct; E of Vitruvius. (Vitruvius A).
Gärtner	60 N	34 E	101	Bay on Frigoris; 'seaward' wall barely traceable.
Gassendi	18 S	40 W	89	Edge of Humorum; 'seaward' wall low; c.p.; many rills on floor.
Gaudibert	11 S	37 E	26	Edge of Nectaris; low walls.
Gauricus	34 S	12 W	64	Pitatus group; irregular outline.
Gauss	36 N	80 E	136	High walls; c.p.; along limb from Humboldtianum.
Gay-Lussac	14 N	21 W	24	N of Copernicus; irregular enclosure.
Geber	20 S	14 E	40	Regular; between Almanon and Abulfeda.
Geissler	3 S	77 E	16	Small crater W of Smythii. (Gilbert D).
Geminus	35 N	57 E	90	Crisium area; broad, terraced walls; central hill.
Gemma Frisius	34 S	14 E	84	N of Maurolycus; high but broken walls.
Gerard	44 N	75 W	87	W of Sinus Roris; fairly distinct.
Gernsback	37 S	100 E	48	Libration zone; beyond Australe.
Gibbs	18 S	84 E	77	Limb; NW of Hekataeus.
Gilbert	3 S	76 E	107	Walled plain; E of Smythii; NW of Kästner.
Gill	70 S	76 E	66	Beyond Rosenberger.
Ginzel	14 N	97 E	55	Libration zone; beyond Marginis.
Gioja	83 N	2 E	35	North polar crater; fairly regular and distinct.
Glaisher	13 N	49 E	16	W border of Crisium; obscure.
Goclenius	10 S	45 E	52	Edge of Fœcunditatis; lava-flooded; low c.p.
Goddard	15 N	89 E	89	Beyond Marginis; dark floor.
Godin	2 N	10 E	43	S of Vaporum; c.,p.; pair with Agrippa.
Goldschmidt	75 N	0	109	E of Anaxagoras; N of Frigoris; low, broken walls.
Golgi	28 N	60 W	4	Small distinct craterlet N of Schiaparelli.
Goodacre	33 N	14 E	48	E of Aliacensis; low c.p.
Gould	19 S	17 W	42	'Ghost' in Nubium; E of Bullialdus.
Graff	42 S	89 W	36	Limb formation; beyond Schickard.
Greaves	13 N	53 E	14	Deep craterlet on Crisium; N of Lick. (Lick D).
Grimaldi	6 S	68 W	193	W of Procellarum; very dark floor; low, irregular walls. No c.p.
Grove	40 N	33 E	26	In Somniorum; bright and deep.
Gruemberger	68 S	10 W	87	Moretus area; high walls; contains a very deep crater.
Gruithuisen	33 N	40 W	10	Bright craterlet on Procellarum.
Guericke	12 S	14 W	53	Fra Mauro area; in Imbrium; broken, irregular walls.
Gum	4 S	87 E	55	Australe area; beyond Marinus; shallow; flooded.
Gutenberg	8 S	41 E	72	Edge of Fœcunditatis, near Goclenius; irregular.
Guthnick	48 S	94 W	36	Libration zone; pair with Rydberg.
Gylden	5 S	1 E	48	N of Ptolemæus; partly lava-filled; crater valley to W.
Hagecius	60 S	46 E	81	Vlacq group; wall broken by craters.
Hahn	31 N	74 E	56	Regular; c.p.; pair with Berosus; Crisium area.
Haidinger	39 S	25 W	12	S of Epidemiarum; inconscpicuous.
Hainzel	41 S	34 W	97	N of Schiller; compound; 2 coalesced rings.
Haldane	2 S	84 E	38	Limb; Smythii area.
Hale	74 S	91 E	84	Libration zone; beyond Boussingault; terraced.
Hall	34 N	37 E	25	Somniorum area; E of Posidonius.
Halley	8 S	6 E	35	Hipparchus group; pair with Hind; regular.
Hamilton	43 S	85 E	57	Limb; beyond Oken; deep and regular.
Hanno	57 S	73 E	64	Australe area; darkish floor.
Hansen	44 S	83 E	36	Regular; similar to Alhazen; Crisium area; near Agarum.
Hansky	10 S	97 E	43	Libration zone; Smythii area; SE of Hirayama.
Hansteen	12 S	52 W	52	Regular; S edge of Procellarum; pair with Billy.
Harding	43 N	70 W	23	Sinus Roris area; low walls
Hargreaves	2 S	64 E	15	Irregular; E of Maclaurin. (Maclaurin S).
Harkhabi	40 N	100 E	282	Libration zone; incomplete; eroded; N of Marginis. Contains Fabry.
Harpalus	53 N	43 W	52	Edge of Frigoris; deep, prominent.
Hartwig	6 S	81 W	80	W of Grimaldi; adjoins Schlüter to the E.
Hase	30 S	63 E	77	S of Petavius; rather irregular.
Hausen	66 S	88 W	167	Bailly area; c.peaks.
Hayn	68 N	85 E	87	Limb; beyond Strabo.
Hedervári	82 S	84 E	69	Libration zone.
Hekataeus	23 S	84 E	180	SE of Vendelinus; irregular walls; c.p.
Heinrich	25 N	15 W	7	In Imbrium; NW of Timocharis; prominent. (Timocharis A).
Heinsius	39 S	18 W	72	Tycho area; irregular; 3 craters on its S wall.
Heis	32 N	32 W	12	Bright craterlet on Imbrium.
Helicon	40 N	23 W	29	In Imbrium; regular; pair with Le Verrier.
Hell	32 S	8 W	31	In W of Deslandres; low c.p.
Helmert	8 S	87 E	26	Libration zone; Smythii area; adjoins Kao.
Helmholtz	72 S	78 E	97	Fairly regular; Boussingault area; S uplands.

Name	Lat	Long	Diam km	Notes
Henry, Paul	24 S	57 W	40	W of Humorum; distinct; pair with Prosper Henry.
Henry, Prosper	24 S	59 W	45	Pair with Paul Henry.
Heraclitus	49 S	6 E	92	Very irregular; Cuvier-Licetus group; S of Stöfler.
Hercules	46 N	39 E	72	Bright walls; deep interior crater; companion to Atlas.
Herigonius	13 S	34 W	16	Bright; c.p.; NE of Gassendi.
Hermann	1 S	57 W	16	On Procellarum; E of Lohrmann; bright.
Hermite	86 N	87 W	110	Well-formed; limb; beyond Anaxagoras.
Herodotus	23 N	50 W	37	Fairly regular; pair with Aristarchus; great valley nearby.
Herschel	6 S	2 W	45	N of Ptolemaeus; terraced walls; large c.p.
Herschel, Caroline	34 N	31 W	13	On Imbrium; group with Carlini and de l'Isle; bright.
Herschel, John	62 N	41 W	145	N of Frigoris; ridge-bordered enclosure.
Hesiodus	29 S	16 W	45	Companion to Pitatus; a major rill runs SW from it.
Hevel	2 N	67 W	122	Grimaldi chain; convex floor; with low c.p. and several rills.
Heyrovsky	40 S	95 W	16	Libration zone; beyond Cordilleras.
Hill	21 N	41 E	16	Edge of Mare; W of Macrobius. (Macrobius B).
Hind	8 S	7 E	26	Hipparchus group; pair with Halley; regular.
Hippalus	25 S	30 W	61	Bay on Humorum; remnant of c.p.; associated with rills.
Hirayama	6 S	94 E	139	Libration zone; irregular; Smythii area.
Hohmann	18 S	94 W	16	Libration zone; small crater in Orientale.
Holden	19 S	63 E	40	S of Vendelinus; deep.
Hommel	54 S	33 E	121	S uplands; 2 large craters on floor.
Hooke	41 N	55 E	43	W of Messala; fairly regular.
Hornsby	28 N	13 E	3	Small craterlet in Serenitatis; between Linne and S. Gallus.
Horrebow	59 N	41 W	32	Pair with Robinson; outside J. Herschel (Frigoris area); deep.
Horrocks	4 S	6 E	29	Within Hipparchus; regular.
Hortensius	6 N	28 W	16	On Procellarum; bright; domes to the N.
Houtermans	9 S	87 E	30	Limb; SE of Kästner.
Hubble	22 N	87 E	81	Limb; SE of Plutarch; partly flooded.
Huggins	41 S	2 W	48	Encroaches on Orontius; irregular.
Humason	31 N	58 W	4	On Procellarum; E of Lichtenberg. (Lichtenberg G).
Humboldt, Wilhelm	27 S	81 E	193	E of Petavius; rather irregular; rills on floor.
Hume	5 S	90 E	23	Libration zone; Smythii area; bordering Hirayama.
Huxley	20 N	5 W	4	Apennine region; E of Imbrium. (Wallace B).
Hyginus	8 N	6 E	6	Depression on Vaporum; in great crater-rill.
Hypatia	4 S	23 E	48	S of Tranquillitatis; triangular; low walls.
Ibn Battuta	7 S	50 E	12	Prominent; NE of Goclenius. (Goclenius A).
Ibn Rushd	12 S	22 E	33	Between Cyrillus and Kant. (Cyrillus B).
Ibn Yunis	14 N	91 E	58	Libration zone; Marginis area; adjoins Goddard.
Ideler	49 S	22 E	44	SE of Maurolycus; distinct.
Idelson	82 S	111 E	60	Libration zone; Demonax area.
Ilyin	18 S	98 W	13	Libration zone; in Orientale.
Inghirami	48 S	70 W	97	Schickard area; regular; c.p.
Isidorus	8 S	33 E	48	Outside Nectaris; deep; pair with Capella.
Ivan	27 N	43 W	4	Small craterlet NE of Prinz. (Prinz B).
Jacobi	57 S	12 E	66	Heraclitus group; fairly distinct.
Jansen	14 N	29 E	26	In Tranquillitatis; low walls; darkish floor.
Jansky	9 N	89 E	73	Limb formation; beyond Neper.
Janssen	46 S	40 E	170	S uplands; great ruin; broken in the N by Fabricius.
Jeans	56 S	91 E	79	Libration zone; beyond Hanno; between Chamberlin and Lyot.
Jehan	21 N	32 W	5	On Imbrium; SW of Euler. (Euler K).
Jenkins	0	78 E	38	W of Smythii.
Jenner	42 S	96 E	71	Libration zone; Australe area.
Joliot	26 N	93 E	143	Libration zone; NE of Marginis; much interior detail.
Joy	25 N	7 E	6	Craterlet in the Haemus foothills.
Julius Caesar	9 N	15 E	71	Vaporum area; low, irregular walls; very dark floor.
Kaiser	36 S	6 E	48	N of Stöfler; well-marked; no c.p.
Kane	63 N	26 E	45	N of Frigoris; fairly regular.
Kant	11 S	20 E	30	W of Theophilus; large c.p. with summit pit.
Kao	7 S	88 E	34	Adjoins Helmert; beyond Kästner and Kiess.
Kapteyn	13 S	70 E	35	E of Langrenus; inconscpicuous.
Kästner	6 S	80 E	129	Smythii area; distinct.
Keldysh	51 N	44 E	33	In Hercules; regular; bright. (Hercules A).
Kepler	8 N	38 W	35	In Procellarum; pair with Encke; major ray-centre.
Kies	26 S	23 W	42	On Nubium; Bullialdus area; very low walls.
Kiess	6 S	84 E	63	Limb formation; beyond Kästner.
Kinau	60 S	15 E	42	Jacobi group; high walls; c.p.
Kirch	39 N	6 W	11	Bright craterlet in Imbrium.
Kircher	67 S	45 W	74	Bettinus chain; Bailly area; very high walls.
Kirchhoff	30 N	39 E	50	Cleomedes area; larger of 2 craters W of Newcomb.
Klaproth	70 S	26 W	119	Darkish floor; contact pair with Casatus (S uplands).
Klein	12 N	3 E	43	In Albategnius; regular; c.p.
Knox-Shaw	5 N	80 E	12	Abuts on Banachiewicz. (Banachiewicz F).
Kopff	17 S	90 W	42	Limb; Orientale area; W of Crüger.

Name	Lat	Long	Diam km	Notes
Krafft	51 N	73 W	51	Pair with Cardanus; on Procellarum; darkish floor.
Kramarov	2 S	99 W	20	Libration zone; crater in Rook Mtns area.
Krasnov	30 S	77 W	41	Irregular; limb; W of Doppelmayer.
Kreiken	9 S	85 E	23	NE of La Peyrouse; trio with Black and Dale.
Krieger	29 N	46 W	23	Aristarchus area; distinct, but walls broken by craterlets.
Krishna	25 N	11 E	3	Craterlet in Serenitatis; S of Linne.
Krogh	9 N	66 E	20	Well-formed; SE of Auzout. (Auzout B).
Krusenstern	26 S	6 E	48	Werner area; not prominent.
Kugler	54 S	104 E	65	Libration zone; beyond Brisbane and Jeans.
Kuiper	10 S	23 W	7	Prominent; on Mare Cognitum; W of Bonpland. (Bonpland E).
Kundt	12 S	12 W	11	Prominent; between Guericke and Davy. (Guericke C).
Kunowsky	3 N	32 W	21	On Procellarum; E of Encke; low central ridge.
La Peyrouse	10 S	78 E	72	E of Fœcunditatis; pair with Ansgarius; well-formed.
Lacaille	24 S	1 E	53	Werner area; irregular.
Lallemand	14 S	84 W	18	W of Rocca. (Kopff A).
Lacroix	38 S	59 W	38	N of Schickard; fairly regular; c.p.
Lade	1 S	10 E	48	Highlands S of Godin; low walls.
Lagalla	45 S	23 W	63	Abuts on Wilhelm I; low-walled; irregular.
Lagrange	33 S	72 W	165	W of Humorum; low walls; rough floor.
Lalande	4 S	8 W	24	Ptolemæus area; low c.p.
Lamarck	23 S	70 W	115	Ruined plain; S of Darwin and W of Byrgius.
Lamb	43 S	101 E	104	Libration zone; Australe area.
Lambert	26 N	21 W	29	In Imbrium; bright; central crater.
Lamé	15 S	65 E	81	On NE wall of Vendelinus; well-formed.
Lamèch	43 N	13 E	95	Eudoxus area; very low, irregular walls.
Lamont	5 N	23 E	62	On Tranquillitatis; Arago area; low walls.
Landsteiner	31 N	15 W	6	Craterlet on Imbrium; E of Carlini.
Langley	51 N	86 W	60	Limb; beyond Repsold; N of Galvani.
Langrenus	9 S	61 E	137	Petavius chain; E edge Fœcunditatis; massive walls; c.p.
Lansberg	0	26 W	42	On Nubium; massive walls; c.p.
Lassell	16 S	8 W	23	In Nubium; Alphonsus area; low walls.
Laue	28 N	97 W	87	Libration zone; beyond Ulugh Beigh; intrudes into Lorentz.
Lauritsen	27 S	96 E	52	Beyond Humboldt; abuts on Curie.
La Voisier	36 N	70 W	71	W of Procellarum; well-formed.
Lawrence	7 N	43 E	24	Flooded; NW of Taruntius. (Taruntius M).
Le Monnier	26 N	31 E	55	Bay on Serenitatis; smooth floor.
Le Verrier	40 N	20 W	25	Distinct crater on Imbrium; pair with Helicon.
Leakey	3 S	37 E	13	Inconspicuous craterlet near Censorinus. (Censorinus F).
Lebesque	5 S	89 E	11	Smythii area; near Warner.
Lee	31 S	41 W	42	S edge of Humorum; damaged by lava.
Legendre	29 S	70 E	74	W. Humboldt area; central ridge.
Legentil	73 S	80 E	140	S of Bailly; distinct.
Lehmann	40 S	56 W	45	Schickard area; irregular.
Lepaute	33 S	34 W	14	Distinct small crater at edge of Epidemiarum.
Letronne	10 S	43 W	113	Bay at S edge Procellarum; low c.p.; N wall destroyed.
Lexell	36 S	4 W	63	Edge of Deslandres; N wall reduced; remnant of c.p.
Liapunov	26 N	89 E	66	Limb formation beyond Seneca.
Licetus	47 S	6 E	74	Cuvier-Heraclitus group; near Stöfler; fairly regular; c.p.
Lichtenberg	32 N	68 W	19	Edge of Procellarum; minor ray-centre.
Lick	12 N	53 E	35	Edge of Crisium; incomplete.
Liebig	24 S	48 W	45	W of Humorum; moderate walls.
Lilius	54 S	6 E	52	Jacobi group; high walls; c.p.
Lindbergh	5 S	53 E	13	Distinct; on Fœcunditatis; SE of Messier. (Messier G).
Lindblad	70 N	99 W	66	Libration zone; beyond Pythagoras; SW of Brianchon.
Lindenau	32 S	25 E	56	Rabbi Levi group; terraced walls.
Lindsay	7 S	13 E	32	In highlands; N of Åndël. (Dollond C).
Linné	28 N	12 E	11	On Serenitatis; craterlet surrounded by light nimbus.
Liouville	3 N	74 E	16	Distinct crater W of Schubert. (Dubiago S).
Lippershey	26 S	10 W	6	Pitatus area (Nubium); distinct craterlet.
Littrow	22 N	31 E	35	Edge of Serenitatis; irregular.
Lockyer	46 S	37 E	48	Intrudes into Janssen; bright walls.
Loewy	23 S	33 W	23	E edge of Humorum; distinct.
Lohrmann	1 S	67 W	45	Between Grimaldi and Hevel; fairly regular; c.p.
Lohse	14 S	60 E	32	On N wall of Vendelinus; deep; c.p.
Lomonosov	23 N	98 E	92	Libration zone; N of Marginis; Joliot group.
Longomontanus	50 S	21 W	145	Clavius area; complex walls; much floor detail.
Lorentz	34 N	97 W	371	Libration zone; huge enclosure; contains Nernst, Röntgen.
Louise	29 N	34 W	1·5	Craterlet between Diophantus and de L'Isle.
Louville	44 N	47 W	35	In Jura Mtns; low walls; darkish floor.
Lovelace	82 N	106 W	54	Libration zone; pair with Froelich; S of Hermite.
Lubbock	4 S	42 E	13	W edge of Fœcunditatis; fairly bright.
Lubiniezky	18 S	24 W	45	In Nubium; Bullialdus area; low walls.
Lucian	14 N	37 E	7	On Tranquillitatis; W of Lyell; (Maraldi B).
Ludwig	8 S	97 E	23	Libration zone; beyond Smythii and Hirayama.
Luther	33 N	24 E	8	Distinct craterlet in N of Serenitatis.
Lyell	14 N	41 E	43	W edge of Somnii; darkish floor.

OUTLINE MAP of the MOON
Patrick Moore
(Drawn by Patricia A. Cullen)
2nd. Edition 1969

NORTH

WEST

33

Name	Lat	Long	Diam km	Notes
Lyot	50 S	84 E	141	Flooded; irregular wall; dark floor; N of Australe.
Maclaurin	2 S	68 E	45	W of Smythii; concave floor; uneven walls.
Maclear	11 N	20 E	18	On Tranquillitatis; darkish floor.
Macmillan	23 N	8 W	8	In Imbrium; SW of Archimedes.
Macrobius	21 N	46 E	68	Crisium area; high walls; compound c.p.
Mädler	11 S	30 E	32	On Nectaris; irregular.
Maestlin	5 N	41 W	50	On Procellarum; near Encke; obscure.
Magelhaens	12 S	44 E	40	Edge of Fœcunditatis; pair with A; darkish floor.
Maginus	50 S	6 W	177	Clavius area; irregular walls; obscure near full moon.
Main	81 N	9 E	48	N polar area; contact twin with Challis.
Mairan	42 N	43 W	40	Jura area; bright; not perfectly regular.
Malapert	85 S	13 E	69	Irregular form; near S pole; E of Cabaeus.
Mallet	45 S	53 E	50	Rheita Valley area; inconspicuous.
Manilius	15 N	9 E	36	On edge of Vaporum; brilliant walls; c.p.
Manners	5 N	20 E	16	Arago area; on Tranquillitatis; bright.
Manzinus	68 S	25 E	90	Boguslawsky area; high, terraced walls.
Maraldi	19 N	35 E	37	In N of Tranquillitatis; distinct.
Marco Polo	15 N	2 W	19	Apennines area; irregular; darkish floor.
Marinus	50 S	75 E	48	Australe area; distinct; with c.p.
Marius	12 N	51 W	42	In Procellarum; pair with Reiner; low c.p.
Markov	53 N	73 W	40	On Sinus Roris; sharp rim.
Marth	31 S	29 W	13	In Epidemiarum; concentric crater.
Maskelyne	2 N	30 E	24	In Tranquillitatis; low c.p.
Mason	43 N	30 E	31	Bürg area; companion to Plana.
Maunder	15 S	94 W	55	Libration zone; on N edge of Orientale; regular; c.p.
Maupertuis	50 N	27 W	43	In Juras; irregular mountain enclosure.
Maurolycus	42 S	14 E	109	E of Stöfler; rough floor; central mountain group.
Maury	37 N	40 E	18	Atlas area; bright and deep.
Maxwell	30 N	99 E	115	Libration zone; beyond Gauss; intrudes into Richardson.
Mayer, C.	63 N	17 E	39	N of Frigoris; rhomboidal.
Mayer, T.	16 N	29 W	35	In Carpathians; c.p.
McAdie	2 N	92 E	45	Libration zone; E edge of Smythii; low walls.
McClure	15 S	50 E	24	Edge of Fœcunditatis; E of Colombo; Regular.
McDonald	30 N	21 W	8	On Imbrium; SE of Carlini. (Carlini B).
McLaughlin	47 N	93 W	79	Libration zone; rather irregular; beyond Galvani.
Mee	44 S	36 W	113	Abuts on Hainzel; low, broken walls.
Mees	14 N	96 W	50	Libration zone; beyond Einstein.
Mendel	49 S	110 W	138	Libration zone.
Menelaus	16 N	16 E	32	On Haemus Mtns; brilliant; c.p.
Menzel	3 N	40 E	3	Craterlet on Tranquillitatis; E of Maskelyne.
Mercator	29 S	26 W	38	Pair with Campanus; dark floor.
Mercurius	56 N	65 E	53	Humboldtianum area; low c.p.
Merrill	75 N	116 W	57	Libration zone; beyond Brianchon.
Mersenius	21 S	49 W	72	W of Humorum; convex floor; rills nearby.
Messala	39 N	60 E	128	Humboldtianum area; oblong; low-walled; broken.
Messier	2 S	48 E	13	On Fœcunditatis; twin with Messier A; 'Comet' ray to W.
Metius	40 S	44 E	81	Janssen group; distinct; pair with Fabricius.
Meton	74 N	25 E	170	N polar area; smooth floor; compound formation.
Milichius	10 N	30 W	13	On Procellarum; bright; dome to the W.
Miller	39 S	1 E	48	Orontius group; fairly distinct.
Mitchell	50 N	20 E	19	Abuts on Aristoteles; distinct.
Moigno	66 N	29 E	37	Contains interior crater; no c.p.; W of Arnold.
Möltke	1 S	24 E	7	N edge of Tranquillitatis; distinct.
Monge	19 S	48 E	32	Edge of Fœcunditatis; rather irregular.
Montanari	46 S	20 W	87	Longomontanus area; distorted.
Moretus	70 S	8 W	105	S uplands; massive walls; very high c.p.
Morley	3 S	65 E	14	NW of Maclaurin. (Maclaurin R).
Moseley	21 N	90 W	90	Libration zone; beyond Einstein.
Mösting	1 S	6 W	26	Medii area; A, to the W, is used as a reference point.
Mouchez	78 N	27 W	82	Ruined plain near Philolaus.
Moulton	61 S	97 E	49	Libration zone; beyond Hanno.
Müller	7 S	3 E	24	Ptolemæus area; fairly regular.
Murchison	5 N	0	56	Edge of Medii; low-walled; irregular.
Mutus	63 S	30 E	81	S uplands; 2 large craters on floor.
Nansen	81 N	95 E	122	Libration zone; beyond Einstein.
Naonobu	5 S	58 E	35	Trio with Biharz, Atwood (Langrenus B).
Nasireddin	41 S	0	48	Orontius group; fairly distinct.
Nasmyth	52 S	53 W	74	Phocylides group; fairly regular; no c.p.
Natasha	20 N	31 W	12	On Imbrium; SW of Euler. (Euler P).
Naumann	35 N	62 W	10	In N of Procellarum; bright walls.
Neander	31 S	40 E	48	Rheita Valley area; well-formed.
Nearch	58 S	39 E	61	Vlacq area; craterlets on floor.
Neison	68 N	28 E	45	Meton area; regular; no c.p.
Neper	7 N	83 E	113	Marginis/Smythii area; deep.
Neumayer	62 S	75 E	80	Boussingault area; distinct.

Name	Lat	Long	Diam km	Notes
Newcomb	30 N	44 E	52	Cleomedes area; S wall broken by crater.
Newton	78 S	20 W	113	Moretus area; unusually deep; irregular.
Nicholson	26 S	85 W	38	In Rook Mtns.
Nicolai	42 S	26 E	43	Janssen area; regular.
Nicollet	22 S	12 W	16	In Nubium; W of Birt; distinct.
Nielsen	32 N	52 W	10	On Procellarum; between Lichtenberg, Gruithuisen. (Wollaston C).
Nièpce	73 N	119 W	57	Libration zone; beyond Brianchon.
Nobile	85 S	53 E	73	Libration zone.
Nobili	0	76 E	42	Trio with Jenkins and X. (Schubert Y).
Noether	67 N	113 W	67	Libration zone; S of Nièpce.
Nöggerath	49 S	45 W	35	Schiller area; low walls.
Nonius	35 S	4 E	32	Stöfler area; fairly regular.
Nunn	5 N	91 E	19	Libration zone; on N edge of Smythii; low walls.
Œnopides	57 N	65 W	68	Limb area; near Sinus Roris; high walls.
Œrsted	43 N	47 E	40	Edge of Somniorum; rather irregular.
Oken	44 S	78 E	80	Australe area; prominent; darkish floor.
Olbers	7 N	78 W	64	Grimaldi area; major ray-centre.
Omar Khayyam	58 N	102 W	70	Libration zone; in Poczobut.
Opelt	16 S	18 W	43	'Ghost' in Nubium; E of Bullialdus.
Oppolzer	2 S	1 W	48	Edge of Medii; low walls.
Orontius	40 S	4 W	84	Irregular; one of a group NE of Tycho.
Palisa	9 S	7 W	32	Alphonsus area on edge of a larger ring.
Palitzsch	28 S	64 E	97×32	Outside Petavius to the E; really a crater-chain.
Pallas	5 N	2 W	47	Medii area; adjoins Murchison; c.p.
Palmieri	29 S	48 W	48	Humorum area; darkish floor.
Paneth	63 N	95 W	65	Libration zone; beyond Xenophanes; N of Smoluchowski.
Parkhurst	32 S	105 E	70	Libration beyond Australe; irregular.
Parrot	15 S	3 E	64	Albategnius area; very irregular compound structure.
Parry	8 S	16 W	42	On Nubium; Fra Mauro group; fairly regular.
Pascal	74 N	70 W	106	Limb formation beyond Carpenter.
Peary	89 N	33 E	74	Walled plain; N polar area; beyond Gioja.
Peek	3 N	87 E	13	Limb; beyond Schubert.
Peirce	18 N	53 E	19	In Crisium; conspicuous.
Peirescius	46 S	71 E	64	Australe area; rather irregular.
Pentland	64 S	12 E	72	S uplands; near Curtius; high walls; c.p.
Petermann	74 N	66 E	73	Limb; beyond Arnold.
Peters	68 N	30 E	15	Plain; no c.p.; between Neison and Arnold.
Petit	2 N	64 E	5	Small crater E of Spumans. (Apollonius W).
Petrov	61 S	88 E	49	Flooded; beyond Pontecoulant.
Pettit	28 S	87 W	35	Limb formation; pair with Nicholson.
Petzval	63 S	110 W	90	Libration zone; beyond Hausen.
Phillips	26 S	78 E	120	W of W. Humboldt; central ridge.
Philolaus	75 N	33 W	74	Frigoris area; pair with Anaximenes; deep; regular.
Phocylides	54 S	58 W	97	Schickard group; much floor detail.
Piazzi	36 S	68 W	90	Schickard area; very broken walls.
Piazzi Smyth	42 N	3 W	10	On Imbrium; bright.
Picard	15 N	55 E	34	Largest crater on Crisium; central hill.
Piccolomini	30 S	32 E	80	End of Altai Scarp; high walls; c.p.
Pickering	3 S	7 E	16	Hipparchus area; distinct.
Pictet	43 S	7 W	48	Closely E of Tycho; fairly regular.
Pilâtre	60 S	86 W	69	Low, irregular walls; beyond Pingré.
Pingré	59 S	74 W	89	Limb formation; no c.p.; beyond Phocylides.
Pitatus	30 S	14 W	80	S edge of Nubium; passes connect it with Hesiodus.
Pitiscus	51 S	31 E	80	S uplands; walls narrow in places.
Plana	42 N	28 E	39	Bürg area; pair with Mason; darkish floor.
Plaskett	82 N	106 E	110	Libration zone; large walled plain; N polar area.
Plato	51 N	9 W	97	Edge of Imbrium; regular; very dark, smooth floor.
Playfair	23 S	9 E	43	Abenezra area; fairly regular.
Plinius	15 N	24 E	48	Between Serenitatis and Tranquillitatis; central craters.
Plutarch	25 N	75 E	64	NE of Crisium; distinct; c.p.
Poczobut	58 N	95 W	209	Libration zone; large, broken by several craters.
Poisson	30 S	11 E	72	Aliacensis area; compound; very irregular.
Polybius	22 S	26 E	32	Theophilus/Catharina area; distinct.
Pomortsev	1 N	67 E	23	Distinct. (Dubiago P).
Poncelet	76 N	54 W	69	Limb; beyond Anaximenes and Philolaus.
Pons	25 S	22 E	32	Altai area; very thick walls.
Pontanus	28 S	15 E	45	Altai area; regular; no c.p.
Pontécoulant	69 S	65 E	97	S uplands; high walls.
Popov	17 N	100 E	65	Libration zone; beyond Marginis.
Porter	56 S	10 W	40	On wall of Clavius; very well-formed; c.p.
Posidonius	32 N	30 E	96	Edge of Serenitatis; narrow walls; much floor detail.
Priestley	57 S	108 E	52	Libration zone; beyond Hanno and Chamberlin.
Prinz	26 N	44 W	50	NE of Aristarchus; incomplete; domes nearby.
Proclus	16 N	47 E	29	W of Crisium; brilliant; low c.p.; minor ray-centre.
Proctor	46 S	5 W	43	Maginus area; fairly regular.

Name	Lat	Long	Diam km	Notes
Protagoras	56 N	7 S	19	In Frigoris; bright and regular.
Ptolemæus	14 S	3 W	148	Trio with Alphonsus, Arzachel; dark floor; contains Ammonius.
Puiseux	28 S	39 W	23	On Humorum; near Doppelmayer; very low walls.
Pupin	24 N	11 W	2	Small but distinct; SE of Timocharis. (Timocharis K).
Purbach	25 S	2 W	120	Walter group; rather irregular.
Purkyně	2 S	95 E	48	Libration zone; beyond Smythii.
Pythagoras	65 N	65 W	113	NW of Iridum; high, massive walls; high c.p.
Pytheas	21 N	20 W	19	On Imbrium; bright; c.p.; minor ray-centre.
Rabbi Levi	35 S	24 E	80	One of a group of 6; SW of Piccolomini.
Raman	27 N	55 W	11	Crater NW of Herodotus; irregular. (Herodotus D).
Ramsden	33 S	32 W	24	Edge of Epidemiarum; rills nearby.
Rankine	4 S	72 E	9	Craterlet E of Gilbert.
Rayleigh	29 N	89 E	107	Limb walled plain; NE of Seneca.
Reaumur	2 S	1 E	45	Medii area; low walls.
Regiomontanus	28 S	0	129×105	Between Walter and Purbach; distorted; c.p.
Regnault	51 N	88 W	47	Limb; near Xenophanes.
Reichenbach	30 S	48 E	48	Rheita area; irregular; crater-valley to the SE.
Reimarus	46 S	55 E	45	Rheita Valley area; irregular.
Reiner	7 N	55 W	32	On Procellarum; pair with Marius; low c.p.
Reinhold	3 N	23 W	48	SW of Copernicus; on Procellarum; pair with B, to the NE.
Repsold	50 N	70 W	140	W of Sinus Roris; one of a group.
Respighi	3 N	72 E	19	SE of Dubiago; distinct. (Dubiago C).
Rhaeticus	0	5 E	45	Medii area; low walls.
Rheita	37 S	47 E	68	Sharp crests; associated with great crater-valley.
Riccioli	3 S	75 W	160	Companion to Grimaldi; low walls; very dark patches on floor.
Riccius	37 S	26 E	80	Rabbi Levi group; broken walls; rough floor.
Richardson	32 N	100 E	161	Libration zone; broken by Maxwell; beyond Gauss, Vestine.
Riemann	40 N	87 E	110	Ruined walled plain; beyond Gauss.
Ritchey	11 S	9 E	26	E of Albategnius; broken walls.
Rittenhouse	74 S	107 E	26	Libration zone; beyond Neumayer; W of Schrödinger.
Ritter	2 N	19 E	31	On Tranquillitatis; c.p.; pair with Sabine.
Ritz	15 S	92 E	51	Libration zone; beyond Ansgarius.
Robinson	59 N	46 W	27	Frigoris area; distinct; similar to Horrebow.
Rocca	15 S	72 W	97	S of Grimaldi; irregular walls.
Rocco	29 N	45 W	4	Craterlet closely E of Krieger. (Krieger D).
Rosenberger	55 S	43 E	80	Vlacq group; darkish floor; c.p.
Ross	12 N	22 E	29	On Tranquillitatis; c.p.
Rosse	18 S	35 E	16	On Nectaris; bright.
Röst	56 S	34 W	55	Schiller area; regular; pair with Weigel.
Rothmann	31 S	28 E	42	Altai area; fairly deep and regular.
Rozhdestvensky	86 N	109 W	170	Libration zone; polar area; contains 2 craterlets.
Rømer	25 N	37 E	37	Taurus area; massive c.p.; with summit pit.
Röntgen	33 N	91 W	126	Libration zone; in Lorentz.
Rümker	41 N	58 W	49	On Sinus Roris; semi-ruined plateau.
Runge	3 S	87 E	38	Smythii area.
Russell	26 N	75 W	103	Extension of Otto Struve.
Ruth	29 N	45 W	3	Craterlet adjoining Krieger to the NE.
Rutherford	61 S	12 W	40	On wall of Clavius; distinct c.p.
Rydberg	46 S	96 W	49	Libration zone; pair with Guthnick.
Rynin	47 N	104 W	75	Libration zone; beyond Galvani and McLaughlin.
Sabatier	13 N	72 E	10	E of Condorcet; low walls.
Sabine	2 N	20 E	31	On Tranquillitatis; c.p.; pair with Ritter.
Sacrobosco	24 S	17 E	84	Altai area; irregular.
Sampson	30 N	17 W	1·5	On Imbrium; distinct craterlet NW of Timocharis.
Santbech	21 S	44 E	70	E of Fracastorius; darkish floor.
Santos-Dumont	27 N	5 E	9	S end of Apennines. (Hadley B).
Sarabhai	25 N	21 E	8	On Serenitatis; NE of Bessel. (Bessel A).
Sasserides	39 S	9 W	97	Irregular enclosure N of Tycho.
Saunder	4 S	9 E	43	E of Hipparchus; low walls.
Saussure	43 S	4 W	50	N of Maginus; interrupts larger ring.
Scheele	9 S	38 W	5	Distinct; on Procellarum; S of Wichmann. (Letronne D).
Scheiner	60 S	28 W	113	Clavius area; high walls; floor craterlet; pair with Blancanus.
Schiaparelli	23 N	59 W	29	On Procellarum; distinct.
Schickard	44 S	54 W	202	Great walled plain; rather low walls.
Schiller	52 S	39 W	180×97	Schickard area; fusion of 2 rings.
Schlüter	6 S	83 W	89	Beyond Grimaldi; terraced walls; prominent.
Schmidt	1 N	19 E	13	In Tranquillitatis; Sabine/Ritter area; bright.
Schömberger	76 N	30 E	65	Regular; Boguslawsky area.
Schönfeld	45 N	98 W	25	Libration zone; regular; beyond Gerard.
Schorr	20 S	90 E	53	Limb formation; beyond Gibbs.
Schrödinger	76 S	134 E	312	Libration zone; beyond Hanno; associated with 310 km valley.

Name	Lat	Long	Diam km	Notes
Schröter	3 N	7 W	32	Medii area; low walls.
Schubert	3 N	70 E	74	In Smythii area; distinct.
Schumacher	42 N	55 E	40	Messala area; fairly distinct.
Schwabe	65 N	46 E	25	NE of Democritus; dark floor.
Schwarzschild	71 N	120 E	235	Libration zone; beyond Petermann; much interior detail.
Scoresby	77 N	15 E	58	Polar uplands; deep and prominent; c.p.
Scott	82 S	45 E	108	Beyond Schömberger; pair with Nansen.
Secchi	2 N	43 E	23	In Fœcunditatis; bright walls; c.p.
Seeliger	2 S	1 E	45	N of Hipparchus; irregular.
Segner	59 S	48 W	74	Schiller area; prominent; well-formed; pair with Zucchius.
Shackleton	90 S	−	19	South polar.
Shaler	33 N	85 W	45	Limb; beyond Lagrange; pair with Wright.
Shapley	10 N	57 E	23	Off N edge of Crisium; dark floor. (Picard H).
Sharp	46 N	40 W	35	In Jura Mtns; deep; small c.p.
Sheepshanks	59 N	17 E	19	N of Frigoris; fairly regular.
Shi Shen	76 N	104 E	43	Libration zone; beyond Nansen.
Short	76 S	5 W	70	Moretus group; deep; high walls.
Shuckburgh	43 N	53 E	47	Cepheus group; fairly regular.
Shuleykin	27 S	93 W	15	Libration zone; just beyond Orientale; Rook Mtns area.
Sikorsky	66 S	103 E	98	Libration zone; crossed by Schrödinger Valley.
Silberschlag	6 N	13 E	13	Near Ariadaeus; bright.
Simpelius	75 S	15 E	80	Moretus area; deep.
Sinas	9 N	32 E	8	On Tranquillitatis; not prominent.
Sirsalis	13 S	60 W	32	Just outside Procellarum; pair with A; associated with rill.
Sklodowska	18 S	96 E	130	Libration zone; well-formed; beyond Hekataeus.
Slocum	3 S	89 E	13	Smythii area.
Smithson	2 N	54 E	6	E of Tranquillitatus. (Taruntius N).
Smoluchowski	60 N	97 W	83	Libration zone; intrudes into Poczobut.
Snellius	29 S	56 E	80	Furnerius area; high walls; c.p.; pair with Stevinus.
Somerville	8 S	65 E	15	Distinct; E of Langrenus. (Langrenus J).
Sömmering	0	7 W	27	Medii area; low walls.
Sosigenes	9 N	18 E	23	Edge of Tranquillitatis; bright; low c.p.
South	57 N	50 W	98	Frigoris area; ridge-bounded enclosure.
Spallanzani	46 S	25 E	25	W of Janssen; low walls.
Spörer	4 S	2 W	35	N of Ptolemæus; partly lava-filled.
Spurr	30 N	1 W	13	In Putredinis; SE of Archimedes.
Stadius	11 N	14 W	70	Pitted ghost-ring E of Copernicus.
Steinheil	50 S	48 E	70	SW of Janssen; contact twin with Watt.
Stevinus	33 S	54 E	70	Furnerius area; high walls; c.p.; pair with Snellius.
Stewart	2 N	67 E	13	Distinct; SW of Dubiago. (Dubiago Q).
Stiborius	34 S	32 E	44	S of Piccolomini; Altai area; c.p.
Stöfler	41 S	6 E	145	SE of Walter; broken by Faraday; darkish floor.
Stokes	53 N	88 W	51	Limb; Regnault-Repsold area.
Strabo	62 N	55 E	47	Near De la Rue; minor ray-centre.
Street	46 S	10 W	50	S of Tycho; fairly regular; no c.p.
Struve	43 N	65 E	18	Adjoins Messala; lies on dark patch.
Struve, Otto	43 N	65 E	160	W edge of Procellarum; pair of two ancient rings.
Suess	4 N	48 W	15	On Procellarum; W of Encke; obscure.
Sulpicius Gallus	20 N	12 E	13	On Serenitatis; very bright.
Sundman	11 N	92 W	40	Libration zone; beyond Vasco da Gama and Bohr.
Sven Hedin	5 N	75 W	98	W of Hevel; irregular; broken walls.
Swasey	6 S	90 E	23	Smythii area.
Swift	19 N	53 E	11	On Crisium; N of Peirce; prominent. (Peirce B).
Sylvester	83 N	80 W	58	Libration zone; beyond Philolaus.
Tacchini	5 N	86 E	40	Limb beyond Banachiewicz. (Neper K).
Tacitus	16 S	19 E	40	Catharina area; polygonal; 2 floor craters.
Tacquet	17 N	19 E	10	Just on Serenitatis; near Menelaus; bright.
Talbot	3 S	85 E	11	Smythii area.
Tannerus	56 S	22 E	24	S uplands; near Mutus; c.p.
Taruntius	6 N	48 E	60	On Fœcunditatis; concentric crater; c.p.; low walls.
Taylor	5 S	17 E	40	Delambre area; rather elliptical.
Tebbutt	10 N	54 E	32	Flooded; foothills of Crisium. (Picard G).
Tempel	4 N	12 E	8	Uplands W of Tranquillitatis; bright.
Thales	59 N	41 E	39	Near Strabo; major ray-centre.
Theaetetus	37 N	6 E	26	On Nebularum; low c.p.
Thebit	22 S	4 W	60	Edge of Nubium; W wall broken by A, which is itself broken by F.
Theiler	13 N	83 E	8	Craterlet W of Marginis.
Theon Junior	2 S	16 E	16	Near Delambre; bright; pair with Theon Senior.
Theon Senior	1 S	15 E	17	Pair with Theon Junior; bright.
Theophilus	12 S	26 E	101	Very deep; massive walls; trio with Cyrillus and Catharina.
Theophrastus	18 N	39 E	9	On Tranquillitatis; NW of Franz. (Maraldi M).
Timaeus	63 N	1 W	34	Edge of Frigoris; bright.
Timocharis	17 N	13 W	35	On Imbrium; central crater; minor ray-centre.
Timoleon	35 N	75 E	130	Adjoins Gauss; fairly distinct.
Tisserand	21 N	48 E	33	Crisium area; regular; no c.p.

Name	Lat	Long	Diam km	Notes
Titus	27 S	101 E	73	Libration zone; beyond W. Humboldt and Lauritsen.
Tolansky	10 S	16 W	13	Between Parry and Guericke; flat floor. (Parry A).
Torricelli	5 S	29 E	19	On Nectaris; irregular; compound structure.
Toscanelli	28 N	48 W	7	Distinct; in highlands N of Aristarchus. (Aristarchus C).
Townley	3 N	63 E	19	SE of Apollonius; distinct. (Apollonius G).
Tralles	28 N	53 E	48	On wall of Cleomedes; very deep.
Triesnecker	4 N	4 E	23	Vaporum area; associated with major rill system.
Trouvelot	49 N	6 E	10	Alpine Valley area; rather bright.
Tucker	6 S	88 E	7	Craterlet in Smythii area.
Turner	2 S	13 W	13	On Nubium; near Gambart; deep.
Tycho	43 S	11 W	84	Terraced walls; c.p.; brightest ray-centre.
Ukert	8 N	1 E	23	Edge of Vaporum; rills nearby.
Ulugh Beigh	29 N	85 W	70	W of Procellarum; high walls; c.p.
Urey	28 N	87 E	38	Limb formation; beyond Seneca.
Väisälä	26 N	48 W	8	Distinct craterlet N of Aristarchus.
Van Albada	9 N	64 E	22	Distinct; closely S of Auzout. (Auzout A).
Van Biesbroeck	29 N	46 W	10	Closely S of Krieger. (Krieger B).
Van Vleck	2 S	78 E	31	Dark floor; N of Kästner. (Gilbert M).
Vasco da Gama	15 N	85 W	80	W of Procellarum; central ridge.
Vashakidze	44 N	93 E	44	Libration zone; outside wall of Harkhebi.
Vega	45 N	63 E	80	Australe area; deep.
Vendelinus	16 S	62 E	165	Petavius chain; broken and irregular.
Vera	26 N	44 W	5	Origin of long rill; Aristarchus area. (Prinz A).
Verne	25 N	25 W	2	Between Euler and Lambert; pair with Artemis.
Very	26 N	25 E	5	Craterlet in Serenitatis; W of Le Monnier. (Le Monnier B).
Vestine	34 N	94 E	96	Libration zone; beyond Gauss.
Vieta	29 S	57 W	52	W of Humorum; low c.p.
Virchow	10 N	84 E	17	Adjoins Neper to the N. (Neper G).
Vitello	30 S	38 W	38	S edge of Humorum; concentric crater.
Vitruvius	18 N	31 E	31	Between Serenitatis and Tranquillitatis; low but rather bright walls.
Vlacq	53 S	39 E	90	One of a group of 6; deep; c.p.; Janssen area.
Vögel	15 S	6 E	60×25	SE of Albategnius; chain of 4 craters.
Volta	58 N	81 W	120	Limb formation; near Repsold.
Von Behring	8 S	72 E	39	Distinct; W of Kästner. (Maclaurin F).
Voskresensky	28 N	88 W	90	Flooded; beyond Otto Struve.
Wallace	10 N	9 W	29	In Imbrium; imperfect ring; very low walls.
Walter	33 S	1 E	129	Massive walls; interior peak and craters; trio with Regiomontanus, Purbach.
Wargentin	50 S	60 W	89	Schickard group; the famous plateau.
Warner	4 S	87 E	35	Regular; Smythii area.
Watt	50 S	51 E	72	SW of Janssen; contact twin with Steinheil.
Watts	9 N	46 E	15	N of Taruntius; low walls; darkish floor. (Taruntius D).
Webb	1 S	60 E	26	Edge of Fœcunditatis; darkish floor; c.p.; minor ray-centre.
Weichart	85 S	104 E	41	Libration zone; polar area; crater on wall.
Weierstrasse	1 S	77 E	33	Fairly regular; E of Maclaurin. (Gilbert N).
Weigel	58 S	39 W	55	Schiller area; pair with Röst; fairly regular.
Weinek	28 S	37 E	30	NE of Piccolomini; darkish floor.
Weiss	32 S	20 W	60	Pitatus group; very irregular.
Werner	28 S	3 E	66	Very regular; high walls; c.p.; pair with Aliacensis.
Wexler	69 S	90 E	52	Libration zone; beyond Neumayer; regular.
Whewell	4 N	14 E	8	Bright craterlet in Tranquillitatis uplands.
Wichmann	8 S	38 W	13	On Procellarum; associated with large 'ghost'.
Widmanstätten	6 S	86 E	46	Limb; E of Maclaurin.
Wildt	9 N	76 E	11	Distinct; W of Neper. (Condorcet K).
Wilhelm I.	43 S	20 W	97	Longomontanus area; uneven walls.
Wilkins	30 S	20 E	64	Rabbi Levi group; irregular.
Williams	42 N	37 E	45	Somniorum area; not prominent.
Wilson	60 S	33 W	74	Bettinus chain; near Bailly; deep and regular; no c.p.
Winthrop	11 S	44 W	18	Ruined crater on W wall of Letronne. (Letronne P).
Wöhler	38 S	31 E	27	E of Riccius; fairly regular.
Wolf	23 S	17 W	32	In S of Nubium; irregular; low walls.
Wollaston	31 S	47 W	11	Bright crater in Harbinger Mtns.
Wright	32 N	87 W	40	Beyond Lagrange; pair with Shaler.
Wrottesley	24 S	57 E	55	Petavius group; twin-peaked central mountain.
Wurzelbauer	34 S	16 W	80	Pitatus group; irregular walls; much floor detail.
Wyld	1 S	98 E	93	Libration zone; beyond Smythii.
Xenophanes	57 N	77 W	108	Limb near Sinus Roris; high walls; c.p.
Yakovkin	54 S	79 W	37	Limb; beyond Phocylides. (Pingré H).
Yangel	17 N	5 E	9	In highlands NW of Manilius. (Manilius F).
Yerkes	15 N	52 E	25	On W edge of Crisium; low walls; irregular.
Young	42 S	51 E	45	Rheita Valley area; irregular.

Name	Lat	Long	Diam km	Notes
Zach	61 S	5 E	52	E of Clavius; fairly deep and regular.
Zagut	32 S	22 E	80	Rabbi Levi group; irregular.
Zähringer	6 N	40 E	11	Deep; W of Taruntius. (Taruntius F).
Zasyadko	4 N	94 E	11	Libration zone; Smythii area; in Babcock.
Zeeman	75 S	105 W	184	Libration zone; beyond Drygalski.
Zeno	45 N	73 E	65	E of Mercurius; deformed.
Zinner	27 N	59 W	4	Distinct craterlet N of Schiaparelli. (Schiaparelli B).
Zöllner	8 S	19 E	40	NW of Theophilus; elliptical.
Zsigmondi	60 N	105 W	65	Libration zone; beyond Poczobut.
Zucchius	61 S	50 W	63	Schiller area; distinct; pair with Segner.
Zupus	17 S	52 W	26	S of Billy; low walls; irregular; very dark floor.

NOMENCLATURE OF THE FAR SIDE OF THE MOON

Names allotted to features on the Moon's far side are listed here. Older maps show, on the Moon's near side, the Leibnitz Mountains and the D'Alembert Mountains; these have proved to be not true ranges, and the names have been deleted so that they could be re-allotted. The name of Porter, in honour of the American astronomer Russell W. Porter, was suggested for a far-side crater at latitude 56°S, longitude 110°W; but following a suggestion by the present writer, it was officially transferred to a crater inside Clavius, on the near side of the Moon. Early maps of the far side showed a mountain range, recorded by the Russian probe Luna 3 in 1959, and named the Soviet Mountains. The name has been deleted, for the excellent reason that the mountains do not exist. (The feature recorded was nothing more than a bright ray.)

Some of the formations which are actually on the far side of the Moon are given in the main list, because libration brings them into view either wholly or in part. The following list includes features which are permanently invisible from Earth:

✳ **The largest named formation never visible from Earth** is Apollo, diameter 520 km. Other very large structures are:

	Diameter km
Hertzsprung	440
Mare Moscoviense	420
Birkhoff	370
Korolev	360
Mare Ingenii	350
Mendeleev	350
Planck	340
Schrödinger	310

The prominent formation Tsiolkovskii, which seems to be a cross between a crater and a mare, has a diameter of 180 km.

✳ **The most recent claim concerning the formation of a large crater** relates to an old report, dating from 18 July 1178, by Gervase of Canterbury. The crescent Moon was seen 'to split in two...a flaming torch sprung up, spewing out over a considerable distance fire, hot coals and sparks. Meanwhile the body of the Moon which was below writhed ... and throbbed like a wounded snake'. This seems to indicate a terrestrial cloud phenomenon, if anything; nevertheless, it has been seriously suggested that the phenomenon was the result of an impact on the far side which led to the formation of the ray-crater Giordano Bruno! This is, of course, absurd; and in any case at this time the altitude of the Moon as seen from Canterbury was less than 5 degrees. Therefore there is no doubt that the claim must be dismissed as merely a 'Canterbury tale'.

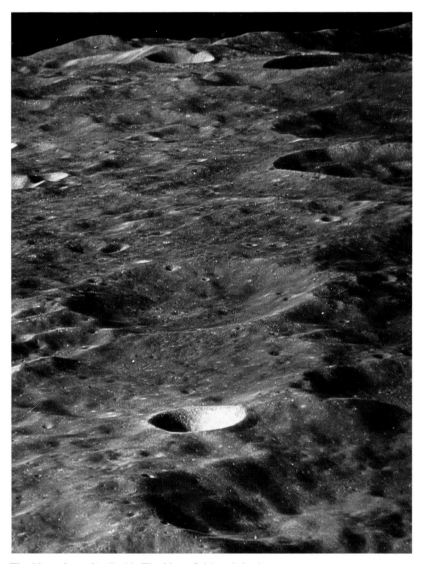

The Moon from Apollo 11. The Mare Crisium is in the centre.

FEATURES ON THE FAR SIDE OF THE MOON

*F*eatures on the Moon's far side are named according to recommendations from the International Astronomical Union.

Name	Lat	Long
Abbe	58 S	175 E
Abul Wafa	2 N	117 E
Aitken	17 S	173 E
Alden	24 S	111 E
Alekhin	68 S	130 W
Alter	19 N	108 W
Amici	10 S	172 W
Anders	42 S	144 W
Anderson	16 N	171 E
Antoniadi	69 S	173 W
Apollo	37 S	153 W
Appleton	37 N	158 E
Artamonov	26 N	104 E
Artemev	10 N	145 W
Ashbrook	81 S	112 W
Avicenna	40 N	97 W
Avogadro	64 N	165 E
Backlund	16 S	103 E
Baldet	54 S	151 W
Barbier	24 S	158 E
Barringer	29 S	151 W
Becquerel	41 N	129 E
Bečvář	2 S	125 E
Beijerinck	13 S	152 E
Bellingshausen	61 S	164 W
Belopolsky	18 S	128 W
Belyayev	23 N	143 E
Bergstrand	19 S	176 E
Berkner	25 N	105 W
Berlage	64 S	164 W
Bhabha	56 S	165 W
Birkeland	30 S	174 E
Birkhoff	59 N	148 W
Bjerknes	38 S	113 E
Blazhko	31 N	148 W
Bobone	29 N	131 W
Boltzmann	55 S	115 W
Bolyai	34 S	125 E
Borman	37 S	142 W
Bose	54 S	170 W
Boyle	54 S	178 E
Brashear	74 S	172 W
Bredikhin	17 N	158 W
Brianchon	77 N	90 W
Bridgman	44 N	137 E
Brouwer	36 S	125 W
Brunner	10 S	91 E
Buffon	41 S	134 W
Buisson	1 S	113 E
Butlerov	9 N	109 W
Buys-Ballot	21 N	175 E
Cabannes	61 S	171 W
Cajori	48 S	168 E
Campbell	45 N	152 E
Cantor	38 N	118 E
Carnot	52 N	144 W
Carver	43 S	127 E
Cassegrain	52 S	113 E
Ceraski	49 S	141 E
Chaffee	39 S	155 W
Champollion	37 N	175 E
Chandler	44 N	171 E
Chang Heng	19 N	112 E
Chant	41 S	110 W
Chaplygin	6 S	150 E
Chapman	50 N	101 W
Chappell	62 N	150 W
Charlier	36 N	132 W
Chaucer	3 N	140 W
Chauvenet	11 S	137 E
Chebyshev	34 S	133 W
Chrétien	33 S	113 E
Clark	38 S	119 E
Coblentz	38 S	126 E
Cockcroft	30 N	164 W
Comrie	23 N	113 W
Comstock	21 N	122 W
Congreve	0	168 W
Cooper	53 N	176 E
Coriolis	0	172 E
Coulomb	54 N	115 W
Crocco	47 S	150 E
Crommelin	68 S	147 W
Crookes	11 S	165 W
Cyrano	20 S	157 E
Dædalus	6 S	180
D'Alembert	52 N	164 E
Danjon	11 S	123 E
Dante	25 N	180
Das	14 S	152 W
Davisson	38 S	175 W
Dawson	67 S	134 W
Debye	50 N	177 W
De Forest	76 S	162 W
Dellinger	7 S	140 E
Delporte	16 S	121 E
Denning	16 S	143 E
Deutsch	24 N	111 E
De Vries	20 S	177 W
Dewar	3 S	166 E
Dirichlet	10 N	151 W
Doppler	13 S	160 W
Douglass	35 N	122 W
Dryden	33 S	157 W
Dufay	5 N	170 E
Dugan	65 N	103 E
Dunér	45 N	179 E
Dyson	61 N	121 W
Ehrlich	41 N	172 W
Eijkman	62 S	141 W
Einthoven	5 S	110 E
Ellerman	26 S	121 W
Elvey	9 N	101 W
Emden	63 N	176 W
Engelhardt	5 N	159 W
Eötvös	36 S	134 E
Esnault-Pelterie	47 N	142 W
Espin	28 N	109 E
Evans	10 S	134 W
Evdokimov	35 N	153 W
Evershed	36 N	160 W
Fechner	58 S	125 E
Feoktistov	31 N	140 E
Fermi	20 S	122 E
Fersman	18 N	126 W
Firsov	4 N	112 E
Fitzgerald	27 N	172 W
Fizeau	58 S	133 W
Fleming	15 N	109 E
Foster	23 N	142 W
Fowler	43 N	145 W
Freundlich	25 N	171 E
Fridman	13 S	127 W
Frost	37 N	119 W
Gadomski	36 N	147 W
Gagarin	20 S	150 E
Galois	16 S	153 W
Gamow	65 N	143 E
Ganswindt	79 S	110 E
Garavito	17 N	131 E
Gavrilov	17 N	131 E
Geiger	14 S	158 E
Gerasmovié	23 S	124 W
Giordano Bruno	36 N	103 E
Glasenapp	2 S	138 E
Golitzyn	25 S	105 W
Golovin	40 N	161 E
Grachev	3 S	108 W
Green	4 N	133 E
Gregory	2 N	127 E
Grigg	13 N	130 W
Grissom	45 S	160 W
Grotrian	66 S	128 E
Gullstrand	45 N	130 W
Guyot	11 N	117 E
Hagen	56 S	135 E
Harriott	33 N	114 E
Hartmann	3 N	135 E
Harvey	19 N	147 W
Hatanaka	29 N	122 W
Hayford	13 N	176 W
Healy	32 N	111 W
Heaviside	10 S	167 E
Hedin	68 N	123 E
Helberg	22 N	102 W
Henderson	5 N	152 E
Hendrix	48 S	161 W
Henyey	13 N	152 W
Hertz	13 N	104 E
Hertzsprung	0	130 W
Hess	54 S	174 E
Heymans	75 N	145 W
Hilbert	18 S	108 E
Hippocrates	71 N	146 W
Hoffmeister	15 N	137 E
Hogg	34 N	122 E
Holetschek	28 S	151 E
Houzeau	18 S	124 W
Hutton	37 N	169 E
Icarus	6 S	173 W
Idelson	81 S	114 E
Ingalls	4 S	153 W
Ingenii, Mare	34 S	163 E
Innes	28 N	119 E
Izsak	23 S	117 E
Jackson	22 N	163 W
Joffe	15 S	129 W
Joule	27 N	144 W
Jules Verne	36 S	146 E
Kamerlingh Onnes	15 N	116 W
Karpinsky	73 N	166 E
Kearons	12 S	113 W
Keeler	10 S	162 E
Kékulé	16 N	138 W
Khwolson	14 S	112 E
Kibaltchitch	2 N	147 W
Kidinnu	36 N	123 E
Kimura	57 S	118 E
King	5 N	120 E
Kirkwood	69 N	157 W
Kleimenov	33 S	141 E
Klute	37 N	142 W
Koch	43 N	150 E
Kohlschütter	15 N	154 E
Kolhörster	10 N	115 W
Komarov	25 N	153 E
Kondratyuk	15 S	115 E
Konstantinov	20 N	159 E
Korolev	5 S	157 W
Kostinsky	14 N	118 E
Kovalevskaya	31 N	129 W
Kovalsky	22 S	101 E
Kramers	53 N	128 W
Krasovsky	4 N	176 W
Krylov	9 S	157 W
Kulik	42 N	155 W
Kuo Shou Ching	8 N	134 W
Kurchatov	38 N	142 E
Lacchini	41 N	107 W
Lamarck	57 S	158 E
Lamb	43 S	101 E
Lampland	31 S	131 E
Landau	42 N	119 W

Name	Lat	Long
Lane	9 S	132 E
Langemak	10 S	119 E
Langevin	44 N	162 E
Langmuir	36 S	129 W
Larmor	32 N	180
Leavitt	46 S	140 W
Lebedev	48 S	108 E
Lebedinsky	8 N	165 W
Leeuwenhoek	30 S	179 W
Leibnitz	38 S	178 E
Lemaître	62 S	150 W
Lenz	3 N	102 W
Leonov	19 N	148 E
Leucippus	29 N	116 W
Levi-Civita	24 S	143 E
Lewis	19 S	114 W
Ley	43 N	154 E
Lobachevsky	10 N	113 E
Lodygin	18 S	147 W
Lomonosov	27 N	98 E
Lorentz	34 N	100 W
Love	6 S	129 E
Lovelace	82 N	107 W
Lovell	39 S	149 W
Lowell	13 S	103 E
Lucretius	9 S	121 W
Lundmark	39 S	152 E
Lütke	17 S	123 E
Lyman	65 S	162 E
Mach	18 N	149 W
Maksutov	41 S	169 W
Malyi	22 N	105 E
Mandelstam	4 N	156 E
Marci	22 N	169 W
Marconi	9 S	145 E
Mariotte	29 S	140 W
McKeller	16 S	171 W
McMath	15 N	167 W
McNally	22 N	127 W
Mees	14 N	96 W
Meggers	24 N	123 E
Meitner	11 S	113 E
Mendeleev	6 N	141 E
Merrill	75 N	116 W
Mesentsev	72 N	129 W
Meshcerski	12 N	125 E
Metchnikiff	11 S	149 W
Michelson	6 N	121 W
Milanković	77 N	170 E
Millikan	47 N	121 E
Mills	9 N	156 E
Milne	31 S	113 E
Mineur	25 N	162 W
Minkowski	56 S	145 W
Mitra	18 N	155 W
Möbius	16 N	101 E
Mohorovičić	19 S	165 W
Moiseev	9 N	103 E
Montgolfier	47 N	160 W
Moore	37 N	178 W
Morozov	5 N	127 E
Morse	22 N	175 W
Moscoviense, Mare	27 N	147 E
Nagaoka	20 N	154 E
Nassau	25 S	177 E
Nernst	36 N	95 W
Neujmin	27 S	125 E
Niépce	72 N	120 W
Nijland	33 N	134 E
Nikolayev	35 N	151 E
Nishina	45 S	171 W
Nobel	15 N	101 W
Nöther	66 N	114 W
Numerov	71 S	161 W
Nušl	32 N	167 E
Obruchev	39 S	162 E
O'Day	31 S	157 E
Ohm	18 N	114 W
Olcott	20 N	117 E

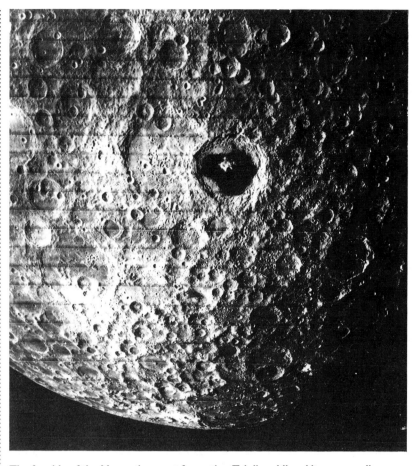

The far side of the Moon: the great formation Tsiolkovskii and its surroundings, taken from Orbiter 3 on 19 Feb. 1967. (NASA)

Name	Lat	Long	Name	Lat	Long
Omar Khayyám	58 N	102 W	Poinsot	79 N	147 W
Oppenheimer	35 S	166 W	Polzunov	26 N	115 E
Oresme	43 S	169 E	Poynting	17 N	133 W
Orlov	26 S	175 W	Prager	4 S	131 E
Östwald	11 N	122 E	Prandtl	60 S	141 E
Pannekoek	4 S	140 E	Priestly	57 S	108 E
Papaleski	10 N	164 E	Quételet	43 N	135 W
Paracelsus	23 S	163 E	Racah	14 S	180
Paraskevopoulos	50 N	150 W	Raimond	14 N	159 W
Parenago	26 N	109 W	Ramsay	40 S	145 E
Parkhurst	34 S	103 E	Rasumov	39 N	114 W
Parsons	37 N	171 W	Rayet	45 N	114 E
Paschen	14 S	141 W	Rayleigh	67 S	179 E
Pasteur	12 S	105 E	Riccó	75 N	177 E
Pauli	45 S	137 E	Riedel	49 S	140 W
Pavlov	29 S	142 E	Riemann	40 N	96 E
Pawsey	44 N	145 E	Rittenhouse	74 S	107 E
Pease	13 N	106 W	Roberts	71 N	175 W
Perelman	24 S	106 E	Robertson	22 N	105 W
Perepelkin	10 S	128 E	Roche	42 S	135 E
Perkin	47 N	176 W	Rowland	57 N	163 W
Perrine	42 N	129 W	Rozhdestvensky	86 N	155 W
Petrie	45 N	108 E	Rumford	29 S	170 W
Petropavlovsky	37 N	115 W	Safarik	10 N	177 E
Petzval	63 S	113 W	Saha	2 S	103 E
Pirquet	20 S	140 E	Sänger	4 N	102 E
Pizzetti	35 S	119 E	St. John	10 N	150 E
Planck	57 S	135 E	Sanford	32 N	139 W
Plaskett	82 N	175 E	Sarton	49 N	121 W
Plummer	25 S	155 W	Scaliger	27 S	109 E
Pogson	42 S	111 E	Schaeberle	26 S	117 E
Poincaré	57 S	161 E	Schjellerup	69 N	157 E

Name	Lat	Long
Schlesinger	47 N	138 W
Schliemann	2 S	155 E
Schneller	42 N	164 W
Schrödinger	75 S	133 E
Schuster	4 N	147 E
Schwarzschild	71 N	120 E
Seares	74 N	145 E
Sechenov	7 S	143 W
Segers	47 N	128 E
Seidel	33 S	152 E
Seyfert	29 N	114 E
Shajn	33 N	172 E
Sharonov	13 N	173 E
Shatalov	24 N	140 E
Shi Shen	76 N	105 E
Siedentopf	22 N	135 E
Sierpinski	27 S	155 E
Sisakian	41 N	109 E
Sklodowska	18 S	159 E
Slipher	50 N	160 E
Sniadecki	22 S	169 W
Sommerfeld	65 N	161 W
Spencer Jones	13 N	166 E
Spiru Haret	59 S	176 W
Stark	25 S	134 E
Stebbins	65 N	143 W
Stefan	46 N	109 W
Stein	7 N	179 E
Steklov	37 S	105 W
Steno	33 N	162 E
Sternberg	19 N	117 W
Stetson	40 S	119 W
Stoletov	45 N	155 W
Stoney	56 S	156 W
Størmer	57 N	145 E
Stratton	6 S	165 E
Strömgren	22 S	133 W
Subbotin	29 S	135 E
Sumner	37 N	109 E
Swann	52 N	112 E
Szilard	34 N	106 E
Teisserenc de Bort	32 N	137 W
ten Bruggencate	9 S	134 E
Terashkova	28 N	147 E
Tesla	38 N	125 E
Thiel	40 N	134 W
Thiessen	75 N	169 W
Thomson	32 S	166 E
Tikhomirov	25 N	162 E
Tikhov	62 N	172 E
Tiling	52 S	132 W
Timiryazev	5 S	147 W
Titov	28 N	150 E
Trümpler	28 N	168 E
Tsander	5 N	149 W
Tsiolkovskii	21 S	129 E
Tsu Chung-chi	17 N	144 E
Tyndall	35 S	117 E
Valier	7 N	174 E
Van de Graaff	27 S	172 E
Van den Bergh	31 N	159 W
Van der Waals	44 S	119 E
Van Gent	16 N	160 E
Van Gu	11 S	139 W
Van Maanen	36 N	127 E
Van Rhijn	52 N	145 E
Van't Hoff	62 N	133 W
Van Wijk	63 S	119 E
Vavilov	1 S	139 W
Vening Meinesz	0	163 E
Ventris	5 S	158 E
Vernadsky	23 N	130 E
Vesalius	3 S	115 E
Vetchinkin	10 N	131 E
Vilev	6 S	144 E
Volterra	57 N	131 E
Von der Pahlen	25 S	133 W
Von Kármán	45 S	176 E
Von Neumann	40 N	153 E

Name	Lat	Long
Von Zeipel	42 N	142 W
Walker	26 S	162 W
Waterman	26 S	128 E
Watson	63 S	124 W
Weber	50 N	124 W
Wegener	45 N	113 W
H. G. Wells	41 N	122 E
Weyl	16 N	120 W
White	48 S	149 W
Wiechart	84 S	165 E
Wiener	41 N	146 E
Wilsing	22 S	155 W
Winkler	42 N	179 W
Winlock	35 N	160 W
Woltjer	45 N	160 W
Wood	44 N	121 W
Yablochkov	61 N	127 E
Yamamoto	59 N	161 E
Zeeman	75 S	135 W
Zelinsky	29 S	167 E
Zernike	18 N	168 E
Zhiritsky	25 S	120 E
Zhukovsky	7 N	167 W
Zinger	57 N	176 E

	Lat	Long
Planck Rima:	65 S	129 E
	to 54 S	125 E
Schrödinger Rima:	62 S	99 E
	to 71 S	114 E

(The name 'Hedin' commemorates the Swedish scientist Sven Hedin. The name of Sven Hedin, used for a large enclosure near Grimaldi which is visible from Earth, has been deleted from some maps, but is retained here because lunar observers continue to use it.)

CRATER DEPTHS

Depth values for lunar craters may carry large standard errors. The following are some typical mean values – ramparts above floor. Formations whose walls are particularly irregular in height are marked §. Depths are given in metres.

4850	Tycho	3010	Menelaus
4730	Maurolycus	3000	Aristarchus
4400	Theophilus	2980	Purbach
4220	Werner	2970	Diophantus
4130	Walter	2890	Lambert
3900	Alpetragius	2830	Theætetus
3850	Theon Junior	2800	Ukert
3830	Alfraganus	2760	Stöfler §
3770	Herschel	2760	Mösting
3770	Copernicus	2760	Triesnecker
3750	Stiborius	2720	Thebit A
3730	Abenezra	2570	Kepler
3650	Aristillus	2530	Pytheas
3620	Arzachel	2510	Halley
3570	Eratosthenes	2450	Almanon
3510	Bullialdus	2400	Ptolemæus §
3470	Theon Senior	2400	Proclus
3430	Autolycus	2320	Plinius
3320	Hipparchus §	2300	Posidonius
3270	Thebit	2200	Hell
3200	Godin	2150	Archimedes
3140	Catharina	2100	Le Verrier
3130	Cayley	2080	Campanus
3120	Kant	2000	Brayley
3110	Timocharus	1960	Fauth
3110	Abulfeda	1860	Aratus
3110	Lansberg	1850	C. Herschel
3050	Manilius	1850	Ammonius

1850	Gassendi	1230	Tempel
1810	Feuillée	1230	Hippalus §
1770	Boscovich §	1180	Agatharchides §
1760	Mercator	1040	Birt
1740	Bessel	850	Kunowsky
1730	Regiomontanus §	750	Encke
1730	Vitello	750	Hyginus
1700	Manners	650	Stadius
1650	Beer	600	Linné
1550	Vitruvius	380	Kies
1490	D'Arrest	310	Spörer
1240	Cassini §	0	Wargentin

ECLIPSES OF THE MOON

Eclipses of the Moon are caused by the Moon's entry into the cone of shadow cast by the Earth. At the mean distance of the Moon, the diameter of the shadow cone is approximately 9170 km; the shadow is 1 367 650 km long on average. Totality may last for up to 1 h 44 m.

Lunar eclipses may be either total or partial. If the Moon misses the main cone, and merely enters the zone of 'partial shadow' or penumbra to either side, there is a slight dimming; but a penumbral eclipse is difficult to detect with the naked eye. Of course, the Moon must pass through the penumbra before entering the main cone or umbra.

During an eclipse the Moon becomes dim, often coppery. The colour and brightness during eclipse depends upon conditions in the Earth's atmosphere; thus the eclipse of 19 March 1848 was so 'bright' that lay observers refused to believe that an eclipse was happening at all. On the other hand, it is reliably reported that during the eclipses of 18 May 1761 and 10 June 1816 the Moon became completely invisible to the naked eye. The French astronomer A. Danjon has given an 'eclipse scale' from 0 (dark) to 4 (bright) and has attempted to correlate this with solar activity, though the evidence is still far from conclusive.

The Danjon scale is as follows:

0 Very dark; Moon almost invisible.
1 Dark; grey or brownish colour, details barely identifiable.
2 Dark or rusty red, with a dark patch in the middle of the shadow; brighter edges.
3 Brick-red; sometimes a bright or yellowish border to the shadow.
4 Coppery or orange-red; very bright, with a bluish cast and varied hues.

Ancient eclipse records are naturally rather uncertain. It has been claimed that an eclipse seen from the Middle East can be dated to 3450 BC; the eclipse of 1361 BC is more definite. Ptolemy gives the dates of observed eclipses as 721 BC and 720 BC. In *The Clouds*, the Greek playwright Aristophanes alludes to an eclipse seen from Athens on 9 October 425 BC. There was certainly a lunar eclipse in August 413 BC which had unfortunate results for Athens, since it caused Nicias, the commander of the Athenian expedition to Sicily, to delay the evacuation of his army; the astrologers advised him to stay where he was 'for thrice nine days'. When he eventually tried to embark his forces, he found that he had been blockaded by the Spartans. His fleet was destroyed, and the expedition annihilated – a reverse which led directly to the final defeat of Athens in the Peloponnesian War. According to Polybius, an eclipse in September 218 BC so alarmed the Gaulish mercenaries in the service of Attalus I of Pergamos that they refused to continue a military advance. On the other hand, Christopher Columbus turned the eclipse of AD 1504 to his advantage. He was anchored off Jamaica, and the local inhabitants refused to supply his men with food; he threatened to extinguish the Moon – and when the eclipse duly took place, the natives were so alarmed that there was no further trouble.

Obviously, a lunar eclipse can happen only at full moon: a solar eclipse, at new moon. In the original edition of the famous novel *King Solomon's Mines*, H. Rider Haggard described a full moon, a solar eclipse and another full moon on successive days. When the mistake was pointed out, he altered the second edition, turning the solar eclipse into a lunar one!

LUNAR ECLIPSES, 1923–1987

* = total. For partial eclipses, the percentage maximum phase is given.

Far side of the Moon: Apollo 11, July 1969.

Year	Date	Phase
1923	March 2	38
1923	August 25	17
1924	February 20	*
1924	August 14	*
1925	February 8/9	74
1925	August 4	75
1927	June 15	*
1927	December 8	*
1928	June 3	*
1928	November 27	*
1930	April 13	11
1930	October 7	3
1931	April 2	*
1931	September 26	*
1932	March 22	97
1932	September 14	98
1934	January 30	12
1934	July 26	67
1935	January 19	*
1935	July 16	*
1936	January 8	*
1936	July 4	27
1937	November 18	15
1938	May 14	*
1938	November 7/8	*
1939	May 3	*
1939	October 28	99
1941	March 13	33

Year	Date	Phase
1941	September 5	6
1942	March 2/3	*
1942	August 26	*
1943	February 20	77
1943	August 15	88
1945	June 25	60
1945	December 18/19	*
1946	June 14	*
1946	December 8	*
1947	June 3	2
1948	April 23	3
1949	April 13	*
1949	October 6/7	*
1950	April 2	*
1950	September 26	*
1952	February 10/11	9

Year	Date	Phase
1952	August 5	54
1953	January 29/30	*
1953	July 26	*
1954	January 18/19	*
1954	July 15/16	41
1955	November 29	13
1956	May 24	97
1956	November 18	*
1957	May 13/14	*
1957	November 7	*
1958	May 3	2
1959	March 24	27
1960	March 13	*
1960	September 5	*
1961	March 2	81
1961	August 26	99

LUNAR ECLIPSES, 1923–1987 (cont.)

1963 July 6/7	71	
1963 December 30	*	
1964 January 24/5	*	
1964 December 19	*	
1965 June 13/14	18	
1967 April 24	*	
1967 October 18	*	
1968 April 13	*	
1968 October 6	*	
1970 February 21	5	
1970 August 17	41	
1971 February 10	*	
1971 August 6	*	
1972 January 30	*	
1972 July 26	55	
1973 December 10	11	
1974 June 4	83	
1974 November 29	*	
1975 May 25	*	
1975 November 18/19	*	
1976 May 13	13	
1977 April 4	21	
1978 March 24	*	
1978 September 16	*	
1979 March 13	89	
1979 September 6	*	
1981 July 17	58	
1982 January 9	*	
1982 July 6	*	
1982 December 30	*	
1983 June 25	34	
1985 May 4	*	
1985 October 28	*	
1986 April 24	*	
1986 October 17	*	
1987 October 1	1	

LUNAR ECLIPSES, 1987–2006

* = total. For partial eclipses, the percentage maximum phase is given.
P = penumbral eclipse, for which the maximum phase is given in brackets.

Date		Type	Time of mid-eclipse GMT		Duration of eclipse Totality		Partial	
			h	m	h	m	h	m
1988	March 3	P(101)	16	14	–	–	–	–
1988	August 27	29	11	05	–	–	1	52
1989	February 20	*	15	37	1	18	3	42
1989	August 17	*	03	09	1	36	3	31
1990	February 9	*	19	13	0	42	3	24
1990	August 6	68	14	12	–	–	2	56
1991	January 30	P(88)	06	00	–	–	–	–
1991	June 27	P(31)	03	16	–	–	–	–
1991	July 26	P(25)	18	09	–	–	–	–
1991	December 21	9	10	34	–	–	1	4
1992	June 15	68	04	58	–	–	3	0
1992	December 9	*	23	45	1	14	3	28
1993	June 4	*	13	02	1	36	3	38
1993	November 29	*	06	26	0	56	3	30
1994	May 25	24	03	32	–	–	1	44
1994	November 18	P(88)	06	45	–	–	–	–
1995	April 15	11	12	19	–	–	1	12
1995	October 8	P(83)	16	05	–	–	–	–
1996	April 4	*	00	11	1	26	3	36
1996	September 27	*	02	55	1	10	3	22
1997	March 24	92	04	41	–	–	3	22
1997	September 16	*	18	47	1	2	3	16
1998	March 13	P(71)	04	42	–	–	–	–
1998	August 8	P(12)	02	26	–	–	–	–
1999	January 31	P(100)	16	20	–	–	–	–
1999	July 28	40	11	34	–	–	2	22
2000	January 21	*	04	45	1	16	3	22
2000	July 16	*	13	57	1	0	3	16
2001	January 9	*	20	22	0	30	1	38
2001	July 5	49	14	57	–	–	1	19
2001	December 30	P(89)	10	30	–	–	–	–
2002	May 6	P(69)	12	05	–	–	–	–
2002	June 24	P(21)	21	29	–	–	–	–
2002	November 20	P(86)	01	47	–	–	–	–
2003	May 16	*	03	41	0	26	1	37
2003	November 9	*	01	20	0	11	1	45
2004	May 4	*	20	32	0	38	1	41
2004	October 28	*	03	05	0	40	1	49
2005	April 24	P(87)	09	57	–	–	–	–
2005	October 17	6	12	04	–	–	0	28
2006	March 14	P(100)	23	49	–	–	–	–
2006	September 7	18	18	52	–	–	0	45

There will be total lunar eclipses on 28 August 2007, 21 February 2008, 21 December 2010, 15 June 2011, 10 December 2011, 15 April 2014, 8 October 2014, 28 September 2015, 31 January 2018, 27 July 2018, 21 January 2019, 26 May 2021, 16 May 2022, 8 November 2022, 14 March 2025, 7 September 2025, 3 March 2026, 31 December 2028, 26 June 2029 and 20 December 2029.

LUNAR PROBES, 1958–1995

AMERICAN, PRE-RANGER

Name	Launch date	Results
Able 1	17 August 1958	Failed after 77 s, at 20 km (explosion of lower stage of launcher).
Pioneer 1	11 October 1958	Reached 113 000 km.
Pioneer 2	9 November 1958	Failed when third stage failed to ignite.
Pioneer 3	6 December 1958	Reached 106 000 km.
Pioneer 4	3 March 1959	Passed within 60 000 km of the Moon on 5 March.
Able 4	26 November 1959	Failure soon after lift-off.
Able 5A	25 October 1960	Total failure.
Able 5B	15 December 1960	Exploded 70 s after lift-off.

AMERICAN RANGERS

Name	Launch date	Landing date	Lat	Long	Results
Ranger 1	23 August 1961	–	–	–	Launch vehicle failure.
Ranger 2	18 November 1961	–	–	–	Launch vehicle failure.
Ranger 3	26 January 1962	–	–	–	Missed the Moon by 37 000 km (28 January). No images obtained.
Ranger 4	23 April 1962	26 April 1962	–	–	Landed on the Moon's night side; instruments and guidance failure.

Name	Launch date	Landing date	Lat	Long	Results
Ranger 5	18 October 1962	–	–	–	Missed the Moon by over 630 km. No lunar data received.
Ranger 6	30 January 1964	2 February 1964	0·2 N	21·5 E	Landed in M. Tranquillitatis. TV failure; no images received.
Ranger 7	28 July 1964	31 July 1964	10·7 S	20·7 W	Landed in M. Nubium. 4306 images returned.
Ranger 8	17 February 1965	20 February 1965	2·7 N	24·8 E	Landed in M. Tranquillitatis. 7137 images returned.
Ranger 9	21 March 1965	24 March 1965	12·9 S	2·4 W	Landed in Alphonsus. 5814 images returned.

AMERICAN SURVEYORS

Name	Launch date	Landing date	Lat	Long	Results
Surveyor 1	30 May 1966	2 June 1966	2·5 S	43·2 W	Landed in M. Nubium, near Flamstead. 11 150 images returned. Transmitted until 13 July; contact regained until January 1967.
Surveyor 2	20 September 1966	22 September 1966	SE of Copernicus		Guidance failure; crash-landed. No data returned.
Surveyor 3	17 April 1967	19 April 1967	2·9 S	23·3 W	Landed in O. Procellarum, 612 km E. of Surveyor 1, close to site of later Apollo 12 landing. 6315 images returned. Soil physics studied.
Surveyor 4	14 July 1967	16 July 1967	0·4 N	1·3 W	Crashed in Sinus Medii. No data returned.
Surveyor 5	8 September 1967	10 September 1967	1·4 N	23·2 E	Landed in M. Tranquillitatis, 25 km from later Apollo 11 site. 18 000 images returned; soil physics studied. Contact lost on 16 December.
Surveyor 6	7 November 1967	9 November 1967	0·5 N	1·4 W	Landed in S. Medii, 29 000 images returned: soil physics studied. Re-started and moved 3 m. Contact lost 14 December.
Surveyor 7	7 January 1968	9 January 1968	40·9 S	11·5 W	Landed on N. rim of Tycho. 21 274 images returned and much miscellaneous data. Contact lost on 20 February 1968.

AMERICAN ORBITERS

Name	Launch date	Landing date	Lat	Long	Results
Orbiter 1	10 August 1966	29 October 1966	6·7 N	162 E	207 images returned. Controlled impact on far side at end of mission.
Orbiter 2	7 November 1966	11 October 1967	4 S	98 E	422 images returned. Controlled impact on far side at end of mission.
Orbiter 3	4 February 1967	9 October 1967	14·6 N	91·7 W	307 images returned. Controlled impact on far side at end of mission.
Orbiter 4	4 May 1967	6 October 1967	Far side		326 images returned. First images of polar regions. Controlled impact on far side at end of mission.
Orbiter 5	1 August 1967	31 January 1968	0	70 W	212 images returned. Controlled impact at end of mission.
Clementine	24 January 1994	5 February 1994	–	–	Lunar mapping. In May left lunar orbit intending to rendezvous with the asteroid 1620 Geographos, but failed to achieve this.

AMERICAN APOLLO MISSIONS

Name	Launch date	Landing date	Lat	Long	Results
Apollo 7	11 October 1968	–	–	–	W. Schirra, D. Eisele, R. Cunningham. Earth orbiter (10 d 20 h).
Apollo 8	21 December 1968	–	–	–	F. Borman, J. Lovell, W. Anders. Flight round Moon (6 d 3 h).
Apollo 9	3 March 1969	–	–	–	J. McDivett, D. Scott, W. Schweickart. LM test in Earth orbit (10 d 2 h).
Apollo 10	18 May 1969	–	–	–	T. Stafford, J. Young, E. Cernan. LM test in lunar orbit (8 d 0 h).
Apollo 11	16 July 1969	20 July 1969	00 67 N	23 49 E	N. Armstrong, E. Aldrin, M. Collins. Landed in M. Tranquillitatis. EVA 2·2 h.
Apollo 12	14 November 1969	19 November 1969	03 11 S	23 23 W	C. Conrad, A. Bean, R. Gordon. Landed in O. Procellarum. EVA 7·6 h (1·4 km).
Apollo 13	11 April 1970	–	–	–	J. Lovell, F. Haise, J. Swigert. Aborted landing. Splashdown 17 April.
Apollo 14	31 January 1971	5 February 1971	03 40 S	17 28 W	A. Shepard, E. Mitchell, S. Roosa. Landed near Fra Mauro. EVA 9·2 h (3·4 km).
Apollo 15	26 July 1971	30 July 1971	26 06 N	03 39 E	D. Scott, J. Irwin, A. Worden. Landed in Hadley-Apennines. EVA 18·3 h (28 km).
Apollo 16	16 April 1972	21 April 1972	08 36 S	15 31 E	J. Young, C. Duke, T. Mattingly. Landed near Descartes. EVA 20·1 h (26 km).
Apollo 17	7 December 1972	11 December 1972	20 12 N	30 45 E	E. Cernan, H. Schmitt, R. Evans. Landed in Taurus-Littrow. EVA 22 h (29 km).

RUSSIAN LUNA AND ZOND PROBES

Name	Launch date	Landing date	Lat	Long	Results
Luna 1	2 January 1959	–	–	–	Passed Moon at 5955 km on 4 January, proving that the Moon has no magnetic field. Contact lost after 62 hours.
Luna 2	12 September 1959	13 September 1959	30 N	1 W	Crash-landed, probably near Archimedes in the Mare Imbrium.
Luna 3	4 October 1959	–	–	–	Went round Moon and imaged far side. Approached Moon to 6200 km.
Luna 4	2 April 1963	–	–	–	Missed Moon by 8529 km; contact lost. In solar orbit.
Luna 5	9 May 1965	12 May 1965	1·5 S	25 W	Unsuccessful soft-lander. Crashed in Mare Nubium.
Luna 6	8 June 1965	–	–	–	Passed Moon at 161 000 km on 11 July; failure. In solar orbit.
Zond 3	18 July 1965	–	–	–	Approached Moon to 9219 km. Photographic probe; 25 images returned, some of the Moon's far side. Images returned on 27 July from 2 200 000 km.
Luna 7	4 October 1965	7 October 1965	9·8 N	47·8 W	Unsuccessful soft-lander. Crashed in Oceanus Procellarum.
Luna 8	3 December 1965	6 December 1965	9·6 N	62 W	Unsuccessful soft-lander. Crashed in Oceanus Procellarum.
Luna 9	31 January 1966	3 February 1966	7·1 N	64·4 W	Successful soft-lander; 100 kg capsule landed in Oceanus Procellarum. Images returned. Contact lost on 7 February.
Luna 10	31 March 1966	–	–	–	Lunar satellite; approached Moon to 350 km; gamma-ray studies of lunar surface layer. Contact lost on 30 May, after 460 orbits.
Luna 11	24 August 1966	–	–	–	Lunar satellite; approached Moon to 159 km. Radiation, meteoritic and gravitational studies. Contact lost on 1 October.
Luna 12	22 October 1966	–	–	–	Lunar satellite; imaged craters to a resolution of 15 m. Contact lost on 19 January 1967.
Luna 13	21 December 1966	23 December 1966	18·9 N	63 W	Soft landing in Oceanus Procellarum. Images returned; soil and chemical studies. Contact lost on 27 December.
Luna 14	7 April 1968	–	–	–	Lunar satellite; approached Moon to 160 km. Valuable data obtained.
Zond 5	14 September 1968	–	–	–	Went round the Moon, approaching to 1950 km, and returned to Earth on 21 September. Plants, seeds, insects and tortoises carried.
Zond 6	10 November 1968	–	–	–	Went round the Moon, approaching to 2420 km, and filmed the far side. Returned to Earth on 17 November.

Apollo 16; John Young drives the Lunar Rover at the Descartes landing site. (NASA)

Name	Launch date	Landing date	Lat	Long	Results
Luna 15	13 July 1969	21 July 1969	17 N	60 E	Unsuccessful sample and return probe. Crashed in Mare Crisium.
Zond 7	7 August 1969	–	–	–	Went round the Moon, approaching to 2000 km, and took colour images of both Moon and Earth. Returned to Earth.
Luna 16	12 September 1970	15 September	0·7 S	56·3 E	Landed in M. Fœcunditatis, secured 100 gr of material, and after 26½ hours lifted off and returned to Earth (24 September).
Zond 8	20 October 1970	–	–	–	Circum-lunar flight; colour pictures of Earth and Moon. Returned to Earth on 27 October, splashing down in Indian Ocean.
Luna 17	10 November 1970	17 November 1970	30·2 N	35 W	Carried Lunokhod 1 to Mare Imbrium.
Luna 18	2 September 1971	15 September 1971	3·6 N	56·5 E	Unsuccessful soft-lander. Contact lost during descent manoeuvre to Mare Fœcunditatis.
Luna 19	28 September 1971	–	–	–	Lunar satellite. Contact kept for 4000 orbits. Studies of mascons, lunar gravitational field, solar flares, etc.
Luna 20	14 February 1972	17 February 1972	3·5 N	56·6 E	Landed near Apollonius, S. of Mare Crisium (120 km N. of Luna 16 site), drilled into the lunar surface, and returned with samples on 25 February.
Luna 21	8 January 1973	16 January 1973	25·9 N	30·5 E	Carried Lunokhod 2 to a site near Le Monnier, 180 km from Apollo 17 site.
Luna 22	29 May 1974	–	–	–	Lunar satellite. TV images, gravitation and radiation studies. Contact maintained until 6 November 1975.
Luna 23	28 October 1974	–	Mare Crisium		Unsuccessful sample and return mission; crashed.
Luna 24	9 August 1976	18 August 1976	12·8 N	62·2 E	Landed in Mare Crisium, drilled down to 2 metres, collected samples, lifting off on 19 August and landing back on Earth on 22 August.

RUSSIAN LUNOKHODS

Name	Carrier	Weight, [kg]	Site	Results
Lunokhod 1	Luna 17	756	M. Imbrium	Operated for 11 months after arrival on 17 November 1970. Area photographed exceeded 80 000 sq. metres. Over 200 panoramic pictures and 20 000 images returned. Distance travelled, 10·5 km.
Lunokhod 2	Luna 21	850	Le Monnier	Operated until mid-May 1973. 86 panoramic pictures and 80 000 television images obtained. Distance travelled, 37 km. On 3 June the Soviet authorities announced that the programme has ended – perhaps prematurely.

JAPANESE LUNAR PROBES

Name	Launch date	
Hagomoro	24 January 1990	Launched by Muses-A vehicle; first Japanese attempt. The satellite Hiten ('Space Flyer') was ejected and put into orbit, 15 February 1992. It had a mass of 180 kg and carried a micrometeoroid detector.
		Hiten crash-landed on the Moon on 10 April 1993, at latitude 34 deg S, longitude 55 deg E (near Furnerius).

MERCURY

DATA ....

Distance from the Sun:
  mean: 57.9 million km (0·387 a.u.)
  max: 69.7 million km (0·467 a.u.)
  min: 45·9 million km (0·306 a.u.)
Sidereal period: 87.969 days
Synodic period: 115·88 days
Rotation period: 58·6461 days
Mean orbital velocity: 47·87 km/s
Axial inclination: negligible
Orbital eccentricity: 0·206
Orbital inclination: 7 deg 00 min 15 sec ·5
Diameter: 4878 km
Surface area: 7·475 × 10⁷ sq.km
Apparent diameter from Earth:
  max: 12 sec ·9
  min: 4 sec ·5
Reciprocal mass, Sun = 1: 6 000 000

Density, water = 1: 5·44
Mass: 3·3 × 10²⁶ g
Mass, Earth = 1: 0·055
Volume, Earth = 1: 0·056
Escape velocity: 4·25 km/s
Surface gravity, Earth = 1: 0·38
Mean surface temperature:
  day +350°C
  night −170°C
Extremes of surface temperature:
  day +427°C
  night −183°C
Oblateness: negligible
Albedo: 0·06
Maximum magnitude: −1.9
Mean diameter of Sun, as seen from Mercury:
  1 deg 22 min 40 sec

✳ *The innermost planet* is Mercury; it is also the *smallest* of the principal planets, with the exception of Pluto. It was once thought that a still closer-in planet must exist, and it was even named (Vulcan), but it is now known that no planet of appreciable size moves within the orbit of Mercury.

✳ *The discovery of Mercury* must have been prehistoric. The oldest observation of the planet which has come down to us is dated 15 November 265 BC, when the planet was one lunar diameter away from a line joining the stars Delta and Beta Scorpii. This information has been given by the last great astronomer of Classical times, Ptolemy (*c.* AD 120–180). Plato (*Republic,* X, 14) commented upon the yellowish colour of Mercury, though most naked-eye observers will describe it as being white. Mercury can actually become brighter than any star, but can never be seen against a really dark sky.

✳ *The phases of Mercury* were probably suspected in the first half of the 17th century by Galileo, Marius, and Martin van den Hove (= Hortensius), but we cannot be sure. They were definitely seen by Giovanni Zupus in 1639, and confirmed by Hevelius in 1644.

ELONGATION DATES, 1991–2005

Western
1991   Jan. 14, May 12, Sept. 7, Dec. 27.
1992   Apr. 23, Aug. 21, Dec. 9.
1993   Apr. 5, Aug. 4, Nov. 22.
1994   Mar. 19, July 17, Nov. 6.
1995   Mar. 1, June 29, Oct. 20.
1996   Feb. 11, June 10, Oct. 3.
1997   Jan. 24, May 22, Sept. 16.
1998   Jan. 6, May 4, Aug. 31, Dec. 20.
1999   Apr. 16, Aug. 14, Dec. 3.
2000   Mar. 28, July 27, Nov. 15.
2001   Mar. 11, July 9, Oct. 29.

2002   Feb. 21, June 21, Oct. 13.
2003   Feb. 4, June 3, Sept. 27.
2004   Jan. 17, May 14, Sept. 9, Dec. 29.
2005   Apr. 26, Aug. 23, Dec. 12.

Eastern
1991   Mar. 27, July 25, Nov. 19.
1992   Mar. 9, July 6, Oct. 31.
1993   Feb. 21, June 17, Oct. 14.
1994   Feb. 4, May 30, Sept. 26.
1995   Jan. 19, May 12, Sept. 9.
1996   Jan. 2, Apr. 23, Aug. 21, Dec.15.
1997   Apr. 6, Aug. 4, Nov. 28.
1998   Mar. 20, July 17, Nov. 11.
1999   Mar. 3, June 28, Oct. 24.
2000   Feb. 15, June 9, Oct. 6.
2001   Jan. 28, May 22, Sept. 18.
2002   Jan. 11, May 4, Sept. 1, Dec. 26.
2003   Apr. 16, Aug. 14, Dec. 9.
2004   Mar. 29, July 27, Nov. 21.
2005   Mar. 12, July 9, Nov. 3.

✳ *The first prediction of a transit of Mercury* across the face of the Sun was made by Kepler, for the transit of 7 November 1631. His prediction enabled Gassendi to observe the transit.

Transits between 1631 and 1995 were on:

1631	Nov. 7	1769	Nov. 9	1881	Nov. 8
1644	Nov. 9	1776	Nov. 2	1891	May 10
1651	Nov. 3	1782	Nov. 12	1894	Nov. 10
1661	May 3	1786	May 4	1907	Nov. 14
1664	Nov. 4	1789	Nov. 5	1914	Nov. 7
1677	Nov. 7	1799	May 7	1924	May 8
1690	Nov. 10	1802	Nov. 9	1927	Nov. 10
1697	Nov. 3	1815	Nov. 12	1937	May 11
1707	May 5	1822	Nov. 5	1940	Nov. 11
1710	Nov. 6	1832	May 5	1953	Nov. 14
1723	Nov. 9	1835	Nov. 7	1957	May 6
1736	Nov. 11	1845	May 8	1960	Nov. 7
1740	May 2	1848	Nov. 9	1970	May 9
1743	Nov. 5	1861	Nov. 12	1973	Nov. 10
1753	May 6	1868	Nov. 5	1986	Nov. 13
1756	Nov. 7	1878	May 6	1993	Nov. 6

The only remaining transit of the present century is that of 15 November 1999, time of mid-transit, 21 h 42 m UT. Then follow transits on 7 May 2003 (7 h 53 m), 8 November

2006 (21 h 42 m), 9 May 2016 (14 h 59 m) and 11 November 2019 (15 h 21 m).

The first transit of Mercury to have been observed at X-ray wavelengths was that of 6 November 1993. The observations were made from the Japanese satellite Yohkoh. Mercury blocked X-ray emissions from the corona, so appearing as a tiny dark hole in the X-ray corona.

Transits can occur only in May and November. May transits occur with Mercury near aphelion; at November transits Mercury is near perihelion, and November transits are the more frequent in the approximate ratio of 7 to 3. The longest transits (those of May) may last for almost 9 hours.

✳ *The last occultation of a planet by Mercury* was on 9 December 1808, when Mercury occulted Saturn. The next will be on 11 August 2079, when Mercury will occult Mars. The following are close planetary conjunctions involving Mercury between 1900 and 2100: all these occur more than 10 degrees from the Sun, and the separations are below 60 seconds of arc.

	Date	UT	Sep sec	Elong
Neptune	1914 Aug. 10	08.11	−30	18 W
Mars	1942 Aug. 19	12.36	−20	16 E
Mars	1985 Sept. 4	21.00	−46	16 W
Mars	1989 Aug. 5	21.54	+47	18 E
Mars	2032 Aug. 23	04.26	+16	13 W
Saturn	2037 Sept. 15	21.32	+18	15 W
Neptune	2039 May 5	09.42	−39	13 W
Neptune	2050 June 4	06.46	−46	17 W
Neptune	2067 July 15	12.04	+13	18 W
Mars	2079 Aug. 11	01.31	Occ.	11 W
Venus	2084 Dec. 24	05.11	+48	11 W

Mercury will occult Jupiter on 27 October 2088 and 7 April 2094, but the elongations will be less than 6 degrees.

✳ *The first serious telescopic observations of Mercury* were made in the late 18th century by Sir William Herschel, who, however, could make out no surface detail. At about the same time Mercury was studied by J. H. Schröter, who recorded some surface patches, and who believed that he had detected high mountains. It seems, however, that these results were illusory.

✳ *The first attempted map of Mercury* was compiled by G. V. Schiaparelli, from Milan, using 0m.218 and 0m.49 refractors between 1881 and 1889. He recorded various dark markings, and believed the rotation period to be synchronous – that is to say, equal to Mercury's revolution period. This would mean that part of the planet would be in permanent daylight and another part in permanent night, with an intervening 'twilight zone' over which the Sun would rise and set. Schiaparelli was also the first observer to study Mercury in daylight, when both it and the Sun were high above the horizon.

MERCURY

✷ *The best pre-Mariner map* was compiled by E. M. Antoniadi, using the 0m.38 Meudon refractor. The map was published in 1934, and various features were named. Like Schiaparelli, Antoniadi believed the rotation period to be synchronous, and he also believed in local obscurations, due to material suspended in a thin Mercurian atmosphere. These conclusions are now known to be wrong.

✷ *The first (and probably only!) observer to draw 'canal-like' features* on Mercury was P. Lowell, at Flagstaff (Arizona) in 1896. These features are completely illusory.

✷ *The first disproof of the synchronous rotation theory* was obtained in 1962 by W. E. Howard and his colleagues at Michigan, who measured the long-wavelength radiations from Mercury and found that the dark side was much warmer than it would be if it never received any direct sunlight. In 1965 the shorter rotation period was confirmed by radar methods (R. Dyce and G. Pettengill, at Arecibo in Puerto Rico). The true period is ⅔ of the revolution period, and when Mercury is best placed for observation from Earth the same face is always presented to us.

✷ *The first (and so far, the only) Mercury Probe,* Mariner 10, made three active passes of the planet: in March and September 1974 and March 1975. It was launched on 3 November 1973, and obtained some photographs of the Moon before by-passing Venus (5 February 1974) and making its first rendezvous with Mercury. *The first crater to be identified on Mercury* during the initial encounter was the bright ray-crater now named Kuiper. Closest approach took place on 29 March, and altogether 647 pictures were obtained. The second closest approach was that of 21 September 1974, and the third occurred on 16 March 1975, by which time the equipment was deteriorating. Contact was finally lost on 24 March 1975, though no doubt the probe is still in solar orbit and continues to make periodical approaches to Mercury.

✷ *The atmosphere of Mercury* was first positively detected from Mariner 10 (earlier spectroscopic reports of an atmosphere, initially by H. Vogel in 1871, were erroneous). The ground density of the Mercurian atmosphere is about 10^{-10} millibar. Helium, hydrogen, oxygen, sodium and potassium have been identified, with helium the main constituent, but the density is so low that the atmosphere is technically an exosphere. It is thought that the solar wind is the main source of the helium and hydrogen, with oxygen, sodium and potassium derived from the surface materials or by meteoritic material.

Crescent Mercury, from Mariner 10. (NASA)

✷ *The greatest temperature range on Mercury* is 610°C. The maximum temperature on any part of the surface is +427°, and the minimum is –183°. The range is greater than for any other planet in the Solar System.

✷ *The magnetic field of Mercury was also detected by Mariner 10.* The encounter yielded a value of 350 gammas (0·3 gauss) at the surface, or about 1 per cent of the Earth's magnetic field. This means that the magnetic field of Mercury is stronger than that of either Venus, Mars or the Moon. The field is dipolar, with two equal magnetic poles of opposite polarity inclined by 11 degrees to the rotational axis of the planet. The polarity is the same as that of the Earth's field (i.e. a compass needle would point north). The magnetic field is just sufficient to deflect the solar wind away from the planet's surface.

✷ *The specific gravity of Mercury* is 5·44. Apart from the Earth (specific gravity 5·5) Mercury is therefore the densest planet in the Solar System. There is an iron core 3600 km in diameter (somewhat larger than the whole of the Moon), containing 80 per cent of the planet's mass. By weight, Mercury is 70 per cent iron and only 30 per cent rocky material, so that it contains twice as much iron per unit volume as any other planet or satellite. There have been suggestions that in the early history of the Solar System, Mercury was struck by a large body and had its outer layers ripped off, leaving behind the heavy iron-rich core. This sounds quite plausible, though there is of course no proof. The core may be partially molten. Above it comes a 600 km mantle and crust composed of silicates. The surface is covered with a layer of porous silicate 'dust' forming the top of a regolith, which goes down to a few metres or a few tens of metres. Compared with that of the Moon, the crust may be rather deficient in iron and titanium-bearing minerals, which block or absorb microwaves.

✳ *The first suggestion of ice on Mercury* came in 1991, following researches by radio astronomers using the VLA (Very Large Array) in New Mexico. There seem to be signs of radar-bright caps at the planet's poles. It was also suggested that there could be ice inside the polar craters, whose floors are permanently shadowed. Results from the Arecibo radio telescope in Puerto Rico were quoted in support. However, there is no definite proof, and the presence of ice on a world such as Mercury would indeed be surprising.

✳ *The Mercurian craters* were first detected from Mariner 10, as were other features: plains (*planitia*), mountains (*montes*), valleys (*valles*), scarps (*dorsa*) and ridges (*rupes*). The south pole of Mercury lies inside the crater Chao Meng Fu. It has been agreed that the 20th-degree meridian of Mercury passes through the centre of the 1·5 km crater Hun Kal, 0 deg ·58 south of the Mercurian equator (the name stands for 20 in the language of the Maya, who used a base-20 number system).

✳ *The most imposing formation on Mercury* is the Caloris Basin (Caloris Planitia), so called because it lies near a 'hot pole', the highest-temperature region of Mercury (there are two 'hot poles'). The Caloris Basin is 1300 km in diameter, and is bounded by a ring of smooth mountain blocks rising 1 to 2 km above the surrounding surface. Unfortunately only half of the Caloris Basin was recorded from Mariner 10; at each encounter the same regions were on view, so that our knowledge of the surface is still very incomplete.

Antipodal to the Caloris Basin is an area which is officially known as 'hilly and lineated terrain', though it is often called 'weird terrain'. It covers 360 000 sq.km, and consists of hills, depressions and valleys

MERCURY

SOUTH POLAR REGION

which have destroyed older features; some of the hills rise to 2 km. It seems that the formation of this terrain has been associated with the event which produced Caloris.

There is an obvious superficial resemblance between the surface of Mercury and that of the Moon, but there are important differences in detail. In particular, about 45 per cent of the surface mapped from Mariner 10 is occupied by 'intercrater plains', which may be very old, and are unlike anything on the Moon. There are also the lobate scarps,

which are cliffs from 20 to over 500 km long and up to 3 km high; they seem to be essentially thrust faults, cutting through other features and displacing older landforms.

Craters are very plentiful, though even the most densely-cratered regions are less cratered than parts of the Moon. Craters less than 20 km in diameter are bowl-shaped, with depths of about ⅕ of their diameters; craters between 20 and 90 km across have flatter floors, with central peaks and terraced walls. The general distribution of the

MERCURY

NORTH POLAR REGION

time of writing (1995). In general craters have been named after persons who have made great contributions to human culture; plains (*planitia*) from the names of Mercury in various languages (for instance, the Suisei Planitia is the Japanese form); valleys (*valles*) after radar installations; ridges (*rupes*) after astronomers who have been particularly associated with the observation of Mercury, and scarps (*dorsa*) after famous ships of exploration and discovery. There are a few exceptions to these general rules. Caloris Basin is 'the hot basin' because it lies near one of the two hot poles; the late Gerard P. Kuiper is honoured by having the first-identified crater on Mercury named after him; and as we have seen, Hun Kal is the Mayan word for 'twenty', since, by definition, the 20th meridian on Mercury passes through its centre.

Apart from the Caloris Basin, the largest circular structure is Beethoven (625 km in diameter). There are various ray-craters; those named are Copley, Kuiper, Mena, Tansen, and Snorri.

No doubt the nomenclature of Mercury will be extended in the near future. Meantime, here is the current list, as approved by the International Astronomical Union.

Craters	Lat	Long	Diam km
Abu Nuwas	17.5	21	115
Africanus Horton	−50.5	42	120
Ahmad Baba	58.5	127	115
Al-Akhtal	59	97	102
Alencar	−63.5	104	85
Al-Hamadhani	39	89.5	170
Al-Jāhiz	1.5	22	95
Amru Al-Qays	13	176	50
Andal	−47	38.5	90
Aristoxenes	82	11	65
Asvaghosa	11	21	80
Bach	−69	103	225
Balagtas	−22	14	100
Balzac	11	145	65
Bartók	−29	135	80
Bashō	−32	170.5	70
Beethoven	−20	124	625
Belinskij	−76	104	70
Bello	−18.5	120.5	150
Bernini	−79.5	136	145
Bjornson	73	110	90
Boccaccio	−80.5	30	135
Boethius	−0.5	74	130
Botticelli	64	110	120
Brahms	58.5	177	75
Bramante	−46	62	130
Brontë	39	126.5	60
Brueghel	50	108	75
Brunelleschi	−8.5	22.5	140
Burns	54	116	45
Byron	−8	33	100
Callicrates	−65	32	65
Camões	−70.5	70	70
Carducci	−36	90	75
Cervantes	−75	122	200
Cézanne	−8	124	75
Chaikovskij	8	50.5	160
Chao Meng-Fu	−87.5	132	150
Chekov	−35.5	61.5	180
Chiang K'ui	14.5	103	40

craters is of lunar type, so that when one crater breaks into another it is almost always the smaller crater which is the intruder. The gently rolling inter-crater plains have a rough texture caused by large numbers of small craters. It seems reasonable to assume that the age of the Mercurian surface is much the same as that of the Moon; it has been estimated that the Caloris Basin is about 4000 million years old, and that extensive vulcanism ended about 3900 million years ago.

NAMED FORMATIONS ON MERCURY

*T*he naming of Mercurian features is not yet complete, and, of course, less than half the total surface was covered from Mariner 10 – our sole source of information at the

Craters	Lat	Long	Diam km
Chŏng Ch'ŏl	47	116	120
Chopin	−64.5	124	100
Chu Ta	2.5	106	100
Coleridge	−54.5	66.5	110
Copley	−37.5	85.5	30
Couperin	30	152	75
Dario	−26	10	160
Degas	37.5	127	45
Delacroix	−44.5	129.5	135
Derzhavin	44.5	35.5	145
Despréz	81	92	40
Dickens	−73	153	72
Donne	3	14	90
Dostoevskij	−44.5	177	390
Dowland	−53	180	80
Dürer	22	119.5	190
Dvořàk	−9.5	12.5	80
Echegaray	43	19	75
Eitoku	−21.5	157.5	105
Equiano	−39	31	80
Fet	−5	180	24
Flaubert	−14	72	95
Futabatei	−15.5	83.5	55
Gainsborough	−36	183	100
Gauguin	66.5	97	75
Ghiberti	−48	80	100
Giotto	12.5	56	150
Gluck	37.5	18.5	85
Goethe	79.5	44	340
Gogol	−28	147	87
Goya	−6.5	152.5	135
Greig	51	14	65
Guido d'Arezzo	−38	19	50
Hals	−55	115	100
Handel	4	34	150
Han Kan	−71.5	145	50
Harunobu	15.5	141	100
Hauptmann	−23	180	120
Hawthorne	−51	116	100
Haydn	−26.5	71.5	230
Heine	33	124.5	65
Hesiod	−58	35.5	90
Hiroshige	−13	27	140
Hitomaro	−16	16	105
Holbein	35.5	29	85
Holberg	−66.5	61	66
Homer	−1	36.5	320
Horace	−68.5	52	48
Hugo	39	47.5	190
Hun Kal	−0.5	20	1.5
Ibsen	−24	36	160
Ictinus	−79	165	110
Imhotep	−17.5	37.5	160
Ives	−32.5	112	20
Janáček	56	154	47
Jókai	72.5	136	85
Judah Ha-Levi	11.5	108	85
Kālidāsā	−17.5	180	110
Keats	−69.5	154	110
Kenkō	−21	16.5	90
Khansa	−58.5	52	100
Kōshō	60	138	65
Kuan Han-ch'ing	29	53	155
Kuiper	−11	31.5	60
Kurosawa	−52	23	180
Leopardi	−73	180	69
Lermontov	15.5	48.5	160
Lessing	−29	90	100
Liang K'ai	−39.5	183.5	105
Li Ch'ing-Chao	−77	73	60
Li Po	17.5	35	120
Liszt	−16	168	85
Lu Hsun	0.5	23.5	95
Lysippus	1.5	133	150
Ma Chih-Yuan	−59	77	170
Machaut	−1.5	83	105
Mahler	−19	19	100
Mansart	73.5	120	75

Craters	Lat	Long	Diam km
Mansur	47.5	163	75
March	31.5	176	55
Mark Twain	−10.5	138.5	140
Martí	−75.5	164	63
Martial	69	178	45
Matisse	−23.5	90	210
Melville	22	9.5	145
Mena	0.5	125	20
Mendes Pinto	−61	19	170
Michelangelo	−44.5	110	200
Mickiewicz	23.5	102.5	115
Milton	−25.5	175	175
Mistral	5	54	100
Mofolo	−37	29	90
Molière	16	17.5	140
Monet	44	9.5	250
Monteverdi	64	77	130
Mozart	8	190.5	225
Murasaki	−12	31	125
Mussorgskij	33	96.5	115
Myron	71	79.5	30
Nampeyo	−39.5	50.5	40
Nervo	43	179	50
Neumann	−36.5	35	100
Nizāmī	71.5	165	70
Ovid	−69.5	23	40
Petrarch	−30	26.5	160
Phidias	9	150	155
Philoxenus	−8	112	95
Pigalle	−37	10.5	130
Po Chü-I	−6.5	165.5	60
Po Ya	−45.5	21	90
Polygnotus	0	68.5	130
Praxiteles	27	60	175
Proust	20	47	140
Puccini	−64.5	46	110
Purcell	81	148	80
Pushkin	−65	24	200
Rabelais	−59.5	62.5	130
Rajnis	5	96.5	85
Rameau	−54	38	50
Raphael	−19.5	76.5	350
Ravel	−12	38	75
Renoir	−18	52	220
Repin	−19	63	95
Riemenschneider	−52.5	100.5	120
Rilke	−44.5	13.5	70
Rimbaud	−63	148	85
Rodin	22	18.5	240
Rubens	59.5	73.5	180
Rublev	−14.5	157.5	125
Rūdakī	−3.5	51.5	120
Rude	−33	80	75
Rūmī	−24	105	75
Sadi	−77.5	56	60
Saikaku	73	177	80
Sarmiento	−28.5	188.5	115
Sayat-Nova	−27.5	122.5	125
Scarlatti	40.5	99.5	135
Schoenberg	−15.5	136	30
Schubert	−42	54.5	160
Scopas	−81	173	95
Sei	−63.5	88.5	130
Shakespeare	48.5	151	350
Shelley	−47.5	128.5	145
Shevchenko	−53	47	130
Sholem Aleichem	51	86.5	190
Sibelius	−49	145	90
Simonides	−29	45	95
Sinan	16	30	140
Snorri	−8.5	83.5	20
Sophocles	−6.5	146.5	145
Sor Juana	49	24	80
Sōseki	39	38	90
Sōtatsu	−48	19.5	130
Spitteler	−68	62	66
Stravinsky	50.5	73	170
Strindberg	54	136	165

Craters	Lat	Long	Diam km
Sullivan	−16	87	135
Sūr Dās	−46.5	94	100
Surikov	−37	125	105
Takanobu	31	108	80
Takayoshi	−37	164	105
Tansen	4.5	72	25
Thākur	−2.5	64	115
Theophanes	−4	143	50
Thoreau	6	133	80
Tintoretto	−47.5	24	60
Titian	−3	42.5	115
Tolstoj	−15	165	400
Ts'ai Wen-chi	23.5	22.5	120
Ts'ao Chan	−13	142	110
Tsurayuki	−62	22.5	80
Tung Yüan	73.5	55	60
Turgenev	66	135	110
Tyagaraja	4	149	100
Unkei	−31	62.5	110
Ustad Isa	−31.5	166	105
Vālmiki	−23.5	141.5	220
Van Dijck	76.5	165	100
Van Eyck	43.5	159	235
Van Gogh	−76	135	95
Velázquez	37	54	120
Verdi	64.5	169	150
Vincente	−56.5	143	85
Vivaldi	14.5	86	210
Vlaminck	28	13	97
Vyāsa	48.5	80	275
Wagner	−67.5	114	135
Wang Meng	9.5	104	170
Wergeland	−37	56.5	35
Whitman	41	111	70
Wren	24.5	36	215
Yakovlev	−40.5	163.5	100
Yeats	9.5	35	90
Yun Sŏn-Do	−72.5	109	61
Zeami	−2.5	148	125
Zola	50.5	178	60

Mountains (Montes)

	Lat	Long	Diam km
Caloris	22–40	180	

Plains (Planitiae)

	Lat	Long	Diam km
Borealis	70	80	
Budh	18	148	
Caloris	30	195	
Odin	25	171	
Sobkou	40	130	
Suisei	62	150	
Tir	3	177	

Ridges (Dorsa)

	Lat	Long	Diam km
Antoniadi	28	30	
Schiaparelli	24	164	

Scarps (Rupes)

	Lat	Long	Diam km
Adventure	−64	63	
Astrolabe	−42	71	
Discovery	−53	38	
Endeavour	38	31	
Fram	−58	94	
Gjöa	−65	163	
Heemskerck	25	125	
Hero	−57	173	
Mirni	−37	40	
Pourquois-Pas	−58	156	
Resolution	−62	52	
Santa Maria	6	20	
Victoria	50	32	
Vostok	−38	19	
Zarya	−42	22	
Zeehaen	50	158	

Valleys (Valles)

	Lat	Long	Diam km
Arecibo	−27	29	
Goldstone	−15	32	
Haystack	5	46.5	
Simeiz	−12.5	65	

VENUS

DATA ....

Distance from the Sun:
mean: 108.2 million km = 0·723 a.u.
max: 109 million km = 0·728 a.u.
min: 107·4 million km = 0·718 a.u.
Sidereal period: 224.701 days
Synodic period: 583·9 days
Rotation period: 243·16 days
Mean orbital velocity: 35·02 km/s
Axial inclination: 178 deg
Orbital inclination: 3 deg 23 min 39·8 sec
Orbital eccentricity: 0·007
Diameter: 12 104 km
Apparent diameter from Earth:
mean: 37 sec ·3
max: 65 sec ·2
min: 9 sec ·5

Reciprocal mass, Sun = 1: 408 520
Density, water = 1: 5·25
Mass: $4·870 \times 10^{27}$ g
Mass, Earth = 1: 0·815
Volume, Earth = 1: 0·86
Escape velocity: 10·36 km/s
Surface gravity, Earth = 1: 0·903
Mean surface temperature: cloud tops −33°C, surface +480°C
Oblateness: 0
Albedo: 0·76
Maximum magnitude: −4.4
Mean diameter of Sun, as seen from Venus: 44 min 15 sec

✳ **The brightest planet** is Venus, second in order of distance from the Sun.

PHENOMENA OF VENUS, 1991–2005

E Elongation		Inferior Conjunction	
1991	June13	1991	Aug. 22
1993	Jan. 19	1993	Apr. 1
1994	Aug. 25	1994	Nov. 2
1996	Apr. 1	1996	June 10
1997	Nov. 6	1998	Jan. 16
1999	June11	1999	Aug. 20
2001	Jan. 17	2001	Mar. 30
2002	Aug. 22	2002	Oct. 31
2004	Mar. 29	2004	June 8
2005	Nov. 3	2006	Jan. 13

W Elongation		Superior Conjunction	
1991	Nov. 2	1992	June 13
1993	June 10	1994	Jan. 17
1995	Jan. 13	1995	Aug. 20
1996	Aug. 19	1997	Apr. 2
1998	Mar. 27	1998	Oct. 30
1999	Oct. 30	2000	June 11
2001	June 8	2002	Jan. 14
2003	Jan. 11	2003	Aug. 18
2004	Aug. 17	2005	Mar. 31

✳ **The first man to find that Venus is at its brightest when in the crescent stage** was Edmond Halley, in 1721. This is because when more of the illuminated hemisphere faces the Earth, Venus is further away from us.

✳ **The discovery of Venus** must, of course, have been prehistoric. The most ancient observations known to us are Babylonian, and are recorded on the Venus Tablet found by Sir Henry Layard at Konyunjik, now to be seen in the British Museum. Homer (*Iliad*, XXII, 318) refers to Venus as 'the most beautiful star set in the sky'.

✳ Venus is **the only planet referred to by Napoleon Bonaparte!** According to the French astronomer F. Arago, Napoleon was visiting Luxembourg when he saw that the crowd was paying more attention to the sky than to him; it was noon, but Venus was visible, and Napoleon himself saw it. Not surprisingly, his followers referred to it as being the star of 'the Conqueror of Italy'.

✳ **The first men to refer to shadows cast by Venus** were the Greek astronomer Simplicius, in his *Commentary on the Heavens of Aristotle*, and the Roman writer Pliny around AD 60.

✳ **The first man to record the phases of Venus telescopically** was Galileo, in 1610. This was an important observation, since according to the old Ptolemaic theory, with the Earth in the centre of the planetary system, Venus could never show a full cycle of phases from new to full. The observation therefore strengthened Galileo's faith in the 'Copernican' or Sun-centred system. (The phases had not previously been mentioned specifically, though exceptionally keen-sighted people can see the crescent form with the naked eye.)

✳ **The first prediction of a transit of Venus across the disk of the Sun** was Johannes Kepler, who in 1627 predicted a transit for 6 December 1631. This transit was not actually observed.

✳ **The first observation of a transit of Venus** was made by two English amateurs, J. Horrocks and W. Crabtree, on 24 November 1639, O.S. (4 December, NS). (It has been claimed that the Eastern scholar Al-Farabi saw a transit in AD 910, from Kazakhstan. This may well be true.)

✳ **The first suggestion of using a transit of Venus to measure the length of the astronomical unit** was Edmond Halley. The method was used during the transits of 1761, 1769, 1874 and 1882, but proved to be inaccurate, and is now obsolete. It de-

pended upon timing the exact moment of the beginning of the transit, but when Venus passes on to the Sun it seems to draw a strip of blackness after it, and when this strip disappears the transit is already in progress. The effect, termed the 'Black Drop', is caused by Venus' atmosphere.

Transits occurred on 13 November 1396, 26 May 1518 and 23 May 1526, but were not observed. Subsequent and future transits are as follows:

Date		Mid-transit UT
1631	Dec. 7	5h 18m
1639	Dec. 4	18h 24m
1761	June 6	5h 19m
1769	June 3	22h 27m
1874	Dec. 9	4h 02m
1882	Dec. 6	17h 04m
2004	June 8	8h 24m
2012	June 6	1h 36m
2117	Dec. 11	2h 51m
2125	Dec. 8	16h 01m

We must then wait until the years 2247 and 2255!

✳ **The only occultation of Mars by Venus to be observed** was that of 3 October 1590, seen by M. Möstlin at Heidelberg.

✳ **The only occultation of Mercury by Venus to be observed** was that of 17 May 1737, seen by J. Bevis at Greenwich.

✳ **The last occultation of a planet by Venus** was on 3 January 1818, when Venus occulted Jupiter. The next occasion will be on 22 November 2065, when Venus will again occult Jupiter, but the elongation will be only 8 deg W. Close planetary conjunctions involving Venus, with separations less than 60 seconds of arc, and elongations of at least 10 degrees from the Sun, are as follows for the 1900–2100 period:

	Date	UT	Sep	Elong
Venus–Neptune	2022 Apr. 27	19.21	−26	43 W
Venus–Neptune	2023 Feb. 15	12.35	−42	28 E
Venus–Uranus	2077 Jan. 20	20.01	−43	11 W
Mercury–Venus	2084 Dec. 24	05.11	+48	17 W

Venus occulted Regulus on 7 July 1959, 14.28 UT, and will do so again on 1 October 2044 at 22.02 UT. On 17 November 1981 Venus occulted the bright star Sigma Sagittarii.

✳ **The first observation of the Ashen Light** was made on 9 January 1643 by G. Riccioli. The Ashen Light is the faint visibility of the night hemisphere of Venus. It was formerly dismissed as a mere contrast effect, but is now thought to be real, and almost certainly due to electrical phenomena in the planet's upper atmosphere.

✳ **The first markings on Venus** were reported in 1645 by F. Fontana. However, Fontana was using a small-aperture, long-focus refractor, and there is no doubt that his 'markings on Venus' were illusory.

✳ **The first map of Venus** was drawn by F. Bianchini in 1727. However, the markings

Venus from Pioneer 10; this is an ultra-violet image, showing the upper clouds. (NASA)

The surface of Venus, from Venera 13, March 1982. Part of the spacecraft is shown.

seen by Bianchini, like those reported by Fontana, were certainly illusory. Bianchini believed that he had charted seas and continents on the surface of Venus!

✳ *The first observer to report an atmosphere around Venus* was the Russian astronomer M. V. Lomonosov, during the transit of 1761.

✳ *The first really serious observer of Venus telescopically* was J. H. Schröter. His series of observations began in 1779, but it was not until 28 February 1788 that he observed markings, which he correctly interpreted as being atmospheric. On 11 December 1798 he believed that he had identified a mountain 43 km high, but this was obviously erroneous. In 1793 he was the first to find that the observed and theoretical phases of Venus do not agree exactly; when Venus is waning, dichotomy is earlier than predicted by a few days; when Venus is waxing dichotomy is late. The phenomenon is now generally called Schröter's Effect (a term due to the writer of this book!). It is due to effects of Venus' atmosphere.

✳ *The first report of a satellite of Venus* was made by G. D. Cassini on 18 August 1686. Other reports followed, the last being that of Montbaron, at Auxerre, on 29 March 1764. It is now certain that no satellite exists, and that the observers were deceived by 'telescopic ghosts'.

✳ *The first man to estimate the rotation period of Venus* was G. D. Cassini in 1666–7; he gave a period of 23 h 21 m. Many other estimates were made in later years, but there was no general agreement.

✳ *The first man to suggest a synchronous rotation period* was G. V. Schiaparelli, in 1890. This would mean that the rotation period and the sidereal period of Venus were identical – 224·7 Earth-days – in which case Venus would keep the same hemisphere turned towards the Sun all the time. However, this did not seem to fit the facts, and in 1954 G. P. Kuiper proposed a period of 'a few weeks'.

✳ *The first estimate of the rotation period by using spectroscopic methods* (the Doppler shift) was made by R. S. Richardson in 1956. Richardson concluded that the rotation was very slow, and retrograde – that is to say, opposite in sense to that of the Earth. This has proved to be correct. The true rotation period is 243·16 days, so that Venus is the only planet with a rotation period longer than its sidereal period. (The 'solar day' on Venus is equal to 118 Earth-days.) Venus is also exceptional in having a genuinely retrograde rotation, though it is true that the axial tilt of Uranus is slightly more than a right angle (98 degrees). However, during the 1960s, French observers established that the upper clouds have a rotation period of only 4 days.

✳ *The first substance to be detected spectroscopically in the atmosphere of Venus* was carbon dioxide, identified by W. S. Adams and T. Dunham at Mount Wilson in 1932.

✳ *The first reliable temperature measurements of the upper clouds of Venus* were made in 1923–8 by E. Pettit and S. B. Nicholson, using a thermocouple attached to the 100 in Hooker reflector at Mount Wilson. They gave a value of –38°C for the day side and –33°C for the dark side, which is in good agreement with modern values.

✳ *The first good infra-red and ultra-violet photographs of Venus* were taken by F. E. Ross in 1923. The infra-red pictures showed no detail, but vague features were shown in

FEATURES ON VENUS
BOLD NUMBERS INDICATE MAP REFERENCES

Terræ	Lat	Long W
Aphrodite	40S–5 N	140–000
Ishtar	52–75 N	080–305

Regiones		
Alpha 1	29–32S	000
Asteria 2	18–30N	228–270
Atla	20N–05S	185–210
Bell	25–35N	045–055
Beta 3	20–38N	292–272
Eisila	10–25N	350–050
Imdr	42S	211
Metis 4	72N	245–255
Mnemosyne	68N	275–285
Ovda	05N–10S	080–108
Phœbe 5	10–20N	275–300
Tellus 6	35N	080
Tethus 7	55N	100
Themis 8	37–40S	275–310
Thetis	02–15S	118–140
Ulfrun	108N–3S	220–230

Planitia		
Aino 9	45S	090
Atalanta 10	54N	162
Guinevere 11	40N	310
Helen 12	55S	255
Lakshmi Planum 23	60N	330
Lavinia 13	45S	350
Leda 14	45N	065
Nava	08S	315
Niobe 15	138N–10S	132–185
Nsomeka	55S	170
Rusa	10N–10S	160–185
Sedna 16	40N	335
Tinatin	15S	015

Pateræ		
Ayrton	06N	228
Cleopatra	67N	009
Ledoux	09S	225
Raskova	51S	223
Tarbell	58S	351
Teasdale	68S	189
Theodora	23N	280

Dorsa		
Chih Nu	73S	195
Juno	32–35S	085–095
Nambi	73S	213
Saule	58S	206
Vaiva	53S	204

Linea		
Antiope	40S	350
Guor	20N	000
Hariasa	19N	015
Hippolyta	42S	345
Kara	44S	306
Lampedo	57N	290
Molpadia	48S	359
Vihansa	54N	020

Chasma		
Artemis	30–42S	121–145
Dali	17–21S	155–180
Devana 22	00	289
Diana	15S	150
Ganis	19–10N	188–200
Hecate	15–18N	230–245
Heng-o	00–10N	350–000
Ix Chel	09–11S	071–078
Juno	32S	102–120
Kuanja	10–16S	097–112
Parga	08–41S	222–321

Chasma	Lat	Long W
Quilla	32–38S	118–135
Virava	13–17S	117–132

Craters		
Baker	63N	041
Cline	22S	317
Colette	65N	322
Dix	37S	329
Earhart	72N	136
Gautier	26N	043
Graham	06S	006
Grey	52S	329
Higgins	08N	241
Hurston	78S	094
Jhirad	17S	106
Lise Meitner	55S	322
Maltby	23S	120
Nevelson	35S	308
Nightingale	62N	132
Pavlova	14N	040
Phryne	46S	315
Ponselle	63S	289
Potter	07N	309
Rhys	09N	299
Sacajawea	63N	335
Sappho	13N	027
Sayers	67S	230
Uvaysi	02N	198
Zenobia	29S	028

Coronae	Lat	Long	Diam km
Artemis	35S	135	2600
Aruru	09N	262	450
Atete	16S	243	600
Beiwe	53N	306	600
Calakomana	06N	043	575
Heng-o	02N	355	1060
Maram	07S	221	600
Poloznitsa	01S	302	675
Zemina	12S	186	530
Zisa	12N	221	850

Volcano	Lat	Long
Theia 21 Beta Regio	29N	285

Montes		
Akna Montes 17	65–72N	310–320
Freyja Montes 18	72N	290–310
Gula	22N	000
Idunn	46S	213
Ilithyia	13S	315
Innini	35S	329
Maat	02S	194
Maxwell Montes	61–67N	355–010
Mbokomu	15S	215
Nokomis	20–18N	189–191
Ozza	05N	200
Rhea 20	31N	285
Talakin	11S	355
Tif	21N	352
Tefnut	39S	304
Tuulikki	10N	278
Ushas	25S	323
Xochiquetzai	03N	270

Terra		
Lada	40S	315–110

Rupes		
Hestia	03–08N	059–075
Ut	48–53N	305–325
Vesta	55–65N	295–355

ultra-violet, indicating high-altitude cloud phenomena.

✳ **The first suggestion of formaldehyde clouds on Venus** was made by R. Wildt in 1937. This remained the favoured theory until the spacecraft results showed it to be incorrect.

✳ **The first suggestion that Venus might be entirely water-covered** was made by F. L. Whipple and D. H. Menzel in 1954; they believed that the clouds were made up of H_2O. The 'marine theory' was disproved by the Mariner 2 results of 1962. Venus is too hot for liquid water to exist on its surface.

✳ **The first radio measurements of Venus at centimetre wavelengths** were made by Mayer and his colleagues in the United States in 1958. They indicated a very high temperature for the surface of Venus, and this has been fully confirmed.

✳ **The first reliable radar contact with Venus was made in 1961** by the team at the Lincoln Laboratory, United States (the success first claimed in 1958 proved to be premature). Several other groups made radar contact with the planet at about the same time.

✳ *The first Venus probe* was Venera 1 (USSR), launched on 12 February 1961. It was not successful, as contact with it was lost at a distance of 7 500 000 km from Earth.

✳ *The first successful Venus probe* was Mariner 2 (USA), launched on 26 August 1962, which bypassed Venus at 35 000 km on 14 December 1962, confirming the high surface temperature and the virtual absence of any magnetic field.

✳ **The first probe to land on Venus** was Venera 3 (USSR), on 1 March 1966. It failed to return any data, because it was crushed by the intense pressure as it parachuted down through the atmosphere of Venus.

✳ **The first probe to transmit information during the descent through Venus' atmosphere** was Venera 4 (USSR), on 18 October 1967.

✳ **The first probe to transmit information from Venus' surface** was Venera 7 (USSR), on 15 December 1970. It survived for 23 minutes before being put out of action by the intensely hostile conditions.

✳ **The first close-range photographs of the clouds of Venus** were obtained from Mariner 10 (USA) on 5 February 1974, as the probe by-passed Venus en route for Mercury.

✳ **The first picture to be sent back from the surface of Venus** was transmitted by Venera 9 (USSR) on 21 October 1975. Venera 10 landed on 25 October 1975, and also sent back one picture.

MISSIONS TO VENUS, 1961–1991

Name	Launch date	Encounter date	Closest approach, km	Capsule landing area Lat	Long		Results
Venera 1	12 Feb. 1961	19 May 1961	100 000	–	–		Contact lost at 7 500 000 km from Earth.
Mariner 1	22 July 1962	–	–	–			Total failure; fell in sea.
Mariner 2	27 Aug 1962	14 Dec. 1962	34 833	–	–		Fly-by. Contact lost on 4 Jan. 1963.
Zond 1	2 Apr. 1964	?	100 000?	–	–		Contact lost in a few weeks.
Venera 2	12 Nov. 1965	27 Feb. 1966	24 000	–	–		In solar orbit.
Venera 3	16 Nov. 1965	1 Mar. 1966	Landed	?	?		Lander crushed during descent.
Venera 4	12 June 1967	18 Oct. 1967	Landed	+19	038	Eisila Regio	Data transmitted during descent.
Mariner 5	14 June 1967	19 Oct. 1967	4100	–	–		Fly-by. Data transmitted.
Venera 5	5 Jan. 1969	16 May 1969	Landed	–03	018	E of Navka Planitia	Lander crushed during descent.
Venera 6	10 Jan. 1969	17 May 1969	Landed	–05	023	E of Navka Planitia	Lander crushed during descent.
Venera 7	17 Aug. 1970	15 Dec. 1970	Landed	–05	351	E of Navka Planitia	Transmitted for 23 min after landing.
Venera 8	26 Mar. 1972	22 July 1972	Landed	–10	335	E of Navka Planitia	Transmitted for 50 min after landing.
Mariner 10	3 Nov. 1973	5 Feb. 1974	5800	–	–		Data transmitted. En route to Mercury.
Venera 9	8 June 1975	21 Oct. 1975	Landed	+31·7	290·8	Beta Regio	Transmitted for 55 min after landing. 1 picture.
Venera 10	14 June 1975	25 Oct. 1975	Landed	+16	291	Beta Regio	Transmitted for 65 min after landing 1 picture.
Pioneer Venus 1	20 May 1978	4 Dec. 1978	145	–	–		Orbiter, 145 to 66 000 km, period 24 h. Contact lost, 9 Oct. 1992.
Pioneer Venus 2	8 Aug. 1978	4 Dec. 1978	Landed (9 Dec.)				Multiprobe. 'Bus' and 4 landers.
			Large Probe	+4·4	304·0	Beta Regio	No transmission after landing.
			North Probe	+59·3	004·8	Ishtar Regio	No transmission after landing.
			Day Probe	+31·7	317·0	Themis Regio	Transmitted for 67 min after landing.
			Night Probe	–28·7	056·7	N of Aino Planitia	No transmission after landing.
			Bus	–37·9	290·9	Themis Regio	Crash-landing.
Venera 11	9 Sept. 1978	25 Dec. 1978	Landed	–14	299	Navka Planitia	Transmitted for 95 min after landing.
Venera 12	14 Sept. 1978	22 Dec. 1978	Landed	–07	303·5	Navka Planitia	Transmitted for 60 min after landing.
Venera 13	30 Oct. 1981	1 Mar. 1982	Landed	–07·6	308	Navka Planitia	Transmitted for 60 min. Soil analysis.
Venera 14	4 Nov. 1981	5 Mar. 1982	Landed	–13·2	310·1	Navka Planitia	Transmitted for 60 min. Soil analysis.
Venera 15	2 June 1983	10 Oct. 1983	1000	–	–		Polar orbiter, 1000–65 000 km. Radar mapper.
Venera 16	7 June 1983	16 Oct. 1983	1000	–	–		Polar orbiter, 1000–65 000 km. Radar mapper.
Vega 1	15 Dec. 1984	11 June 1985	8890	–	–		Fly-by; en route to Halley's Comet. Lander
			Lander	+08·5	176·9	Rusalka Planitia	transmitted for 20 min after arrival. Balloon dropped into Venus' atmosphere.
Vega 2	20 Dec. 1984	15 June 1985	8030	–	–		Fly-by; en route to Halley's Comet. Lander
			Lander	–07·5	179·8	Rusalka Planitia	transmitted for 21 min after arrival. Balloon dropped into Venus' atmosphere.
Magellan	5 May 1989	10 Aug. 1990	294	–	–		Orbiter, 294–8450 km. Radar mapper. Burned away in Venus' atmosphere 11 Oct. 1994.
Galileo	18 Oct. 1989	10 Feb. 1990	16 000	–	–		Fly-by. En route to Jupiter.

Venus is an intensely hostile planet. The first pictures from the surface, sent back by Venera 9 and 10, were obtained under a pressure of about 90 000 millibars (90 times as great as that of the Earth's atmosphere at sea-level) and an intolerably high temperature – which had been expected, since the atmosphere consists mainly of carbon dioxide which acts in the manner of a greenhouse, shutting in the solar heat. The Venera 9 landscape was described as 'a heap of stones', several dozen centimetres in diameter and with sharp edges; the Venera 10 landing-site was smoother, as though it were an older plateau. Wind velocities were sluggish, and the light level was described as similar to that at noon in Moscow on a cloudy winter day.

✳ *The first attempts at analysis of the surface materials* were made in March 1982 by Veneras 13 and 14, which landed in the general area of Phœbe Regio. Venera 13 dropped a lander which continued to transmit for a record 127 minutes after arrival; the temperature was 457°C, and the pressure 89 atmospheres.

Venera 14 came down in a plain near Navka Planitia (Phœbe Regio area); there were fewer of the sharp, angular rocks of the Venera 13 site. The temperature was 465°C, and the pressure 94 atmospheres. In each case, highly alkaline potassium basalts were much in evidence.

The eastern edge of Alpha Regio, lat 30S, long 11·8E. Seven circular, domelike hills, up to 25 km in diameter and 750 m high, dominate the scene: they are due to thick lava flows emerging from openings in relatively level ground. Magellan image. (NASA)

✳ *The first reliable information about Venus' atmosphere* was obtained by the Venera probes, subsequently confirmed by other spacecraft, both Russian and American. Below an altitude of 80 km, over 99·9 per cent of the atmosphere is composed of carbon dioxide and nitrogen; 96·5 per cent carbon dioxide and 3·5 per cent N_2, with traces of helium, neon, argon and krypton. The clouds are rich in sulphuric acid, and at some levels there must be sulphuric acid 'rain', which, however, evaporates before reaching the planet's surface.

Venus has no detectable overall magnetic field (it must be at least 25 000 times weaker than that of the Earth), but the dense atmosphere and magnetic eddy currents induced in its ionosphere produce a well-marked bow shock and prevent the solar wind particles from reaching the surface.

According to the space-probe results, the top of Venus' atmosphere lies at 400 km above the surface. The upper clouds lie at 70 km, and at a height of 63 km the temperature is 13°C, with an outside pressure of 0·5 atmosphere. At an altitude of 50 km above the ground the temperature is 20°C; below lies a clear layer, and then a layer of denser cloud. Beneath this layer, at 47 km, there is a second clear region. The cloud-deck ends at 30 km above the surface. The wind structure of the atmosphere is remarkable, remembering that the upper clouds have a rotation period of 4 days as against 243 days for the solid surface – in each case, retrograde (east to west) for reasons which are unknown. Winds decrease close to the surface, from about 100 m/s at the cloud-top level to only 50 m/s at 50 km, and only a few metres per second at the surface, though in that thick atmosphere even a slow wind will have tremendous force.

✳ *The first balloon to be dropped into Venus' atmosphere* was released from the Vega 1 spacecraft, which flew past Venus on its

Perspective view of Gula Mons. (NASA)

Gula Mons on Venus (Magellan) measures 400 × 250 km and is elliptical; it rises to 3.2 km above its surroundings. The flanks slope away from the summit at angles of between 0·25 and 1·4 degrees. Lava flows extend for hundreds of kilometres across the fractured plains. Another volcano, Sif Mons, adjoins Gula. (NASA)

way to Halley's Comet; a small lander was also dropped. The balloon entered Venus' atmosphere at 11 km/s on 10 June 1985, over the night side, and was tracked for 46 hours as it drifted over into the day side. A second balloon, from Vega 2 (15 June 1985) was equally successful. One revelation was that at the level of the balloons, there were stronger vertical atmospheric up-currents than had been expected. On 10 February 1990 the Galileo probe flew past Venus, en route for Jupiter, and scanned the night side; it appeared that the cloud deck, 50 to 58 km high, was very turbulent. It may be transporting heat upward from below via very large convection cells.

✳ *The first accurate maps of the surface of Venus* were obtained by radar equipment in the Pioneer probe, in 1978. Further accurate maps were obtained by Veneras 15 and 16, in 1983.

✳ *The most accurate maps to date* are those obtained by radar from the Magellan probe, which on 15 May 1991 completed one 'Venus year' of mapping and was fully operative until it burned away in Venus' atmosphere on 11 October 1994. Magellan could resolve features down to 120 metres; the orbital period was 3·2 hours, and the high inclination of the orbit meant that the polar zones could be studied. The main dish (3·7 metres across) sent down a pulse at an oblique angle to the spacecraft, striking the surface below much as a beam of sunlight will do on Earth. The surface rocks modified the pulse before it was reflected back to the antenna; rough areas are radar-bright, smooth areas are radar-dark. A smaller antenna sent down a vertical pulse, and the time-lapse between transmission and return gave the altitude of the surface below to an accuracy of 30 metres.

Venus is a world of plains, highlands and lowlands. A huge rolling plain covers some 65 per cent of the surface, with lowlands accounting for 27 per cent and highlands for only 8 per cent. About 60 per cent of the surface lies within 500 metres of Venus' mean radius, with only 5 per cent more than 2 km above it; the total range of elevations is 13 km. *The highest mountains* are the Maxwell Mountains (Maxwell Montes), which rise to 11 km above the mean radius, or 8·2 km above the adjoining plateau in Ishtar Terra. *The lowest point* is Diana Chasma, in the Aphrodite area, 2 km below the mean radius.

There are two large highland areas, Ishtar Terra in the northern hemisphere and Aphrodite, which is mainly in the south but does cross the equator. Ishtar is about the size of Australia (2900 km in diameter); the western part, Lakshmi Planum, is a high, smooth plateau covered with lava which has flowed from the caldera-like structure Colette; Colette itself has collapsed to 3 km below the adjacent surface. Ishtar Terra is characterized by mountain belts and parallel linear faults. Maxwell Montes occupies its eastern end.

Aphrodite Terra is larger – 9700 × 3200 km – and consists of eastern and western mountains separated by a lower area; one end has been nicknamed the Scorpion's Tail (officially known as Atla Regio) which is thought to be one of the main volcanic areas. A smaller highland, Beta Regio, shows a huge rift valley between two massive peaks, Theia Mons (a shield volcano) and Rhea Mons.

Volcanic activity has left its mark all over Venus; there are structures ranging from small domes up to huge shield volcanoes such as Sapas Mons, Maat Mons and others. Current activity is probable. It may well be that the crust of Venus is less mobile than that of the Earth, so that terrestrial-type plate tectonics does not apply; if so, a volcano which forms over a 'hot spot' will not drift away (as Mauna Kea and Hawaii has done), but will remain active for a long period. Volcanic features do tend to concentrate along the equatorial regions (Beta Regio, Atla, Themis Regio for example).

There has been extensive flooding by lavas sent out both from volcanic craters and from fissures at lower levels probably from 500 to 300 million years ago. Faulting occurs almost everywhere, and several major zones of large-scale faulting can be identified. Volcanic structures include what are termed coronæ (circular, up to 2 km high and 400 km in diameter, surrounded by ridges and troughs); 'arachnoids', so named because of their outward resemblance to spiders' webs, which are circular structures surrounded by complex troughs; and strange features which look rather like pancakes, and are explained as low domes.

The idea of active vulcanism is supported by reports of the amount of variability of sulphur dioxide in the atmosphere; for example there was a sudden increase in 1978, possibly following a major eruption, followed by a decline.

It is notable that over 70 per cent of all shield fields, coronæ and large volcanoes fall within one-half of the surface. There are also areas characterized by intersecting ridges and grooves. These were once called 'parquet' regions, but the term was abandoned as being insufficiently scientific, to be replaced by 'tesseræ'. They seem to represent blocks of thickened crust, and occur as elevated plateaux with steep sides; their surfaces are very rough. Like the coronæ and arachnoids, they are unique to Venus – at least as far as we know. *The largest contiguous area* covers 10 000 000 km², stretching from Ovda Regio to the westward end of Aphrodite. *The largest circular lowland area* is Atalanta Planitia, east of Ishtar; it is on average 1.4 km below the level of the mean radius, and is about as large as the Gulf of Mexico. *The longest channel* is Hildr Chasma – the length is 6800 km, longer than the Nile. (This is a lava-channel.)

Craters are widespread, though there are fewer small craters than on the Moon or Mercury. Many are surrounded by radar-dark haloes. If the craters are due to impact, as is generally believed, it follows that the impactor will be preceded by a shock wave which slams down and pulverizes the upper few metres of rock, producing a surface which is relatively smooth and therefore radar-dark. *The largest crater* is Mead, diameter 280 km.

It has been laid down that all names of features on Venus must be female (the Maxwell Mountains were named before this edict was passed). Many of the names are familiar, though some are more obscure – for example Quetzalpetlatl (an Aztec fertility goddess), Titibu (a Japanese poet) and Erxleben (a German scholar).

If Venus and the Earth are so alike in size and mass, why have they developed so differently? The cause must be Venus' lesser distance from the Sun. It seems that in the early history of the Solar System, the Sun was less luminous than it is now, so that the two planets began to evolve in similar fashion; but as the Sun became more powerful, the oceans of Venus evaporated, the carbonates were driven out of the rocks, and there was a 'runaway greenhouse' effect leading to the intolerable conditions of today. If this is so, then Venus has had a tragic history; any life which may have gained a foothold there was ruthlessly destroyed.

EARTH

DATA ....

Distance from the Sun:
  mean: 149·5979 million km (1 a.u.)
  max: 152 million km (1·0167 a.u.)
  min: 147 million km (0·9833 a.u.)
Perihelion (1992): 3 January
Aphelion (1992): 3 July
Equinoxes (1992): 20 March, 08 h 48 m; 22
  September, 18 h 43 m
Solstices (1992): 21 June, 03 h 14 m; 21
  December, 14 h 43 m
Obliquity of the ecliptic (1992): 23 deg ·44032
Sidereal period: 365·256 days
Rotation period: 23 h 56 m 04 s
Mean orbital velocity: 29·79 km/s

Orbital inclination: 0 deg (by definition)
Orbital eccentricity: 0·01671
Diameter:
  equatorial: 12 756 km
  polar: 12 714 km
Circumference: 40 075 km (equatorial)
Surface area: 510 065 500 sq. km
Oblateness: 1/298·25
Mass: $5·976 \times 10^{24}$ g
Reciprocal mass, Sun = 1: 328 946
Density, water = 1: 5·517
Escape velocity: 11·18 km/s
Albedo: 0·36
Mean surface temperature: 22°C

*T*he Earth is the largest and most massive of the inner group of planets. Mainly from seismic studies we have a good knowledge of its structure. The crust has an average depth of 10 km below the oceans, but 50 km below the continents; the base of the crust is marked by the Mohorovičič Discontinuity (the 'Moho'), named after the Yugoslav scientist Andrija Mohorovičič, who discovered that the velocity of seismic waves changes abruptly at this depth, indicating a sudden change in density. Below the crust comes the mantle, which extends down to 2890 km and contains 67 per cent of the Earth's mass. Partial melting of mantle material produces basalt, which issues from volcanic vents on the ocean floors. The base of the mantle is marked by the Gutenberg Discontinuity, where the rock composition changes from silicate to metallic and its state from solid to liquid. The outer liquid core extends down to 5150 km, and contains 31 per cent of the Earth's mass. The inner core, down to the centre of the globe, is solid and has 1·7 per cent of the total mass; it has been said to 'float' in the outer liquid core. The core is iron-rich. The solid inner core, approximately 2400 km in diameter, is thought to have a central temperature of about 4530°C, with a density of $13·1$ g/cm³. Currents in the liquid core, involving iron, are responsible for the Earth's magnetic field. The magnetosphere is extensive, and the Earth is associated with belts of charged particles, discovered by equipment on the first American satellite, Explorer 1, in 1958, and named the Van Allen Belts, after the chief investigator.

MARS

DATA

Distance from the Sun:
mean: 227·94 million km (1·524 a.u.)
max: 249·1 million km (1·666 a.u.)
min: 206·7 million km (1·381 a.u.)
Sidereal period: 686.980 days (= 668·60 sols)
Synodic period: 779.9 days
Rotation period: 24 h 37 m 22·6 s (= 1 sol)
Mean orbital velocity: 24·1 km/s
Axial inclination: 23 deg 59 min
Orbital inclination: 1 deg 50 min 59 sec
Orbital eccentricity: 0·093
Diameter:
equatorial: 6794 km
polar: 6750 km

Apparent diameter from Earth:
max: 25 sec ·7
min: 3 sec ·5
Reciprocal mass, Sun = 1: 3 098 700
Density, water = 1: 3·94
Mass, Earth = 1: 0·107
Mass: $6·421 \times 10^{26}$ g
Volume, Earth = 1: 0·150
Escape velocity: 5·03 km/s
Surface gravity, Earth = 1: 0·380
Mean surface temperatures: −23°C
Oblateness: 0·009
Albedo: 0·16
Maximum magnitude: −2·8
Mean diameter of Sun, as seen from Mars: 21 min

OPPOSITION DATES, 1992–2005

Date	Closest Approach to Earth	Minimum distance millions of km	Apparent diameter sec	Mag.	Constellation
1993 Jan. 7	1993 Jan. 3	94	14·9	−1·2	Gemini
1995 Feb. 12	1995 Feb. 11	101	13·8	−1·0	Leo
1997 Mar. 17	1997 Mar. 20	99	14·2	−1·1	Virgo
1999 Apr. 24	1999 May 1	87	16·2	−1·5	Virgo
2001 June 13	2001 June 21	67	20·8	−2·1	Sagittarius
2003 Aug. 28	2003 Aug. 27	56	25·1	−2·7	Capricornus
2005 Nov. 7	2005 Oct. 30	69	20·2	−2·1	Aries

Between 1992 and 2005 Mars is at perihelion on 17 May 1992, 4 Apr. 1994, 20 Feb. 1996, 7 Jan. 1998, 25 Nov. 1999, 12 Oct. 2001, 30 Aug. 2003 and 17 July 2005; at aphelion on 25 Apr. 1993, 14 Mar. 1995, 29 Jan. 1997, 16 Dec. 1998, 2 Nov. 2000, 21 Sept. 2002 and 7 Aug. 2004.

✷ **The only red planet** is Mars, which comes fourth in order of distance from the Sun, and is the first planet beyond the orbit of the Earth.

Close planetary conjunctions involving Mars for the 1900–2100 period:

Date	UT	Sep sec	Elong
Mercury–Mars 1942 Aug. 19	12.36	−20	16 E
Mars–Uranus 1947 Aug. 6	01.59	+43	48 W
Mercury–Mars 1985 Sept. 4	21.00	−46	16 W
Mars–Uranus 1988 Feb. 22	20.48	+40	63 W
Mercury–Mars 1989 Aug. 5	21.54	+47	18 E
Mercury–Mars 2032 Aug. 23	04.26	+16	13 W
Mercury–Mars 2079 Aug. 11	01.31	Occ.	11 W

✷ **The closest oppositions** occur with Mars near its perihelion. During this period the closest approach is that of 1988, with a minimum distance from Earth of 36 300 000 miles or 58 400 000 km. The closest approach during the past half-century has been that of August 1971 (minimum distance 56 200 000 km).

✷ **The greatest distance** between Mars and the Earth, with Mars at superior conjunction, may amount to 400 000 000 km.

✷ **The longest interval between successive oppositions** is 810 days; the **shortest interval** is 764 days. The mean synodic period (see Data Table) is 779.9 days.

✷ **The least favourable oppositions** occur with Mars at aphelion, as in 1980 (least distance from Earth 101 320 000 km).

✷ **The minimum phase of Mars** as seen from Earth is 85 per cent. (At opposition, the phase is, of course, virtually 100 per cent.)

✷ **Martian Seasons:** The seasons are of the same general type as those of the Earth, since the axial tilt is very similar and the Martian day (sol) is not a great deal longer (1 sol = 1·029 days). The lengths of the seasons are as follows:

	Days	Sols
S spring (N autumn)	146	142
S summer (N winter)	160	156
S autumn (N spring)	199	194
S winter (N summer)	182	177
	687	669

Southern summer occurs near perihelion. Therefore, climates in the southern hemisphere of Mars show a wider range of temperature than for the north. The effect is much greater than in the case of the Earth, partly because there is no sea on Mars and partly because of the greater eccentricity of the Martian orbit. At perihelion, Mars receives 44 per cent more solar radiation than at aphelion.

✷ **The discovery of Mars** was unquestionably prehistoric. The planet was recorded in Egypt, China and Assyria. Later, its redness led to its being named in honour of the War-God, Ares (Mars); the study of the Martian surface is still officially known as 'areography'.

✷ **The first observation of an occultation of Mars by the Moon** was made by Aristotle (384–322 BC). The exact date of the phenomenon is not known.

✷ **The first precise observation of the position of Mars** dates, according to Ptolemy, back to 17 January 272 BC, when Mars was observed to be close to the star Beta Scorpii.

✷ **The best pre-telescopic observations of the movements of Mars** were made by the Danish astronomer Tycho Brahe, from his island observatory at Hven between 1576 and 1596. It was these observations which enabled Kepler, in 1609, to publish his first Laws of Planetary Motion, showing that the planets move round the Sun in elliptical orbits.

✷ **The first telescopic observations of Mars** were made by Galileo in 1610. No surface details were seen.

✷ **The first detection of the phase of Mars** was also due to Galileo in 1610, and recorded by him in a letter written to Father Castelli on 30 December of that year.

✷ **The first telescopic drawing of Mars** was made by F. Fontana, in Naples, in 1636. Mars was shown as spherical, and 'in its centre was a dark cone in the form of a pill'. This feature was, of course, an optical effect. Fontana's second drawing (24 August 1638) was similar.

✷ **The first marking to be recorded on Mars** was the Syrtis Major, by C. Huygens on 28 November 1659, at 7 p.m. The Syrtis Major is easily recognizable (though exaggerated in size). The sketch has been very useful in confirming the constancy of Mars' rotation period.

✷ **The first reasonably accurate estimate of the rotation period of Mars** was also due to Huygens. On 1 December 1659 he recorded that the period was about 24 hours.

✷ **The first really accurate measurement of the rotation period** was made by G. D. Cassini in 1666; his value was 24 h 40 m.

✷ **The first record of the polar caps** was also due to Cassini in 1666. It has been suggested that a cap was seen by Huygens in 1656, but his surviving drawing is very inconclusive. (However, Huygens undoubtedly saw the south polar cap in 1672.)

✷ **The discovery that the polar caps do not coincide with the geographical (or areographical!) poles** was made by G.

Summit caldera of Olympus Mons, 13 June 1977, from Viking Orbiter 1. Olympus Mons is the highest volcano on Mars (27·4 km). The summit caldera is complex, with varied floor levels due to different eruptions. The deepest crater (top centre) is 25 km in diameter. At the lower left, a lava-flow extends from the caldera rim down the flank of the volcano. (NASA)

Maraldi, in 1719. In 1704 Maraldi had studied the caps, and had also given a rotation period of 24 h 39 m, which was very near the truth.

✳ *The first known panic due to Mars* was that of 1719. Mars was at perihelic opposition, and was so bright that people mistook it for a red comet which might well collide with the Earth.

✳ *The first identification of the polar caps with ice and snow* was suggested by W. Herschel, from his observations made between 1777 and 1783. Herschel also measured the rotation period, and his observations were later re-worked by W. Beer and J. H. Mädler, yielding a period of 24 h 37 m 23 s ·7, which is only one second in error.

✳ *The first comment upon the Martian atmosphere* was made by Herschel in 1783. From observations of the close approach of a star to Mars, he rightly concluded that the atmosphere could not be very extensive.

✳ *The first good determination of the axial inclination of Mars* was also due to Herschel at the same period. His value was 28 degrees, which is only 4 degrees too great. The Martian north polar star is at present Deneb (Alpha Cygni), but the inclination varies between 35 degrees and 14 degrees in a period of about 50 000 years, so that precessional effects are important in any long-term consideration of the changing Martian climate.

✳ *The first suggestion of signalling to the inhabitants of Mars* was made by the great German mathematician K. F. Gauss, about 1802; the plan was to draw vast geometrical patterns in the Siberian tundra. In 1819 J. von Littrow, of Vienna, proposed to use signal fires lit in the Sahara. Later (1874) Charles Cros, in France, put forward a scheme to focus the Sun's heat on to the Martian deserts by means of a huge burning-glass; the glass could be swung around to write messages in the deserts!

✳ *The first prize offered for communicating with extraterrestrial beings* was the Guzman Prize, announced in Paris on 17 December 1900. The sum involved was 100 000 francs – but Mars was excluded, because it was felt that communication with the Martians would be too easy!

In 1909 the well-known physicist R.W. Wood suggested building a large array of cylinders, blacked on one side, and setting them up in a desert so that they could be turned around to send signals to Mars. Before 1909 the astronomer David Todd and balloonist L. Stevens planned to take a sensitive receiver up in a balloon, so that detecting Martian signals would be easier. Then in 1906, when Marconi set up a telegraph station at Cape Clear, British operators reported a strange regular signal of three dots (the Morse letter S) which they could not explain. And in 1921 Marconi himself reported receiving the Morse letter V at 150 000 m.

A concerted effort was made in August 1924, when Mars was at its closest. Radio transmitters in various parts of the United States were temporarily shut down so that signals from Mars could be picked up, and ready to translate them was W.F. Friedman, head of the code section of the US army Signals Corps. At Dulwich, in Outer London, radio listeners reported possible Martian signals at 30 000 m. And in 1926 a Dr Mansfield Robinson went to the Central Telegraph Office in London and dispatched a telegram to Mars – for which he was charged the standard 18 pence per word. Tactfully, the postal authorities noted it as 'Reply not guaranteed'.

In 1992 Mr C. Cockell fought the General Election as a Forward to Mars Party candidate. The constituency he selected was Huntingdon, where the sitting member was the Prime Minister, Mr John Major. Alas, Mr Cockell lost his deposit!

✳ *The first suggestion of clouds in the atmosphere of Mars* was made by the French astronomer H. Flaugergues in 1811.

✳ *The first reasonably good charts of Mars* were made by W. Beer and J. H. Mädler in 1830–32, from Berlin. They were also the first to report a dark band round the periphery of a shrinking polar cap. This band was later seen by many observers, and in the late 19th century P. Lowell attributed it, wrongly, to moistening of the ground owing to the melting polar ice.

✳ *The first report that the southern polar cap has a greater range of size than the northern* was made by the French astronomer F. Arago in 1853. (Flaugergues, in 1811, had suggested that this should be so, because the temperature range is more extreme in the south.)

Mars, from three frames taken at intervals of 9 seconds from Viking Orbiter 1 on 18 June 1976. (NASA)

* *The first observations of 'white' clouds on Mars* were made by A. Secchi in 1858.

* *The first suggestion that the dark areas are vegetation tracts rather than seas* was made in 1860 by E. Liais, a French astronomer who spent much of his life in Brazil. This was also the view of G. V. Schiaparelli, who in 1863 pointed out that the dark areas did not show the Sun's reflection as they would be expected to do if they were sheets of water. The vegetation theory was regarded as probable up to the space-probe era.

* *The first reported detection of water vapour in the Martian atmosphere* was by J. Janssen in 1867. Taking his equipment to the top of Mount Etna, to get above the densest and wettest part of the Earth's atmosphere, Janssen compared the spectrum of Mars with that of the airless Moon, and believed that he could see differences. It is now known that these results were spurious.

* *The first system of Martian nomenclature* was that of R. A. Proctor in 1867. The features were named in honour of past and contemporary observers of the planet. The system was superseded in 1877 by that of Schiaparelli.

* *The first detection of the alleged canal network* was due to G. V. Schiaparelli in 1877, when he recorded 40 canals. Streaks had been recorded earlier by various observers, including Beer and Mädler (1830–1840) and the Rev W. R. Dawes (1864), but Schiaparelli's work marked the beginning of the 'canal controversy'. He also produced a new map, and as the zero for Martian longitudes he adopted the feature which had also been selected by Beer and Mädler; it is now called Sinus Meridiani, or the Meridian Bay. Following the space-probe results from 1965, the zero was fixed as the centre of a small crater in the Sinus Meridiani; this has been named Airy-0, in honour of the 19th-century Astronomer Royal, Sir George Airy (the zero for terrestrial longitudes passes through the Airy transit circle at the old Royal Observatory in Greenwich Park). On this map Schiaparelli introduced a new system of nomenclature, which has only recently been modified in view of the Mariner and Viking results.

* *The first reports of the twinning or 'gemination' of some canals* were made by Schiaparelli in 1879. The first observers to support Schiaparelli's canal network were Perrotin and Thollon, at the Nice Observatory in France, in 1886. Subsequently canals became fashionable, and P. Lowell, who built the observatory which bears his name at Flagstaff and observed Mars between 1896 and 1916, believed the network to be

artifical. Only since the close-range photography of Mars by space-probes has it been shown that the canal system is completely illusory.

* *The first astronomer to use the term 'deserts' for the bright regions of Mars* was W. H. Pickering, in 1886. In 1892 Pickering observed features in the dark regions as well as the 'deserts', and this more or less killed the theory that the dark areas were seas.

* *The first observations of craters on Mars* seem to have been made by E. E. Barnard in 1892, using the Lick Observatory 36-in refractor. However, Barnard did not publish his results, for fear of ridicule. Craters were also seen by J. Mellish in 1917, with the 40-in Yerkes refractor, but these observations also remained unpublished. The first prediction of craters on Mars was made by D. L. Cyr in 1944, but it was not until the flight of Mariner 4, in 1965, that the existence of craters was proved.

* *The first suggestion that the polar caps were made of solid carbon dioxide* was due to A. C. Ranyard and Johnstone Stoney in 1898. The theory was more or less disregarded until the early space-probe results, beginning with Mariner 4 in 1965, but was then favoured until the missions of Vikings 1 and 2 in 1976. It is now known that the caps are of ice, many metres thick. Each cap has a coating of carbon dioxide ice; in the case of the northern cap, this coating disappears seasonally, by subliming (i.e. turning straight from a solid into a gas as the temperature rises), exposing the residual water ice cap beneath; this does not happen with the southern cap, where the carbon dioxide coating never completely dissipates. The cause of the difference is that during northern spring and summer the Martian atmosphere is comparatively clear and transparent, so that the CO_2 ice is exposed to solar radiation and dissipates, while during southern spring and summer the atmosphere is more dust-laden, protecting the upper CO_2 cap.

* *The first good measurements of the surface temperature of Mars* were made in 1909, by Nicholson and Pettit at Mount Wilson and by Coblentz and Lampland at Flagstaff. They found that Mars has a mean surface temperature of –28°C, as against +15°C for the Earth.

* *The first report of the 'Violet Layer'* was made by Lampland in 1909. This was supposed to block out the short-wavelength radiations, and prevent them from reaching the Martian surface, except on occasions when the layer cleared away. It could not be seen visually.

* *The first analysis of the Martian atmos-*

phere by using the Doppler method was made by W. S. Adams and T. Dunham in 1933. When Mars is approaching the Earth, the spectral lines should be shifted toward the short-wave end of the band; when Mars is receding, the shift should be to the red. It was hoped that in this way any lines due to gases in the Martian atmosphere could be disentangled from the lines produced by the same gases in the Earth's atmosphere. From their results, Adams and Dunham concluded that the amount of oxygen over Mars was less than 1/10 of 1 per cent of the amount existing in the atmosphere of the Earth.

* *The first reasonably good determination of the atmospheric pressure on Mars* was made by the Russian astronomer N. Barabaschev in 1934. He gave a value of 50 millibars. Later estimates increased this to between 80 and 90 millibars. However, the Mariner and Viking results have shown that the pressure is nowhere as high as 10 millibars.

* *The first reported spectroscopic detection of carbon dioxide in the Martian atmosphere* was due to G. P. Kuiper in 1947. However, it was then thought that the atmosphere must consist chiefly of nitrogen. The Mariner and Viking results show that this is wrong. The atmosphere of Mars is made up of 95 per cent carbon dioxide (CO_2), 2·7 per cent nitrogen (N_2), 0·03 per cent oxygen (O_2), 1·6 per cent argon, and smaller amounts of gases such as krypton and xenon.

* *The first reported detection of organic matter in the spectra of the Martian dark areas* was announced by W. M. Sinton in 1959. The results were later found to be spurious.

* *The first attempt to send a probe to Mars* was made by the Russians in 1962 (Mars-1). The first successful probe was the US Mariner 4, in 1965.

* *The latest nomenclature* for Martian features involves naming the craters in honour of men and women who have been associated with the study of Mars. Other classes of features are as follows:

Catena: crater-chain; a line or chain of craters (e.g. Tithonia Catena).

Cavi: steep-sided hollow (e.g. Angusti Cavi).

Chaos: distinctive area of broken terrain (e.g. Aromatum Chaos).

Chasma: very large linear chain (e.g. Ophir Chasma).

Colles: hills (e.g. Deuteronilus Colles).

Dorsum: ridge; irregular, elongated elevation (e.g. Solis Dorsum).

Fossa: ditch – long, shallow, narrow depression (e.g. Claritas Fossæ).

MARS PROBES, 1962–1991

Name	Nationality	Launch date	Encounter date	Closest approach of orbiter, km	Landing site of capsule	Results
Mars 1	USSR	1 Nov. 1962	?	190 000?	–	Contact lost at 106 000 km.
Mariner 3	USA	5 Nov. 1964	–	–	–	Shroud failure. In solar orbit but contact lost soon after launch.
Mariner 4	USA	28 Nov. 1964	14 July 1965	9789	–	Returned 21 images, plus miscellaneous data. Contact lost on 20 Dec. 1967.
Zond 2	USSR	30 Nov. 1964	Aug. 1965?	?	–	Contact lost on 2 May 1965.
Mariner 6	USA	24 Feb. 1969	31 July 1969	3392	–	Returned 76 images; flew over Martian equator. In solar orbit.
Mariner 7	USA	27 Mar. 1969	4 Aug. 1969	3504	–	Returned 126 images, mainly over the S. hemisphere. In solar orbit.
Mariner 8	USA	8 May 1971	–	–	–	Total failure: fell in the sea.
Mars 2	USSR	19 May 1971	27 Nov. 1971	In orbit, 2448 × 24 400	44 S, 213 W (Eridania)	Capsule landed, with Soviet pennant but no images received.
Mars 3	USSR	28 May 1971	2 Dec. 1971	In orbit, 1552 × 212 800	45 S, 158 W (Phæthontis)	Orbiter returned data. Contact with lander lost 20 sec after arrival.
Mariner 9	USA	30 May 1971	13 Nov. 1971	In orbit, 1640 × 16 800	–	Returned 7329 images. Contact lost on 27 Oct 1972.
Mars 4	USSR	21 July 1973	10 Feb. 1974	Over 2080	–	Missed Mars; some fly-by data returned. Failed to orbit.
Mars 5	USSR	25 July 1973	12 Feb. 1974	In orbit, 1760 × 32 500	–	Failure; contact lost.
Mars 6	USSR	5 Aug. 1973	12 Mar. 1974	?	?24 S, 25 W (Erythræum)	Contact lost during landing sequence.
Mars 7	USSR	9 Aug. 1973	9 Mar. 1974	1280	–	Failed to orbit; missed Mars.
Viking 1	USA	20 Aug. 1975	19 June 1976	In orbit	22·4 N, 47·5 W (Chryse)	Landed 20 July 1976.
Viking 2	USA	9 Sept. 1975	7 Aug. 1976	In orbit	48 N, 226 W (Utopia)	Landed 3 Sept. 1976.
Phobos 1	USSR	7 July 1988	?	?	–	Contact lost, 29 Aug. 1988.
Phobos 2	USSR	12 July 1988	–	In orbit, 850 × 79 750	–	Contact lost, 27 Mar. 1989. Some images and data from Mars and Phobos returned.
Mars Observer	USA	25 Sept. 1992	24 Aug. 1993	–	–	Contact lost, 25 Aug. 1993.

Labyrinthus: valley complex; network of linear depressions (e.g. Noctis Labyrinthus, the only really major example).

Mensa: mesa. Small plateau or table-land (e.g. Nilosyrtis Mensæ).

Mons: mountain or volcano (e.g. Olympus Mons).

Montes: mountains (e.g. Tharsis Montes).

Patera: saucer-like volcanic structure with fluted or scalloped edges (e.g. Alba Patera).

Planitia: smooth, low-lying plain (e.g. Hellas Planitia).

Planum: plateau; smooth, high area (e.g. Solis Planum).

Rupes: cliff (e.g. Ogygis Rupes).

Scopulus: irregular, degraded scarp (e.g. Nilokeras Scopulus).

Sulci: intricate network of linear depressions and ridges (e.g. Gigas Sulci).

Terræ: lands, names often given to classical albedo features (e.g. Sirenum Terra).

Tholus: domed hill (e.g. Uranius Tholus).

Vallis: valley. Sinuous channel, often with tributaries (e.g. Valles Marineris).

Vastitas: extensive plain (e.g. the north-polar circumpolar plain, Vastitas Borealis).

There are two major bulges in the Martian

lithosphere: those of Tharsis and Elysium. Following the zero contour for Martian datum (taken as the 6·1 millibar at atmospheric pressure level) around the planet, it seems that it describes a great circle inclined to the equator; north of this line the surface is low and lightly cratered, while south of this line it is higher and more cratered. The datum line is inclined to the equator by 30 degrees. The heavily-cratered area is more extensive than the smoother regions; the area ratio is about 2 to 1. The Tharsis bulge,

with a mean elevation of 8 km, straddles the boundary between the lower, smoother northern hemisphere and the higher, more heavily cratered south. North of Tharsis is the curious Alba Patera, a different kind of volcanic structure; it is only a few kilometres high, but over 1500 kilometres across.

The Elysium bulge is much smaller than Tharsis: it is around 2000 km across, with an average altitude of 4 km. It is centred at latitude 25°N, longitude 210°. The associ-

View of Utopia, from the Viking 2 lander in August 1976. The view extends for over 3 km to the horizon. (NASA).

ated volcanoes are decidedly smaller than those of Tharsis.

There is no definite evidence of Earth-type plate tectonics on Mars, so that the volcanoes may have remained over their 'hot spots' for very long periods and grown to great size. It may be that some of the volcanoes remained active for up to 2500 million years. Lava-flows are extensive, and the equatorial plains have been described as enormous flows of lava and ash.

✴ *The highest volcano on Mars* is Olympus Mons, formerly known as Nix Olympica (it is visible from Earth as a bright point). It is 25 km high, and has a base measuring 600 km, so that it is the largest known volcano in the Solar System; by contrast, the Hawaiian volcanoes on Earth, which also are shield volcanoes, have a height of only 9 km at most above their bases on the ocean floor. (Heights on Mars are bound to be somewhat arbitrary, since there is no sea-level to act as a standard.)

✴ *The largest caldera on a Martian volcano* is that of Arsia Mons; the caldera is 110 km in diameter. Other large calderas crown the volcanoes Olympus Mons (85 km), Ascræus Mons (65 km) and Pavonis Mons (45 km); all these volcanoes lie in the Tharsis area. Arsia Mons has a height of 14 km.

✴ *The youngest of the major volcanoes* is Olympus Mons. The estimated ages of some of the volcanoes, in thousands of millions of years, are:

Tempe Patera	3·4
Ceraunius Tholus	2·4
Uranius Tholus	2·3
Elysium Mons	2·2
Alba Patera	1·7
Hecates Tholus	1·7
Tharsis Tholus	1·4
Arsia Mons	0·7
Pavonis Mons	0·3
Ascræus Mons	0·1
Olympus Mons	0·03

✴ *The deepest basin on Mars* is Hellas, which measures 2200 km by 1800 km and is

about 3 km deep (reckoned from the level of Mars where the atmospheric pressure is 6·1 millibars).

✴ *The longest valley system on Mars* is that of the Valles Marineris, in the Coprates-Tithonius region and not very far from the giant volcanoes of the Tharsis ridge. The system can be traced for a length of 4500 km, with a maximum width of 600 km, and a greatest depth of about 7 km below the rim. There are various other sinuous channels over 1000 km long, giving every impression of having been cut by running water.

✴ *The greatest canyon system on Mars* is Noctis Labyrinthus (formerly known as

Chryse area of Mars; Viking 1, 1976. The area is centred on lat. 17N, long 55W, near Chryse Planitia. Just to the west of the area are the plains of Lunæ Planum. The channels must certainly indicate past floods. (NASA)

Noctis Lacus). The canyons are from 10 to 20 km wide, with slopes of from 10 to 15 degrees. The whole system covers 120 000 km².

✴ *The most conspicuous dark area on Mars* is the Syrtis Major, formerly known as the Hourglass Sea or Kaiser Sea (or even the Atlantic Canal, though there is no connection with Lowell's canal network). The Syrtis Major is V-shaped, and was drawn by Huygens more than 300 years ago. It was formerly thought to be a vegetation-filled sea-bed, but is now known to be a plateau. It lies in the equatorial region of Mars; the most prominent dark marking further north is the Acidalia Planitia, formerly known as Mare Acidalium.

✴ *The brightest area on Mars*, excluding the polar caps, may often be Hellas, once thought to be a snow-covered plateau and now known to be a basin. However, Hellas is very variable in brightness.

✴ *The most widespread dust-storms* occur when Mars is near perihelion, and the surface winds are strongest. What seems to happen is that if the wind-speed exceeds a certain critical value (50 to 100 metres per second), grains of surface material, about 100 microns across, are whipped up and given a 'skipping' motion, known technically as saltation. On striking the surface they propel smaller grains, a few microns across, into the atmosphere, where they remain suspended for weeks. Over 100 localized storms occur every Martian year, and at times a storm may become global, so that for a while all surface details are hidden; this happened, for instance, in late 1971, when Mariner 9 reached the neighbourhood of Mars. Other major storms occurred in 1909, 1911, 1924 and 1972.

✴ *The maximum windspeeds* may, it is estimated, reach 400 km/h, but winds in that thin atmosphere will have little force. Windspeeds measured from the Vikings were light, little exceeding 20 km/h.

✴ *The highest atmospheric pressure so far measured on Mars* is 8·9 millibars, on the floor of Hellas. The pressure at the top of Olympus Mons is below 3 millibars. When Viking 1 landed in Chryse, the pressure was approximately 7 millibars. A decrease of 0·012 millibar per sol was subsequently measured, due to carbon dioxide condensing out of the atmosphere to be deposited on the south polar cap, but this is a seasonal phenomenon, and ceased while Vikings 1 and 2 were still operating.

✴ *The maximum temperature of the aeroshell of Viking 1 during descent to Mars* was 1500°C; that of Viking 2 was similar.

✴ *The first analysis of the surface material* was made from the lander of Viking 1: 44 per cent silica (SiO_2), 5·5 per cent alumina (Al_2O_3), 18·0 per cent iron, 0·9 per cent titanium, 0·3 per cent potassium. Results from Viking 2 were similar. The general composition seems to be not very different from that of the lunar maria.

COMPOSITION OF THE MARTIAN ATMOSPHERE AT THE SURFACE
ppm = parts per million

Gas	Proportion
Carbon dioxide, CO_2	95·32%
Nitrogen, N_2	2·7%
Argon, Ar	1·6%
Oxygen, O_2	0·03%
Carbon monoxide, CO	0·07%
Water vapour, H_2O	0·03%
Neon, Ne	2·5 ppm
Krypton, Kr	0·3 ppm
Xenon, Xe	0·08 ppm
Ozone, O_3	0·03 ppm

✴ *The first temperature measurements* from the lander of Viking 1 ranged between −86°C after dawn to a maximum of −31°C near noon. The results from Viking 2 were similar.

✴ *The minimum summer temperature of the north polar cap* was found to be −73°C; at this temperature carbon dioxide would evaporate. At this time the temperature of the southern cap was found to be −137°C.

✴ *The first attempts to detect Martian life from the surface* were made by Viking 1. There were three main experiments (pyrolitic release, labelled release, and gas exchange) but all were inconclusive; no certain evidence of Martian life has been found. The experiments were repeated from Viking 2, with equally inconclusive results.

All our ideas about Mars have been revolutionized by the Mariner and Viking results. The dark areas are no more than albedo features, and are not depressed seabeds; neither are they vegetation-covered. The crust of Mars may be about 200 km deep; below this comes the mantle, and then the core. No magnetic field has been definitely detected. Much of the water on Mars is locked up in the permanent polar caps. If all the water in the atmosphere were condensed, it would cover the Martian surface with a layer only 1/100 mm deep, but the release of all the water in the ice-caps would produce a layer 10 m deep.

The atmospheric pressure is at present too low for liquid water to exist on the surface, though the presence of apparently water-cut channels indicates that the pressure was much greater some tens of thousands of years ago; evidently there are marked climatic changes, due possibly to the fact that when the axial inclination is at a suitable value the volatiles in the caps may be evaporated, leaving no permanent residual ice-caps such as those which exist today. Traces of carbon monoxide, ozone and hydrogen have been detected in the atmosphere; the height of the Martian tropopause is perhaps 120 km, with carbon dioxide or H_2O wave clouds rising to 24 km, water ice clouds to a maximum of 8 km, and dust-storms up to over 40 km (though even during major storms, the tops of the Tharsis volcanoes often protrude above the dusty level). No doubt future probes to Mars will produce further unexpected results!

FEATURES ON MARS

Craters	Lat	Long	Diam km
Achar	+45·7	236·8	6
Adams	+31	197	100
Agassiz	−70	089	110
Airy	+00·5	000	57
Albany	+23·3	049	2
Albi	−41·9	034·7	8
Alexei Tolstoy	−48	234	–
Alga	−24·6	026·5	19
Amsterdam	+23·3	047	2
Annapolis	+23·4	047·8	1
Antoniadi	+22	299	380
Apt	+40·1	009·5	10
Arago	+10	330	150
Arandas	+42·6	015·1	22
Arrhenius	−40	237	120
Aspen	−21·6	022·9	16
Azul	−42·5	042·3	20
Azusa	−05·5	040·5	41
Bacht	+18·9	257·5	7
Bahn	−03·5	043·5	10
Bakhuysen	−23	344	140
Balboa	−03·5	034	20
Baldet	+23	295	190
Balta	−24·1	026·4	15
Baltisk	−42·6	054·5	48
Bamba	−03·5	041·7	21
Bamberg	+39·9	033·1	55
Banff	+17·4	030·5	3
Banh	+19·7	055·5	13
Bar	−25·6	019·3	2
Barabashov	+47	069	130
Barnard	−61	298	122
Basin	+18·1	253·2	16
Batoka	−07·6	036·8	14
Batos	+21·7	029·5	16
Baykonur	+46·7	227·2	4
Becquerel	+22	008	175
Beer	−15	008	80
Belz	+21·8	043·2	10
Bend	−22·6	027·5	2
Bentong	−22·6	018·9	10
Berseba	−04·4	037·7	36
Bhor	+42	225·5	6
Bianchini	−64	096	50
Bigbee	−25·0	034·6	18
Bise	+20·5	056·5	9
Bison	−26·6	029·0	15
Bjerknes	−43	189	85
Bled	+21·8	031·4	7
Blitta	−26·3	020·8	11
Blois	+23·9	055·8	11
Bluff	+23·7	250·1	6
Boeddicker	−15	197	110
Bogra	−24·4	028·6	18
Bok	+20·8	031·6	7
Bole	+25·6	054·9	8
Bole	+25·4	053·7	8
Bond	−33	036	120
Bor	+18·1	033·8	4

Mars, a photomosaic based on Viking data. The immense canyon of Valles Marineris is shown. (NASA)

Craters	Lat	Long	Diam km
Bordeaux	+23·4	048·9	2
Boru	−24·6	027·7	11
Bouguer	−19	333	100
Bozkir	−44·5	031·9	77
Brashear	−54	120	115
Bremerhaven	+23·5	048·6	2
Briault	−10	270	90
Bridgetown	+22·2	047·1	2
Bristol	+22·4	047·0	3
Broach	+23·8	056·9	11
Brush	+21·9	248·8	5
Bulhar	+50·6	225·6	19
Burroughs	−72	243	110
Burton	−14	156	120
Buta	−23·5	032·2	12
Butte	−05·1	39·1	12
Byrd	−64	232	120
Byske	−05	031·1	5
Cadiz	+23·4	049·1	1
Cairns	+23·8	047·5	10
Calbe	−25·5	028·7	14
Camiling	−00·8	038·1	21
Camiri	−45·1	041·9	20
Campbell	−54	195	133
Campos	−22·0	027·6	7
Can	+48·4	014·6	6
Canaveral	+47·1	224·0	3
Canberra	+47·5	227·2	3
Cartago	−23·6	017·8	33
Cassini	+24	328	440
Cave	+21·9	035·6	8
Cerulli	+32	338	125
Chamberlin	−66	124	120
Changsong	+23·8	057·3	32
Chapais	−22·6	020·4	33
Charleston	+22·9	047·9	2
Charlier	−69	169	110
Chauk	+23·6	055·9	9
Cheb	−24·5	019·3	8
Chekalin	−24·5	026·6	80
Chimbote	−01·5	039·8	65
Chincoteague	+41·5	236	35
Chinju	−04·6	042·3	65
Chinook	+22·8	055·4	17
Chive	+21·9	056·0	8
Choctaw	−41·5	037·0	20
Chom	+38·7	002·2	6
Chur	+16·7	029·2	4
Circle	−22·4	025·4	11
Clark	−56	134	95
Clogh	+20·8	047·7	10
Cluny	−24·1	027·1	9
Cobalt	−26·1	026·8	10
Coblentz	−55	091	95
Colon	+23·0	047·1	2
Columbus	−29	166	110
Comas Solà	−20	158	130
Conches	−04·2	034·3	18
Concord	+16·6	034·1	20
Copernicus	−50	169	280
Corby	+43·1	222·4	7
Cost	+15·3	256·1	10
Cray	+44·4	016·2	5
Creel	−06·1	039·0	8
Crewe	−25·2	019·4	3
Crommelin	+05	101	107
Cruls	−43	197	78
Cruz	+38·6	001·7	6
Curie	+29	005	115
Cypress	−47·6	047·0	11
Daet	−07·4	041·9	12
Daly	−66	022	90
Dana	−73	032	90
Dank	+22·2	253·2	7
Darwin	−57	020	173
da Vinci	+02	039	100
Dawes	−09	322	190
Deba	−24·3	017·1	7

Craters	Lat	Long	Diam km
Dein	+38·5	002·4	24
Delta	−46·3	039·0	5
Denning	−18	326	150
Dese	−45·8	030·3	12
Dessau	−43·1	053·0	9
Dia-Cau	−00·3	042·8	28
Dingo	−24·0	017·3	13
Dison	−25·4	016·3	20
Dokuchaev	−61	127	–
Douglass	−52	070	90
Dromore	+20·1	049·6	14
du Martheray	−06	266	100
du Toit	−72	046	200
Eads	−28·9	029·8	2
Eagle	+44·0	008·2	12
Echt	−22·2	028·0	2
Edam	−26·6	019·9	20
Eddie	+12	218	100
Eger	−48·7	051·8	12
Eil	+42·0	009·7	4
Eiriksson	−19	174	50
Elath	+46·1	013·7	13
Ely	−23·9	027·1	10
Escalante	000	245	80
Esk	+45·4	007·0	4
Eudoxus	−45	147	90
Evpatoriya	+47·3	225·5	6
Faith	+43·2	011·9	5
Falun	−24·2	024·4	10
Fastov	−25·4	020·2	12
Fesenkov	+22	087	90
Flammarion	+26	312	179
Flat	−25·8	019·4	3
Flaugergues	−17	341	230
Flora	−45	051	15
Focas	+34	347	70
Fontana	−63	073	80
Fournier	−04	287	118
Freedom	+43·6	009·0	12
Funchal	+23·2	049·5	2
Gaan	+38·5	003·2	3
Gagra	−20·9	021·9	14
Gah	−45·1	032·3	2
Gale	−06	222	160
Gali	−44·1	036·9	24
Galilaei	+06	027	133
Galle	−51	031	210
Galu	−22·3	021·5	9

Craters	Lat	Long	Diam km
Gandu	−45·8	047·0	5
Gardo	−27·0	024·6	11
Garm	+48·4	009·2	5
Gastre	+24·8	247·6	5
Gatico	−21·2	020·9	18
Gilbert	−68	274	117
Gill	+16	354	85
Glazov	−20·8	026·4	20
Gledhill	−53	274	73
Glide	−08·2	043·3	10
Globe	−24·0	027·1	45
Goba	−23·5	020·9	11
Goff	+23·5	255·2	6
Gol	+47·4	010·7	7
Gold	+20·2	031·3	9
Golden	−22·2	033·3	19
Goldstone	+48·1	225·3	1
Gori	−23·2	028·6	6
Graff	−21	206	160
Green	−52	008	160
Grojec	−21·6	030·6	37
Groves	−04·1	044·6	10
Guaymas	+26·2	045·0	20
Guir	−21·8	020·4	12
Gulch	+16·1	251·2	7
Gusev	−14	184	170
Gwash	+38·9	003·0	5
Hadley	−19	203	110
Haldane	−53	231	80
Hale	−36	036	150
Halley	−49	059	78
Ham	−45·0	032·0	1
Hamaguir	+49	227·3	3
Hamelin	+20·4	032·8	10
Harad	−27·8	027·8	8
Hartwig	−39	016	100
Heaviside	−71	095	88
Heinlein	−65	244	83
Helmholtz	−46	021	110
Henry	+11	336	170
Herschel	−14	230	320
Hilo	−44·8	035·5	20
Hipparchus	−45	151	90
Hit	+47·3	221·5	7
Holden	−26	034	175
Holmes	−75	292	116
Honda	−22·7	016·2	9
Hooke	−45	044	140

Craters	Lat	Long	Diam km	Craters	Lat	Long	Diam km	Craters	Lat	Long	Diam km
Hope	+45·1	010·3	6	Kong	−05·4	038·7	10	Mitchel	−68	284	179
Houston	+48·5	223·9	2	Korolev	+73	196	92	Mohawk	+43·2	005·4	11
Hsuanch'eng	+47·0	227·2	2	Kourou	+47·0	227·1	2	Molesworth	−28	211	175
Huancayo	−03·7	039·9	25	Kristofovich	−48	262	–	Moreux	+42	315	135
Huggins	−49	204	80	Kribi	−43·4	043·4	8	Moss	+19·4	250·7	8
Hussey	−54	127	100	Kuba	−25·6	019·5	25	Muller	−26	232	120
Hutton	−72	255	90	Kufra	+40·6	239·7	37	Murgoo	−24·0	022·3	24
Huxley	−63	259	98	Kumara	+43·3	231·4	12	Mut	+22·7	035·7	7
Huygens	−14	304	495	Kuiper	−57	157	80	Naar	+23·2	042·1	12
Ibragimov	−25·5	059	–	Kunowsky	+57	009	60	Naic	+24·7	252·8	8
Innsbrück	−06·5	030·1	15	Kushva	−44·3	035·2	39	Nain	+41·7	233·2	7
Inta	−24·6	024·9	14	Lachute	−04·3	039·9	13	Naju	+45·3	237·1	8
Irbit	−24·6	024·7	13	Laf	+48·3	005·9	6	Nan	−27·1	019·8	2
Jal	−26·6	028·6	5	Lagarto	+50·0	008·4	19	Nansen	−50	141	90
Janssen	+03	322	160	Lamas	−27·4	020·5	21	Nardo	−27·8	032·7	23
Jarry-Desloges	−09	276	100	Lambert	−20	335	90	Naukan	+21·4	030·6	6
Jeans	−70	206	80	Lamont	−59	114	235	Navan	−26·2	023·2	25
Jen	+40·1	010·6	8	Lampland	−36	079	80	Nazca	−44·0	038·2	13
Jezza	−48·8	037·7	6	Land	+48·4	008·8	5	Nema	+20·9	052·1	14
Jijiga	+25·4	053·9	14	La Paz	+21·2	049·0	1	Nepa	−25·3	019·5	15
Jodrell	+47·8	227·6	3	Lar	−26·1	028·8	6	New Bern	+21·8	049·2	2
Johannesburg	+48·2	226·7	2	Lassell	−21	063	90	Newcomb	−24	358	250
Joly	−75	042	80	Lasswitz	−09	222	110	New Haven	+22·3	049·3	2
Jones	−19	020	90	Lau	−74	107	110	Newport	+22·5	049·0	2
Kagoshima	+47·6	224·1	1	Lebu	−20·6	019·4	20	Newton	−40	158	280
Kagul	−24·0	018·9	8	Lemgo	−42·9	034·5	11	Nicholson	00	165	100
Kaid	−04·6	044·8	7	Le Verrier	−38	343	120	Niesten	−28	302	110
Kaiser	−46	340	205	Lexington	+22·1	048·7	5	Nif	+20·2	056·3	7
Kaj	−27·4	029·2	2	Liais	−75	253	122	Nitro	−21·5	023·8	28
Kakori	−41·9	029·6	25	Libertad	+23·3	029·4	31	Noma	−25·7	024·0	38
Kaliningrad	+48·8	224·9	2	Li Fan	−47	153	100	Nordenskjöld	−53	159	–
Kampot	−42·1	045·4	9	Linpu	+18·3	247·0	17	Nune	+17·7	038·6	9
Kanab	−27·6	018·8	14	Lins	+15·9	029·8	6	Nutak	+17·6	030·3	11
Kansk	−20·8	017·1	34	Lisboa	+21·5	047·6	1	Ochakov	−42·5	031·6	30
Kantang	−24·8	017·3	48	Liu Hsin	−53	172	122	Oglala	−03·2	038·3	15
Karpinsk	−46·0	031·8	28	Livny	−27·5	028·9	10	Okhotsk	+23·2	047·4	2
Karshi	−23·6	019·2	22	Locana	−03·5	038·3	7	Omura	−25·7	025·0	8
Kartabo	−41·2	052·3	17	Lockyer	+28	199	80	Oraibi	+17·4	032·4	31
Kashira	−27·5	018·1	67	Lod	+21·2	031·6	7	Ore	+16·9	033·9	7
Kasmimov	−25·0	022·8	85	Lohse	−43	016	160	Orinda	+45·6	233·0	9
Kaup	+22·5	033·6	3	Loja	+41·4	223·8	10	Ostrov	−26·9	027·5	70
Kaw	+16·7	255·9	10	Lomonosov	+65	008	135	Ottumwa	+24·9	055·7	55
Keeler	−61	152	55	Longa	−20·9	025·7	10	Oudemans	−10	092	120
Kem	−45·3	032·7	2	Lorica	−19·9	028·1	68	Pabo	−27·3	022·9	9
Kepler	−47	219	238	Lota	+46·5	011·9	13	Paks	−07·8	042·2	7
Keul	+46·3	237·7	6	Loto	−22·2	022·3	22	Pasteur	+19	335	125
Kholm	−07·3	042·1	10	Lowell	−52	081	200	Peixe	+20·8	047·6	9
Kifri	−46·0	054·1	12	Luck	+17·4	036·9	7	Perepelkin	+52	065	80
Kimry	−20·5	016·2	17	Luga	−44·6	047·2	42	Peridier	+26	276	100
Kin	+20·4	033·4	7	Lutsk	+38·5	002·9	5	Pettit	+12	174	90
Kingston	+22·4	047·1	2	Luzin	+26	328·5	115	Philadelphia	+22	048	2
Kipini	+24·8	032·0	35	Lyell	−70	015	120	Phillips	−67	045	175
Kirs	−26·7	019·3	3	Lyot	+50	331	220	Phon	+15·8	257·3	8
Kirsanov	+22·4	025·0	15	Mädler	−11	357	100	Pickering	−34	133	110
Kita	−23·1	017·0	11	Madrid	+48·7	224·4	3	Pinglo	−03	037	18
Knobel	−06	226	125	Mafra	−44·4	053·0	12	Playfair	−78	125	40
Kok	+15·7	028·1	6	Magelhæns	−32	174	110	Plum	−26·4	018·9	3
				Maggini	+28	350	140	Podor	−44·6	043·0	25
				Maidstone	−41·9	054·1	9	Poona	+24·0	052·3	20
				Main	−77	310	115	Polotsk	−20·1	026·1	23
				Manah	−04·7	033·8	9	Port-au-Prince	+21·3	048·2	1
				Manti	−03·7	037·7	15	Porter	−50	114	105
				Manzi	−22·3	027·3	7	Porth	+21·5	255·9	10
				Maraldi	−62	032	115	Portsmouth	+22·8	049·1	2
				Marbach	+17·9	249·2	20	Porvoo	−43·7	040·6	9
				Mariner	−35	164	160	Priestley	−54	228	120
				Marth	+13	003	100	Princeton	+21·9	049·1	2
				Martz	−34	217	88	Proctor	−48	330	160
				Maunder	−50	358	84	Ptolemæus	−46	158	160
				McLaughlin	+22	022	90	Punsk	+20·7	041·2	10
				Mega	−01·5	037·1	13	Pylos	+16·9	030·1	29
				Mellish	−73	024	99	Quénisset	+34	319	130
				Mendel	−59	199	73	Quick	+18·5	048·8	8
				Mie	+48	220	100	Quorn	−05·6	33·8	5
				Mila	−27·5	020·6	10	Rabe	−44	325	108
				Milankovič	+55	147	120	Radau	+17	005	120
				Millman	−54·4	150	82	Rakke	−04·7	043·5	18
				Millochau	−21	275	118	Rana	−26·0	021·6	12
				Misk	−01	035·5	10	Raub	+42·6	224·9	7

Surface of Mars; Viking 1 lander (Chryse) 1976. (NASA)

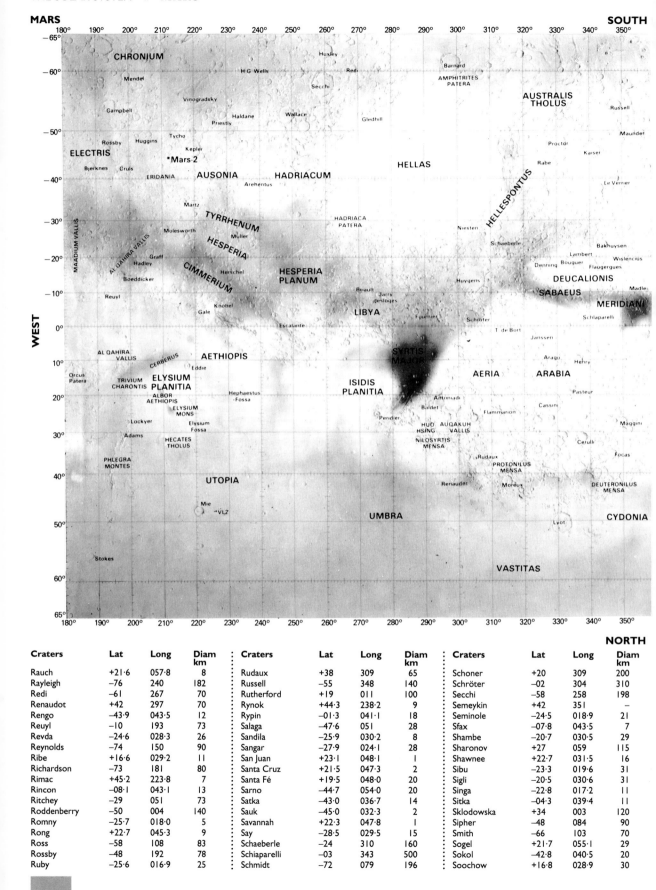

MARS

SOUTH

NORTH

WEST

Craters	Lat	Long	Diam km	:	Craters	Lat	Long	Diam km	:	Craters	Lat	Long	Diam km
Rauch	+21·6	057·8	8	:	Rudaux	+38	309	65	:	Schoner	+20	309	200
Rayleigh	−76	240	182	:	Russell	−55	348	140	:	Schröter	−02	304	310
Redi	−61	267	70	:	Rutherford	+19	011	100	:	Secchi	−58	258	198
Renaudot	+42	297	70	:	Rynok	+44·3	238·2	9	:	Semeykin	+42	351	−
Rengo	−43·9	043·5	12	:	Rypin	−01·3	041·1	18	:	Seminole	−24·5	018·9	21
Reuyl	−10	193	73	:	Salaga	−47·6	051	28	:	Sfax	−07·8	043·5	7
Revda	−24·6	028·3	26	:	Sandila	−25·9	030·2	8	:	Shambe	−20·7	030·5	29
Reynolds	−74	150	90	:	Sangar	−27·9	024·1	28	:	Sharonov	+27	059	115
Ribe	+16·6	029·2	11	:	San Juan	+23·1	048·1	1	:	Shawnee	+22·7	031·5	16
Richardson	−73	181	80	:	Santa Cruz	+21·5	047·3	2	:	Sibu	−23·3	019·6	31
Rimac	+45·2	223·8	7	:	Santa Fé	+19·5	048·0	20	:	Sigli	−20·5	030·6	31
Rincon	−08·1	043·1	13	:	Sarno	−44·7	054·0	20	:	Singa	−22·8	017·2	11
Ritchey	−29	051	73	:	Satka	−43·0	036·7	14	:	Sitka	−04·3	039·4	11
Roddenberry	−50	004	140	:	Sauk	−45·0	032·3	2	:	Sklodowska	+34	003	120
Romny	−25·7	018·0	5	:	Savannah	+22·3	047·8	1	:	Sipher	−48	084	90
Rong	+22·7	045·3	9	:	Say	−28·5	029·5	15	:	Smith	−66	103	70
Ross	−58	108	83	:	Schaeberle	−24	310	160	:	Sogel	+21·7	055·1	29
Rossby	−48	192	78	:	Schiaparelli	−03	343	500	:	Sokol	−42·8	040·5	20
Ruby	−25·6	016·9	25	:	Schmidt	−72	079	196	:	Soochow	+16·8	028·9	30

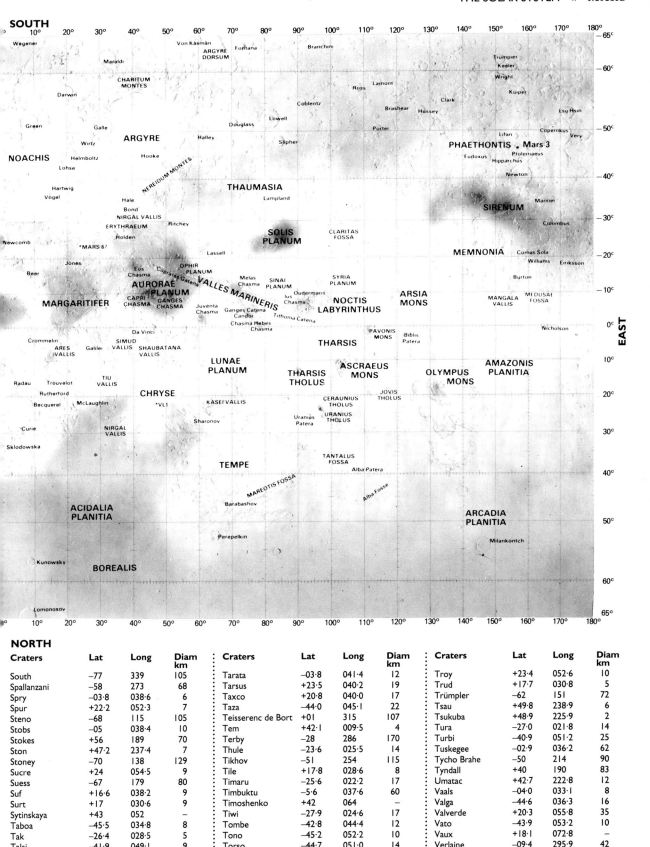

SOUTH

ARGYRE

NOACHIS

NEREIDUM MONTES

THAUMASIA

SIRENUM

PHAETHONTIS · Mars 3

MEMNONIA

SOLIS PLANUM

CLARITAS FOSSA

OPHIR PLANUM

AURORAE PLANUM

VALLES MARINERIS

Eos Chasma

Coprates Catena

Melas Chasma

SINAI PLANUM

SYRIA PLANUM

ARSIA MONS

MANGALA VALLIS

MEDUSAE FOSSA

MARGARITIFER

CAPRI CHASMA

GANGES CHASMA

Juventa Chasma

Ius Chasma

Oudemans

NOCTIS LABYRINTHUS

Candor

Ganges Catena

Tithonia Catena

Chasma Hebes Chasma

PAVONIS MONS

Biblis Patera

SIMUD VALLIS

SHAUBATANA VALLIS

ARES VALLIS

THARSIS

Nicholson

LUNAE PLANUM

THARSIS THOLUS

ASCRAEUS MONS

OLYMPUS MONS

AMAZONIS PLANITIA

TIU VALLIS

CHRYSE

KASEI VALLIS

CERAUNIUS THOLUS

JOVIS THOLUS

Sharonov

Uranius Patera

URANIUS THOLUS

NIRGAL VALLIS

TANTALUS FOSSA

Alba Patera

TEMPE

MAREOTIS FOSSA

Alba Fosse

Barabashov

ACIDALIA PLANITIA

ARCADIA PLANITIA

Perepelkin

Milankontch

BOREALIS

EAST

NORTH

Craters	Lat	Long	Diam km	Craters	Lat	Long	Diam km	Craters	Lat	Long	Diam km
South	−77	339	105	Tarata	−03·8	041·4	12	Troy	+23·4	052·6	10
Spallanzani	−58	273	68	Tarsus	+23·5	040·2	19	Trud	+17·7	030·8	5
Spry	−03·8	038·6	6	Taxco	+20·8	040·0	17	Trümpler	−62	151	72
Spur	+22·2	052·3	7	Taza	−44·0	045·1	22	Tsau	+49·8	238·9	6
Steno	−68	115	105	Teisserenc de Bort	+01	315	107	Tsukuba	+48·9	225·9	2
Stobs	−05	038·4	10	Tem	+42·1	009·5	4	Tura	−27·0	021·8	14
Stokes	+56	189	70	Terby	−28	286	170	Turbi	−40·9	051·2	25
Ston	+47·2	237·4	7	Thule	−23·6	025·5	14	Tuskegee	−02·9	036·2	62
Stoney	−70	138	129	Tikhov	−51	254	115	Tycho Brahe	−50	214	90
Sucre	+24	054·5	9	Tile	+17·8	028·6	8	Tyndall	+40	190	83
Suess	−67	179	80	Timaru	−25·6	022·2	17	Umatac	+42·7	222·8	12
Suf	+16·6	038·2	9	Timbuktu	−5·6	037·6	60	Vaals	−04·0	033·1	8
Surt	+17	030·6	9	Timoshenko	+42	064	−	Valga	−44·6	036·3	16
Sytinskaya	+43	052	−	Tiwi	−27·9	024·6	17	Valverde	+20·3	055·8	35
Taboa	−45·5	034·8	8	Tombe	−42·8	044·4	12	Vato	−43·9	053·2	10
Tak	−26·4	028·5	5	Tono	−45·2	052·2	10	Vaux	+18·1	072·8	−
Talsi	−41·9	049·1	9	Torso	−44·7	051·0	14	Verlaine	−09·4	295·9	42
Tara	−44·4	052·7	27	Troika	+17·1	255·0	13	Very	−50	177	140
Tarakan	−41·6	030·1	37	Trouvelot	+16	013	150	Viana	+19·5	255·3	29

71

Craters	Lat	Long	Diam km
Vils	+39·2	011·7	6
Vinogradsky	−56	217	150
Vishniac	−77	276	78
Vivero	+49·2	241·1	27
Vögel	−37	013	115
Volgorad	+48·4	224·8	2
Volsk	+23·2	051·3	8
von Kármán	−64	059	80
Voo	−27·3	019·8	2
Wabash	+21·5	033·7	42
Wahoo	+23·5	033·7	63
Wallace	−53	249	150
Wallops	+46·9	227·2	2
Warra	+20·8	037·5	11
Waspam	+20·7	056·6	40
Wassamu	+25·5	053·0	17
Wau	−45·2	042·4	3
Weert	+19·8	051·4	9
Wegener	−65	004	70
Weinbaum	−66	245	82
Wells	−60	238	80
Wer	+45·6	036·2	3
Wicklow	−02·0	040·7	21
Wien	−11	220	120
Williams	−18	164	120
Wilmington	+21·9	047·5	1
Windfall	−02·1	043·5	20
Wink	−06·6	041·5	8
Wirtz	−49	026	70
Wislencius	−18	349	150
Woomera	+48·4	227·3	3
Wright	−59	151	115
Yakima	+43·3	003·2	12
Yala	+17·5	038·5	19
Yar	+22·5	039·1	5
Yat	+18·3	029·1	8
Yegros	−22·5	023·5	14
Yorktown	+23·1	048·7	8
Yoro	+23·0	028·0	8
Yungay	−44·3	044·6	17
Yuty	+22·4	034·1	18
Zir	+18·7	036·6	6
Zulanka	−02·3	042·3	47
Zuni	+19·3	029·6	25

Catena	Lat	Long	Length km
Acheron	+35 to 42	097–103	558
Coprates	−14 to 16	067–058	505
Ganges	−02 to 03	071–067	233
Phlegethon	+35 to 41	095–105	875
Tithonia	−06 to 05	087–080	400
Tractus	+20 to 35	100–104	935

Cavi	Lat	Long	Diam km
Angusti	−73 to 84	055–100	730
Anseris	−30	264·5	35
Novi	−65 to 70	320–340	535
Sisyphi	−76 to 82	340–000	405
Ultimi	−73 to 76	195–214	320

Chaos	Lat	Long	Diam km
Aram	0 to +05	019–024	275
Aromatum	−01	043·5	–
Arsinoes	−08	021	–
Aureum	−02 to 07	030–024	365
Eos	−16 to 19	043–050	–
Hellas	−46	293	900
Hydaspes	+01 to 04	024–030	350
Hydraotes	−01 to 03	032–040	415
Iani	−05 to 02	020–013	405
Margaritifer	−07 to 13	017–025	430
Pyrrhæ	−11	029	–

Chasma	Lat	Long	Length km
Australe	−80 to 89	284–257	501
Borealis	+85 to 81	050–030	318

Chasma	Lat	Long	Length km
Candor	−04 to 08	078–070	400
Capri	−15 to 03	053–031	1275
Coprates	−10 to 16	069–053	975
Echus	−01 to 05	078–032	310
Eos	−16 to 19	043–050	–
Gangis	−06 to 09	055–043	575
Hebes	−02 to 00	079–073	320
Ius	−10 to 14	092–077	890
Juventai	−02 to 06	054–060	250
Melas	−08 to 14	077–069	505
Ophir	−03 to 09	075–064	660
Thyles	−69 to 73	230–235	235
Tithonium	−03 to 07	092–077	880

Colles (hills)	Lat	Long
Ariadnes	−35	187
Deuteronilus	+41	338

Dorsum (Dorsa)	Lat	Long	Length km
Arena	−11 to 15	291–292	205
Argentea	−76 to 80	015–040	385
Brevia			
Cerberus	−09 to 18	252–258	–
Eumenides	0 to −10	155–158	625
Felis	−20 to 27	063–070	575
Gordii	−01 to 07	142–146	400
Melas	−15 to 25	070–075	560
Solis	−17 to 30	078–085	825
Xanthe	+20 to 25	047–051	280
Zea	−47 to 53	277–283	300

Fossa (Fossæ)	Lat	Long	Length km
Aganippe	−05 to 11	125–127	325
Alba	+38 to 49	117–109	1300
Amenthes	+03 to 14	252–263	655
Calydon	−7·5	087–093	–
Ceraunius	+38 to 20	112–105	1110
Claritas	−12 to 40	108–100	1794
Coloe	−32 to 41	294–307	–
Elysium	−30 to 21	224–217	550
Erythræa	−27 to 28	029–033	200
Hephæstus	+25 to 18	242–233	725
Ismeniæ	−37 to 42	315–337	–
Mareotis	−35 to 53	090–062	1487
Medusæ	0 to −06	160–165	350
Memnonia	−23 to 18	160–147	750
Nia	−13 to 15	070–074	–
Nili	−17 to 26	287–279	625
Oceanidum	−60 to 63	027–031	255
Oti	−07 to 12	115–118	–
Sacra	−12 to 21	070–073	545
Sirenum	−40 to 25	177–139	2165
Tantalus	−33 to 47	108–090	1582
Tempe	−35 to 50	082–062	1179
Thaumasia	−33 to 45	100–085	802
Tithoniæ	−01 to 03	034–047	–

Labyrinthus	Lat	Long	
Adamas	+30 to 42	247–257	–
Noctis	−04 to 14	110–095	1025

Mensæ	Lat	Long	Diam km
Æolis	−06 to 02	213–224	700
Candor	−06	028	–
Ceti	−06	76·3	55
Cydonia	−30 to 43	008–020	890
Deuteronilus	−40 to 48	344–325	754
Hebes	−01	032	–
Nepenthes	−08 to 13	239–251	775
Nilosyrtis	−30 to 45	304–280	1225
Protonilus	−49 to 38	325–303	1032

Mons	Lat	Long	Diam km
Arsia	−09	121	500
Ascræus	−12	104	370

Mons	Lat	Long	Diam km
Elysium	+25	213	180
Labeatis	+38	076	28
Olympus	+18	133	540
Pavonis	+01	113	340

Montes	Lat	Long	Length km
Charitum	−50 to 59	060–027	1279
Erebus	+30 to 42	170–180	–
Geryon	−07 to 08	074–084	–
Hellespontus	−35 to 50	319–310	854
Libya	+15 to −02	253–282	2015
Nereidum	−50 to 38	060–030	1626
Phlegra	+30 to 46	195	919
Tartarus	+03 to 28	181–193	893
Tharsis	−12 to +16	125–101	2175

Patera	Lat	Long	Diam km
Alba	+40	110	130
Amphitrites	−61	299	113
Apollinaris	−09	186	200
Biblis	+02	124	55
Hadriaca	−31	267	70
Meroe	+07	291	–
Nepenthes	+12	271	70
Nili	+09	293	–
Orcus	+14	181	370
Tyrrhena	−22	253	55
Ulysses	+03	121	60
Uranius	+26	093	125

Planitia	Lat	Long	Diam km
Acidalia	+55 to 14	060–000	2615
Amazonis	00 to +40	168–140	2416
Arcadia	+55 to 40	195–110	3052
Argyre	−45 to 36	043–051	741
Chryse	+19 to 30	051–037	840
Elysium	−10 to 30	260–205	5312
Hellas	−60 to 30	313–272	1955
Isidis	−04 to 20	279–255	800
Syrtis Major	+20 to −01	298–283	1262
Utopia	+35 to 50	310–195	3276

Planum	Lat	Long	Diam km
Auroræ	−09 to 15	053–043	565
Bathys	−38 to 48	100–112	626
Dædalia	−17 to 23	115–125	–
Hesperia	−10 to 35	258–242	2125
Lunæ	+05 to 23	075–060	1050
Ophir	−06 to 12	063–054	550
Planum Australe	−80 to 90	Circumpolar	1130
Planum Boreum	+80 to 90	Circumpolar	1130
Planum Erythræum	−18 to 37	356–065	1750
Sinai	−09 to 20	097–070	1495
Solis	−20 to 30	098–088	1000
Syria	−10 to 20	112–097	900
Thaumasia	−22	065	930

Rupes	Lat	Long	Length km
Amenthes	00	248–252	–
Argyre	−61 to 65	065–073	347
Arena	+11 to 16	288–291	375
Bosporus	−39 to 47	053–060	525
Morpheos	−36 to 39	231–240	–
Ogygis	−32 to 35	053–055	185
Olympus	−15 to 24	129–013	2010
Phison	+25 to 28	308–310	160
Ultima	−65 to 73	198–204	495

Scopulus	Lat	Long	Length km
Eridana	−47 to 60	208–235	1344
Frigoris	+78 to 81	045–323	948
Hyperboreus	+80 to 85	060–160	1059
Hypernotius	−75 to 81	244–320	1918

Scopulus	Lat	Long	Length km
Nilokeras	+30 to 42	051–063	1231
Œnotria	–15 to 10	270–292	–
Promethei	–75 to 83	190–320	2508

Sulci			
Cyane	+23 to 28	126–130	300
Gigas	+06 to 14	125–136	750
Lycus	+18 to 34	130–148	1004
Sulci Gordii	+16 to 20	125–127	250

Tholus	Lat	Long	Diam km
Albor	+19	210	115
Australis	–59	323	40
Ceraunius	+24	097	135
Charitum	–55	041	30
Hecates	+32	210	152
Iaxartes	+72	015	53
Jovis	+18	117	50
Kison	+73	358	50
Ortygia	+70	008	100
Tharsis	+14	091	150
Uranius	+26	098	65

Valllis (Valles)	Lat	Long	Length km
Al-Qahira	–19 to 14	200–194	420
Arda	–20 to 22	030–034	156
Ares	+12 to 02	023–017	915
Arnus	–8 to +8	290	–
Auqakuh	+30 to 27	300–297	195
Axius	–53 to 58	285–295	435

Valllis (Valles)	Lat	Long	Length km
Bahram	+20 to 22	055–057	135
Brazes	–2·5 to 6	192–197	–
Clota	–25 to 27	020–022	115
Dao	–23 to 40	265–275	430
Drava	–17 to 23	115–125	–
Drinus	–22 to 24	165–195	–
Evros	–12 to 13	343–349	–
Granicus	–27 to 30	225–231	–
Harmakhis	–33 to 40	265–275	698
Himera	–21 to 23	022–023	–
Hrad	+38 to 45	225–240	825
Huo Hsing	+34 to 28	299–292	662
Indus	–17 to 21	317–322	–
Kasei	+27 to 18	075–056	1090
Ladon	–02 to 25	027–030	280
Licus	–01 to 05	233–234	–
Locras	+07 to 11	311–313	265
Loire	–17·5 to 22	014–017	–
Louros	+08 to 09	033–041	–
Ma'adim	–28 to 16	184–181	955
Mad	–58	183	460
Maja	+16 to 21	049–057	398
Mamers	+32 to 39	338–345	628
Mangala	–04 to 09	150–152	272
Marineris	+01 to –18	024–113	5272
Maumee	+18 to 20	054–056	130
Mawrth	+14 to 20	019–027	500
Naktong	000	330–321	–
Nanedi	00 to +09	047–051	570
Nirgal	–27 to 30	037–047	665
Oltis	–24 to 26	021–022	105

Valllis (Valles)	Lat	Long	Length km
Osuga	–15 to 17	038–040	–
Parana	–23 to 26	009–012	240
Ravi	–02 to +02	039–043	365
Reull	+37 to 43	248–265	1065
Rubicon	+43 to 37	115–117	230
Scamander	–14 to 18	330–331	–
Shalbatana	00 to +15	040–045	995
Simud	–03 to 14	035–042	690
Tagus	–07	245	35
Tiu	+03 to 14	030–035	680
Uzboi	–27 to 37	035–037	290
Vedra	+19 to 20	054–056	122
Warrego	–42 to 44	091–095	160

Vastitas	Lat	Long	Diam km
Borealis	+55 to 70	Circumpolar	9999

Terræ	Lat	Long	Diam km
Aonia	–25 to 65	060–120	2917
Arabia	00 to +43	024–280	5625
Cimmeria	–45	210	–
Margaritifer	+02 to –27	012–045	1924
Meridiani	–05	000	–
Noachis	–15 to 83	040–300	1025
Promethei	–30 to 65	240–300	2967
Sabæa	–01	325	–
Sirenum	–50	150	–
Tempe	+24 to 54	050–093	1628
Tyrrhena	–10	280	–
Xanthe	+19 to –13	015–065	2797

THE SATELLITES OF MARS

DATA

	Phobos	Deimos
Mean distance from centre of Mars, km	9270	23 400
Mean angular distance from Mars, mean opposition:	24 sec ·6	1 min 01 sec ·8
Mean sidereal period, days:	0·3189	1·2624
Mean synodic period:	7 h 39 m 26 s·6	1 d 6 h 21 m 15 s·7
Orbital inclination to Martian equator:	1 deg ·1	1 deg ·8
Orbital eccentricity:	0·0210	0·0028
Diameter, km:	27 × 22 × 18	15 × 12 × 10
Density, water = 1:	2·0	1·7
Mass, grams:	1·08 × 10^{19}	1·8 × 10^{18}
Escape velocity, km/s:	0·016	0·008
Magnitude, mean opposition:	11·6	12·8
Maximum apparent magnitude, seen from Mars:	–3·9	–0·1
Apparent diameter, seen from Mars:		
max:	12 min ·3	2 min
min:	8 min	1 min ·7

*T*he first mention of possible Martian satellites was fictional – by Jonathan Swift, in Gulliver's *Voyage to Laputa* (1727). Swift described two satellites, one of which had a revolution period shorter than the rotation period of its primary; but there was no scientific basis at that time for such an object. Two satellites were also described in another novel, Voltaire's *Micromégas* (1750). The reasoning was, apparently, that since the Earth had one satellite and Jupiter was known to have four, Mars could not possibly manage with less than two!

✳ *The first systematic search for Martian satellites* was made by W. Herschel in 1783. The result was negative. H. D'Arrest at Copenhagen, in 1862 and 1864, was similarly unsuccessful.

✳ *The first satellite to be discovered was Deimos*, by Asaph Hall at Washington on 10 August 1877. Hall discovered Phobos on 16 August 1877. The names Phobos and Deimos were suggested to Hall by Mr Madan of Eton.

✳ *The first colour estimates of the satellites* were made by E. M. Antoniadi, who in 1930 reported that Phobos was white and Deimos bluish. These results were certainly spurious, even though Antoniadi was using the 83 cm Meudon refractor. It is now known that the satellites are dust-covered, with very low albedos.

✳ *The first proof that Phobos is irregular in shape* was obtained in 1969, when Mariner 7 photographed the shadow of Phobos cast on to Mars and revealed that it was elliptical. *The first accurate size measurements* were obtained by Mariner 9 in 1971–2. Both satellites are irregular, so that diameter values are bound to be somewhat arbitrary.

✳ *The first detection of craters on the satellites* was also due to Mariner 9, in 1971–2.

✳ *The largest crater on Phobos* is Stickney, 5 km in diameter. (Stickney was the maiden name of Asaph Hall's wife.) The principal craters on Deimos have been named Swift and Voltaire.

The Martian satellites are very dark (albedo c. 5 per cent) and dense (probably about 2 g/cm^2), and their material may be similar to that of carbonaceous chondrites. Each is covered with a regolith a few mm thick; that of Deimos is thicker than Phobos', so that its surface is more subdued. Phobos also shows strange parallel grooves inclined by 30 degrees to the equator, around 100–200 m wide, and 10–20 m deep. It may well be that they are captured asteroids rather than bona-fide satellites, though there are admittedly difficulties in the way of this hypothesis. It has been calculated that Phobos is slowly spiralling downward at the rate of about 18 metres per century – in which case it may impact Mars in some 40 000 000 years hence. Phobos would be invisible to a Martian observer above latitude 69 degrees; the limiting latitude for Deimos would be 82 degrees. To a Martian observer, Phobos would be less than half the apparent diameter of the Moon as seen from Earth, and would give only about as much light as Venus does to us; from Phobos, Mars would subtend an angle of 42 degrees! The apparent diameter of Deimos as seen from Mars would be only about twice the maximum apparent diameter of Venus as seen from Earth, and with the naked eye the phases would be almost imperceptible. Deimos would remain above the Martian horizon for 2½ sols consecutively; Phobos would cross the sky in only 4½ hours, moving from west to east, and the interval between successive risings would be only just over 11 hours. This curious behaviour is due to the fact that Phobos has a revolution period less than the rotation period of its primary. Total solar eclipses

could never occur; Phobos would transit the Sun 1300 times a year, taking 19 seconds to cross the disk, while Deimos would show an average of 130 transits, each taking 1 minute 48 seconds. To a Martian observer, eclipses of the satellites by the shadow of Mars would be very frequent.

The rotation periods of the satellites are synchronous: that is to say they would keep the same faces turned permanently towards the planet, though there would be marked 'libration' effects similar to those of the Moon.

CRATERS ON PHOBOS

	Lat	Long W
D'Arrest	43S	185
Hall	0	50
Roche	56N	186
Sharpless	24S	170
Stickney	0	50
Todd	5S	165
Wendell	0	145

(There is also one named ridge: Kepler Dorsum)

CRATERS ON DEIMOS

	Lat	Long W
Swift	10N	358
Voltaire	23N	2

Deimos; Viking Orbiter 2, showing resolution down to 300 metres. The area covered measures 1·2 × 1·5 km. (NASA)

Phobos, from Viking Orbiter 1 on 18 February 1977, from a range of 480 km; c. 75 per cent illuminated. The south pole of Phobos is inside the crater Hall (5 km in diameter) at the bottom centre, where the pictures overlap. (NASA)

Map of Deimos

Map of Phobos

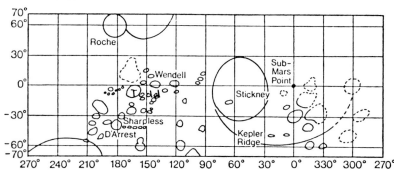

THE MINOR PLANETS

*T*he Minor Planets are also known as asteroids or planetoids. Over 3500 have had their orbits satisfactorily computed. It is estimated that the total membership of the swarm may exceed 40 000.

THE FIRST KNOWN MINOR PLANETS

q = perihelion, in astronomical units.
a = aphelion distance, in astronomical units.
P = period, in years.
e = orbital eccentricity.
i = orbital inclination.

T = type.
A = albedo.
D = diameter, in kilometres.
R = rotation period, in hours.
Mag = apparent magnitude, at mean opposition.

No.	Name	Discoverer	q	a	P	e	i	T	A	D	R	Mag
1	Ceres	Piazzi, 1 Jan. 1801	2·55	2·77	4·60	0·078	10·60	C	0·054	940	9·08	7·4
2	Pallas	Olbers, 28 Mar. 1802	2·12	2·77	4·62	0·234	34·80	CU	0·074	580 × 470	7·81	8·0
3	Juno	Harding, 1 Sept. 1804	1·98	2·67	4·36	0·258	13·00	S	0·151	288 × 230	7·21	8·7
4	Vesta	Olbers, 29 Mar. 1807	2·15	2·37	3·63	0·090	7·14	V	0·229	576	5·34	6·5
5	Astræa	Hencke, 8 Dec. 1845	2·08	2·57	4·13	0·190	5·36	S	0·140	120	16·81	9·8
6	Hebe	Hencke, 1 July 1847	1·94	2·43	3·77	0·202	14·79	S	0·164	204	7·28	8·3
7	Iris	Hind, 13 Aug. 1847	1·84	2·39	5·51	0·229	5·51	S	0·154	208	7·14	7·8
8	Flora	Hind, 18 Oct. 1847	1·86	2·20	3·27	0·156	5·89	S	0·144	162	12·35	8·7
9	Metis	Graham, 26 Apr. 1848	2·10	2·39	3·69	0·121	5·59	S	0·139	158	5·08	9·1
10	Hygeia	De Gasparis, 12 Apr. 1849	2·76	3·13	5·54	0·120	3·84	C	0·041	430	17·50	10·2
11	Parthenope	De Gasparis, 11 May 1850	2·06	2·45	3·84	0·100	4·63	S	0·126	156	7·83	9·9
12	Victoria	Hind, 13 Sept. 1850	1·82	2·33	3·56	0·220	8·38	S	0·114	136	8·65	9·9
13	Egeria	De Gasparis, 2 Nov. 1850	2·36	2·58	4·14	0·087	16·50	C	0·041	244	7·05	10·8
14	Irene	Hind, 19 May 1851	2·16	2·59	4·16	0·016	9·11	S	0·162	150	9·35	9·6
15	Eunomia	De Gasparis, 29 July 1851	2·15	2·64	4·30	0·185	11·76	S	0·155	260	6·08	8·5
16	Psyche	De Gasparis, 17 Mar. 1851	2·53	2·92	5·00	0·133	3·09	M	0·093	248	4·20	9·9
17	Thetis	Luther, 17 Apr. 1852	2·13	2·47	3·88	0·138	5·60	S	0·103	98	12·28	10·7
18	Melpomene	Hind, 14 June 1852	1·80	2·30	3·48	0·217	10·14	S	0·144	162	11·57	8·9
19	Fortuna	Hind, 22 Aug. 1852	2·05	2·44	3·82	0·159	1·57	C	0·032	198	7·45	10·1
20	Massalia	De Gasparis, 19 Sept. 1852	2·06	2·41	3·74	0·145	0·70	S	0·164	134	8·09	9·2
21	Lutetia	Goldschmidt, 15 Nov. 1852	2·04	2·44	3·80	0·161	3·07	M	0·093	108	8·17	10·5
22	Calliope	Hind, 16 Nov. 1852	2·62	2·91	4·96	0·098	13·70	M	0·130	174	4·15	10·6
23	Thalia	Hind, 15 Dec. 1852	2·02	2·63	4·26	0·231	10·15	S	0·164	116	12·31	10·1
24	Themis	De Gasparis, 5 Apr. 1853	2·71	3·13	5·54	0·134	0·76	C	0·030	228	8·38	11·8
25	Phocæa	Chacornac, 6 Apr. 1853	1·79	2·40	3·72	0·254	21·59	S	0·184	72	9·95	10·3
26	Proserpina	Luther, 5 May 1853	2·41	2·67	2·41	0·089	3·56	S	±0·2	88	12·00	11·3
27	Euterpe	Hind, 8 Nov. 1853	1·95	2·35	3·60	0·171	1·59	S	0·147	116	8·50	9·9
28	Bellona	Luther, 1 Mar. 1854	2·37	2·78	4·64	0·148	9·40	S	0·132	124	15·70	10·8
29	Amphitrite	Marth, 1 Mar. 1854	2·37	2·56	4·08	0·073	6·09	S	0·140	200	5·39	10·0
30	Urania	Hind, 22 July 1854	2·07	2·37	3·63	0·126	2·09	S	0·144	94	13·69	10·6
31	Euphrosyne	Ferguson, 1 Sept. 1854	2·43	3·15	5·58	0·228	26·35	C	0·030	270	5·53	11·0
32	Pomona	Goldschmidt, 26 Oct. 1854	2·37	2·59	4·16	0·084	5·52	S	0·140	92	9·44	11·3
33	Polyhymnia	Chacornac, 28 Oct. 1854	1·89	2·86	4·84	0·341	1·89	S	±0·2	62	18·60	11·0
34	Circe	Chacornac, 6 Apr. 1855	2·40	2·69	4·40	0·107	5·49	C	0·039	112	15·00	12·3
35	Leucothea	Luther, 19 Apr. 1855	2·34	3·00	5·20	0·220	8·04	C	?	±67	?	12·4
36	Atalantë	Goldschmidt, 5 Oct. 1855	1·91	2·74	4·55	0·305	18·49	C	0·024	120	9·93	11·6
37	Fides	Luther, 5 Oct. 1855	2·17	2·64	4·29	0·177	3·07	S	0·186	96	7·33	10·7
38	Leda	Chacornac, 12 Jan. 1856	2·33	2·74	4·54	0·151	6·95	CU	±0·2	±118	13·00	12·2
39	Lætitia	Chacornac, 8 Feb. 1856	2·45	2·77	4·60	0·114	10·38	S	0·169	156	5·13	10·3
40	Harmonia	Goldschmidt, 31 Mar. 1856	2·16	2·67	3·41	0·047	4·26	S	0·123	116	9·14	10·5
41	Daphne	Goldschmidt, 22 May 1856	2·03	2·77	4·61	0·269	15·79	C	0·056	204	5·99	10·3
42	Isis	Pogson, 23 May 1856	1·89	2·43	3·81	0·226	8·54	S	0·125	94	13·59	10·2
43	Ariadne	Pogson, 15 Apr. 1857	1·83	2·20	3·27	0·169	3·47	S	0·113	84	5·75	10·2
44	Nysa	Goldschmidt, 27 May 1857	2·06	2·42	3·77	0·151	3·71	E	0·377	68	6·42	9·7
45	Eugenia	Goldschmidt, 27 June 1857	2·49	2·72	4·49	0·083	6·60	C	0·030	244	5·70	11·5
46	Hestia	Pogson, 16 Aug. 1857	2·09	2·52	4·01	0·172	2·33	F	0·028	164	21·04	11·1
47	Aglaia	Luther, 15 Sept. 1857	2·50	2·88	4·89	0·133	4·98	C	0·027	158	13	12·1
48	Doris	Goldschmidt, 19 Sept. 1857	2·90	3·11	5·49	0·068	6·54	C	±0·03	246	11·90	11·9
49	Pales	Goldschmidt, 19 Sept. 1857	2·35	3·08	5·41	0·236	3·18	C	?	176	10·4	11·4
50	Virginia	Ferguson, 4 Oct. 1857	1·88	2·65	4·31	0·289	2·84	C	?	±88	24	11·6
51	Nemausa	Laurent, 22 Jan. 1858	2·21	2·37	3·63	0·066	9·97	CU	0·050	152	7·8	10·8
52	Europa	Goldschmidt, 4 Feb. 1858	2·79	3·11	5·48	0·103	7·46	C	0·035	292	5·6	11·1
53	Calypso	Luther, 4 Apr. 1858	2·09	2·62	4·24	0·203	5·16	C	?	110	26·6	11·7
54	Alexandra	Goldschmidt, 10 Sept. 1858	2·12	2·71	4·47	0·197	11·78	C	0·030	176	7·0	10·9
55	Pandora	Searle, 10 Sept. 1858	2·36	2·76	4·59	0·145	7·20	CMEU	?	112	4·8	11·6
56	Melete	Goldschmidt, 9 Sept. 1859	2·00	2·60	4·19	0·232	8·09	P	0·026	144	16·0	11·5
57	Mnemosyne	Luther, 22 Sept. 1859	2·78	3·15	5·59	0·118	15·12	S	0·140	116	12·3	12·2
58	Concordia	Luther, 24 Mar. 1860	2·59	2·70	4·44	0·043	5·07	C	0·030	104	?	13·1
59	Elpis	Chacornac, 12 Sept. 1860	2·40	2·71	4·47	0·117	8·64	C	0·05	104	13·7	11·4
60	Echo	Ferguson, 14 Sept. 1860	1·96	2·40	3·70	0·183	3·60	S	0·154	52	26·2	11·5
61	Danaë	Goldschmidt, 9 Sept. 1860	2·50	3·00	5·16	0·162	18·21	S	?	88	11·5	11·7
62	Erato	Förster, 14 Sept. 1860	2·53	3·11	5·49	0·186	2·28	C	?	64	?	13·0
63	Ausonia	De Gasparis, 10 Feb. 1861	2·09	2·40	3·71	0·126	5·78	S	0·128	92	8·8	10·2
64	Angelina	Tempel, 4 Mar. 1861	2·35	2·69	4·40	0·125	1·31	E	0·342	60	8·8	11·3
65	Cybele	Tempel, 8 Mar. 1861	3·07	3·43	6·36	0·105	3·55	CPF	0·022	308	6·1	11·8
66	Maia	Tuttle, 9 Apr. 1861	2·19	2·65	4·30	0·174	3·05	C	0·029	92	?	12·5
67	Asia	Pogson, 17 Apr. 1861	1·97	2·42	3·77	0·187	6·01	S	0·157	60	15·89	11·6

No.	Name	Discoverer	q	a	P	e	i	T	A	D	R	Mag
68	Leto	Luther, 29 Apr. 1861	2·27	2·78	4·64	0·185	7·96	S	0·126	128	14·84	10·8
69	Hesperia	Schiaparelli, 29 Apr. 1861	2·47	2·98	5·13	0·170	8·56	M	?	108	5·66	11·3
70	Panopæa	Goldschmidt, 5 May 1861	2·13	2·61	4·23	0·184	11·59	C	0·039	152	14·0	11·5
71	Niobe	Luther, 13 Aug. 1861	2·27	2·75	4·57	0·175	23·29	S	0·140	106	28·8	11·3
72	Feronia	Safford, 29 May 1861	1·99	2·67	3·41	0·120	5·42	U	0·032	96	8·1	12·0
73	Clytia	Tuttle, 7 Apr. 1862	2·55	2·66	4·35	0·042	2·38	?	?	±56	?	13·4
74	Galatea	Temple, 29 Aug. 1862	2·11	2·78	4·63	0·239	4·07	C	?	±108	9	12·0
75	Eurydice	Peters, 22 Sept. 1862	1·86	2·67	4·37	0·304	4·99	M	?	±48	8·9	11·1
76	Freia	D'Arrest, 21 Oct. 1862	2·82	3·40	6·28	0·171	2·11	P	0·03	196	9·8	12·3
77	Frigga	Peters, 12 Nov. 1862	2·31	2·67	4·36	0·133	2·43	M	0·113	66	9·0	12·1
78	Diana	Luther, 15 Mar. 1863	2·09	2·62	4·25	0·205	8·66	C	?	144	7·2	10·9
79	Eurynome	Watson, 14 Sept. 1863	1·97	2·44	3·28	0·193	4·62	S	0·137	80	5·8	10·8
80	Sappho	Pogson, 3 May 1864	1·84	2·30	3·48	0·200	8·67	SU	0·113	84	14·1	12·8
81	Terpsichore	Tempel, 30 Sept. 1864	2·26	2·86	4·82	0·209	7·81	C	?	122	?	12·1
82	Alcmene	Luther, 27 Nov. 1864	2·15	2·76	4·58	0·223	2·84	S	0·138	66	13·0	11·5
83	Beatrix	De Gasparis, 26 Apr. 1865	2·23	2·43	3·79	0·084	4·98	C	0·030	118	10·2	12·0
84	Clio	Luther, 25 Aug. 1865	1·80	2·36	3·63	0·236	9·32	C	0·037	88	?	11·2
85	Io	Peters, 19 Sept. 1865	2·14	2·65	4·32	0·193	11·96	C	0·042	148	6·8	11·2
86	Semele	Tietjen, 4 Jan. 1866	2·45	3·11	5·48	0·213	4·79	C	?	112	16·6	12·6
87	Sylvia	Pogson, 16 May 1866	3·19	3·48	6·50	0·083	10·87	P	?	282	5·2	12·6
88	Thisbe	Peters, 15 June 1866	2·31	2·77	4·61	0·163	5·21	C	0·045	210	6·0	10·7
89	Julia	Stéphan, 6 Aug. 1866	2·09	2·55	4·08	0·180	16·12	S	0·086	168	11·4	10·3
90	Antiope	Luther, 1 Oct, 1866	2·64	3·15	5·59	0·162	2·24	C	?	128	?	12·4
91	Ægina	Stéphan, 4 Nov. 1866	2·32	2·59	4·17	0·106	2·12	C	0·031	104	6·0	12·2
92	Undina	Peters, 7 July 1867	2·91	3·20	5·71	0·087	9·88	U	±0·03	184	15·9	12·0
93	Minerva	Watson, 24 Aug. 1867	2·37	2·76	4·56	0·141	8·56	C	0·039	168	6·0	11·5
94	Aurora	Watson, 6 Sept. 1867	2·90	3·16	5·62	0·081	8·01	C	0·029	190	7·2	12·5
95	Arethusa	Luther, 23 Nov. 1867	2·63	3·07	5·38	0·143	12·93	C	0·019	228	8·7	12·1
96	Ægle	Coggia, 17 Feb. 1868	2·62	3·05	5·32	0·140	16·01	U	0·04	112	?	12·4
97	Clotho	Tempel, 17 Feb. 1868	1·98	2·67	4·36	0·256	11·76	MP	0·121	108	35·0	10·5
98	Ianthe	Peters, 18 Apr. 1868	2·18	2·68	4·40	0·188	15·56	C	?	106	?	12·6
99	Dike	Borrelly, 28 May 1868	2·14	2·66	4·35	0·195	13·88	C	?	80	30·0	12·1
100	Hekate	Watson, 11 July 1868	2·62	3·10	5·46	0·155	6·39	SU	?	84	10·0	12·1

SOME FURTHER MINOR PLANET DATA

The following asteroids are of note in some way or other – either for orbital peculiarities, physical characteristics, or because they are the senior members of 'families' – or even because they have very fast or very slow rotation periods, for example.

No.	Name	Discoverer	q	a	P	e	i	T	D	R	Mag
128	Nemesis	Watson, 25 Nov. 1872	2·40	2·75	4·56	0·126	6·25	CEU	116	39·0	11·6
132	Æthra	Watson, 13 June 1873	1·61	2·61	4·22	0·384	25·07	SU	38	?	11·9
135	Hertha	Peters, 18 Feb. 1874	1·93	2·42	3·78	0·204	2·29	M	80	8·4	10·5
153	Hilda	Palisa, 2 Nov. 1875	3·40	3·97	7·91	0·142	7·83	P	222	?	13·3
158	Koronis	Knorre, 4 Jan. 1876	2·71	2·87	4·86	0·053	1·00	S	36	14·2	14·0
221	Eos	Palisa, 18 June 1882	2·72	3·01	5·23	0·098	10·87	CEU	112	10·4	12·4
243	Ida	Palisa, 29 Sept. 1884	2·74	2·86	4·84	0·042	1·14	S	24 × 21	4·6	14·6
279	Thule	Palisa, 25 Oct. 1888	4·22	4·27	8·23	0·011	2·33	D	130	7·4	15·4
288	Glauke	Luther, 20 Feb. 1890	2·18	2·76	4·58	0·210	4·33	S	30	1500	13·2
311	Claudia	Charlote, 11 Jan. 1891	2·89	2·90	4·94	0·003	3·23	S	28	11·5	14·9
324	Bamberga	Palisa, 23 Feb. 1892	1·77	2·68	4·39	0·341	11·14	C	252	29·4	9·2
349	Dembowska	Charlote, 9 Dec. 1892	2·66	2·92	5·00	0·092	8·26	R	164	4·7	10·6
423	Diotima	Charlote, 7 Dec. 1896	2·97	3·07	5·38	0·031	11·22	C	208	4·6	12·4
434	Hungaria	Wolf, 11 Sept. 1898	1·80	1·94	2·71	0·074	22·51	E	11	26·5	13·5
444	Gyptis	Coggia, 31 Mar. 1899	2·29	2·77	4·61	0·173	10·26	CS	166	6·2	11·8
451	Patientia	Charlote, 4 Dec. 1899	2·85	3·06	5·36	0·070	15·23	C	280	9·7	11·9
511	Davida	Dugan, 30 May 1903	2·61	3·18	5·66	0·177	15·93	C	324	5·2	10·5
532	Herculina	Wolf, 20 Apr. 1904	2·28	2·77	4·62	0·176	16·35	S	220	9·4	10·7
704	Interamnia	Cerulli, 2 Oct. 1910	2·61	3·06	5·36	0·148	17·30	F	338	8·7	11·0
747	Winchester	Metcalf, 7 Mar. 1913	1·97	3·00	5·19	0·342	18·17	C	204	9·4	10·7
878	Mildred	Nicholson, 6 Sept. 1916	1·82	2·36	3·63	0·231	2·02	?	26	?	15·0
903	Nealley	Palisa, 13 Sept. 1918	3·16	3·24	5·84	0·025	11·69	?	42	?	15·1
944	Hidalgo	Baade, 31 Oct. 1920	2·01	5·84	14·15	0·656	42·40	MEU	28	10·0	14·9
951	Gaspra	Neujmin, 30 July 1916	1·82	2·20	3·28	0·173	4·10	S	20	20·0	14·1
1269	Rollandia	Neujmin, 20 Sept. 1930	2·76	3·90	7·69	0·097	2·76	D	110	55	15·3
1300	Marcelle	Reisa, 10 Feb. 1934	2·76	2·78	4·64	0·009	9·54	?	21	?	15·9
1578	Kirkwood	Cameron, 10 Jan. 1951	3·02	3·94	7·82	0·231	0·81	D	56	?	15·8

The Trojans

These asteroids move in virtually the same orbit as Jupiter, but are about 60 degrees from Jupiter, so that there is no fear of collision. They are named after characters of the Trojan War. Of the first fifteen Trojans, ten lie east of Jupiter; the western group includes Patroclus, Priamus, Æneas, Anchises and Troilus. (In this table, Trojans indicated by an * are east of Jupiter; the others, west.)

THE BRIGHTEST TROJANS

No.	Name	Discoverer	q	a	P	e	i	T	D	R	Mag
588	Achilles*	Max Wolf, 22 Feb. 1906	4·40	5·17	11·77	0·149	10·33	D	116	58	15·3
617	Patroclus	Kopff, 17 Oct. 1906	4·50	5·23	11·97	0·139	22·04	P	164	?	15·2
624	Hector*	Kopff, 10 Feb. 1907	5·05	5·17	11·76	0·023	18·23	D	232	6·9	15·2

No.	Name	Discoverer	q	a	P	e	i	T	D	R	Mag
659	Nestor*	Max Wolf, 23 Mar. 1908	4·66	5·24	12·01	0·111	4·51	C	110	?	15·8
884	Priamus	Max Wolf, 22 Sept. 1917	4·53	5·12	11·70	0·122	8·91	D	94	?	16·0
911	Agamemnon*	Reinmuth, 19 Mar. 1919	4·85	5·20	11·87	0·068	21·84	D	154	7	15·1
1143	Odysseus*	Reinmuth, 28 Jan. 1930	4·76	5·25	12·01	0·093	3·14	D	174	?	15·6
1172	Æneas	Reinmuth, 17 Oct. 1930	4·62	5·16	11·72	0·104	16·72	D	162	?	15·7
1173	Anchises	Reinmuth, 17 Oct. 1930	4·56	5·30	12·21	0·137	6·91	C	116	?	16·0
1208	Troilus	Reinmuth, 31 Dec. 1931	4·72	5·20	11·85	0·091	33·64	C	124	24	16·0
1404	Ajax	Reinmuth, 17 Aug. 1936	4·66	5·27	12·09	0·011	18·04	?	92	?	16·6
1437	Diomedes	Reinmuth, 3 Aug. 1937	4·87	5·10	11·51	0·045	20·59	C	172	18·0	15·7
1583	Antilochus*	Arend, 20 Sept. 1950	4·83	5·11	11·55	0·054	28·55	D	158	?	16·3
1647	Menelaus*	Nicholson, 23 June 1957	5·12	5·25	12·03	0·025	5·64	?	50	?	18·1
1749	Telamon*	Reinmuth, 23 Sept. 1949	4·64	5·24	11·99	0·011	6·07	?	57	?	17·5

The next numbered Trojans are:

No.	Name	No.	Name	No.	Name
1868	Thersites*	2207	Antenor	2674	Pandorus
1869	Philocteles*	2223	Sarpedon	2759	Idomeneus*
1870	Glaukos	2241	Alcathous	2797	Teucer*
1871	Astyanax	2260	Neptolemus*	2893	Peiroos
1872	Helenes	2357	Phereclos	2895	Memnon
1873	Agenor	2363	Cebriones	2920	Automedon*
2146	Stentor*	2456	Palamedes*	3063	Makhaon
2148	Epeios*	2594	Acamas	3317	Paris
				3391	Sinon

CLOSE-APPROACH MINOR PLANETS

Minor planets which invade the inner part of the Solar System are conventionally divided into several classes. *Aten* asteroids have an average distance from the Sun of less than 1 a.u., though some may cross the Earth's orbit. *Apollo* asteroids have average distances from the Sun of more than 1 a.u., but cross the Earth's orbit. *Amor* asteroids cross the orbit of Mars, but not that of the Earth. Many of these close-approach asteroids are now known, and more are being found regularly; the following list is merely representative.

Aten Class

No.	Name	Discoverer	q	a	P	e	i	D	R
2062	Aten	Helin, 1976	0·79	0·97	0·95	0·183	18·9	0·9	?
2100	Ra-Shalom	Helin, 1978	0·47	0·83	0·76	0·450	15·8	2·4	9·7
2340	Hathor	Kowal, 1976	0·46	0·84	0·77	0·450	5·9	0·5	?
3362	Khufu	Dunbar, Barocci, 1984	0·52	0·99	0·98	0·469	9·9	1·4	?
3554	Amun	Shoemaker, 1986	0·70	1·24	0·96	0·280	23·4	2	?
3753	1986 TO	Waldron, 1986	0·48	1·51	0·99	0·515	19·8	3	?

Apollo Class

No.	Name	Discoverer	q	a	P	e	i	D	R	
—	Hermes	Reinmuth, 1937	0·62	2·66	2·10	0·624	6·2	1	?	Lost
1566	Icarus	Baade, 1949	0·19	2·40	1·12	0·827	22·0	1·4	2·3	
1620	Geographos	Wilson, Minkowski, 1951	0·82	1·66	1·39	0·335	13·3	2	5·2	
1685	Toro	Wirtanen, 1948	0·77	1·37	1·60	0·436	9·4	7·6	10·2	
1862	Apollo	Reinmuth, 1932	0·65	1·47	1·78	0·560	6·3	1·4	3·1	
1863	Antinoüs	Wirtanen, 1948	0·89	2·26	3·40	0·607	18·4	3	4·0	
1864	Daedalus	Gehrels, 1971	0·56	1·46	1·77	0·614	22·2	3·2	8·6	
1865	Cerberus	Kohoutek, 1971	0·58	1·08	1·12	0·467	16·1	1·6	6·8	
1866	Sisyphus	Wild, 1972	0·87	1·89	2·61	0·539	41·1	7·6	?	
1981	Midas	Kowal, 1973	0·62	1·78	2·37	0·650	39·8	1·6	?	
2063	Bacchus	Kowal, 1977	0·70	1·01	1·11	0·349	9·4	1·2	?	
2101	Adonis	Delporte, 1936	0·44	1·87	2·57	0·764	1·4	2	?	
2102	Tantalus	Kowal, 1978	0·90	1·29	1·47	0·298	64·0	2	?	
2135	Aristæus	Helin, 1977	0·79	1·60	2·02	0·503	23·0	0·8	?	
2201	Oljato	Giclas, 1947	0·63	2·17	3·20	0·711	2·5	2·8	24·0	
2212	Hephaistos	Chernykh, 1978	0·36	2·16	3·18	0·835	11·9	8·8	?	
2329	Orthos	Schuster, 1976	0·82	2·40	3·73	0·659	24·4	3·4	?	
3103	Eger	Lovas, 1982	0·91	1·41	1·67	0·355	20·9	5	5·7	
3200	Phæthon	IRAS, 1983	0·14	1·27	1·43	0·890	22·0	5	4	
3360	1991 VA	Helin, Dunbar, 1981	0·62	2·46	3·85	0·744	22·0	1·4	?	
3361	Orpheus	Torres, 1982	0·82	1·21	1·33	0·322	2·7	0·8	?	
3752	Camillo	Helin, Narucci, 1985	0·99	1·84	1·68	0·303	55·6	2	?	
4015	Comet Wilson-Harrington	Helin, 1979	0·99	4·29	4·29	0·623	2·8	3·2	?	
4034	1988 PA	Helin, 1986	0·59	1·53	1·09	0·444	11·2	1	?	
4660	Nereus	Helin, 1982	0·95	1·49	1·6	0·360	1·4	0·5	?	
4679	Castalia	Helin, 1989	0·55	1·06	1·2	0·483	8·9	0·6	?	

Amor Class

No.	Name	Discoverer	q	a	P	e	i	D	R	Notes
433	Eros	Witt, 1898	1·13	2·46	1·76	0·223	10·83	22	5·3	Close approaches 1931, 1975.
719	Albert	Palisa, 1911	1·19	2·58	4·15	0·540	10·82	2·6	?	Lost.
997	Alinda	Wolf, 1918	1·01	2·49	3·97	0·558	9·25	4	74·0	
1036	Ganymed	Baade, 1924	1·23	2·66	4·35	0·537	26·45	40	10·3	
1221	Amor	Delporte, 1932	1·09	1·92	2·66	0·435	11·90	1	?	Well placed every 8 years.
1580	Betulia	Johnson, 1950	1·12	2·20	3·26	0·049	52·01	1	6·1	No close approach before 2000.
1627	Ivar	Hertzsprung, 1929	1·12	1·86	2·54	0·397	8·44	6	4·8	

No.	Name	Discoverer	q	a	P	e	i	D	R	
1915	Quetzalcoatl	Wilson, 1953	1·07	2·53	4·03	0·577	20·50	0·4	4·9	Recovered in 1974.
1916	Boreas	Arend, 1953	1·25	2·27	3·43	0·450	12·84	3	?	Recovered in 1974.
1917	Cuyo	Cesco and Samuel, 1968	1·06	2·15	3·15	0·505	23·99	3	?	
1943	Anteros	Gibson, 1973	1·06	1·43	1·71	0·256	8·70	4	?	
1951	Lick	Wirtanen, 1949	1·30	1·39	1·63	0·062	39·09	2·2	?	
1980	Tezcatlipoca	Wirtanen, 1980	1·09	1·70	2·23	0·365	26·85	6	?	
2059	Babuquivari	Goethe, 1963	1·26	2·65	4·31	0·526	10·99	4	?	
2061	Anza	Giclas, 1960	1·15	3·38	3·40	0·537	3·74	2·4	11·6	
2202	Pele	Lemola, 1972	1·12	2·23	3·47	0·512	1·12	1·2	?	
2368	Baltrovata	Wild, 1977	1·23	2·10	3·06	0·413	5·26	5	?	
2608	Seneca	Schuster, 1978	1·03	2·48	3·90	0·586	15·63	1·4	18·5	
3199	Nefertiti	Shoemaker, 1982	1·12	1·57	1·98	0·283	32·97	4·5	3·1	
3271	1982 RB	Shoemaker, 1982	1·27	2·10	3·05	0·394	25·00	2	?	
3288	Seleucus	Schuster, 1982	1·10	2·03	2·90	0·457	5·93	4	75	
3352	McAuliffe	Thomas, 1981	1·18	1·88	2·58	0·369	4·78	2·4	?	
3671	Dionysius	Shoemaker, 1984	1·01	3·34	3·26	0·540	13·61	2	?	
3757	1982 XB	Helin, 1982	1·02	2·66	2·45	0·446	3·87	0·4	?	

All the close-approach asteroids are so faint that it is difficult to keep track of them. Apollo, discovered in 1932, was 'lost' for many years before its recovery in 1973; it passed within 0·01 astronomical units of Venus in 1950 and within 0·07 astronomical units of Venus in 1968; it also by-passed Mars at 0·05 units in 1946. It approached the Earth to within 0·06 units in November 1980 and again in May 1982.

Toro was within 0·14 units of Earth on 8 August 1972. It makes periodical fairly close approaches, and there were popular reports that it ranked as a minor Earth satellite. This is, of course, incorrect; it is an ordinary asteroid in orbit round the Sun, though admittedly its mean distance from the Sun is very nearly the same as that of the Earth. (It is not, however, moving in the Earth's orbit, since its path is considerably more eccentric than ours.)

* The only numbered asteroid to have been permanently lost is 719 Albert, seen only at the return of 1911.

* The only named but unnumbered asteroid to have been permanently lost is Hermes (1937 UB). It was discovered by K. Reinmuth, from Heidelberg, by-passing the Earth at a mere 800 000 km on 28 October 1937; it reached the 8th magnitude, and moved at 5 degrees per hour, so that it crossed the sky in 9 days. It has not been seen again, and is unlikely to be recovered.

On 18 January 1991, D. Rabinowitz from Kitt Peak discovered 1991 BA, which passed the Earth at 170 000 km – less than half the distance of the Moon. Even so, the magnitude was only 17·5; the diameter was estimated as 9 m. Even closer was 1994 XM₁, which passed by in December 1994 at a mere 104 000 km. The diameter was probably between 6 and 13 m.

On 5 December 1991 J. V. Scotti (University of Arizona) was carrying out a routine patrol with the 0 m.9 Spacewatch Telescope when he discovered a most peculiar object,

1991 VG. It passed within 460 000 km of the Earth, but varied in brightness; the estimated diameter is 6 m. The orbit has an inclination of 0 deg ·4, and the period is only slightly longer than that of the Earth. At present it is uncertain whether it is an asteroid, or a piece of space débris from an Earth-launched rocket. If it is an asteroid, it is among the smallest so far found. However, it now seems almost certain that this object was a piece of space débris of terrestrial origin.

The dimmest asteroid ever recorded is 1993 KA₂, absolute magnitude +29.

The first (and, so far, the only) Martian 'Trojan' is 1990 MB, discovered by Holt and Levy, from Kitt Peak, on 19 June 1990. Its period is 1·88 years; the orbital eccentricity is 0·065 and the inclination 20·2 degrees, while the distance from the Sun ranges between 1·42 and 1·53 astronomical units.

RECORD CLOSE APPROACHES

Name	Date	Min. dist. a.u.
1994 XM₁	9 Dec. 1994	0·0007
1993 KA₂	20 May 1993	0·001
1991 BA	18 Jan. 1991	0·001
4581 Åsclepius	Mar. 1989	0·004
Hermes	30 Oct. 1937	0·006
1988 TA	29 Sept. 1988	0·009
2340 Hathor	20 Oct. 1976	0·008
2101 Adonis	7 Feb. 1936	0·015
1986 JK	29 May 1986	0·028
3671 Dionysius	19 June 1984	0·031
3361 Orpheus	13 Apr. 1982	0·032
1991 JX	8 May 1991	0·032
2102 Tantalus	26 Dec. 1975	0·04
1466 Icarus	14 June 1967	0·04
1915 Quetzalcoatl	5 Mar. 1953	0·05
1950 DA	12 Mar. 1950	0·06
1620 Geographos	27 Aug. 1969	0·06
2061 Anza	7 Oct. 1960	0·06
1862 Apollo	25 May 1932	0·07
1973 NA	2 July 1973	0·08

The last asteroid to be recovered after having been missing for many years is 878 Mildred. It was discovered in 1916 by Nicholson and Shapley, and named after Harlow Shapley's young daughter. It was then lost, but finally rediscovered in April

1990, and subsequently identified on plates taken in 1977, 1984 and 1985.

Asteroids are divided into various types according to their surface characteristics, as derived from their spectral affinities. The main classes are:

C (Carbonaceous). These are the most numerous, increasing in number from 10 per cent at a distance of 2·2 astronomical units from the Sun out to 80 per cent at 3 astronomical units. They are of low albedo – often below 5 per cent (darker than coal) – and have flat, featureless spectra similar to those of carbonaceous chondrites.

S (Silicaceous). These are most numerous in the inner part of the main zone, making up 60 per cent of the total at 2·2 astronomical units, but only 15 per cent at 3 astronomical units. Their albedoes range from 15 to 25 per cent, and they seem to resemble the metal-bearing meteorites known as chondrites: generally they are reddish.

M (Metallic). Moderate albedoes, and may be the metal-rich cores of large 'parent asteroids' which have been broken up by collision. Asteroid 16 (Psyche) is almost pure nickel-iron alloy in composition.

E (Enstatite). Relatively rare; high albedo (sometimes over 40 per cent). They may resemble some types of chondrites, in which enstatite ($MgSiO_3$) is a major constituent.

D Low albedo; reddish. Their surfaces seem to be 90 per cent clays, with magnetite and carbon-rich substances. Most of the D-type asteroids are remote from the Sun, including many Trojans, but also a few in the main belt, such as 336 Lacadiera.

A Almost pure olivine. The best examples are 246 Asporina, 289 Nenetta and 446 Æternitas.

P Low albedo; spectra not unlike those of type M. 87 Sylvia is the best-known example.

Q Close-approach asteroids; in composition they seem to resemble chondrites (see Meteorites).

V Igneous rock surfaces. Very rare; Vesta is the only large example.

U Asteroids which are regarded as unclassifiable.

In 1987 D. P. Cruikshank and R. H. Brown, using the NASA Infra-Red Telescope at the high-altitude Mauna Kea Observatory in Hawaii, reported organic compounds in the spectrum of asteroid 130, Electra. The compounds seemed to resemble those found in meteorites of the carbonaceous chondrite type. The results showed a 3·4 micrometre absorption band in the spectrum of Electra. These observations indicate widespread distribution of organic matter throughout the Solar System.

✳ *The largest asteroid* is Ceres, which has one-third the total mass of all the asteroids combined. In 1977 L. Lebofsky detected an absorption band at 3 microns, due to H_2O embedded in the mineral structure of the asteroid. It is a spherical body, grey in colour, and of type C. Second in size is Vesta, which is probably triaxial, and whose surface seems to be coated with formerly molten igneous rock: the spectrum shows indications of pyroxene. Colour changes show that the surface is not uniform. Pallas, formerly believed to be larger than Vesta, is triaxial, measuring $580 \times 530 \times 470$ km. Its spectrum is not very unlike that of Ceres.

Asteroid diameters are not easy to measure. One method depends upon infra-red studies. The apparent brightness of a body in sunlight will depend partly on its size and partly on its albedo; everyone knows that in sunlight a white car will feel cooler than a black one! The amount of infra-red emission gives a clue to the albedo, and hence to the diameter. Another method is to observe the occultations of stars by asteroids. Some of the larger asteroids, such as Ceres, seem to be almost spherical; others, such as Pallas and Herculina, are ellipsoids; Juno, one of the four 'original' asteroids, seems to be spheroidal.

According to the latest results, the largest asteroids are:

Asteroid	Diam km
1 Ceres	940
4 Vesta	576
2 Pallas	525 ($580 \times 530 \times 470$)
10 Hygeia	430
511 Davida	384
704 Interamnia	338
65 Cybele	308
52 Europa	292
87 Sylvia	282
451 Patientia	280
31 Euphrosyne	270
15 Eunomia	260
324 Bamberga	252
16 Psyche	248
3 Juno	248 (288×230)

Masses, and hence densities, are derived mainly from mutual perturbations. Ceres is the giant of the asteroid family; all the rest make up only 1·5 times the mass of Ceres, and Ceres, Pallas and Vesta combined make up 55 per cent of the total mass of the asteroids – the total is 3 per cent of the mass of the Moon, or 0·04 per cent the mass of the Earth. The derived values are:

Mass, g	Density, water = 1
Ceres 1×10^{24}	2·7
Vesta $2·8 \times 10^{23}$	3·3
Pallas $2·2 \times 10^{23}$	2·6

The remaining asteroids have a total mass of $1·30 \times 10^{23}$ g.

✳ *The smallest known asteroid* is 1991 BA, with an estimated diameter of 9 m. Next comes 2340 Hathor (500 m). Masses of some small asteroids have been given as $5·0 \times 10^{12}$ kg (Icarus), $5·0 \times 10^{15}$ kg (Eros) and $5·0 \times 10^{10}$ kg (Adonis), but with great uncertainty.

✳ *The asteroid with the smallest known orbit* is Ra-Shalom, discovered by R. McNaught in December 1994. It is the 19th Aten asteroid to be found. Its mean distance from the Sun is 0·67 astronomical unit (100 000 000 km) perihelion distance 0·32 a.u. (48 000 000 km, just inside Mercury's orbit). Its orbital eccentricity is high, and so is its orbital inclination (28°). The diameter is perhaps 300 m. The orbital period is a mere 6½ months!

✳ *The brightest asteroid* is Vesta, which is the only member of the swarm ever visible with the naked eye. There are only 10 asteroids which can exceed the ninth magnitude, and one of these (Eros) does so only during its rare close approaches, as in 1931 and 1975. The list is as follows (mean opposition magnitudes):

4 Vesta	6·4	
1 Ceres	7·3	
2 Pallas	7·5	
7 Iris	7·8	
3 Juno	8·1	
6 Hebe	8·3	
433 Eros	8·3	(maximum at close approach)
15 Eunomia	8·5	
8 Flora	8·7	
18 Melpomene	8·9	

✳ *The most reflective asteroid* so far measured is 44 Nysa (albedo about 0·4; one estimate gives 0·377, another as much as 0·428). Another high-albedo asteroid is 434 Hungaria.

✳ *The darkest asteroid* of the first hundred to be discovered is 95 Arethusa (albedo 0·019, blacker than a blackboard). Asteroid 324 Bamberga is also very dark (albedo 0·032), but there are many very dark asteroids of smaller size, such as 1273 Helma and 1749 Telamon.

✳ *The shortest known rotation periods* (based upon variations in magnitude) are for 1566 Icarus (2 h 16 m) and 321 Florentina

(2 h 52 m) (magnitude range 10·6 to 11·4). *The slowest known spinner* is usually said to be 288 Glauke (1500 hours), though it has recently been suggested that Glauke may be a binary asteroid rather than a single slow-rotating body.

✳ *The greatest magnitude range* for a 'normal' asteroid is shared by two. 1864 Dædalus has a range of 0·9 magnitude and a period of 8 h 34 m; 1220 Crocus has also a range of 0·9 magnitude and a period of 30·7 days. However, 624 Hektor has a range of 1·1 magnitude in a period of 6·923 hours; it may be either elliptical in shape, or even double. 433 Eros has a range of 1·50 magnitude and a period of 5·270 hours, but is known to be very elliptical in form.

✳ *Asteroids to be detected by radar* include the 'earth-gazers' Eros, Icarus, Toro, Quetzalcoatl and the binary asteroid 4179 Toutatis. Excellent results were obtained from Eros during its 1975 approach; the surface was found to be rough on a scale of centimetres. The longest and shortest diameters are 36 and 15 km. In 1981 the Arecibo radio telescope was used to contact asteroids by radar: Psyche, Klotho, Apollo and Quetzalcoatl. At that time Apollo was 12 800 000 km from Earth and Psyche 40 000 000 km. Psyche is *the largest known iron asteroid*, with an estimated mass of 50 000 million million tonnes.

✳ *The first systematic search for a planet between the orbits of Mars and Jupiter* was initiated in 1800, by six astronomers meeting at Lilienthal – at Johann Schröter's observatory. They based their search on Bode's Law (actually first described by Titius, but popularized by Bode). This so-called Law may be summed up as follows:

Take the numbers 0, 3, 6, 12, 24, 48, 96, 192 and 384, each of which (apart from the first) is double its predecessor. Add 4 to each. Taking the Earth's distance from the Sun as 10, the remaining figures give the mean distances of the planets with considerable accuracy out to as far as Saturn, the outermost planet known when Bode popularized the relationship in 1772. Uranus, discovered in 1781, fits in well. Neptune and Pluto do not, but of course these planets were not discovered until considerably later.

Planet	Distance by Bode's Law	Actual distance
Mercury	4	3·9
Venus	7	7·2
Earth	10	10·0
Mars	16	15·2
–	28	–
Jupiter	52	52·0
Saturn	100	95·4
Uranus	196	191·8
Neptune	–	300·7
Pluto	388	394·6

Whether the Law is anything more than a coincidence is not known, but it was enough for the 'Celestial Police' to hunt for the missing planet corresponding to Bode's number 28. Schröter became president of the association; the secretary was the Hungarian astronomer Baron Franz Xavier von Zach. They very logically concentrated their searches in the region of the ecliptic.

✳ *The first asteroid to be discovered* was Ceres, by G. Piazzi at Palermo on 1 January 1801. (He was working on a star catalogue, and was not a member of the 'Police', though he joined later.) Ceres was found to have a distance of 27·7 on Bode's scale. Four asteroids were known by the time that the 'Police' disbanded in 1815; the fifth, Astræa, was not found until 1845 as a result of painstaking solo searches by a German amateur, Hencke.

✳ *The first asteroid to be discovered photographically* was 323 Brucia, on 20 December 1891 by Max Wolf.

✳ *The 19th-century astronomer to have discovered the greatest number of asteroids to which permanent numbers have been given* is K. Reinmuth, with 246 discoveries; Max Wolf had 232, and J. Palisa 121. By 1900, 452 asteroids were known.

Only one numbered asteroid has been lost: 719 Albert. Asteroid 330, Adalberta, never existed at all; it was recorded by Max Wolf in 1892, but was photographed only twice, and it was then found that the images were of two separate stars.

✳ *The first asteroid known to come well within the orbit of Mars* was Eros, discovered by Witt at Berlin in 1898. Its variability was detected by von Oppolzer in 1900; and in 1931 Finsen and van den Bos, at Johannesburg, saw its elongated shape. Eros can come to within 23 000 000 km of the Earth, and at the approach of 1931 was intensively studied in an attempt to improve our knowledge of the length of the astronomical unit – though the final value (derived by H. Spencer Jones) is now known to have been rather too great. The last close approach occurred in 1975.

✳ *The first numbered asteroid whose orbit lies wholly between those of the Earth and Mars* is 1951 Lick.

✳ *The first asteroid found to approach the Sun within the orbit of Mercury* is 1566 Icarus, discovered by W. Baade in 1949. It has a period of 409 days, and can approach the Sun to a distance of 28 000 000 km, so that its surface must then be at a temperature of approximately 500°C. The next asteroid found to invade these torrid regions was 3200 Phæthon, discovered in 1983 by J. Davis and S. Green from data supplied by the IRAS satellite. Phæthon is about 5 km in diameter, and its distance from the Sun ranges between 21 000 000 km and 390 000 000 km. It seems to be darkish, with a rotation period of about 4 hours. Its orbit is not unlike that of the Geminid meteor stream, and it may well be a dead comet. Other asteroids which pass close to the Sun are 5786 Talos (to 0·187 astronomical units), 1989 VA (0·295 units), 1990 UO (0·298 units) and 1991 VE (0·299 units).

✳ *The first asteroid found to have a very tenuous 'atmosphere'* is Ceres. In 1990 M. A'Hearn detected a thin cloud of OH ions, so that there must be enough water ice on the surface to produce this cloud when warmed.

✳ *The named asteroid with the smallest orbit* is Ra-Shalom, discovered by Mrs Helin on 25 October 1976. The period is 283 days. The distance from the Sun ranges between 69 000 000 and 126 000 000 km, so that the orbit lies well within that of the Earth. 2340 Hathor also has a very small orbit; the period is almost the same as that of Ra-Shalom.

✳ *The greatest perihelion distance for a numbered asteroid* (excluding Chiron and Pholus) is that of 588 Achilles; 5·98 astronomical units.

✳ *The greatest aphelion distance for a named asteroid* (again excluding probable planetesimals) is that of 944 Hidalgo (9·61 astronomical units, slightly greater than the mean distance between the Sun and Saturn). Hidalgo also has the *largest orbit*, and the *longest revolution period* (over 14 years). The magnitude ranges between 10 and 19. This range is greater than for any other asteroid. The greatest aphelion distance for a 'normal' asteroid is that of 1038 Tuckia (4·87 a.u.).

✳ *The greatest orbital eccentricity* is for 5145 Pholus (0·9). Then follow 3200 Phæthon (0·89), 2212 Hephaistos (0·834), and 5786 Talos and 1566 Icarus (each 0·827).

✳ *The least orbital eccentricty* is for 311 Claudia (0·0031). Then follow 903 Nealley (0·0039) and 1300 Marcelle (0·0050).

✳ *The greatest orbital inclination* is for 1973 NA (66·8 degrees). The greatest inclination for a named asteroid is for 1580 Betulia (52·0 degrees).

✳ *The least orbital inclination* is for 1383 Limburgia (0·014 degrees). Then follows 1486 Marilyn (0·084 degrees).

✳ *Gaps in the asteroid zone* were predicted by D. Kirkwood in 1857, and confirmed in 1866; they are due to the perturbing influence of Jupiter. For example, the Hilda asteroids have periods of approximately two-thirds that of Jupiter, so that the perturbations are cumulative. Families include:

Family	Distance from Sun, astronomical units	
Hungaria	1·9	Just inside the main belt. Most are of class E, with inclinations of over 16°.
Flora	2·2	Most populous family; well over 150 members. Most are reddish.
Phocæa	2·4	Rather ill-defined.
Koronis	2·9	In the outer belt, with low eccentricities and inclinations. Mostly of Class S.
Eos	3·0	Over 60 members, mainly of Class S with a few of Class M.
Themis	3·1	Major family in the outer ring; over 100 members including 24 Themis and 90 Antiope. Mainly of Class C.
Hilda	4·0	Over 30 members: Classes P, C and D.
Thule	4·2	Isolated, but perhaps the brightest member of a family?

✳ *The first known binary asteroid* is 4769 Castalia, discovered by Eleanor Helin on 9 August 1989. Radar images show that it is dumbbell-shaped, with two distinct lobes about ¾ km across. There is a narrow 'waist' at the mid-section, from 100 to 150 m thick, and it seems that the two lobes were once separate bodies which came together in a gentle collision. *The first contact binary asteroid known* is 4179 Toutatis; radar images were obtained from NASA's Deep Space Network on 8 December 1992. The two objects are heavily cratered, and are about 4 and 2·5 km across respectively; they rotate in a period of approximately 10·5 days. Here too it may be that two separate objects came gently together.

✳ *The first asteroid suspected to have a satellite* is 532 Herculina. On 7 June 1978 it occulted a star (SAO 120 774) and a second occultation indicated the presence of a 50 km satellite (Herculina itself is elliptical, with a mean diameter of 220 km). Other asteroids suspected to have satellites have been Melpomene, Kleopatra and Metis. However, precise confirmation is lacking. The Trojan asteroid 624 Hektor has been suspected of either having a satellite, or else being dumbell-shaped. On 11 March 1990 an occultation observation by M. Kidger, R. Casals and J. Sanchez, from the Canary Islands, indicated that asteroid 444 Gyptis may be double.

✳ *The first asteroid known to have a satellite* is 243 Ida. On 28 August 1993 the Galileo spacecraft imaged Ida and discovered the satellite, Dactyl, though the information did not reach Earth until the following February because of the problem with Galileo's high-gain antenna. Ida measured 56 × 24 × 21 km, and crater density is five times that of Gaspra, so that Ida's surface is presumably older. Dactyl measures $1.2 \times 1.4 \times 1.6$ km, and is 90 km from Ida.

✳ *The first asteroid to be named* was 1 Ceres, discovered by Piazzi from Palermo and named by him after the patron goddess of Sicily. All early names were mythological, but when the supply of deities began to be exhausted, more 'modern' names were introduced. The first was that of 125 Liberatrix, which seems to have been given either in honour of Joan of Arc or else of Adolphe Thiers, first President of the French Republic. Asteroids may be named after countries (136 Austria, 1125 China, 1279 Uganda, 1197 Rhodesia), people (1123 Shapleya, in honour of Dr Harlow Shapley; 1462 Zamenhof, after the inventor of Esperanto; 1486 Marilyn, after the daughter of Paul Herget, director of Cincinnati Observatory). The author feels very honoured that Asteroid 2602 has been named after himself!

✳ *The first masculine name* was that of 433 Eros.

✳ *The first asteroid to be named after an electric calculator* is 1625 The NORC – the Naval Ordnance Research Calculator at Dahlgren, Virginia. Plants are represented (978 Petunia, 973 Aralia); universities (694 Ekard – 'Drake' spelled backwards; the orbit was computed by students at Drake University), musical plays (1047 Geisha), foods (518 Halawe – the discoverer, R. S. Dugan, was particularly fond of the Arabian sweet *halawe*), social clubs (274 Philagoria – a club in Vienna), and even shipping lines (724 Hapag – Hamburg Amerika Linie).

One of the more unusual choices for an asteroid name is for No. 2309. Mr Spock – named after a ginger cat which was itself named after the Vulcanian space-traveller in the fictional starship *Enterprise*. One name has been expunged: No. 4267 was named Anvlad in honour of a Soviet philanthropist, Anatoli Vladimirov, but was withdrawn when Vladimirov was exposed as a financial swindler!

✳ *The only name to be auctioned* was that of 250 Bettina, whose discoverer, Palisa, offered to sell his right of naming for £50. The offer was accepted by Baron Albert von Rothschild, who named the asteroid after his wife. (The Rothschilds are well represented; 719 is named Albert, and 977 Philippa after Philip von Rothschild.) 443 Photographica commemorates the new method of asteroid discovery, and 1581 Abanderada – someone who carries a banner – is said to be associated with Eva Perón, wife of the former ruler of Argentina! Finally, 1000 is, appropriately enough, named Piazzia.

✳ *The first picture of an asteroid from a spacecraft* was obtained on 13 November 1991, when the Galileo probe, en route for Jupiter, imaged asteroid 951 Gaspra from a range of 16 000 km. Gaspra proved to be wedge-shaped (not unlike a distorted potato) with a darkish rocky surface pitted with craters. There were also indications of three areas from which pieces of it had been broken off. It has been suggested that Gaspra is the survivor from a series of catastrophic collisions. Gaspra was found to have a magnetic field strong enough to create a 'bubble' in the solar wind.

Craters on Gaspra have been named after famous spas. They include:

	Lat	Long	Diam km
Aix	47·9	160·3	14
Bath	13·4	009·7	20
Beppu	03·9	058·4	10
Brookton	27·7	103·3	12
Carlsbad	29·7	088·8	10
Charax	08·6	0	22
Helwan	22·4	118·9	12
Ixtapan	11·9	086·9	10
Lisdoonvarna	16·4	358·2	20
Ramlösa	15·0	004·9	20
Rio Hondo	31·7	20·7	14
Rotorua	18·8	030·7	12
Yalta	57·6	261·3	10

There are three regiones: Dunne (lat 15·0, long 15·0); Neujmin (002·0, 80·0) and Yeates (65·0, 75·0).

✳ *The second asteroid imaged by the Galileo probe* was 243 Ida. One crater on it has been named: Afon, lat 6·4S, long 0, diam 5·2 km.

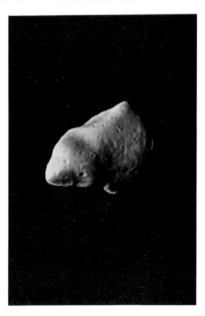

Asteroid 951 Gaspra, imaged by the Galileo spacecraft on 29 October 1991 from a distance of 16 200 km. Sunlight is coming from the left. (NASA)

Kuiper Belt object 1992 FW, imaged by David Jewitt and Jane Luu (Hawaii) and reproduced by their kind permission.

KUIPER BELT OBJECTS

*I*n 1977 C. Kowal, using the Schmidt telescope at Palomar, discovered an extraordinary object which has been named Chiron. It has been given an asteroidal number (2060) but it is far from being an ordinary asteroid, since its path lies mainly between those of Saturn and Uranus. At perihelion it comes to within 1278 million km of the Sun, and about one-sixth of its orbit lies within that of Saturn; but at aphelion it recedes to 2827 million km, greater than the minimum distance between the Sun and Uranus. The orbital inclination of 7 degrees prevents any close encounters with Saturn or Uranus at the present epoch, but in 1664 BC Chiron approached Saturn to within about 16 000 000 km, which is not much greater than the distance between Saturn and its outermost satellite, Phœbe. If Chiron's orbit is somewhat unstable, it may at some time approach Saturn within the orbit of Phœbe. Chiron was last at perihelion in 1895 and 1945; it will again reach perihelion in 1996. At discovery the magnitude was 18, but at perihelion it will rise to 15.

Preliminary spectroscopic results indicated a fairly low albedo with a dusty or rocky surface; the diameter was given as around 200 km – larger than that of Phœbe, and decidedly large by asteroidal standards. Kowal's discovery was made on 1 November 1977, from plates taken on 18 and 19 October of that year, but Chiron's image has since been traced back on plates taken as long ago as 1895.

There was a major surprise in 1988, when it was found to be brightening – not spectacularly, but appreciably. Inevitably there were suggestions that it might be a comet rather than an asteroid, and this idea was strengthened in April 1990, when K. Meech and J.S. Belton, using electronic equipment on the 4 m reflector at Kitt Peak in Arizona, photographed Chiron and found that it appeared to be 'fuzzy'; in other words it had developed a coma. Using the 2·24 m telescope on Mauna Kea on 29 January 1990, D. Jewitt and J. Luu found that the coma extends for 80 000 km, and the true length may be as much as 2 000 000 km; moreover, it is elongated away from the Sun in comet-like fashion. The ejected material has been thought to be vaporized carbon monoxide carrying away dust grains.

This is certainly comet-like behaviour. On the other hand, Chiron seems to be much too large to be a comet. *The first occultation*

of a star by Chiron was observed on 7 November 1993, at 13h 16m UT, by R. Marcialis and R. Hill (University of Arizona), using a 36 cm telescope. The fade lasted for 7 seconds, so that the diameter of Chiron must be at least 166 km and is probably rather more.

On 9 March 1994 there was another occultation, of the star Ch08, and from this it was found that Chiron's diameter must be between 166 and 312 km; probably 250 km is a good estimate. Probably the coma or atmosphere is due to outbursts from discrete jets on the surface, and there are several definite dust structures, one of which is fan-shaped; the outbursts sometimes make Chiron brighten by a full magnitude, and since 1988 activity has been almost continuous. There could well be an analogy with the geysers detected on Neptune's satellite Triton by Voyager 2.

Chiron is the brightest member of the group now known as the Centaurs; Pholus and Damocles are other examples – though Chiron itself may well have escaped from what is now known as the Kuiper Belt. Note too that Chiron is 40 times larger and 50 000 times more massive than a typical comet.

On 31 August 1992 David Jewitt and Jane Luu, using the 2·2 m telescope on Mauna Kea, found a 23rd-magnitude object which was given the provisional designation of 1992 QB₁. Its orbit keeps it well beyond that of Neptune, and it seems to come from a ring or disk of such bodies (the Kuiper Belt) ranging between 35 and 45 astronomical units from the Sun. By 1995 seventeen Kuiper

Belt objects had been found. They seem to be primordial, ice-rich planetesimals which were never incorporated into planets at the time when the Solar System as we now know it was being formed. The total mass may be as much as that of the Earth , and the total number of objects may be of the order of 20 000 to 40 000 – perhaps more. It is now generally believed that the Kuiper Belt is the source of these asteroidal bodies and at least some of the Centaurs, as well as short-period comets, though longer-period comets are still believed to come from the much more remote Oort cloud.

The Kuiper Belt objects seem to have diameters of from 100 to 300 km, and to be red. *The reddest object* is the Centaur asteroid 5145 Pholus; it has been suggested that the surface is covered with organic compounds produced when mixtures of methane and other simple compounds are bombarded by ultra-violet radiation from the Sun.

Pluto is probably a Kuiper Belt object. The first known Neptune-crossing object from the Kuiper Belt is 1993 SB; its orbital eccentricity is 0·32. Of the Centaurs, the orbit of Pholus crosses those of Saturn, Uranus and Neptune; Damocles crosses Mars, Jupiter, Saturn and Uranus; Hidalgo crosses Mars and Jupiter; Chiron crosses Saturn and Uranus. Note also that the orbit of Halley's Comet crosses those of all the principal planets apart from Mercury; its distance from the Sun ranges between 0·6 and 35·3 astronomical units. Unlike the objects listed above, Halley's Comet has retrograde motion.

In January 1995 four Centaur bodies were known. They are:

Name	Period years	Absolute magnitude	Eccentricity	Inclination degrees
2060 Chiron	50·9	6·0	0·38	6·9
5145 Pholus	92·1	7·3	0·57	24·7
1993 HA₂	124	9·5	0·52	15·6
1994 TA	73·1	11·5	0·39	5·4

Absolute magnitude indicates how bright the object would appear to an observer standing on the Sun, if the object were 1 astronomical unit away.

Some orbital data are as follows:

Name	Distance from Sun perihelion	aphelion	Orbital inclination	Period years	Diam km
Centaurs					
944 Hidalgo	2·0	5·9	42·4	14·1	28
5335 Damocles	1·6	22·2	60·9	40·9	15
2060 Chiron	8·5	18·5	6·9	50·7	±200 Mag. 15
5145 Pholus	8·7	32·0	25·0	93·0	±200
Kuiper Belt objects					
1993 SB	26·8	52·0	1·9	247·5	188
Pluto	29·7	49·3	17·1	247·7	2324 Mag. 14
1993 SC	32·4	46·6	5·2	248·2	319
1992 QB₁	40·8	46·9	2·2	290·2	283
1993 FW	42·1	45·6	7·7	291·2	256

JUPITER

DATA ....

Distance from the Sun:
mean: 778·34 million km (5·203 a.u.)
max: 815·7 million km (5·455 a.u.)
min: 740·9 million km (4·951 a.u.)
Sidereal period: 11·86 Years = 4332·59 days
Synodic period: 398·9 days
Rotation period:
System I (equatorial) 9 h 50 m 30 s;
System II (rest of planet) 9 h 55 m 41 s;
System III (radio methods) 9 h 55 m 30 s
Mean orbital velocity: 13·06 km/s
Axial inclination: 3 deg 04 min
Orbital inclination: 1 deg 18 min 15 sec ·8
Orbital eccentricity: 0·048
Diameter:
equatorial 143 884 km
polar 133 708 km

Apparent diameter from Earth:
max 50 sec ·1
min: 30 sec ·4
Reciprocal mass, Sun = 1: 1047·4
Density, water = 1: 1·33
Mass, Earth = 1: 317·89
Volume, Earth = 1: 1318·7
Escape velocity: 60·22 km/s
Surface gravity, Earth = 1: 2·64
Mean surface temperature: −150°C
Oblateness: 0·06
Albedo: 0·43
Maximum magnitude: −2·9
Mean diameter of Sun, as seen from Jupiter:
6 min 09 sec

Belt		Lat
N Equatorial	N edge	+18
	S edge	+7
S Equatorial	N edge	−7
	S edge	−21
S Temperate		−30
SS Temperate		−41

✳ *The first man to find that System I rotates faster than System II* was G. D. Cassini, in 1690.

We now have a sound knowledge of the way in which the Jovian winds blow. In the equatorial region, they blow west–east at some 100 m/s relative to the core, reaching maximum speed 6 to 7 deg north and south of the equator. In the northern hemisphere the east wind decreases with increasing latitude, until at 18 degrees N the clouds are moving westward at 25 m/s. North of this the wind speed falls to zero, and then shifts in an eastward direction, reaching a maximum of about 170 m/s at 24 deg N. In the southern hemisphere conditions are not quite the same, due probably to the presence of the Great Red Spot.

✳ *The most famous marking on Jupiter* is the Great Red Spot. It may be identical with a feature recorded by Robert Hooke in 1664, and it (or a similar feature) was certainly seen by Cassini in 1665, but it became famous in 1878, when it was described as very prominent and brick-red. Subsequently it has shown variations in both intensity and colour, and at times (as in 1976–7) it has been invisible, but it always returns after a few years, so that during telescopic times it has been to all intents and purposes a permanent feature – unlike any other of the spots. At its greatest extent is may be 40 000 km long and 14 000 km wide, so that its surface area is then greater than that of the Earth. On 5 September 1831 H. Schwabe found that it encroached into the region of the southern part of the South Equatorial Belt, making up what is now always termed the Red Spot Hollow.

Though the latitude of the Red Spot varies little, it drifts about in longitude. Over the past century the total longitude drift amounts to approximately 1200 degrees! The latitude is generally very close to −22 degrees.

It was once thought that the Red Spot might be a solid or semi-solid body floating in Jupiter's outer gas, in which case it would be expected to disappear if its level sank for any reason (possibly a decrease in the density of the outer gas). In 1963 R. Hide suggested that it might be the top of a 'Taylor Column', a sort of standing wave above a mountain or depression below the gaseous layer. However, the Pioneer and Voyager results have shown that ideas of this kind are untenable. The Red Spot is a phenom-

*T*he largest and most massive planet is Jupiter, which is fifth in order of distance from the Sun.

Jupiter is well placed for observation for several months in every year, and the opposition magnitude has a range of only about 0^m·5. Generally speaking it 'moves' about one constellation per year; thus the 1993 opposition occurred in Virgo and the 1994 opposition in Libra. Perihelion dates: 1987, 1999. Aphelion: 1981, 1993, 2005.

✳ *The last occasion when Jupiter was occulted by a planet* was on 3 January 1818, when Venus occulted Jupiter. This will happen again on 22 November 2065, 12.46 UT, but the elongation from the Sun will be only 8 degrees W. Close planetary conjunctions involving Jupiter for the 1900–2100 period:

	Date	UT	Sep sec	Elong
Jupiter–Uranus	1955 May 10	20.39	−46	65 E
Venus–Jupiter	2065 Nov. 22	12.46	Occ.	8 W
Mercury–Jupiter	2088 Oct. 27	13.44	Occ.	5 W
Mercury–Jupiter	2094 Apr. 7	10.49	Occ.	2 W

✳ *The last occasion when Jupiter occulted a planet* was on 15 August 1623 when it occulted Uranus.

On average Jupiter is the brightest of the planets apart from Venus, though for relatively brief periods during perihelic oppositions Mars may outshine it. Note that in the period between 1977 and 2000, the only years in which Jupiter does not come to opposition are 1978 and 1990. This is, of course, because oppositions occurred in the last part of December in 1977 and 1989.

✳ *The shortest 'day' in the Solar System* is that of Jupiter, insofar as the principal planets are concerned (some of the asteroids rotate much more quickly). It is not possible to give an overall value for the axial rotation period, because Jupiter does not rotate in the way that a solid body would do. The equatorial zone has a shorter rotation than the rest of the planet. Conventionally, System I refers to the region between the north edge of the South Equatorial Belt and the south edge of the North Equatorial Belt; the mean rotation period is 9 h 50 m 30 s, though individual features may have periods which differ perceptibly from this. System II, comprising the rest of the planet, has a period of 9 h 55 m 41 s, though again different features have their own periods; that of the Great Red Spot varies between 9 h 55 m 36 s to 9 h 55 m 42 s. In addition there is System III, which relates not to optical features but the bursts of decametre radio radiation; the period seems to be 9 h 55 m 29 s·4.

There are no true 'seasons' on Jupiter, because the axial inclination to the perpendicular of the orbital plane is only just over 3 degrees – less than for any other planet.

OPPOSITION DATES, 1994–2005

Date	Diameter seconds of arc	Declination at opposition, deg	Mag.
1994 Apr. 30	44·5	−13	−2·0
1995 June 1	45·6	−21	−2·1
1996 July 4	47·0	−23	−2·2
1997 Aug. 9	48·6	−17	−2·4
1998 Sept. 16	49·7	−4	−2·5
1999 Oct. 23	49·8	+10	−2·5
2000 Nov. 28	48·5	+20	−2·4
2002 Jan 1.	47·1	+23	−2·3
2003 Feb. 2	45·5	+18	−2·0
2004 Mar. 4	44·5	+8	−2·0
2005 Apr. 3	44·2	−4	−2·0

The latitudes of the various belts and zones are also variable, over a limited range. The following values are therefore approximate only:

Belt	Lat
NN Temperate	+38
N Temperate	+27

enon of Jovian meteorology – an anticyclonic vortex. The present dimensions are 23 000 × 12 400 km, so that the Spot has decreased in size over the past century or so; it may or may not be a permanent feature, and its origin is uncertain (suggestions include a very large rising convective cell. To the south it is bounded by a gentle east wind but to the north side it is bounded by an extremely strong west wind. This means that as the winds are reflected round the Spot, they set up a rotation of about 7 days in an anti-clockwise direction. The Spot has a profound effect upon all neighbouring features – notably the famous white ovals which lie some 10 degrees to its south, but are variable in longevity and size.

✱ **The longest-observed 'disturbance' on Jupiter** was the South Tropical Disturbance, discovered by P. Molesworth on 28 February 1901 and last indicated by many observers during the apparition of 1939–40. It took the form of a shaded zone between white spots. The rotation period of the disturbance was shorter than that of the Red Spot, so that periodically the Disturbance caught the Spot up and passed it; the two were at the same latitude, and the interactions were of great interest. Nine conjunctions were observed, and possibly the beginning of a tenth in 1939–40, though by then the Disturbance had practically vanished. Its average rotation period was 9 h 55 m 27 s·6. It has not reappeared, but there have been several smaller, shorter-lived Disturbances like it; the most notable of these lasted from 1979 to 1981.

✱ **The most prominent belt on Jupiter** is generally the North Equatorial, while the South Equatorial Belt is often double and becomes obscure at times. However, during the apparition of 1988 the South Equatorial Belt was fully equal to the North Equatorial, and this seems also to have been the case at times before 1900, though the records are far from complete. Revivals of the South Equatorial Belt are the most spectacular phenomena seen on Jupiter; they involve sudden outbreaks of bright and dark clouds, with intense turbulence and many spots moving on rapid currents. Other belts are also subject to marked variations in intensity. Around latitude 16 degrees N may be seen brown ovals which are known as 'barges'. They are low-lying, and it may be that their colour is caused by the fact that they are slightly warmer than adjacent regions, so that some of the ammonia ice particles melt.

✱ **The first proof that Jupiter is not a miniature Sun** was given by H. Jeffreys in his classic papers published in 1923 and 1924. Jeffreys proposed a model in which Jupiter

The Great Red Spot, imaged by Voyager 1. Also shown is an oval, with its wake of vortices. (NASA)

would have a rocky core, a mantle composed of solid water and carbon dioxide, and a very deep, tenuous atmosphere.

✱ **The first identification of methane and ammonia in Jupiter's atmosphere** was due to R. Wildt in 1932; it followed that much of the planet must be composed of hydrogen. In 1934 Wildt proposed a model giving Jupiter a rocky core 60 000 km in diameter, overlaid by an ice shell 27 000 km thick above which lay the hydrogen-rich atmosphere. (This was certainly more plausible than a strange theory proposed by E. Schoenberg in 1943. Schoenberg believed Jupiter to have a solid surface, with volcanic rifts along parallels of latitude; heated gases rising from these rifts would produce the belts!)

✱ **The first 'hydrogen models'** were proposed independently in 1951 by W. Ramsey in England and W. DeMarcus in America. On Ramsey's theory, the 120 000 km diameter core was composed of hydrogen, so compressed that it assumed the characteristics of a metal. This core was overlaid by an 8000 km deep layer of ordinary solid hydrogen, above which came the atmosphere.

✱ **The first suggestion that Jupiter could be mainly liquid** seems to have been made by G. W. Hough as long ago as 1871. Hough, incidentally, believed the Red Spot to be a kind of floating island.

✱ **The latest models of Jupiter** are based on one proposed by W. De Marcus in 1959. The most up-to-date versions are by J. D. Anderson and W. B. Hubbard in the United States. There is a relatively small, rocky core

made up of iron and silicates, and at a temperature of about 20 000°C (perhaps rather more). Around this is a thick shell of liquid metallic hydrogen. At about 46 000 km from the centre of the planet there is a sudden transition from liquid metallic hydrogen to liquid molecular hydrogen; in the transition region the temperature is assumed to be about 11 000°C, with a pressure about three million times that of the Earth's air at sea-level. Above the liquid molecular hydrogen comes the gaseous atmosphere, which is about 1000 km deep. The upper clouds consist of ammonia ice; below comes a layer of ammonium hydrosulphide, and then a layer of water ice or liquid water droplets.

Gases warmed by the internal heat of the planet rise into the upper atmosphere, and cool, forming clouds of ammonia crystals floating in gaseous hydrogen. The clouds form the bright zones on Jupiter, which are both colder and higher than the dark belts. The colours of the various features have yet to be satisfactorily explained, but are certainly due to characteristics of Jovian chemistry.

Jupiter radiates 1·7 times more energy than it would do if it depended only upon radiation received from the Sun. On the liquid-planet model, this excess heat must be nothing more than the remnant of the heat generated when Jupiter was first formed. It had been suggested that Jupiter might be slowly contracting, and that this con-traction was the cause of the excess radiation emitted; but this is not now generally favoured.

JUPITER PROBES 1972–95

Name	Launch date	Encounter date	Nearest approach km	Remarks
Pioneer 10	2 Mar. 1972	3 Dec. 1973	131 400	Complete success; images and data. Now on its way out of the Solar System. Will probably be tracked into the late 1990s.
Pioneer 11	5 Apr. 1973	2 Dec. 1974	46 400	Passed quickly over the Jovian equator to avoid radiation damage. Complemented the results of Pioneer 10. Went on to a rendezvous with Saturn (1 Sept. 1979). Now on its way out of the Solar System. Contact probably until into the late 1990s.
Voyager 1	5 Sept. 1977	5 Mar. 1979	350 000	More detailed information about Jupiter and the Galilean satellites Io, Ganymede and Callisto; volcanoes on Io discovered. Went on to rendezvous with Saturn, 12 Nov. 1980. Now on its way out of the Solar System.
Voyager 2	20 Aug. 1977	9 Jul. 1979	714 000	Complemented Voyager 1. Went on to fly by Saturn (1981), Uranus (1986) and Neptune (1989). Now on its way out of the Solar System; should be tracked well into the next century.
Galileo	18 Oct. 1989	1995	Entry	Orbiter and entry probe; fly-bys of Venus (10 Feb. 1990) and Earth (8 Dec. 1990 and 8 Dec. 1992); images of asteroids Gaspra (10 Oct. 1989) and Ida (8 Aug. 1993).
Ulysses	6 Oct. 1990	8 Feb. 1992	378 000	Studies of Jupiter's magnetosphere, radiation zones and general environment. Went on to study poles of the Sun.

✳ **The first detection of radio radiation from Jupiter** was due to B. F. Burke and K. L. Franklin, in the United States, in 1955. (It must be admitted that the discovery was accidental!) The radio emissions are concentrated in wavelengths of tens of metres (decametric emissions) and tenths of metres (decimetric). Efforts to correlate the radio burst with visual features, such as the Red Spot, were not successful, but it seems that the decametric radiation is affected by the orbital position of Io, the innermost of the large satellites of Jupiter. From these early results it was surmised – correctly – that Jupiter has an extremely powerful magnetic field.

✳ **The first probe to Jupiter** was Pioneer 10. It was launched from Cape Canaveral on 2 March 1972, and by-passed Jupiter on 3 December 1973 at a distance of 131 400 km. Studies of the Jovian magnetosphere and atmosphere were carried out, and more than 300 pictures returned. Pioneer 10 was the first probe to leave the Solar System; it will never return. In case it should ever be found by some alien civilization, it carries a plaque to give a clue as to the planet of its origin – though whether any other beings would be able to decipher the message on the plaque has been questioned!

✳ **The second Jupiter probe** was Pioneer 11, launched on 5 April 1973 and which by-passed Jupiter at 46 400 km on 2 December 1974. It also studied Jupiter's magnetosphere and atmosphere, and sent back excellent pictures; it was then put into a path en route

for a rendezvous with Saturn, after which it too will leave the Solar System permanently.

✳ **The third Jupiter probe**, Voyager 1, was actually launched a few days later than its twin Voyager 2, but travelled in a more economical path. (To confuse matters still further, initial faults detected in the first spacecraft caused a switch in numbers, so that the original Voyager 1 became Voyager 2, and vice versa!) The Voyager was much more elaborate than either of the Pioneers, and the results obtained were of far higher quality. In addition to Jupiter itself, Voyager 1 also obtained good images of Io, Ganymede and Callisto.

✳ **The fourth Jupiter probe**, Voyager 2, followed much the same programme, but was sent close to Europa, the only Galilean satellite not satisfactorily observed by its predecessor.

The Voyagers confirmed the Pioneer data concerning Jupiter's magnetic field, and also confirmed the association between the radiation zones and the active satellite Io. It also seems that at times some of the trapped particles are ejected as bursts of cosmic rays. Jupiter and Io are connected by a flux tube of electrons and ions carrying a current of five million amperes; material from Io produces an ionized plasma ring or torus round Jupiter in Io's orbit.

✳ **The Red Spot** was found to be in rotation; it spins anti-clockwise, with a period of 12 days at its outer edge and 9 days somewhere inside. Material rises in its centre,

and spirals outward toward the edge. The centre is some 8 km above the surrounding clouds, and the Spot is colder than the adjacent regions. Adjacent spots may be 'sucked in', though during the Voyager 2 encounter there seemed to be some sort of barrier east of the Red Spot which had earlier been absent. The colour of the Red Spot may be due to phosphorus, produced by the action of sunlight upon phosphine sent up from the planet's interior.

✳ **The Jovian atmosphere** was found to consist of 89 per cent hydrogen and 11 per cent helium, with smaller amounts of other substances. Hydrogen compounds include ammonia, water, and methane.

✳ **Auroræ on Jupiter** were first recorded by Voyager 1 during its passage around the night side of the planet. It may be, however, that the main cause of Jovian auroræ is not the solar wind, as with Earth, but material from the volcanoes of Io. Lightning was detected, and is probably very violent; Jovian 'noise' was recorded, due to high-energy particles in motion, so that at some wavelengths Jupiter is a noisy world as well as a turbulent one. The findings were confirmed by Voyager 2.

✳ **The ring system of Jupiter** was also discovered from Voyager 1:

Name	Distance from centre of Jupiter km	Width km	Thickness km
Halo	100 000–122 800	22 800	20 000
Main	122 800–129 200	6400	30
Gossamer	129 200–214 200	850 000	?

The Hale and Gossamer rings are extremely tenuous.

The only ring which is at all substantial is the 'Main', but it too is very dark. The rings reach to over 50 000 km above the cloud-tops. There are two main sections, a 5000 km wide segment and a brighter, outer 800 km segment. The thickness of the rings is probably not much more than 1 km and the tiny particles which make up the system orbit Jupiter in from 5 to 7 hours. The smallness of the ring particles is shown by the fact that the rings are best seen when 'backlit' – that is to say, when the spacecraft's camera is behind them with respect to the Sun, so that the sunlight is reaching the camera via the ring. It has been suggested that the particles may slowly spiral downward to the planet; if so, they must be continually replenished, and once more the active Io may be involved. Clearly, Jupiter's ring is very different from the glorious ring system of Saturn, and it may not even be a permanent feature. Note also that the two small inner satellites, Metis and Adrastea, move in the same region, and certainly have profound effects upon the rings in spite of their small size and mass.

✳ *The Jupiter probe Galileo* was launched in 1989. Its orbiter is designed to enter a closed path round the planet and survey both Jupiter and its main satellites, while the entry probe, previously separated, should plunge into the clouds and send back data for perhaps an hour before being destroyed. The faulty deployment of the high-gain antenna has, however, caused severe problems. In November 1991 Galileo obtained the first close-range image of an asteroid (Gaspra).

The solar probe Ulysses flew past Jupiter on 8 February 1992, using Jupiter's gravity to put it into an orbit taking it well away from the ecliptic so that it could subsequently survey the poles of the Sun. It carried no camera, but carried out valuable studies of the Jovian environment.

✳ *The magnetic field of Jupiter* is stronger than for any other planet. Except near the planet, the field is dipolar, like that of the Earth, but the direction of the field is opposite to ours – so that if a terrestrial compass were taken to Jupiter, it would point south instead of north. The axis of the dipole field is inclined to the axis of rotation by 10·8 degrees. Inside about three Jupiter radii the field is so complex that it cannot be regarded as truly dipolar. The main dipole field has a strength of about 4 gauss (as against 0·3 to 0·8 gauss at the surface of the Earth). Pioneer 10 crossed the shock wave (i.e. the boundary between the Jovian magnetosphere and the general interplanetary field) on 26 November 1973 – the first probe to do so. Energetic particles were detected as far out as 300 Jovian radii from the planet, and during its passage through the radiation belts it received 200 000 rads from electrons and 50 000 rads from energetic ions; the lethal dose for a man is never 500 rads. The Jovian radiation belts – that is to say, the regions in which ions and high-energy electrons are trapped by the magnetic field – are about 10 000 times more intense than the Van Allen belts associated with the Earth. The electrical instruments on Pioneer 10 were 95 per cent saturated, and if the probe had approached much closer to Jupiter the mission would have failed. Pioneer 11 was put into a different trajectory, passing quickly over the equatorial zone so as to avoid the worst of the radiation and during the passage of Ulysses through the radiation zone on 8 February 1992 some of the sensitive instruments were switched off to guard against damage. The Jovian magnetosphere is very extensive, and the outer part does not share in the planet's rotation. The magnetic 'tail' extends so far that when suitably placed, Saturn may lie within it.

✳ *The first observed impact of a comet on a planet* occurred in July 1994, when Comet Shoemaker-Levy 9 impacted Jupiter. What was described as 'a squashed comet' was detected on 26 March 1993 by Eugene and Carolyn Shoemaker on a photograph taken with the Schmidt telescope at Palomar. It was found to be in orbit round Jupiter – it had probably been so ever since around 1970 or even earlier – and calculations showed that in mid-1992 it had passed within 21 000 km of the Giant Planet, and had been disrupted. A chain of fragments resulted, and further calculations showed that these would hit the planet in turn from 16 to 22 July 1994. This is precisely what happened. The first impact (lettered A) occurred at 20 h 11 min on 16 July, and the last (W) at 8 h 5 min on 22 July. All the fragments landed in roughly the same latitude – around 50 degrees S, well south of the Red Spot – but there was a wide distribution in longitude. Unfortunately, all the impacts occurred on the side of Jupiter turned away from the Earth, but the planet's quick rotation brought the affected areas into view after only a few minutes, and the results of the impacts were very marked indeed. Great dark spots were produced; the most impressive was lettered G, which impacted at 7 h 32 min on 18 July. Fragment G, which may have been 3 to 4 km across, created a fireball at least 3000 km high, and left a multi-ringed 'scar' on Jupiter's surface. It was estimated that had Fragment G hit the Earth, it would have made a crater 60 km in diameter. The large impacts such as G created vast clouds of 'smoke' which remained visible from Earth for many months.

The impacts were observed with the Hubble Space Telescope, and also from the Galileo spacecraft. Hubble results showed the presence of ammonia (NH_3), sulphur (S_2) and hydrogen sulphide H_2S) in the G fireball as it cooled, but apparently the fragment did not penetrate into Jupiter as far as the layer of water ice cloud.

The effects upon Jupiter were temporary only (for example, there was strong indication of enhanced auroral activity), but the impact provided a great deal of information. Relatively small though the fragments were, they hit Jupiter at a relative speed of 60 km/s.

SATELLITES OF JUPITER

*T*he four large satellites are known as the Galileans, though they were observed at about the same time independently by

Simon Marius. There is, however, considerable evidence that a naked-eye observation of one of the satellites (probably Ganymede, or even two satellites close together) was made in China by Gan De as long ago as 364 BC. The four outermost satellites (Ananke, Carme, Pasiphaë and Sinope) have retrograde orbits. All the satellites beyond Callisto are so far from Jupiter, and so subject to solar perturbation, that their orbits are not even approximately circular, and orbital elements are subject to change. Indeed, Pasiphaë was 'lost' after its discovery in 1908, found again in 1922, lost once more until 1938, and again between 1941–55.

✳ *The last satellite to be discovered by visual means* was Amalthea, by E. E. Barnard in 1892; all subsequent discoveries have been photographic or by space-probe.

✳ *The first attempted maps of the surfaces of the Galilean satellites* were due to A. Dollfus and his colleagues at the Pic du Midi Observatory, in 1961. Some features were recorded, but it is hardly to be expected that the results could have been reliable.

Since the apparent diameter of the Sun as seen from Jupiter is less than 6 minutes of arc, Amalthea and all the Galileans can produce total eclipses. To a Jovian observer, Io would appear much the largest of the satellites.

The satellites beyond Callisto may well be in the nature of captured asteroids, and do not in any way resemble the massive Galileans. Their diameters are very uncertain; only Himalia can be as much as 100 km in diameter. It is believed that they are darkish bodies, but at present it must be admitted that very little is known about them.

Metis and Adrastea, the two satellites closest to Jupiter, are less than 60 000 km above the cloud-tops, and are associated with the form of Jupiter's ring. They seem to have rather low albedoes, in which case they will be rocky rather than icy, but we have no further information.

Amalthea, on the other hand, has been known since 1892; it was found by E. E. Barnard during a deliberate search with the great Lick refractor. Barnard commented that it was 'much more difficult to see than the satellites of Mars'. Voyager 1 approached it to within 420 100 km and Voyager 2 to 558 270 km, and both took pictures. The gravitational pull of Jupiter has drawn Amalthea out into an irregular shape, and, as expected, the longest axis (270 km) points toward Jupiter. In view of its irregular shape, Amalthea may be fairly dense, and probably the surface is covered with a thick regolith.

The surface is very red, due probably to sulphur contamination from Io. The albedo is generally low, but the bright, rather greenish patches have a reflectivity of up to 20 per cent, and may be due to recently-exposed material which contains a lesser quantity of sulphur-rich glass. The two main craters are bowl-shaped. Pan is 90 km in diameter and well over 8 km deep. Gaea, 75 km in diameter, has a depth of between 10 and 20 km. If the latter figure is correct, then the slope angle of the wall must be 30°, therefore making them the highest known scarps in the Solar System. (It would be interesting to watch a piece of material fall from the crest to the floor; the descent time would be about 10 minutes!) Both Pan and Gaea are deeper, relatively, than craters of similar size on the Moon. Between them from long 0 to 60 deg W is a complex region of troughs and ridges, tens of kilometres long and sometimes at least 20 km wide. The two bright patches, Ida and Lyctos, are about 15 km in diameter, and are presumably mountains.

Thebe was also found on Voyager pictures during an examination to confirm the existence of Adrastea. Thebe moves between the orbits of Amalthea and Io. All we know about it is that its revolution period is 16 h 16 m, and that with a diameter of about 100 km it is decidedly larger than either Metis or Adrastea.

IO

*I*o is slightly larger than our Moon, and is the densest of the four Galilean satellites. Before the space missions it was tacitly assumed to be a rocky, cratered world, but in the event nothing could have been further from the truth. Io is bright orange; it is covered with sulphur, and there are active volcanoes.

✳ *The first indication of vulcanism* was found on 9 March 1979 by Linda Morabito, a member of the Voyager team who was looking for a faint star, AGK-10021, to check on Io's position. The volcanic plume rose to 280 km. Other volcanoes were then found; altogether eight were detected from Voyager 1. There are many volcanic craters, generally looking like black spots a few tens of kilometres in diameter and frequently surrounded by irregular haloes nearly as black as the central spot. The volcanic centres are very hot; for example the lava lakes associated with the volcanoes Pele and Loki were measured at at least 150°C hotter than their surroundings. The heights of the volcanic plumes during the Voyager 1 flyby were considerable; 300 km for Pele, 190 km for Loki, 120 km for Marduk, and between

SATELLITES OF JUPITER

No.	Name	Discoverer	Mean distance from Jupiter, km	Mean angular distance from Jupiter, at mean opposition distance min sec
XVI	Metis	Synott 1979	127 960	
XIV	Adrastea	Jewitt and Danielson 1979	128 980	
V	Amalthea	Barnard 1892	181 300	0 59·4
XV	Thebe	Synott 1979	221 900	
I	Io	Galileo	421 600	2 18·4
II	Europa	and	670 900	3 40·1
III	Ganymede	Marius	1 070 000	5 51·2
IV	Callisto	1610	1 880 000	10 17·6
XIII	Leda	Kowal 1974	11 094 000	60 45
VI	Himalia	Perrine 1904	11 480 000	62 45
X	Lysithea	Nicholson 1938	11 720 000	64 05
VII	Elara	Perrine 1905	11 737 000	64 10
XII	Ananke	Nicholson 1951	21 200 000	116
XI	Carme	Nicholson 1938	22 600 000	123
VIII	Pasiphaë	Melotte 1908	23 500 000	129
IX	Sinope	Nicholson 1914	23 700 000	130

	Sidereal period days	Mean synodic period d h m s	Apparent diameter as seen from Jupiter min sec
Metis	0·295		
Adrastea	0·298		
Amalthea	0·498	0 11 57 27·6	7 24
Thebe	0·675		
Io	1·769	1 18 28 35·9	35 40
Europa	3·551	3 13 17 53·7	17 30
Ganymede	7·155	7 03 59 35·9	18 06
Callisto	16·689	16 18 05 06·9	9 30
Leda	238·7	254	0 0·15
Himalia	250·6	266	0 8·2
Lysithea	259·2	276	0 0·03
Elara	259·7	276	0 1·4
Ananke	631	551	0 0·2
Carme	692	597	0 0·2
Pasiphaë	735	635	0 0·2
Sinope	758	645	0 0·2

	Orbital inclination degrees	Orbital eccentricity	Diameter km	Density water = 1	Escape velocity km/s
Metis	0	0	40	3?	0·02?
Adrastea	0	0	26×20×16	3?	0·01?
Amalthea	0·45	0·003	262×146×143	3?	0·16?
Thebe	0·9	0·013	110×90	3?	0·8?
Io	0·04	0·004	3660×3637×3631	3·55	2·56
Europa	0·47	0·009	3130	3·04	2·10
Ganymede	0·21	0·002	5268	1·93	2·78
Callisto	0·51	0·007	4806	1·81	2·43
Leda	26·1	0·148	10	3?	0·005?
Himalia	27·6	0·158	170	3?	0·1?
Lysithea	29·0	0·107	24	3?	0·01?
Elara	24·8	0·207	80	3?	0·05?
Ananke	147	0·17	20	3?	0·01?
Carme	164	0·21	30	3?	0·02?
Pasiphaë	145	0·38	36	3?	0·02?
Sinope	153	0·28	28	3?	0·01?

	Reciprocal mass Jupiter = 1	Albedo	Magnitude at mean opposition distance		Reciprocal mass Jupiter = 1	Albedo	Magnitude at mean opposition distance
Metis	?		17·4	Himalia		?	14·8
Adrastea	?		18·9	Lysithea		?	18·4
Amalthea		0·05	14·1	Elara		?	16·7
Thebe	?		15·5	Ananke		?	18·9
Io	21 300	0·63	5·0	Carme		?	18·0
Europa	39 000	0·64	5·3	Pasiphaë		?	17·7
Ganymede	12 700	0·43	4·6	Sinope		?	18·3
Callisto	17 800	0·17	5·6				
Leda	?		20·2				

IO

60 and 90 km for the rest. Oddly enough Pele, the most violent of the volcanoes, had become inactive by the time of the Voyager 2 pass a few months later, but there is no reason to assume that it has become extinct; Loki and Prometheus were more active than before, with the Loki plume rising to over 200 km. Using large Earth-based infra-red telescopes and also the Hubble Space Telescope, it is now possible to monitor some of the Ionian volcanoes with reasonable resolution, and, as expected, activity is continuous.

It has recently been found that much of Io's heat comes not from its lava-filled calderas, which erupt at well over 300°C, but from broad areas spanning thousands of square kilometres, averaging around 30°C. The lavas may be mainly molten silicate rock or molten sulphur.

Why is Io so active? According to one theory, the crust may be a 'sea' of sulphur and sulphur dioxide about 4 km deep, with only the uppermost kilometre solid. Heat escapes from the interior in the form of lava, erupting beneath the sulphur ocean, and the result is a violent outpouring of a mixture of sulphur, sulphur dioxide gas, and sulphur dioxide 'snow'. It has been proposed that the interior is constantly flexed by tidal effects produced by Jupiter; the orbit of Io is slightly eccentric (due mainly to the perturbations produced by the other Galileans) and so the tidal effects vary.

WEST

What may be termed Io's 'atmosphere' is so tenuous that it corresponds to what we normally call an efficient vacuum, but the volcanoes are sending out material all the time; the ejected material can escape from Io, but not from Jupiter, so that it produces a torus all round Io's orbit. The flux tube connecting Io with Jupiter amounts to 600 000 volts across Io's diameter, and currents of up to 5 000 000 ampères flow along

Ionian vulcanism: The plume rising from the limb (Loki) is striking. Taken from Voyager 1.

SOUTH

it. The flux tube was identified as a small emission feature in infra-red images of Jupiter's polar regions obtained with equipment on Mauna Kea; the feature marks the foot of the flux tube where charged particles from Io impact on Jupiter's upper atmosphere. Io is indeed a strange, colourful place; but since it moves within Jupiter's radiation zones, it must be just about the most lethal world in the entire Solar System.

FEATURES ON IO

Bold numbers indicate map references

Volcanoes	Lat	Long W
Amirani (Plume 5) **1**	27 N	119
Loki (Plume 2) **2**	19 N	305
Marduk (Plume 7) **3**	28 S	210
Masubi (Plume 8) **4**	45 S	053
Maui (Plume 6) **5**	19 N	122
Pele (Plume 1) **6**	19 S	257
Prometheus (Plume 3) **7**	03 S	153
Surt **8**	46 N	336
Volund (Plume 4) **9**	22 N	177
Pateræ		
Amaterasu	38 N	307
Asha	09 N	226
Atar	30 N	279

Impacts of comet Shoemaker-Levy 9 on Jupiter (Hubble Space Telescope). Eight sites are visible; from left to right: the E/F complex, the star-shaped H site, the impact sites for tiny N, Q1, small Q2 and R, and on the far right the D/G complex. The smallest features in this image are about 200 km across.
(NASA)

Volcanoes	Lat	Long W
Aten	48 S	311
Babbar	40 S	272
Bochica	61 S	022
Creidne	52 S	345
Culann	20 S	150
Dædelus	19 N	175
Dazhbog	54 N	302
Emakong	00	110
Fuchi	28 N	328
Galai	11 S	289
Gibil	15 S	295
Heno	57 S	312
Hephæstus	02 N	290
Hiruko	65 S	331
Horus	10 S	340
Huo Shen	15 S	329
Inti	68 S	349
Kane	48 S	015
Kibero	12 S	306
Loki	13 N	310
Maasaw	40 S	341
Mafuike	15 S	261
Malik	34 S	128
Manua	35 N	322
Masaya	22 S	350
Maui	20 N	125
Mbali	32 S	007
Mihr	16 S	306
Nina	40 S	165
Nusku	63 S	007
Nyambe	00	345
Ra	08 S	325
Reiden	24 S	236
Ruwa	00	002
Sengen	33 S	304
Shakuru	23 N	267
Shamash	36 S	152
Shoshu	19·5 S	324·5
Svarog	48 S	267
Talos	26 S	356
Taw	33·5 S	000·5
Tohil	28 S	157
Ülgen	41 S	288
Uta	35 S	027
Vahagn	27 S	359
Viracocha	62 S	284
Other Features		
Apis Tholus	11 S	349
Bactria Regio 10	45 S	125
Chalybes Regio 11	55 N	085
Colchis Regio 12	10 N	170
Dodona Planum 13	60 S	350
Hæmus Mons	70 S	050
Inachus Tholus	29 S	354
Lerna Regio 14	65 S	300
Mazda Catena	08 S	315
Media Regio 15	00	070
Mycenæ Regio 16	35 S	170
Nemea Planum	80 S	270
Silpium Mons	62 S	282
Tarsus Regio	30 S	055

EUROPA

*E*uropa was not closely studied by Voyager 1, which by-passed it at 732 230 km, so that our main information comes from Voyager 2, where the minimum distance was only just over 204 000 km. The whole scene is utterly different from that on Io. The overall hue of the disk is white, with no volcanoes, and virtually no craters; only three have been identified, none more than about 20 km in diameter. There is no doubt

EUROPA   **NORTH**

SOUTH

that the surface is made up of ice. The observed features are unlike any previously found anywhere, and new terms have had to be introduced: 'linea' (a dark or bright elongated marking, either straight or curved); 'flexus' (a very low curved ridge with a scalloped pattern) and 'macula' (a dark, sometimes irregularly-shaped patch).

Europa is a smooth world, with an almost complete absence of vertical relief. It has been said that the satellite is as smooth as a billiard-ball. It has also been likened to a cracked eggshell, and that the dark lines look as though they have been drawn on a white ball with a felt-tipped pen.

We know little about the interior, but there may well be an ice crust perhaps 100 km deep, overlying either an ocean of water or else a region of 'slushy ice' below which is a silicate core.

Europa is a map-maker's nightmare; one region looks very much like another. If craters ever existed, they have been obliterated, perhaps by water seeping out from below and then freezing. Yet unless the surface is

very young, some craters would be expected. We also have to ask why Europa is so completely unaffected by the forces which have moulded the surface of Io; after all, Europa is less than twice Io's distance from Jupiter. In its way, Europa is as puzzling a world as we have yet found in the Solar System.

FEATURES ON EUROPA

Bold numbers indicate map references

Maculæ	Lat	Long W
Thera Macula 16	45 S	178
Thrace Macula 17	44 S	169
Tyre Macula 18	34 N	144
Other Features		
Adonis Linea 1	38–60 S	112–122
Agenor Linea 2	44–35 S	175–221
Argiope Linea 3	20–1 S	171–204
Asterius Linea 4	29 S–29 N	251–314
Belus Linea 5	14–26 N	170–226
Cadmus Linea 6	29–26 N	170–251
Libya Linea 7	66–41 S	153–263
Minos Linea 8	45–31 N	119–150
Pelorus Linea 9	26 S–90 N	110–195
Phineus Linea 10	29–26 S	314–340
Sarpedon Linea 11	23–51 S	80–94
Thasus Linea 12	58–76 S	176–217
Cicilia Flexus 13	58–70 S	153–200
Gortyna Flexus 14	62–50 S	146–182
Sidon Flexus 15	70–64 S	150–215
Cilix Crater	1 N	182

GANYMEDE

NORTH

SOUTH

Ganymede, from 1 200 000 km: Voyager 2, 7 July 1979. The large dark area to the upper right is Galileo Regio. (NASA)

Ganymede is the largest satellite in the Solar System, and is larger than the planet Mercury, though less massive. Its overall density is less than twice that of water, and it seems to be composed of ice and rock in about equal proportions. Both Voyagers obtained good views of it, Voyager 1 from 112 030 km and Voyager 2 from only 59 530 km, so that 80 per cent of the surface has been mapped down to a resolution of 5 km or less.

The surface is certainly icy, and gives every impression of being ancient. There are dark areas, of which the largest, Galileo Regio, has a diameter of 4000 km, so that it is nearly as large as the continental United States. There are also brighter, presumably younger regions distinguished by 'sulci', i.e. grooves or furrows, with ridges which may rise to a kilometre or so. The two types of terrain intermingle, so that the overall picture is highly complex. There are many craters, some of which are ray-centres.

✳ *The largest crater is Gilgamesh*, diameter 150 km; it is surrounded by rough, blocky terrain. *The most prominent ray-crater* is Osiris.

Probably Ganymede has an icy crust less than 100 km deep, below which is a convecting mantle of water or slushy ice 400–800 km deep; there is presumably a silicate core. The surface could once have been active enough to produce what may be termed tectonic blocks – icy versions of the Earth's tectonic plates – but Ganymede today is completely inert.

An ingenious attempt to detect an atmosphere was made from Voyager 1. Ganymede occulted a star, and the event was carefully monitored to see if any trace of atmosphere could be found. The results were negative, and any atmosphere must have a density of less than one hundred thousand millionth of the Earth's air.

FEATURES ON GANYMEDE (MAP P.92)
Bold numbers indicate map references

Regiones	Lat	Long W
Barnard 10	22 N	010
Galileo 11	35 N	145
Marius 12	10 S	200
Nicholson 13	20 S	000
Perrine 14	40 N	030

Sulci		
Anshar 15	15 N	200
Apsu 16	40 N	230
Aquarius 17	50 N	010
Dardanus 18	20 S	013
Harpagia 19	00	317
Kishar 20	15 S	220
Mashu 21	22 N	200

GANYMEDE

WEST

Regiones	Lat	Long W
Mysia **22**	10 N	340
Nun **23**	50 N	320
Philus **24**	37 N	215
Phrygia **25**	20 N	005
Sicyon **26**	44 N	003
Tiamat **27**	03 S	210
Uruk **28**	00	157

Craters		
Achelous **1**	66 N	004
Adad	62 N	352
Adapa	83 N	022
Ammura	36 N	337
Anu	68 N	332
Asshur	56 N	325
Aya	67 N	303
Ba'al	29 N	326
Daniel	04 N	021
Diment	29 N	346
Enkidu	17 S	322
Enlil	52 N	301
Eshmun **2**	22 S	187
Etana	78 N	310
Gilgamesh **3**	58 S	124
Gula	68 N	001
Hathor	70 S	265
Isis **4**	64 S	197
Keret	22 N	034

Regiones	Lat	Long W
Khumbam	15 S	332
Kishar	78 N	330
Melkart **5**	13 S	182
Mor	35 N	323
Nabu	36 S	002
Namtar	49 S	343
Nigitsu	48 S	308
Nut **6**	61 S	268
Osiris **7**	39 S	161
Ruti	15 N	304
Sapas	59 N	031
Sebek **8**	65 N	348
Sin	56 N	349
Tanit	59 N	032
Teshub	02 N	016
Tros **9**	20 N	028
Zaqar	60 N	031

CALLISTO

*C*allisto is slightly smaller, slightly less massive and slightly less dense than Ganymede, and it also has a lower albedo or reflecting power, so that as seen from Earth it is actually the faintest of the four Galileans. Voyager 1 passed it at 123 950 km and Voyager 2 at 212 510 km, so that the surface has not been mapped as accurately as Ganymede's.

The crust has often been described as being made up of 'dirty ice', and may go down to a depth of 200 to 300 km; below this there may be a 1000 km deep mantle of water or soft ice overlying the silicate core. The surface temperatures are very low; about –118°C during the daytime, –193°C just before dawn (remembering that Callisto has a long 'day'; like all the Galileans its rotation is synchronous, so that its rotation period of 16·7 Earth-days is the same as the time which it takes to orbit Jupiter).

Callisto is the most cratered world known to us, though in general the craters are fairly shallow. The geology seems to be much more straightforward than that of Ganymede; apparently there has been vir-

EAST

Callisto, from Voyager 1. (NASA)

tually no activity since a very early stage in the history of the satellite, so that we may well be looking at the oldest planetary surface in the Solar System. There are two huge ringed basins, Valhalla and Asgard. Valhalla has a brighter circular region 600 km in diameter, and is surrounded by concentric rings, the largest of which has a diameter of 3000 km; Asgard is similar, though rather smaller. It is surely appropriate to name the ancient features of Callisto after the dwelling-places and the gods of ancient Norse mythology!

The lack of past tectonic activity may indicate that Callisto's slight inferiority in size and mass compared with Ganymede meant that the core temperature never rose sufficiently for major movements to begin.

Dead though Callisto is, it may well be the first world in the Jovian system to be visited by future astronauts, because it lies outside the main radiation zones of Jupiter itself.

FEATURES ON CALLISTO
(MAP P. 94–5)
Bold numbers indicate map references

	Lat	Long W
Adal	77 N	079
Adlinda **1**	58 S	020
Ägröi	42 N	012
Akycha	74 N	325
Alfr **2**	09 S	222
Ali	57 N	058
Anarr **3**	43 N	003
Aningan	51 N	011
Asgard **4**	30 N	140
Askr	53 N	327
Balkr	27 N	012
Bavorr	48 N	023
Beli	61 N	079
Bragi	77 N	069
Brami	26 N	018
Bran **5**	25 S	207
Buga	22 N	326
Buri	43 S	044
Burr **6**	40 N	136
Dag	56 N	074
Danr	61 N	075
Dia	73 N	056
Dryops	77 N	029
Durinn	66 N	087
Egdir	31 N	035

	Lat	Long W
Erlik	66 N	358
Fadir	56 N	015
Fili	65 N	349
Finnr	14 N	014
Freki	82 N	010
Frodi	69 N	136
Fulla	74 N	102
Geri	66 N	353
Gipul Catena	65 N	055
Gisl	56 N	035
Gloi **7**	48 N	246
Goll	58 N	323
Gondul	59 N	115
Grimr **8**	43 N	214
Gunnr	64 N	100
Gymir	61 N	055
Habrok	77 N	129
Haki **9**	26 N	315
Har	06 N	357
Hepti	64 N	027
Hodr **10**	69 N	087
Hoenir	36 N	261
Hogni	15 S	005
Igaluk **12**	05 N	315
Ivarr	06 S	322
Jumo	62 N	015
Kari	47 N	103
Lodurr **13**	52 S	270
Loni **14**	04 S	215

CALLISTO

NORTH

WEST

SOUTH

	Lat	Long W		Lat	Long W		Lat	Long W
Losy	68 N	329	Oski	56 N	266	Sudri 19	53 N	137
Mera	63 N	073	Ottar	60 N	100	Sumbur	69 N	332
Mimir	30 N	054	Pekko	17 N	006	Tindr	05 S	355
Mitsina	57 N	097	Reginn 17	42 N	088	Tornarsuk 20	25 N	130
Modi	67 N	115	Rigr 18	69 N	240	Tyn 21	68 N	229
Nama	57 N	336	Sarakka	08 S	053	Valfodr 22	03 S	246
Nar	04 S	045	Sequinek	55 N	027	Valhalla 23	10 N	055
Nerivik	22 S	055	Sholmo	52 N	018	Vali	09 N	327
Nidi	66 N	093	Sigyn	33 N	027	Vestri	42 N	054
Nori 15	46 N	347	Skoll	57 N	317	Vitr	23 S	347
Nuada 16	62 N	269	Skuld	06 N	037	Ymir	51 N	097

THE OUTER SATELLITES

Of the eight known satellites moving round Jupiter beyond the orbit of Callisto, only Himalia is of any size; its diameter is thought to be about 170 km, while none of the others attain 100 km. They fall into two main groups. Leda, Himalia, Lysithea and Elara move round Jupiter between 11 and 12 million km, while Ananke, Carme, Pasiphaë and Sinope move between 20 and 24 million km. The outer four have retrograde orbits, which strengthens the suggestion that these small satellites may be captured asteroids – particularly since they seem to have fairly low albedoes, in which case they are rocky rather than icy. Neither of the Voyagers went close enough to obtain surface pictures of any of the eight, and therefore we know little about them. One, Sinope, was discovered by accident. On 21 July 1914 S. B. Nicholson set out to photograph Pasiphaë, which had been discovered six years earlier; when he developed the plate, he found not only Pasiphaë but also the newcomer. No doubt there are more faint satellites awaiting discovery.

SATURN

DATA ....

Distance from the Sun:
  mean: 1427.0 million km (9·539 a.u.)
  max: 1507 million km (10·069 a.u.)
  min: 1347 million km (9·008 a.u.)
Sidereal period: 10 759·20 days = 29·46 years
Mean synodic period: 378·1 days
Rotation period (equatorial):
  10 h 13 m 59 s
  (Internal) 10 h 39 m 25 s
Mean orbital velocity: 9·6 km/s
Axial inclination: 26 deg 44 min
Orbital inclination: 2 deg 29 min 21 sec ·6
Orbital eccentricity: 0·056
Diameter:
  equatorial: 120 536 km
  polar: 108 728 km

Apparent diameter seen from Earth:
  max: 20 sec ·9
  min: 15 sec ·0
Reciprocal mass, Sun = 1: 3498·5
Density, water = 1: 0·71
Mass, Earth = 1: 95·17
Volume, Earth = 1: 744
Escape velocity: 32·26 km/s
Surface gravity, Earth = 1: 1·16
Mean surface temperature: −180°C
Oblateness: 0·1
Albedo: 0·61
Opposition magnitude: max −0·3, min +0·8
Mean diameter of Sun, as seen from Saturn:
  3 min 22 sec

*T*he most remote planet known in ancient times was Saturn, which is sixth in order of distance from the Sun.

OPPOSITION DATES, 1994–2005

Date		Mag.	Declination at opposition deg	Diam sec of arc
1994	Sept. 1	+0·7	−10	19·1
1995	Sept. 14	+0·8	−6	19·4
1996	Sept. 26	+0·7	−1	19·6
1997	Oct. 10	+0·4	+4	19·9
1998	Oct. 23	+0·2	+9	20·1
1999	Nov. 6	0·0	+13	20·3
2000	Nov. 19	−0·1	+17	20·5
2001	Dec. 3	−0·3	+20	20·6
2002	Dec. 17	−0·3	+22	20·7
2003	Dec. 31	−0·3	+23	20·7
2005	Jan. 13	−0·2	+24	20·6

The opposition magnitude is affected both by Saturn's varying distance and by the angle of presentation of the rings. The last edgewise presentations were in 1980 and in 1995. The intervals between successive edgewise presentations are 13 years 9 months and 15 years 9 months. During the shorter interval, the south pole is sunward, the southern ringface is seen, and Saturn passes through perihelion. Perihelion fell in 1944 and 1974: the next will be in 2003. The aphelion dates are 1959, 1988 and 2018.

On 15 September 2037, 21.32 UT, there will be a close conjunction of Mercury and Saturn; separation 18 seconds of arc.

RING DIAMETERS

Ring A: outer 272 300 km; inner 239 600 km.
Ring B: outer 234 200 km; inner 181 100 km.
Ring C: inner 149 300 km.

✻ **The first observations of Saturn** must date back before recorded history, since the planet is a bright naked-eye object (at its most brilliant it surpasses any of the stars apart from Sirius and Canopus).

✻ **The first recorded observations of Saturn** seem to have been those made in Mesopotamia in the mid-7th century BC. About 650 BC there is a record that Saturn 'entered the Moon', which is presumably a reference to an occultation of the planet.

✻ **The first observation of Saturn by Copernicus** was made on 26 April 1514, when Saturn lay in a line with the stars 'in the forehead of Scorpio'. Copernicus made three more recorded observations of Saturn: on 5 May 1514, 13 July 1520 and 10 October 1527.

✻ **The first observation of Saturn by Tycho Brahe** was made on 18 August 1563, when Saturn was in conjunction with Jupiter.

✻ **The first telescopic observation of Saturn** was made in July 1610 by Galileo, using a magnification of 32 on his largest telescope. He recorded that 'the planet Saturn is not one alone, but is composed of three, which almost touch one another and never move nor change with respect to one another. They are arranged in a line parallel to the Zodiac, and the middle one is about three times the size of the lateral ones.' Galileo's

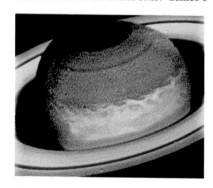

Saturn, from Voyager 1. (NASA)

telescope was inadequate for him to see the ring-system in its true guise, and subsequently he lost sight of the 'companions', as the rings were edgewise-on in December 1612; later he recovered them.

✻ **The earliest-dated drawing of Saturn to be published** was made by Pierre Gassendi on 19 June 1633, though it, together with other drawings made by Gassendi, was not actually published until 1658.

✻ **The first true explanation of the nature of the ring system** was given by Christiaan Huygens in 1656, from his observations made in 1655. His explanation was challenged, and was not finally accepted by all astronomers until 1665.

✻ **The first indications of the Crêpe Ring (Ring C)** may have been given by Campani in 1664, though they were not interpreted as being such. Campani also appears to have recorded a belt on the planet's disk. *The discovery of the Cassini Division*, separating Rings A and B, was made by G. D. Cassini in 1675 (claims that W. Ball had observed the Division ten years earlier have been discounted). Cassini also recorded the South Equatorial Belt on the planet's disk.

✻ **The first suggestion that the rings are not solid** was made by J. Cassini in 1705. Theoretical confirmation was provided by James Clerk Maxwell in 1875, who showed that no solid ring could exist; it would be disrupted by the gravitational pull of Saturn.

✻ **The discovery of the polar flattening of Saturn** was made by W. Herschel in 1789. Herschel gave the ratio of the equatorial to the polar diameter as 11:10, which is approximately correct.

✻ **The first well-defined spots to be detected on Saturn** were observed by J. H. Schröter and his assistant, K. Harding, at Schröter's observatory at Lilienthal, near Bremen, in 1796.

✻ **The discovery of Encke's Division in Ring A** was made by J. F. Encke at Berlin on 28 May 1837. *The discovery of the Crêpe Ring, Ring C*, was made by W. Bond at Harvard in November 1850, using the 15 in Merz refractor there. Independent confirmation was provided by Lassell and Dawes in England.

✻ **The transparency of Ring C was discovered in 1852**, independently by Lassell and by C. Jacob at Madras.

✻ **The first explanation of the divisions in the ring system** was given by D. Kirkwood in 1866. Kirkwood attributed the Divisions to the perturbing effects of Saturn's inner satellites.

✻ **The discovery of dark absorption bands in the spectrum of Saturn** was made by A. Secchi in 1863.

✳ *The first brilliant white spot to be recorded on Saturn* was discovered by A. Hall on 8 December 1876.

✳ *The first observation of the globe of Saturn seen through the Cassini Division* was made in 1883 by A. Hall and C. A. Young.

✳ *The first spectroscopic proof of the meteoritic nature of the rings* was given by J. E. Keeler in 1895. Keeler showed that the rings did not rotate as a solid mass would do; the inner sections had the fastest rotation. This was proved by their Doppler shifts.

✳ *The first report of a dusky ring exterior to Ring A* was made by G. Fournier, using an 11 in refractor, at Mont Revard (France) in 1907. (There are in fact three exterior rings, detectable only by space-probes.)

✳ *The most famous observation of the passage of a star behind the rings* was made by M. A. Ainslie and J. Knight, independently, on 9 February 1917. The dimming of the star as it passed behind the rings and divisions provided valuable information about the transparencies of the ring sections. The star concerned was B. D. +21°1714. Ainslie used a 9 in refractor at Blackheath; Knight a 5 in refractor at Rye in Sussex.

✳ *The first theoretical demonstration that Saturn has a cold surface, and is therefore not a 'miniature sun',* was given by H. Jeffreys in 1923.

✳ *The spectroscopic detection of methane and ammonia in Saturn's atmosphere* was made by T. Dunham at Mount Wilson, California, in 1932. Bands had been seen by V. M. Slipher in 1909, but had not previously been identified.

✳ *The first prominent white spot on Saturn to be seen during the present century* was discovered by W. T. Hay on 3 August 1933. Hay was using a 6 in refractor, and is the only famous comedian to have discovered a spot on Saturn – he is best remembered as Will Hay! The spot remained identifiable until 13 September 1933. W. H. Wright considered it to have been of an 'eruptive nature'.

An equally brilliant white spot, also in the equatorial region of Saturn, was discovered on 24 September 1990 by S. Wilber, in America. It rapidly spread out, and by 10 October was no longer a 'spot', but extra outbreaks were seen in it, clearly denoting uprush of material from below.

✳ *The first modern-type model of Saturn* was described by R. Wildt in 1938. Wildt believed that there was a rocky core, overlaid by a thick layer of ice and then by the gaseous atmosphere. The model has now been rejected, as has an alternative model proposed in 1951 by W. Ramsey, who con-

sidered that hydrogen was the main constituent but that near the core it acquired metallic characteristics.

✳ *The first serious observation of Saturn with the Hale reflector at Palomar* was made by G. P. Kuiper in 1954. Kuiper concluded that the Cassini Division was the only genuine gap, Encke's Division being merely a 'ripple' and the other reported minor divisions illusory.

✳ *The first observations of infra-red thermal emission from Saturn* were made in 1969, from equipment carried in an aircraft.

✳ *The first radar reflections from the rings* were observed in 1972. It was concluded that the ring particles were icy, with diameters of between 4 and 30 cm.

Much of our present knowledge of Saturn has been drawn from the space-probes, of which there have so far been three. The first was Pioneer 11. Its prime target had been Jupiter; but the success of its predecessor, Pioneer 10, and the fact that there was adequate power reserve meant that could be swung back across the Solar System to a rendezvous with Saturn in 1979. The results were preliminary only, but involved several important discoveries. Perhaps the most significant fact was that Pioneer was not destroyed by a ring particle collision; at that time nobody knew what conditions near Saturn would be like, and estimates of the probe's chances of survival ranged from 1 per cent to 99 per cent! In fact there was a near-encounter with the hitherto-unknown small satellite, but Pioneer emerged unscathed.

Details of the three probes are as follows:

its smaller mass and lower density Saturn bears a strong resemblance to Jupiter. The outer clouds contain more hydrogen but less helium than with Jupiter (the helium content for Saturn's clouds is 6 per cent, as against 11 per cent for Jupiter's). Below the clouds comes liquid hydrogen, at first molecular and then, below a depth of 30 000 km, metallic; the rocky core seems to be little larger than the Earth, though three times as massive. The core temperature is of the order of 15 000 degrees.

Like Jupiter, Saturn sends out more energy than it would do if it depended entirely upon what it receives from the Sun, but the cause is different. Saturn must now have lost all its heat produced during its formation stage, whereas the more massive Jupiter has not had time to do so. It is likely that the excess energy of Saturn is gravitational, produced by helium droplets as they fall through the lighter hydrogen toward the centre of the planet. This also explains why Saturn has less high-altitude helium than with Jupiter. The rotation period of the interior of the globe was found to be 10 h 39.4 m, longer than that of the upper clouds.

✳ *The magnetic field* is 1000 times stronger than that of the Earth, though 20 times weaker than Jupiter's. The magnetic axis is within one degree of the rotational axis, in which respect Saturn differs from both Jupiter and the Earth. The magnetic field is stronger at the north pole than at the south (0.69 gauss, against 0.53 gauss), and the centre of the field is displaced some 2400 km northward along the planet's axis. The magnetosphere is decidedly variable; its average boundary lies near the orbit of

Name	Launch date	Encounter date	Nearest approach, km	Remarks
Pioneer 11	5 Apr. 1973	1 Sept. 1979	20 880	Preliminary results.
Voyager 1	5 Sept. 1977	12 Nov. 1980	124 200	Complete success. Good images also of Titan, Dione, Rhea and Mimas.
Voyager 2	20 Aug. 1977	25 Aug. 1981	101 300	Success apart from one temporary failure after rendezvous. Good images of Iapetus, Hyperion, Tethys and Enceladus.

At this planning stage, Titan was regarded as almost as important a target as Saturn itself. Voyager 1 was programmed to encounter it. Had Voyager 1 failed, then Voyager 2 would have been reprogrammed to bypass Titan – but in this case it would have been unable to go on to rendezvous with Uranus and Neptune. There was therefore tremendous relief when Voyager 1 succeeded, leaving its follower free to carry out the full programme.

✳ *The latest models of Saturn,* supported by the Voyager results, indicate that despite

Titan, so that Titan is sometimes inside the magnetosphere and sometimes outside it. Radiation zones are present, though they cannot compare with those of Jupiter. The numbers of electrons fall off quickly at the outer edge of the main ring system (i.e. the outer edge of Ring A), because the electrons are absorbed by the ring particles; the region between Ring A and the globe is the most radiation-free area in the Solar System. Auroræ were recorded by the Voyagers, but are ten times weaker than those of Jupiter.

Saturn as seen from the Hubble Space Telescope on 9 Nov. 1990, showing a white 'zone' – previously a white spot.

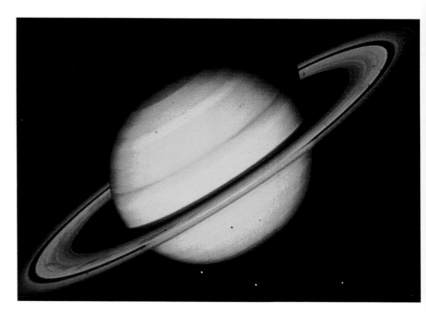

Saturn, like Jupiter, has a violently disturbed cloud surface. Features over 1000 km in diameter are not common, and even the largest ovals are no more than half the size of Jupiter's Great Red Spot. A red spot on Saturn was detected from Voyager 1 in August 1980, and was still present in September 1981; it too may owe its colour to phosphorus, but it is not likely to be so long-lived as the Great Red Spot on Jupiter. Smaller reddish and brownish spots were also found on Saturn. The lower temperature as compared with Jupiter means that ammonia crystals form at higher levels, covering the planet with 'haze' and giving it a somewhat bland appearance.

One surprise was that the wind patterns are different from those of Jupiter. The wind zones do not follow the light and dark bands, but are symmetrical with the equator; between latitudes 35N and 35S there is an eastward flow or jetstream with winds up to 1500 km/h, four times faster than any winds found on Jupiter. One prominent 'ribbon' at latitude 47N was taken to be a wave pattern in a particularly unstable jet-stream.

✳ *Seasonal effects occur.* The Sun crossed into Saturn's northern hemisphere in 1980, but there is a definite lag effect, and during the Voyager missions the northern hemisphere was still the colder of the two; the temperature difference between the north and south poles amounted to 10 degrees.

THE RINGS

*D*espite the fascinating information about the globe, it is fair to say that the main emphasis was upon the ring-system. As the results came into Mission Control at Pasadena, California, there was a feeling not only of excitement, but also incredulity. The rings were so completely different from anything which had been envisaged. Instead of being more or less homogenous, they proved to be immensely complex, made up of thousands of small ringlets and narrow gaps; it was even said that the rings had more grooves than any gramophone record. Also, the system turned out to be much more extensive than expected. The so-called Ring D is not a true ring, and is unobservable from Earth, but there were three rings outside the main system (one of which, Ring F, had been found by Pioneer 11). Even the Cassini Division contained ringlets, and, most remarkable of all, some of the rings were eccentric instead of perfectly circular.

Details, with the satellites involved in the system, are as follows:

Feature	Distance from centre of Saturn km	Period hours
Cloud-tops	60 330	10·66
Inner edge of Ring D	67 000	4·91
Huygens Gap	71 000	5·2
Inner edge of Ring C	73 200	5·61
Maxwell Gap	87 500	6
Inner edge of Ring B	92 200	7·93
Outer edge of Ring B	117 500	11·41
Middle of Cassini Division	119 000	11·75
Inner edge of Ring A	121 000	11·92
Pan	133 600	13·7
Centre of Encke Division	135 700	13·82
Keeler Gap	136 530	14
Outer edge of Ring A	136 800	14·14
Atlas	137 670	14·61
Prometheus	139 350	14·71
Centre of Ring F	140 600	14·94
Pandora	141 700	15·07
Epimetheus	151 420	16·65
Janus	151 470	16·68
Centre of Ring G	170 000	19·90
Inner edge of Ring E	180 000	21
Mimas	185 540	22·60
Brightest part of Ring E	230 000	31·3
Enceladus	238 040	32·88
Tethys	294 670	45·3
Dione	377 420	65·7
Outer edge of Ring E	480 000	80
Rhea	527 040	108

✳ *The system is very varied.* The D 'ring' has no sharp inner edge, and the particles may go down close to the cloud-tops. Ring C, the Crêpe Ring, is 19 000 km broad, and has various gaps, including two with widths of 200 and 300 km respectively; the outer gap contains a narrow, dense and slightly eccentric bright ring, 90 km wide when at its closest. The C-ring particles seem, on average, to be about 2 m in diameter.

The main ring is, of course, Ring B, where the particles range in size from 10 cm to about a metre. The particles are redder than those of the C Ring and D region, and are made up mainly of water ice, as had already been suspected from Earth-based radar work. The ring temperatures ranged between –180°C in sunlight down to –200° in shadow. It was also found that there is a rarefied cloud of neutral hydrogen extending to 60 000 km above and below the ring-plane. Strange radial spokes were seen in the B ring – incidentally, shown many years earlier on drawings made by E. E. Barnard and E. M. Antoniadi – and were decidedly puzzling, because logically no such features should form; the difference in rotation period between the inner and outer edges of the ring is well over three hours – and yet the spokes persisted for hours after emerging from the shadow of the globe; when they were distorted and broken up, new spokes emerged from the shadow to replace them. Presumably they are due to particles of a certain definite size elevated away from the ring-plane by magnetic or electrostatic forces. They are confined wholly to Ring B. Another point finally cleared up concerned the thickness of the ring. The maximum cannot be more than 150 m, and is more probably as little as 100 m.

The 4000 km Cassini Division is not empty, but contains many rings a few hundred kilometres wide; the particles are less red than those of Ring B, and more closely resemble those of Ring C. At least one of the rings in the Cassini Division is markedly eccentric.

Ring A is made up of particles ranging from fine 'dust' to larger blocks around 10 m in diameter. The Encke Division in Ring A is 270 km wide, and contains ringlets which are both discontinuous and 'kinky'. (The

Encke Division has been referred to as the Keeler Gap, but it has now very properly been decided that the old, familiar name should be retained; three other narrow gaps in the ring system have been named Huygens, Maxwell and Keeler.) Atlas moves close to the outer edge of Ring A, and is responsible for the sharp border to the ring.

Outside the main system we come to Ring F. It is faint, but of special interest because it seems to be made up of several rings which are intertwined; the effect was startling in the Voyager 1 pictures, but less so in Voyager 2. Ring F is stabilized by two small satellites, Prometheus and Pandora. The satellite slightly closer-in to Saturn will be moving faster than the ring particles, and will speed a particle up if it strays away from the ring region; the outer satellite, moving more slowly, will drag any errant ring particles back and return them to the central region. For obvious reasons, these two tiny bodies are known as the 'shepherd' satellites. It was thought for a while that the numerous divisions in the main rings could be similarly explained, and Voyager 2 was re-programmed to search for any satellites moving in Rings A and B, but none was found.

Beyond Ring F and its shepherds come the two co-orbital satellites, Epimetheus and Janus. Next there is the very tenuous Ring G. Still further out comes the familiar satellite Mimas. The final ring, E, is nearly 300 000 km wide, but very tenuous indeed; its brightest part is slightly closer-in than the orbit of Enceladus, the second of the well-known satellites, and it has even been suggested that material ejected from Enceladus may be concerned in the formation of the ring. Enceladus has an icy surface, and there can no longer be any doubt that the ring-particles are made up simply of water ice. Ring E extends out even beyond the orbit of Tethys and Dione.

An ingenious experiment was carried out from Voyager 2. As seen from the space-craft, the third-magnitude star δ Scorpii was occulted by the rings. It had been expected that there would be regular 'winks', as the starlight would shine through any gaps but would be blocked by ring material. In fact thousands of 'blips' were recorded, but it seems that there is very little empty space anywhere in the main ring system.

What can be the cause of this amazing complexity? The Cassini Division had been explained as due to Mimas; a particle moving in the Division will have a revolution period exactly half that of Mimas, and cumulative perturbations will drive it away from the 'forbidden zone', leaving the Division

swept clear. Yet the Voyagers have shown that this explanation is clearly inadequate, and it now seems more likely that we are dealing with a density wave effect. A satellite such as Mimas can alter the orbit of a ring particle and make it elliptical. This causes 'bunching' in various areas; a spiral density wave will be created, and particles in it will collide, moving inward toward the planet and leaving a gap just outside the resonance orbit. This theory, due to Peter Goldreich and Scott Tremaine, does at least sound plausible, but it would be wrong to claim that we really understand the mechanics of the ring system.

The data for the minor satellites are extremely uncertain, and different authorities give different values. As the Sun's apparent diameter as seen from Saturn is only 3 min 22 sec, total eclipses may be produced on Saturn by Mimas, Enceladus, Tethys, Dione, Rhea and Titan.

At the opposition of 1966, when the rings were edgewise-on to the Earth, A. Dollfus reported the detection of a new satellite, closer-in than Mimas, the quondam nearest satellite. It was also glimpsed by several others (including myself) and was even given a name: Janus. However, it is now clear that there is no satellite moving in the orbit worked out for Dollfus' satellite, and the name Janus has been given to the satellite known formerly as 1980 S–1. There seems little doubt that the object seen in 1966 was either the present Janus or else its co-orbiter, Epimetheus.

Phœbe has retrograde motion, and may possibly be a captured asteroid, particularly as preliminary results show that its surface is darkish and reddish – quite unlike the familiar icy satellites.

Wavelike patterns in the rings, not unlike the water-wave wake of a speedboat, led M. W. Showalter (USA) to suspect that a small satellite might be orbiting inside Encke's Division. In 1991, following the close inspection of the Voyager images, he was able to confirm the existence of the satellite, now named Pan. It may be responsible for the presence of Encke's Division. It is very small, and nothing is known about its physical nature, though presumably it is icy.

THE SATELLITES OBSERVED BY THE VOYAGERS

*A*part from Phœbe, all the satellites were reasonably well observed from the Voyageur. Minimum distances were as follows:

Satellite	Min. distance, km	
	Voyager 1	Voyager 2
Atlas	219 000	287 170
Prometheus	300 000	246 590
Pandora	270 000	107 000
Epimetheus	121 000	147 010
Janus	297 000	222 760
Mimas	88 440	309 990
Enceladus	202 040	87 140
Tethys	415 760	93 000
Calypso	237 332	284 396
Telesto	432 295	153 518
Dione	161 520	502 250
Helene	230 000	318 200
Rhea	73 980	645 280
Titan	6 490	665 960
Hyperion	880 440	470 840
Iapetus	2 470 000	909 070
Phœbe	13 537 000	1 473 000

SATELLITES OF SATURN

Satellite	Discoverer	Mean distance from Saturn km	Mean angular distance from Saturn, at mean opposition distance	
			min	sec
Pan	M. Showalter, 1990, from Voyager images	133 600		
Atlas	From Voyager	137 670		
Prometheus	photographs	139 350		
Pandora	1980: S. Collins, D. Karlson	141 700		
Janus	Fountain and	151 470		
Epimetheus	Larson 1978	151 420		
Mimas	W. Herschel 1789	185 540	0	30·0
Enceladus	W. Herschel 1789	238 040	0	38·4
Tethys	G. D. Cassini 1684	294 670	0	47·6
Telesto	Group led by	294 670	0	47·6
Calypso	B. Smith 1980	294 670	0	47·6
Dione	G. D. Cassini 1684	377 420	1	01·1
Helene	P. Lacques and J. Lecacheaux 1980	377 420	1	01·1
Rhea	G. D. Cassini 1672	527 040	1	25·1
Titan	C. Huygens 1655	1 221 860	3	17·3
Hyperion	W. Bond 1848	1 481 100	3	59·4
Iapetus	G. D. Cassini 1671·	3 561 300	9	35
Phœbe	W. H. Pickering 1898	12 954 000	34	51

	Sidereal period days	Mean Synodic period d	h	m	s	Apparent diameter seen from Saturn min	sec
Pan	0·57						
Atlas	0·602						
Prometheus	0·613						
Pandora	0·629						
Janus	0·695						
Epimetheus	0·694						
Mimas	0·942	0	22	37	12.4	10	54
Enceladus	1·370	1	8	53	22	10	36
Tethys	1·888	1	21	18	55	17	36
Telesto	1·888						
Calypso	1·888						
Dione	2·737	2	17	42		12	24
Helene	2·737						
Rhea	4·518	4	12	18		10	42
Titan	15·945	15	23	16		17	10
Hyperion	21·277	21	7	39		0	43
Iapetus	79·331	79	22	05		1	48
Phoebe	550·4	523	13			0	3·2

	Orbital inclination degrees	Orbital eccentricity	Diameter km	Density water=1	Escape velocity km/s
Pan	Low	Low	20?		
Atlas	0·3	0·002	37 × 34 × 27		
Prometheus	0·0	0·004	148 × 100 × 68		
Pandora	0·1	0·004	110 × 88 × 62		
Janus	0·1	0·007	194 × 190 × 154	0·6	
Epimetheus	0·3	0·009	138 × 110 × 110	0·6	
Mimas	1·52	0·020	421 × 395 × 385	1·4	0·1
Enceladus	0·02	0·004	512 × 495 × 488	1·2	0·2
Tethys	1·86	0·000	1046	1·1	0·4
Telesto	2	0·0	30 × 25 × 15		
Calypso	2	0·0	30 × 16 × 16		
Dione	0·02	0·002	1120	1·4	0·9
Helene	0·2	0·005	35		
Rhea	0·35	0·001	1528	1·3	0·6
Titan	0·33	0·029	5150	1·9	2·47
Hyperion	0·43	0·104	360 × 280 × 225	3?	0·2
Iapetus	7·52	0·028	1436	1·2	0·67
Phoebe	175	0·163	230 × 220 × 210	3?	0·1

	Reciprocal mass Saturn=1	Albedo	Magnitude at mean opposition distance
Atlas		0·4?	18·1
Prometheus		0·6?	16·5
Pandora		0·6?	16·3
Janus		0·5?	14·5
Epimetheus		0·5?	15·5
Mimas	15 000 000	0·7	12·9
Enceladus	7 000 000	1·0	11·8
Tethys	910 000	0·8	10·3
Telesto		0·8?	19·0
Calypso		0·6?	18·5
Dione	490 000	0·6	10·4
Helene		0·5?	18·5
Rhea	250 000	0·6	9·7
Titan	4150	0·2	8·4
Hyperion	5 000 000	0·2	14·2
Iapetus	300 000	0·5 to 0·05	10–12
Phoebe		0·1	16·5

Atlas is the 'A-ring shepherd'; Prometheus and Pandora are the 'F-ring shepherds'. Additional satellites have been reported from time to time. In April 1861 H. Goldschmidt announced that he had found a satellite moving between the orbits of Hyperion and Iapetus, and named it Chiron; in 1904 W.H. Pickering claimed that he had detected a satellite between the orbits of Titan and Hyperion, and named it Themis. Neither of these satellites was confirmed, and neither seems to exist, but there is much better evidence of the existence of several new small satellites reported in 1994 by British astronomers at Queen Mary and Westfield College. Confirmation is awaited.

Little is known of the physical nature of the first three satellites, except that they are irregular in shape. However, Epimetheus and Janus are of interest because they move in almost identical orbits, and there seems little doubt that they represent two parts of a former larger satellite which broke up. Their behaviour reminds one of a game of cosmic musical chairs. At present (1995) Epimetheus is slightly closer to Saturn, and is moving faster than Janus, so that it will eventually catch its companion up. As they approach each other there will be mutual interactions, so that Epimetheus will be slowed down and Janus speeded up. The end result is that the satellites will exchange orbits – something which happens every four years or so. Obviously they cannot collide, as otherwise they would not continue to exist as separate bodies, but they may come within a few kilometres of each other.

A few craters have been found and named. There are two named craters on Epimetheus (Hilairea and Pollux) and four on Janus (Castor, Idas, Lynceus and Phoibe).

MIMAS

Mimas, first of the satellites known before the Space Age, is only slightly denser than water, and may be made up mainly of ice all

MIMAS

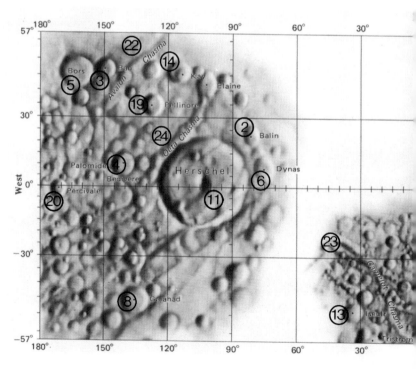

the way through to its centre, though there must be some rock as well. The surface is dominated by a huge crater, named Herschel, which has a diameter of 130 km – one-third that of the satellite itself – with walls which rise 5 km above the floor, the lowest part of which is 10 km deep. Herschel has a massive central mountain, with a base measuring 30 × 20 km; it rises to 6 km above the floor. There are many other craters (such as Modred and Bors) but few as much as 50 km across. Grooves (chasma) are also much in evidence, and some are more or less parallel, indicating that the crust has been subjected to considerable strain. If Herschel were formed by a gigantic impact there would have been a serious risk of breaking up the entire satellite. It was originally to be named 'Arthur', and the other main features of Mimas have been named after characters in the Arthurian legend.

FEATURES ON MIMAS
Bold numbers indicate map references

Craters	Lat	Long W
Accolon	68 S	166
Arthur **1**	35 S	190
Balin **2**	22 N	082
Ban **3**	47 N	149
Bedivere **4**	10 N	145
Bors **5**	45 N	165
Dynas **6**	08 N	075
Elaine	44 N	102
Gaheris **7**	46 S	187
Galahad **8**	47 S	135
Gareth **9**	44 S	280
Gawain	60 S	254
Gwynevere **10**	12 S	312
W. Herschel **11**	00	104
Igraine **12**	40 S	225

Craters	Lat	Long W
Iseult **13**	48 S	035
Kay **14**	46 N	116
Lamerok	65 S	283
Launcelot **15**	10 S	317
Lot	30 S	227
Mark	28 S	297
Merlin **16**	38 S	215
Modred **17**	05 N	213
Morgan **18**	25 N	240
Palomides	04 N	157
Pellinore **19**	35 N	128
Percivale **20**	01 S	171
Tristram	58 S	026
Uther **21**	35 S	244
Chasma		
Avalon **22**	20–57 N	160–120
Camelot **23**	15–60 S	000–045
Oeta **24**	10–35 N	130–105
Ossa **25**	10–30 S	305–280
Pangea **26**	25–55 S	340–290
Pelion **27**	20–25 S	235–200
Tintagel **28**	43–60 S	235–190

A very small satellite has been suspected moving in virtually the same orbit as Mimas, but nothing definite is known about it.

ENCELADUS

*E*nceladus, less than twice the size of Mimas, is completely different. Instead of one huge crater and many others of fair size, we find several completely different types of terrain. Craters exist in many areas, but give the impression of being young and fairly sharp, while there is an extensive plain which is almost crater-free and is instead dominated by long grooves. Like Mimas, Enceladus is not much denser than

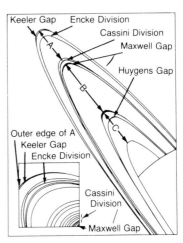

Nomenclature of Saturn's rings.

water, and its albedo is virtually 100 per cent, making it the most reflective body in the Solar System.

It is suggested that despite its small size, Enceladus may be an active world. We have to explain the paucity of craters over wide areas, and it may be that the interior is 'flexed' by the tidal forces of Saturn and of the outer, much more massive satellite Dione, whose revolution period is twice that of Enceladus. If so, then there may be periods when soft ice or even water wells up over the surface, obliterating features there. The surface formations have been given names from the Arabian Nights legend.

Another interesting fact is that Enceladus is exceptionally cold even by Saturnian standards. Equipment on Voyager 2 gave an average daytime temperature of –200°C.

FEATURES ON ENCELADUS
Bold numbers indicate map references

Craters	Lat	Long W
Ahmad	58 N	304
Aladdin	63 N	017
Ali Baba **1**	55 N	011
Dalilah **2**	53 N	244
Duban	58 N	176
Dunyazad **3**	34 N	200
Gharib	81 N	245
Julnar **4**	54 N	340
Musa	73 N	003
Peri-Banu	63 N	315
Salih	06 S	000
Samad	61 N	353
Shahrazad **5**	49 N	200
Shahryar **6**	58 N	222
Sindbad	66 N	210
Fossæ		
Bassorah **7**	40–50 N	023–345
Daryabar **8**	05–10 N	020–335
Isbanir **9**	10 S–20 N	000–350
Planitia		
Diyar **10**	00	250
Sarandib **11**	05 N	300
Sulci		
Harran **12**	5 S–35 N	270–210
Samarkand **13**	10 S–75 N	300–340

TETHYS

Most of our detailed knowledge of Tethys comes from Voyager 2 (Voyager 1 did not make a close approach), and it is unfortunate that some eagerly-awaited images of the satellite were lost when Voyager 2 developed a fault after it had started on its journey from Saturn to Uranus. Tethys seems to be made up of almost pure ice. There is one huge crater, Odysseus, with a diameter of 400 km – larger than Mimas! It is not very deep, and is presumably old, but it is certainly puzzling, since it may well be that an impact violent enough to produce such a formation would have shattered Tethys completely. On the other hand there can be nothing in the way of conventional vulcanism on a globe like Tethys. There are many other craters, such as Penelope and Telemachus (the names given are from the Trojan War).

However, the main feature is Ithaca Chasma, a tremendous trench 2000 km in length, running from near the north pole across the equator and along to the south polar region, Nothing similar is known in the Solar System. It was presumably caused when the water inside Tethys froze, expanding as it did so. It extends three-quarters of the way round Tethys; its average width is 100 km, and it is 4 to 5 km deep, with a rim rising to half a kilometre above the outer terrain.

The orbit of Tethys is shared by two small satellites, Calypso and Telesto. They are irregular in shape, with longest diameters of less than 40 km. We have as yet no positive information about their surface features.

FEATURES ON TETHYS
Bold numbers indicate map references

Craters	Lat	Long W
Ajax **1**	30 S	285
Anticleia **2**	55 N	038
Antinoüs	62 S	275
Arete	04 S	300
Circe **3**	08 S	049
Elpenor **4**	54 N	268
Eumæus	27 N	047
Eurycleia **5**	56 N	247
Lærtes **6**	50 S	060
Melanthius	62 S	204
Mentor **7**	03 N	039
Nausicaa	80 N	352
Nestor **8**	57 S	058
Odysseus **9**	30 N	130
Penelope **10**	10 S	252
Phemius **11**	12 N	290
Polyphemus **12**	05 S	285
Teiresias	62 N	005
Telemachus **13**	50 N	338

Chasma		
Ithaca **14**	60 S–35 N	030–340

ENCELADUS

TETHYS

DIONE

Dione is only a little larger than Tethys, but it is much more massive, and may be the densest of all Saturn's satellites apart from Titan. There are suspicions that it may have an effect upon Saturn's radio emission, since there seems to be a cycle of 2·7 days – which is also the revolution period of Dione. Moreover, as has been noted, Dione may play a rôle in the flexing of the interior of Enceladus, whose revolution period is almost exactly half that of Dione.

The surface is not uniform. The trailing hemisphere is relatively dark, with an albedo are, in the main, associated with Roman history). Valleys exist, though some of them, notably Latium Chasma, appear to be essentially craterlet-chains.

Dione has one small co-orbital satellite, Helene, and a second has been suspected.

FEATURES ON DIONE (Map p. 104–5)
Bold numbers indicate map references

Craters

Craters	Lat	Long W
Adrastus	64 S	040
Æneas **1**	26 N	047
Amata **2**	07 N	287
Anchises **3**	35 S	063
Antenor **4**	06 S	008
Butes	68 N	050
Caieta **5**	25 S	080
Cassandra **6**	42 S	245
Catillus **7**	01 S	275
Coras **8**	03 N	268
Creusa	48 N	078
Dido **9**	22 S	015
Halys	60 S	045
Ilia **10**	03 N	344
Italus **11**	20 S	076
Latagus	16 N	026
Lausus **12**	38 N	023
Magus	20 N	024
Massicus **13**	36 S	052
Palinurus	05 S	062
Remus **14**	10 S	030
Ripheus	56 S	029
Romulus **16**	08 S	024
Sabinus **17**	44 S	190
Turnus **18**	21 N	342

Chasma

Chasma	Lat	Long W
Larissa **19**	20–48 N	015–065
Latium **20**	03–45 N	064–075
Palatine	55–73 S	075–320
Tibur **21**	48–80 N	060–080

Linea

Linea	Lat	Long W
Carthage **22**	20–10 N	337–310
Padua **22**	05 N–40 S	245–190
Palatine **24**	10–55 S	285–320

RHEA

With Rhea, considerably larger than Tethys or Dione, we come to yet another heavily-cratered world, with features named after characters in Far Eastern legends. The most prominent crater is Izanagi, but there are few large craters anywhere on the surface; moreover, the craters tend to be irregular. As with Dione, the trailing hemisphere is darkish, with wispy features reminiscent of those on Dione although much less prominent; the leading hemisphere is brighter, with one suspected ray-centre. It has been found that there are two separate types of cratered area, and it has been suggested that the first cratering period – whether internal or impact-produced – gave rise to some large craters; there was then a period of resurfacing, presumably by material welling up to the surface, followed by a second cratering era in which the structures were smaller. Rhea seems to be made up of ice

of 0·3, while the brightest features on the leading hemisphere have albedo 0·6. The most prominent feature is Amata, 240 km in diameter, which may be either a crater or a basin; its precise nature is uncertain, but it is associated with a system of bright wispy features which extend over the trailing hemisphere, and are accompanied by narrow

linear troughs and ridges. It is believed that these wispy features have been produced by bright, new ice which has seeped out from the interior, so that Dione shows considerably more past activity than Tethys. There are some large, well-marked craters with central peaks, of which Æneas is a good example (the names of the features and rock in about equal amounts. It has certainly been inactive for a very long time indeed, and parts of the surface look not unlike the heavily-cratered surface of Callisto in Jupiter's system.

An interesting observation was made on 8 April 1921 by six English observers independently: A. E. Levin, P. W. Hepburn, L. J. Comrie, E. A. L. Attkins, F. Burnerd and C. J. Spencer. Rhea was eclipsed by the shadow of Titan, and according to the first three observers Rhea vanished completely for over half an hour.

FEATURES ON RHEA
Bold numbers indicate map references

Craters

Craters	Lat	Long W
Aananin	39 N	330
Adjua	46 N	126
Aguana	70 N	065
Ameta	59 N	014
Arunaka	14 S	021
Atum	45 S	000
Bulagat 1	35 S	014
Bumba	70 N	040
Burkhan	69 N	288
Con	24 S	010
Djuli 2	26 S	046
Ellyay	78 N	095
Faro 3	52 N	121
Haik 4	34 S	026
Heller 5	09 N	310
Haoso	09 N	008
Iraca	45 N	120
Izanagi 6	49 S	298
Izanami 7	46 S	310
Jumo	56 N	065
Karora	07 N	016
Khado	45 N	349
Kiho 8	10 S	354
Kumpara	11 N	321
Leza 9	19 S	304
Lowa	45 N	009
Malunga	74 N	049
Manoid	33 N	002
Melo 10	51 S	006
Mubai	61 N	011
Num	23 N	093
Ormazd	62 N	052
Pan Ku	72 N	115
Pedn	48 N	340
Qat 11	23 S	347
Sholmo	13 N	340
Taaroa	14 N	099
Thunapa 12	51 N	015
Tika	25 N	087
Tore	00	335
Torom	68 S	345
Uku	85 N	115
Whanin	74 N	121
Wuraka	28 N	357
Xamba 13	04 N	347
Xu	61 N	070
Yu-ti 14	55 N	085

Chasma

Chasma	Lat	Long W
Kun Lun 15	37–50 N	275–300
Pu Chou 16	10–35 N	085–115

TITAN

*T*he first suggestion of an atmosphere round Titan was made in 1903 by the Spanish astronomer J. Comas Solá, who noted

DIONE

RHEA

(LEADING HEMISPHERE)

that the satellite appeared brighter at its centre than near the limb. The existence of an atmosphere was proved spectroscopically in 1944 by G. P. Kuiper, but it was tacitly assumed that the main constituent would be methane, and it was thought that the atmospheric density must be low; after all, the escape velocity of Titan is a mere 2·5 km/s.

As Voyager 1 approached Saturn's system, astronomers were still divided as to whether any surface details would be seen on Titan. In the event, they were not. Titan is permanently shrouded by its orange clouds. The northern hemisphere was the darker of the two, and there were vague indications of 'banding', but that was all.

tion period of the solid globe is presumably synchronous (15·945 days) or virtually so; however, this is little more than speculation as yet.

Titan's globe is probably made up of about 55 per cent rock and the rest ice. According to this picture there is a rocky core surrounded by a mantle of liquid water with some dissolved ammonia and methane, above which comes the crust. The surface temperature is −168°C. This is near the triple point of methane, which freezes at −182°C and boils at −155°C. On Titan, methane could therefore exist as a solid, liquid or gas. It is even possible that methane-nitrogen droplets actually rain down to higher portions of the surface.

It has been suggested that Titan was completely covered with an ocean of ethane (C_2H_6); earlier., C. Sagan had proposed a methane ocean at least 350 metres deep. However, recent radar studies indicate that there cannot be an ethane ocean covering Titan completely; this would be a poor radar reflector, and results show that the reflectivity is very variable, so that there must be solid masses. D. Muhleman has suggested that the solid areas are made up of water ice together with solid carbon dioxide, silicates and tars. Images taken in infra-red with the Hubble Space Telescope show a certain amount of detail, and in 1994 recorded a large region, about the size of Australia, 10 to 15 per cent brighter than its surroundings, but whether it can be regarded as a 'continent' is a very open question.

Certainly, Titan is unlike any other body in the Solar System. We now await the controlled landing of the Huygens probe, part of the Cassini mission, due to be launched in 1997.

HYPERION

*H*yperion moves between the orbits of Titan and Iapetus. (In 1904 W. H. Pickering reported another satellite in this region, and it was even given a name – Themis – but it has never been confirmed, and in all probability Pickering mistook a faint star for a satellite.) Hyperion is irregular in shape, with a longest diameter of 360 km and a shortest diameter of 225 km. This is strange, because an object as large as this ought to be regular in form; and to make matters still stranger, the longest axis does not point directly at Saturn, as dynamically it ought to do. It has been suggested that Hyperion is part of a larger satellite which broke up – but if so, then where is the rest of it? The rotation period is 'chaotic', and it has been

However, important measurements were made. Titan's atmosphere proves to be mainly nitrogen, with a ground pressure 1·3 times that of the Earth's air at sea-level. There is a little helium and hydrogen. Voyager found that the atmosphere contains a wide variety of hydrocarbons, including methane and ethane; there seem to be patchy methane clouds from below 10 to at least 30 km above the surface, and methane rain is likely. On 2 July 1990 Titan occulted the star 28 Sagittarii, and a central 'flash' due to the effects of Titan's atmosphere was observed, leading to the suggestion that the atmosphere may be oblate (i.e. flattened at the poles) by 1·7 per cent. B. Sicardy has also suggested that the upper clouds may have a rotation period of 26 hours, though the rota-

IAPETUS

said that Hyperion is 'tumbling through space'. The 1983 rotation period was about 13 days.

Hyperion is less reflective than the inner icy satellites, but it too is probably composed largely of ice; there may be a 'dirty' layer covering wide areas. Several craters are seen, with diameters of up to 120 km, and there is one long ridge or scarp, extending for 300 km, which has been named Bond-Lassell in honour of the discoverers of Hyperion in 1848. Bond actually found the satellite first, but Lassell's confirmation shortly afterwards was independent.

FEATURES ON HYPERION

Craters	Lat N	Long W
Bahloo	036	196
Helios	071	132
Jarilo	061	183
Meri	003	151
Dorsum		
Bond-Lassell	012–084	171

IAPETUS

*I*apetus was the second of Saturn's satellites to be discovered, by G. D. Cassini in 1671. Cassini soon found that the magnitude is very variable. When west of Saturn in the sky, Iapetus is bright enough to be seen with a small telescope; when east, it becomes much fainter. The magnitude range is from about 9 to below 12. Assuming that the rotation is synchronous, this would mean that one side must be bright and the other side dark.

The Voyager results confirmed this view, though neither probe made a close fly-by, and the surface of Iapetus is less well known than those of the other icy satellites. The leading hemisphere is as black as a blackboard, with an albedo of 0·04 to 0·05; the

trailing hemisphere is bright, with an albedo of 0·5. The line of demarcation is not abrupt; there is a 200–300 km 'transition zone'.

Some craters in the bright region have dark floors, but unfortunately we do not know whether the dark floor material has the same composition as the dark hemisphere coating.

In view of the low density of Iapetus, it seems certain that the satellite itself is bright and icy, so that the dark material represents a deposit of some kind of material whose nature is unknown. There have been suggestions that the material has been wafted on to Iapetus from the outermost satellite, Phœbe, but this seems highly unlikely, partly because Phœbe is so small and distant and partly because some of the craters on the bright trailing hemisphere of Iapetus are dark-floored. Presumably, then, the dark material has welled out from the interior of Iapetus. We know nothing about its depth. Carl Sagan has suggested that it may be thick, and made up of organics; others believe it to be no more than a millimetre or two deep. Obviously we know little about the dark areas, but the bright regions contain the usual craters, named from the Charlemagne period.

It is worth noting that future space-travellers will see Saturn well from Iapetus, because the rings will not be edgewise-on. The orbital inclination is nearly 15°, and Saturn will indeed be a glorious object in the Iapetan sky. All the inner satellites move practically in the plane of the rings.

FEATURES ON IAPETUS
Bold numbers indicate map references

Craters	Lat	Long W
Almeric	53 N	274
Baligant **1**	15 N	225
Basan **2**	30 N	197
Berenger	59 N	220

Craters	Lat	Long W
Besgun	72 N	296
Charlemagne **3**	54 N	266
Geboin **4**	56 N	175
Godefroy	78 N	253
Grandoyne **5**	18 N	215
Hamon **6**	10 N	271
Lorant	64 N	165
Marsilion **7**	41 N	177
Milon	75 N	270
Ogier **8**	42 N	274
Oliver	61 N	203
Othon **9**	24 N	344
Roland	78 N	030
Turpin **10**	43 N	000
Regio		
Cassini **11**	48 S–55 N	210–340
Terra		
Roncevaux **12**	30 S–90 N	300–130

PHŒBE

*T*he outermost satellite, Phœbe, differs from all the rest. It moves in a retrograde orbit, and its rotation period is not synchronous; Phœbe takes over 550 days to complete one journey round Saturn, but only 9·4 hours to spin round. Apparently it is a darkish body, and may be of the same type as the carbonaceous asteroids. Indeed, there is every chance that Phœbe itself is a captured asteroid rather than a bona-fide satellite. Investigations carried out by D. Cruikshank at the high-altitude Mauna Kea Observatory have shown that the dark material of Phœbe differs from that on the leading hemisphere of Iapetus, but its nature is unknown, and since neither Voyager made a close fly-by of Phœbe our knowledge of its surface is still very incomplete. There are suggestions that Phœbe may be of the same type as the exceptional Chiron. As we have noted, Chiron approached Saturn in the year 1664 BC to a distance of 16 million km – only about three million km beyond the orbit of Phœbe.

URANUS

DATA ....

Distance from the Sun:
mean: 2869.6 million km (19·181 a.u.)
max: 3004 million km (20·088 a.u.)
min: 2735 million km (18·275 a.u.)
Sidereal period: 84·01 years = 30 684·9 days
Synodic period: 369·7 days
Rotation period: 17·24 hours
Mean orbital velocity: 6·90 km/s
Axial inclination: 98 deg
Orbital inclination: 0 deg ·773
Orbital eccentricity: 0·047
Diameter:
equatorial: 51 118 km
polar: 49 946 km

Apparent diameter seen from Earth:
max: 3 sec ·7
min: 3 sec ·1
Reciprocal mass, Sun = 1: 22 800
Density, water = 1: 1·27
Mass, Earth = 1: 14·6
Volume, Earth = 1: 67
Escape velocity: 22·5 km/s
Surface gravity, Earth = 1: 1·17
Mean surface temperature: −214°C
Oblateness: 0·24
Albedo: 0·35
Maximum magnitude: +5·6
Mean diameter of Sun, as seen from Uranus:
1 min 41 sec

*T*he first planet to be discovered in modern *(telescopic) times* was Uranus, which comes seventh in order of distance from the Sun.

OPPOSITION DATES, 1994–2005

	Declination at opposition
1994 July 17	−22
1995 July 21	−21
1996 July 25	−20
1997 July 29	−19
1998 Aug. 3	−18
1999 Aug. 7	−17
2000 Aug. 11	−16
2001 Aug. 15	−15
2002 Aug. 20	−13
2003 Aug. 24	−12
2004 Aug. 27	−11
2005 Sept. 1	−9

During this period Uranus passes from Sagittarius into Capricornus.

In June 1989 Uranus reached its greatest southerly declination (−23 deg ·7). (Greatest northerly declination had been reached in March 1950.)

Close planetary conjunctions involving Uranus for the 1900–2100 period:

	Date	UT	Sep sec	Elong
Mars–Uranus	1947 Aug. 6	01.49	+43	48 W
Jupiter–Uranus	1955 May 10	20.39	−56	65 E
Mars–Uranus	1988 Feb. 22	20.48	+40	63 W
Venus–Uranus	2077 Jan. 20	20.01	−43	11 W

Between January and March 1610 Uranus was within 3 degrees of Jupiter. This was when Galileo was making his first telescopic observations of Jupiter, but Uranus was beyond the limits of his telescopic field.

✳ *The most recent aphelion passage* was that of 1 April 1925. Uranus was at its maximum distance from the Earth (21·09 a.u.) on 13 March of that year. Aphelion will next be reached on 27 February 2009.

✳ *The most recent perihelion passage* was that of 20 May 1966. Uranus was at its closest to the Earth (17·29 a.u.) on 9 March of that year. Perihelion will next be reached on 13 August 2050.

The unique axial inclination of Uranus (98°) means that the rotation is technically retrograde, though not generally reckoned as such; all the satellites move virtually in the equatorial plane of Uranus, and in the same sense as the rotation of the planet. From Earth, the equator of Uranus is regularly presented (as in 1923 and 1966); at other times a pole is presented (the south pole in 1901 and 1985, the north pole in 1946 and 2030).

✳ *The first record of Uranus* seems to have been an observation by the Astronomer Royal, the Rev. John Flamsteed, on 23 December 1690, when the planet was in Taurus; Flamsteed took it for a star (34 Tauri). Altogether, 22 pre-discovery observations have been listed, as follows:
Flamsteed 1690, 1712, four times in 1715.
J. Bradley, 1748 and 1750.
P. Le Monnier, twice in 1750, 1764, twice in 1768, six times in 1769, 1771.
T. Mayer, 1756.

It is interesting that Le Monnier failed to identify Uranus because of its movement. He observed it eight times in four weeks (27 December 1768–23 January 1769) without realizing that it was anything but a star. He has often been ridiculed for this, but it seems that when he made his observations Uranus was near its stationary point – it is hardly surprising that Le Monnier failed to identify it.

Dust in the rings of Uranus; Voyager 2, Jan. 1986. When taking the picture, Voyager was in the shadow of Uranus, at a distance of 236 000 km; resolution about 33 km. Lanes of fine dust particles are shown. Exposure 96 seconds, so that there is a noticeable 'smear' as well as streaks due to trailed stars. (NASA)

✳ *The discovery of Uranus* was made on 13 March 1781 by William Herschel, using a 6·2 in reflector of 7 ft focal length and a magnification of 227. Herschel recognized that the object (in Gemini) was not a star, and he believed it to be a comet. His communication to the Royal Society was indeed entitled 'An Account of a Comet'.

✳ *The first recognition of the new body as a planet* seems to have been due, independently but at about the same time (May 1781) by the French amateur astronomer J. de Saron – who was guillotined in 1794 during the Revolution – and by the Finnish mathematician Anders Lexell. Lexell calculated an orbit, finding that the distance of the planet from the Sun was 19 a.u. – only slightly too small. He gave a period of between 82 and 83 years, and stated that the apparent diameter was between 3 sec and 5 sec, making the planet larger than any other apart from Jupiter and Saturn.

✳ *The first proposal to name the planet 'Uranus'* was made in 1781 by the German astronomer J. E. Bode. Other names were proposed – for instance 'Hypercronius' (J. Bernoulli, 1781) and 'the Georgian Planet' (by Herschel himself in 1782, in honour of his patron, King George III). Others called it simply 'Herschel'. Until 1850 the *Nautical Almanac* continued to call it the Georgian Planet, but in that year the famous mathematician John Couch Adams suggested changing over to 'Uranus'. This was done, and the name became universally accepted.

✳ *The first attempt to measure the apparent diameter of Uranus* was made by Herschel in 1781. His value (4 sec ·18) was rather too great. In 1788 he gave the diameter as 34 217 miles (55 067 km), with a mass 17·7 times that of the Earth; these values were also slightly too high. In 1792–4 Herschel made an attempt to measure the polar flattening, and from his results concluded that Uranus must have a rapid rotation.

✳ *The first observation of an occultation of Uranus by the Moon* was made on 6 August 1824 by Captain (later Rear-Admiral) Sir John Ross, using a power of 500 on a reflector of 25 ft (7·26 m) focal length. On 4 October 1832 Thomas Henderson observed an occultation of Uranus by the Moon from the Cape of Good Hope.

✳ *The first estimate of the axial rotation period* was made by J. Houzeau, in France in 1856. Houzeau stated that the rotation period must be between 7¼ hours and 12½ hours.

✳ *The first reports of markings on the disk of Uranus* were made by J. Buffham on 25 January 1870; the telescope used was a 9 in refractor, magnification 320. Two bright round spots were recorded, and on 19 March a light streak was described. It seems rather improbable that these markings were genuine features, though one cannot be sure.

✳ *The first observations of dark lines in the spectrum of Uranus* were made by A. Secchi in 1869, from Italy. Further lines were detected by the English amateur W. Huggins in 1871, and in 1889 Huggins obtained the first photographs of these lines.

✳ *The first really good tables of the motion of Uranus* were compiled in America by Simon Newcomb, in 1875.

✳ *The first spectroscopic confirmation of the retrograde rotation* was obtained by the French astronomer H. Deslandres, in 1902. Final confirmation was obtained by P. Lowell and V. M. Slipher, at the Flagstaff Observatory in Arizona, in 1911. For many years the rotation period was thought to be 10 h 49 m, but is now known to be 17·24 hours.

✳ *The first identification of methane in the atmosphere of Uranus* was made by R. Mecke, at Heidelberg, in 1933 (it had been suggested, on theoretical grounds, by R. Wildt in 1932). Confirmation was obtained by V. M. Slipher and A. Adel, at Flagstaff, in 1934.

✳ *The first widely-accepted model of Uranus* was that of R. Wildt, who proposed in 1934 that the planet must have a rocky core, overlaid by a thick layer of ice which was in turn overlaid by the atmosphere. In 1951 W. R. Ramsey, of Manchester University, proposed an alternative model, according to which Uranus was made up largely of methane, ammonia and water.

✳ *The first photographs of Uranus from a balloon-borne telescope* were obtained in 1970, from a height of 24 km; the balloon was *Stratoscope II*. The resolution was 0 sec ·15, ten times better than can be obtained by a telescope on the ground.

✳ *The first (and so far, the only) spaceprobe to by-pass Uranus* is Voyager 2, which made its closest approach (80 000 km) on 24 January 1986. Much information was obtained, including the discovery of ten new satellites.

It now seems that Uranus contains more heavy elements than Jupiter or Saturn, and is intermediate in type between the hydrogen- and helium-rich Jupiter and Saturn on the one hand, and the rocky, metal- and oxygen-rich inner planets on the other. According to the so-called 'three-layer' model, there is a rocky core surrounded by a liquid ocean which is in turn overlaid by the atmosphere; on the 'two-layer' model, which from the Voyager results is the more probable, there is a dense atmosphere in which gases are mixed with 'ices', i.e. substances which would be frozen at the low temperatures at the surface. These ices are mainly water, ammonia and methane. It seems that the bulk of the atmosphere is made up of water, and that water, ammonia and methane condense in that order to form thick, icy cloud-layers. Methane freezes at the lowest temperature, and so forms the top layer, above which comes the predominantly hydrogen atmosphere with 15 per cent helium. Methane absorbs red light, which is why Uranus appears bluish-green.

Features at different latitudes have rotation periods from 14 to 17 hours, so that we are observing winds blowing in an east–west direction; cloud bands were also seen concentric with the south pole, which was at this time turned toward the Sun and the Earth. The equator was in twilight, and the north pole in darkness, but the temperature at the poles and the equator was found to be about the same.

(Incidentally, there has been some confusion about Uranus' 'poles'. The International Astronomical Union has decreed that all poles above the ecliptic (i.e. the plane in which the Earth orbits) are north poles, and all poles below the ecliptic are south poles. In this case it was Uranus' south pole which was in sunlight during the Voyager pass. However, the Voyager team members reversed this, and referred to the sunlit pole as the *north* pole. Take your pick!)

Images of Uranus taken in 1994 with the Hubble Space Telescope did at least show two high-latitude clouds in the southern hemisphere, between latitudes 20 and 35 degrees south, together with a polar cap of atmospheric haze.

Uranus seems to have little internal heat compared with the other giant planets. Jupiter radiates 1·7 times as much energy as it would do if it depended entirely upon what it receives from the Sun, and the figures for Saturn and Neptune are much the same; for Uranus the upper limit is only 1·06, but with a possible error of 1 – so that there may be no excess energy at all. Moreover, the temperatures of Uranus and Neptune as measured from Earth are almost equal, though Neptune is so much further from the Sun.

The magnetic field of Uranus proved to be surprising. It is appreciable – the equatorial field strength, in gauss, is 0·23, as against 0·69 for Saturn and 4·2 for Jupiter (the value for the Earth is 0·305). However, the magnetic axis is displaced by 58·6 degrees to the rotational poles, and the magnetic axis does not pass through the centre of the globe; it is displaced by 30 per cent of Uranus' radius. The polarity is opposite to that of the Earth.

The magnetosphere extends to 590 000 km on the day side and 6 000 000 km on the night side.

The reason for the tilted magnetic axis is unknown. It was originally thought that Uranus might be experiencing a 'magnetic reversal', but subsequently it was found that Neptune had similar displacement – to assume that two reversals are occurring simultaneously would be too much of a coincidence.

Ultra-violet observations showed strong emissions on the day side, producing what is termed the electroglow. It may be caused by electrons exciting the hydrogen molecules in the upper atmosphere, though it is not clear how these electrons get their energy.

On Uranus, sunlight would be relatively strong, ranging between 1068 and 1334 times that of full moonlight on Earth. Saturn would be fairly bright when well placed (every 45½ years); Neptune would just be visible with the naked eye when near opposition; Jupiter would have an apparent magnitude of 1·7, but would remain inconveniently close to the Sun in the Uranian sky.

RING-SYSTEM OF URANUS

Ring	Distance from Uranus km	Eccentricity	Inclination deg	Width km
6	41 837	1·01	0·06	1–3
5	42 234	1·90	0·05	2–3
4	42 570	1·06	0·03	2
α	44 718	0·76	0·01	4–11
β	45 661	0·44	0·005	7–11
η	47 175	0·004	0·001	2
γ	47 630	0·0	0·11	1–4
δ	48 320	0·0	0·04	3–9
λ	50 023	0·0	0·0	1–2
ε	51 149	7·93	0·002–0·008	20–96

* *The rings of Uranus* were discovered on 10 March 1977. It had been predicted that the planet would occult the star SAO 158687, magnitude 8·9, and the occultation would give a good opportunity for measuring the diameter of Uranus. Calculations made by Gordon Taylor at the Royal Greenwich Observatory indicated that the occultation would be visible only from a restricted area in the southern hemisphere, and observations were made by J. Elliott, T. Dunham and D. Mink, flying at 12·5 km above the southern Indian Ocean in the Kuiper Airborne Observatory (KAO), which is in fact a modified C–141 aircraft carrying a 36 in reflecting telescope. Close watches were also being kept at ground-based observatories, including those in South Africa. Thirty-five minutes before occultation the KAO observers saw the star 'wink' five times, so that it was apparently being temporarily obscured by material in the vicinity of Uranus. The occultation by Uranus began at

20·52 h UT, and lasted for 25 minutes. After emersion there were more winks, and it was later found that these were symmetrical with the first set, indicating a system of rings. The post-emersion winks were also recorded from Cape Town by J. Churms.

Subsequent occultations confirmed the existence of the ring system. The first Earth-based picture of them was obtained in 1978 by G. Neugebauer and his colleagues at Palomar, using the Hale reflector, and in 1984 D. A. Allen, at Siding Spring in Australia, obtained clear pictures of them in infra-red.

The rings were surveyed in detail by Voyager 2; details are given in the accompanying

THE SATELLITES OF URANUS

	Discoverer	Mean distance from Uranus km	Angular distance from Uranus at mean opposition distance sec
Cordelia	⎫	49 471	
Ophelia		53 796	
Bianca		59 173	
Cressida	From	61 777	
Desdemona	⎬ Voyager 2 1986	62 676	
Juliet		64 352	
Portia		66 085	
Rosalind		69 941	
Belinda	⎭	75 258	
Puck	Voyager 2 1985	86 000	
Miranda	Kuiper 1948	129 400	9·9
Ariel	Lassell 1851	191 000	14·5
Umbriel	Lassell 1851	266 300	20·3
Titania	W. Herschel 1787	435 000	33·2
Oberon	W. Herschel 1787	583 500	44·5

	Sidereal period, days	Mean synodic period d h m s	Apparent diameter as seen from Uranus min sec
Cordelia	0·330		
Ophelia	0·372		
Bianca	0·433		
Cressida	0·463		
Desdemona	0·475		
Juliet	0·493		
Portia	0·513		
Rosalind	0·558		
Belinda	0·622		
Puck	0·762		
Miranda	1·414	1 9 55 31	17 54
Ariel	2·520	2 12 29 39·0	30 54
Umbriel	4·144	4 3 28 25·8	14 12
Titania	8·706	8 17 00 1·2	15 00
Oberon	13·463	13 11 15 36·5	9 48

	Orbital inclination degrees	Orbital eccentricity	Diameter km	Density water = 1	Escape velocity km/s
Cordelia			26		
Ophelia			30		
Bianca			42		
Cressida			62		
Desdemona			54		
Juliet			84		
Portia	Very low	Very low	108	Low?	Very low
Rosalind			54		
Belinda			66		
Puck			154		
Miranda	0·0	0·017	472	1·3	0·5?
Ariel	0·0	0·003	1158	1·6	1·2?
Umbriel	0·0	0·004	1169	1·4	1·2?
Titania	0·0	0·002	1578	1·6	1·6?
Oberon	0·0	0·001	1523	1·5	1·5?

	Reciprocal mass Uranus = 1	Albedo	Magnitude at mean opposition distance
Miranda	1 000 000	0·27	16·5
Ariel	67 000	0·40	14·4
Umbriel	67 000	0·19	15·3
Titania	20 000	0·28	14·0
Oberon	30 000	0·24	14·2

Ariel, from Voyager 2; 24 Jan. 1986; range 170 000 km. The southern hemisphere is shown; resolution 3 km. The craters, fault scarps, and graben (fault-bounded valleys) are well shown. (NASA)

table. The outermost ring, the ε-ring, is not symmetrical, and is narrowest when closest to Uranus; the newly-discovered satellites Cordelia and Ophelia are 'shepherds' to it. There is little obvious dust in the main rings, but Voyager 2 took a final picture, when on its outward journey from Uranus when the planet hid the Sun, showing 200 very diffuse, nearly transparent bands of microscopic dust surrounding the main rings. The rings seem to be made up of particles a few metres in diameter, with not many centimetre- and millimetre-sized objects; they are very dark – as dark as coal. It has been suggested, though without proof, that they may be relatively young, and perhaps not even permanent features of the Uranian system. Their maximum thickness is 0·1–1 km.

The rings are not alike: 6, 5 and 4 show significant internal structure. Rings α and β lack sharp edges; a faint ring-like feature was suspected just outside Ring β, but has not been confirmed. Ring η does not show sensible eccentricity or inclination to Uranus' equatorial plane, and is made up of two components – a sharp inner feature, and a much fainter one which extends about 55 km beyond the sharp feature. Both edges of the γ ring are very sharp, while with the δ ring there is a faint component inside the main ring. Ring λ was discovered between the δ and ε rings; it is very thin, and apparently circular. Ring ε is of variable width, with sharp edges. In addition to the ten rings, there is a broad sheet of material closer-in than Ring 6, extending from 37 000 to

39 500 km from Uranus. In 1994 the Hubble Space Telescope sent back excellent images of Uranus and its ring system, together with the small inner satellites which had not been seen since the Voyager 2 pass.

As the diameter of the Sun as seen from Uranus is below 2 min, all the satellites known before the Voyager 2 pass could produce total eclipses. To an observer on Uranus, Ariel would appear much the largest of the satellites. As the satellites move round Uranus in the same sense as the planet rotates, their movements are technically retrograde, though not generally regarded as such.

✳ *The first satellites to be discovered were Oberon and Titania*, by William Herschel in 1787. Both were seen on 11 January, though Herschel delayed making any announcement until he was certain of their nature. In 1797 Herschel announced the discovery of four more satellites, but three of these are certainly non-existent, and although the fourth may have been Umbriel there is considerable uncertainty. Ariel and Umbriel were discovered on 24 October 1851 by the English amateur W. Lassell, with his 24 in reflector (previous observations by Lassell in 1847 had been inconclusive). In 1894, 1897 and 1899 W. H. Pickering had searched unsuccessfully for new satellites.

✳ *The last satellite to be discovered visually* was Miranda, on 16 February 1948, by G. P. Kuiper, with the 82 in reflector at the McDonald Observatory in Texas. The ten

URANUS from Voyager 2. The disk appears remarkably featureless!

inner satellites were discovered during the Voyager 2 pass in 1985–6.

✳ *The names of the first four satellites* were suggested by Sir John Herschel – two Shakespearean; one (Ariel) from both Shakespeare and Pope's *Rape of the Lock*, and Umbriel from *Rape of the Lock*. The name 'Miranda' was suggested by Kuiper, and the Shakespearean names of the ten inner satellites were adopted by the International Astronomical Union in 1987.

✳ *The nine innermost satellites* are very small (between 16 and 108 km in diameter). Cordelia and Ophelia act as 'shepherds' to the ring. Nothing definite is known about the physical characteristics of these satellites.

MIRANDA

PUCK

*P*uck was the first of the inner satellites to be discovered from Voyager 2 (on 30 December 1985). On 24 January 1986 a single image was obtained of it, from a range of 500 000 km. The resolution is of the order of 10 km. Puck is roughly spherical, with a low albedo (7 per cent). Three craters were recorded: Bogle, Lob and Butz.

MIRANDA

*A*s with the other satellites, only one half of Miranda was available for study from Voyager 2; the northern hemisphere was in darkness. Miranda was imaged from close range, and has proved to have an incredibly varied landscape. There are several distinct types of terrain: old, cratered plains, brighter areas with cliffs and scarps, and 'ovoids', large trapezoidal-shaped regions 200 to 300 km across (Arden Corona, Inverness Corona and Elsinore Corona). The cliffs, presumably of ice, may be as much as 20 km high. There are few really large craters.

There have been suggestions that Miranda was broken up early in its history, and that the broken pieces subsequently re-formed. This may or may not be true, but it would go some way to explaining Miranda's remarkably varied surface.

FEATURES ON MIRANDA

	Lat S	Long E
1 Alonso	47	345
2 Arden Corona	10–60	30–120
3 Argier Rupes	40–50	310–340
4 Dunsinane Regio	20–75	345–65
5 Elsinore Corona	10–42	215–305
6 Ferdinand	36	208
7 Francisco	70	246
8 Gonzalo	13	75
9 Inverness Corona	38–90	0–360
10 Mantua Regio	10–90	75–300
11 Prospero	35	323
12 Silicia Regio	10–50	295–340
13 Stephano	36	239
14 Trinculo	67	168
15 Verona Rupes	10–40	340–350

ARIEL

*A*riel was imaged from 130 000 km, giving a resolution down to 2·4 km. There are plenty of craters, in terrain transected by fault scarps and graben, but the dominant features are broad, branching, smooth-floored valleys such as Korrican Chasma and Kewpie Chasma, and although Ariel is not totally inert it has clearly experienced tremendous tectonic activity earlier in its history. Its surface certainly appears to be younger than those of Umbriel, Titania or Oberon.

FEATURES ON ARIEL

		Lat S	Long E			Lat S	Long E
1	Abans	16	251	15	Korrigan Chasma	25–46	328–353
2	Agape	47	336	16	Kra Chasma	32–36	355–2
3	Ataksak	53	225	17	Laica	22	44
4	Befanak	17	32	18	Leprachaun Vallis	5–15	350–25
5	Berylune	23	328	19	Mab	39	353
6	Brownie Chasma	5–21	325–357	20	Melusine	53	9
7	Deive	23	23	21	Onagh	22	244
8	Djadek	12	251	22	Pixie Chasma	18–25	350–20
9	Domovoy	72	339	23	Rima	18	260
10	Finvara	16	19	24	Sprite Vallis	12–17	332–355
11	Gwyn	78	23	25	Sylph Chasma	45–50	328–15
12	Huon	39	33	26	Yangoor	68	260
13	Kachina Chasma	24–40	210–280				
14	Kewpie Chasma	15–42	307–335				

UMBRIEL

UMBRIEL

Umbriel has a much darker, more sub-dued surface than Ariel, and there are no ray-craters. The most prominent crater, Skynd, is 110 km in diameter, with a bright central peak. Wunda, diameter 140 km, lies near Umbriel's equator at around longitude 270 deg; its nature is not known, but it is the most reflective feature shown, and may be a crater.

FEATURES ON UMBRIEL

		Lat S	Long E
1	Alberich	31	43
2	Fin	36	41
3	Gob	9	26
4	Kanaloa	11	351
5	Malingee	22	13
6	Peri	9	6
7	Setibos	31	350
8	Skynd	1 (N)	335
9	Vuver	2	311
10	Wokolo	33	7
11	Wunda	6	274
12	Zlyden	24	330

TITANIA

TITANIA

Like Ariel, though to a lesser extent, Titania has clearly been the site of past tectonic activity. There are many craters, together with ice-cliffs; there are prominent fault valleys, such as Messina Chasmata, up to 1500 km in length. The 200 km crater Ursula is cut by a younger fault valley over 100 km wide. The largest crater recorded, Gertrude, is well over 200 km in diameter. In size and mass Titania and Oberon are virtual twins, but Titania shows evidence of considerably more activity in the past.

FEATURES ON TITANIA

		Lat S	Long E
1	Adriana	20	4
2	Belmont Chasma	4–25	25–35
3	Bona	55	351
4	Calpurnia	42	292
5	Elinor	44	334
6	Gertrude	15	288
7	Imogen	25	318
8	Iras	19	339
9	Jessica	55	286
10	Katherine	51	333
11	Lucetta	9	277
12	Marina	15	316
13	Messina Chasma	8–28	325–5
14	Mopsa	11	302
15	Phrynia	24	309
16	Rousillon Rupes	7–25	17–38
17	Ursula	13	44
18	Valeria	34	4

OBERON

OBERON

Oberon shows less obvious signs of past tectonic activity than Titania, and differs also inasmuch as some of the major craters such as Hamlet, Othello and Falstaff have dark material inside them; this may be a mixture of ice and carbonaceous material erupted from the interior. Oberon was imaged at a minimum distance of 660 000 km, with a resolution down to 12 km. One interesting feature is the high mountain, 6 km in altitude, seen on the limb near Macbeth.

FEATURES ON OBERON

		Lat S	Long E
1	Anthony	28	65
2	Cæsar	27	61
3	Coriolanus	11	345
4	Falstaff	22	19
5	Hamlet	46	45
6	Lear	5	31
7	Macbeth	59	112
8	Mommur Chasma	16–20	240–343
9	Othello	65	44
10	Romeo	28	88

NEPTUNE

DATA

Distance from the Sun:
mean: 4496·7 million km (30·058 a.u.)
max: 4537 million km (30·316 a.u.)
min: 4456 million km (29·800 a.u.)
Sidereal period: 164·8 years = 60 190·3 days
Synodic period: 367·5 days
Rotation period: 16·1 hours
Mean orbital velocity: 5·43 km/s
Axial inclination: 28 deg 48 min
Orbital inclination: 1 deg 34 min 19 sec ·8
Orbital eccentricity: 0·009
Diameter:
equatorial: 50 538 km
polar: 49 600 km

Apparent diameter from Earth:
max: 2 sec ·2
min: 2 sec ·0
Reciprocal mass, Sun = 1: 19 300
Density, water = 1: 2·0643
Mass, Earth = 1: 17·2
Volume, Earth = 1: 57
Escape velocity: 23·9 km/s
Surface gravity, Earth = 1: 1·2
Mean surface temperature: −220°C
Oblateness: 0·02
Albedo: 0·35
Maximum magnitude: +7·7
Mean diameter of Sun, as seen from Neptune: 1 min 04 sec

*T*he first planet to be discovered by mathematical calculation was Neptune, which comes eighth in order of distance from the Sun.

OPPOSITION DATES, 1991–2005

1991 July 8
1992 July 9
1993 July 12
1994 July 14
1995 July 17
1996 July 18
1997 July 21
1998 July 23
1999 July 26
2000 July 27
2001 July 30
2002 Aug. 2
2003 Aug. 4
2004 Aug. 6
2005 Aug. 8

Close planetary conjunctions involving Neptune for the 1900–2100 period:

	Date	UT	Sep sec	Elong
Mercury–Neptune	1914 Aug. 10	08.11	−30	18 W
Venus–Neptune	2022 Apr. 27	19.21	−26	43 W
Venus–Neptune	2023 Feb. 15	12.35	−42	28 E
Mercury–Neptune	2039 May 5	09.42	−39	13 W
Mercury–Neptune	2050 June 4	06.46	−46	17 W
Mercury–Neptune	2067 July 15	12.04	+13	18 W

* The first telescope observation of Neptune seems to have been made by Galileo on 27 December 1612. While drawing Jupiter and its four satellites, he recorded a 'star' which was almost certainly Neptune. He again saw it twice in January 1613, and noted movement, but, not surprisingly, took it for a star. The telescope had a magnification of ×18 and a resolving power of 10 seconds of arc, with a field of view 17 minutes of arc in diameter. Neptune's magnitude was 7·7, and Galileo often plotted stars fainter than that. Jupiter actually occulted Neptune in 1613.

The next telescopic observation was made in May 1795 by J. J. de Lalande, but again

Neptune was mistaken for a star. Further observations were made in 1845–6 by John Lamont (often referred to as Johann von Lamont; he was Director of the Munich Observatory).

* The first suggestion of an unknown planet beyond Uranus was made by the Rev. T. J. Hussey, Rector of Hayes, in a letter written to the Astronomer Royal (Airy) on 17 November 1834. Uranus had been observed to depart from its predicted position, and in 1832 Airy had himself pointed out that the best available tables of Uranus, compiled by Alexis Bouvard, were in error by half a minute of arc. Various theories were later advanced to explain this (an unknown satellite of Uranus, braking due to 'cosmic fluid', and even a cometary collision!) but Hussey's suggestion was supported in 1835 by J. E. B. Valz, and subsequently by others.

* The first plan to search for the trans-Uranian planet was announced by F. W. Bessel in 1840. Bessel hoped to collaborate with his pupil F. W. Flemming, but Flemming died soon after; Bessel himself became ill, and was never able to follow the problem up (he died in 1846).

* The first estimated position of the new planet was calculated by John Couch Adams, of Cambridge, in 1845. Adams had made up his mind to tackle the problem while still graduating, in 1841. On 21 October 1845 Adams sought an interview with Airy, who had given him little encouragement, but was unsuccessful. No search was instigated. Meanwhile, U. J. J. Le Verrier, in France, had begun working on the problem (1845), and reached similar conclusions. Le Verrier's memoir reached Airy in December 1845, and finally, on 9 July 1846, Airy requested Challis, professor of astronomy at Cambridge, to begin searching for a new planet on the basis of Adams' calculations.

Challis used the 11·75 in 'Northumberland' refractor, but had no proper maps of the area, and his searches were laborious – and, it must be admitted, carried out with a strange lack of energy.

* The first identification of Neptune was made by J. Galle and H. D'Arrest, at Berlin, on 23 September 1846, on the basis of Le Verrier's calculations. It was found that Le Verrier's predicted position was in error by only 55 minutes of arc. The telescope used by Galle and D'Arrest was a 9 in refractor, which was powerful enough to show the planet as a tiny disk. Subsequently, Challis found that he had seen the planet twice – the second occasion being on 12 August – but had failed to identify it. Following the announcement from Berlin the planet was widely observed; in England J. R. Hind, using a 7 in refractor, saw it on 30 September.

* The first suggested name for the new planet was 'Janus', by Galle. Le Verrier suggested 'Neptune', but then changed his mind and proposed to have the planet named after himself; Challis proposed 'Oceanus', but before long the name 'Neptune' became universally accepted.

* The first announcement of Adams' independent work (almost as accurate as Le Verrier's) was made by Sir John Herschel on 3 October 1846. The announcement caused deep resentment in France, and led to acrimonious disputes in which, however, neither Adams nor Le Verrier took much part; they remained on cordial terms for the rest of their lives. Nowadays they are recognized as co-discoverers of Neptune.

* A ring of Neptune was reported by the English amateur W. Lassell on 14 October 1846, but later observations showed that this was an optical effect. Neptune does indeed have rings, but they were not discovered until 1989, by Voyager 2, and have no connection with the spurious Lassell ring. Lassell did, however, discover Triton, the largest satellite, soon after the identification of Neptune itself.

Neptune is a twin of Uranus – but it is a non-identical twin. It is very slightly smaller, but appreciably denser and more massive. Unlike Uranus, it has considerable internal heat. It sends out 2·6 times more energy than it would do if it depended entirely upon what it receives from the Sun; the corresponding figures for the other giants are 1·7 for Jupiter, 1·8 for Saturn and an upper limit of only 1·06 for Uranus. Neptune does not share Uranus' unusual axial tilt; at the time of the Voyager pass it was Neptune's south pole which was in sunlight.

Almost all our detailed knowledge of Neptune comes from the Voyager 2 encounter.

TRITON

NAMED FORMATIONS ON TRITON

	Lat	Long		Lat	Long		Lat	Long
Abatos Planum 1	35–8 S	35–81 E	Jumna Fossæ	6·5–20 S	38–50 E	Ravgga	3 S	71·5 E
Akupara Maculæ	24–31 S	61–65 E	Kasu Patera	39 N	14 E	Raz Fossæ	1–11 N	15–27 E
Amarum	26 N	24·5 E	Kasyapa Cavus	7·5 N	358 E	Rem Maculæ	11·5–15 N	348–351 E
Andvari	20·5 N	34 E	Kibu Patera	10·5 N	43 E	Ruach Planitia	24–31 N	19·5–28 E
Apep Cavus	20 N	301·5 E	Kikimora Maculæ	28–33·5 S	73·5–82 E	Ryugu Planitia 8	3–7 S	25–28·5 E
Awib Dorsa	1–13 S	73–87 E	Kormet Sulci	3 S–33 N	314–345 E	Set Catena	21–23 N	35–40 E
Bheki Cavus	16 N	308 E	Kraken Catena	12–15 N	32·5–41 E	Sipapu Planitia	2–6·5 S	32–39 E
Bin Sulci 2	28–48 S	351–14·5 E	Kulilu Cavus	41 N	4 E	Slidr Sulci	5 S–35 N	312·5–13 E
Boynne Sulci 3	18·5–4 S	301–21 E	Kurma	65·5 N	61 E	Tangaroa	25 S	65·5 E
Bubembe Regio 4	25–43 S	285–24·5 E	Leipter Sulci	6–8 N	0–17 E	Tano Sulci	23–41 N	327–359·5 E
Cay	12 S	44 E	Leviathan Patera	16–19 N	27–30 E	Tuonela Planitia 9	36–42 N	7–19·5 E
Cipango Planum	1–26 N	23–46 E	Lo Sulci	3–4·5 N	316–326 E	Uhlanga Regio 10	60–0 S	285–85 E
Dagon Cavus	29 N	354 E	Mah Cavus	38 N	6 E	Ukupanio Cavus	35 N	23 E
Dilolo Patera	26 N	24·5 E	Mahilani	49–50·5 S	352·5–1 E	Vimur Sulci	3–12 S	47–77 E
Doro Macula	27–28 S	31–32·5 E	Mangwe Cavus	7 S	343 E	Viviane Macula	29·5–32 S	34·5–38·5 E
Gandvik Patera	28 N	5·5 E	Mazomba	18·5 S	63·5 E	Vodyanoy	17 S	28·5 E
Hekt Cavus	26 N	342 E	Medamothi Plan. 5	16·5 S–17 N	50–90 E	Yasu Sulci	2 S–5 N	332·5–0 E
Hili	56–57 S	24–38 E	Monad Regio 6	30 S–45 N	330–90 E	Yenisey Fossa	18 S–17 N	52–56 E
Hirugo Cavus	14·5 N	345 E	Namazu Macula	24–28 S	12–16 E	Zin Maculæ	21–27·5 S	65–71·5 E
Ho Sulci	0·5–2·5 N	293–317·5 E	Ob Sulci 7	19 S–14 N	325·5–327 E			
Ilomba	14·5 S	57 E	Ormet Sulci	18 S–29·5 N	328–351 E			

At 06.50 GMT on 25 August 1989, Voyager passed over the darkened north pole, at a minimum relative velocity of 17·1 km/s. At that time the spacecraft was 4 425 000 000 km from the Earth, so that the 'light-time' was 4 hours 6 minutes.

Neptune proved to be a much more dynamic world than Uranus. There is an appreciable magnetic field, though it is weaker than those of the other giants; the field strength at the surface is 1·2 gauss in the southern hemisphere, but only 0·06 gauss in the northern. Surprisingly, the inclination of the magnetic axis relative to the axis of rotation is 47 deg, so that in this respect Neptune is not unlike Uranus; moreover the magnetic axis does not pass through the centre of the globe, but is displaced by

10 000 km or 0·4 Neptune radii. This indicates that the dynamo electric currents must be closer to the surface rather than near the core. The magnetosphere has been described as comparatively 'empty'; it gyrates dramatically as the planet rotates, and the satellites are also involved. Predictably, Neptune was found to be a radio source. Lightning was detected by Voyager's plasma wave equipment, though not visually. Auroræ were confirmed, though they are of course most prominent near the magnetic poles rather than the poles of rotation.

The interior of Neptune, like that of Uranus, is dominated by planetary 'ices', primarily water. A silicate iron core out to 0·2 of the planet's radius may be present, with a mantle out to 0·8 per cent of the radius, but the core does not seem to be strongly differentiated from the ice components.

Radio measures fixed the rotation period of the core as 16 h 7 m. The rotation of the surface is somewhat complicated, with a greater difference between the various latitudes than is the case with the other giants. The most conspicuous surface feature is a huge oval, the Great Dark Spot, at latitude 8·28 deg S, and has been compared with the Great Red Spot on Jupiter, though it is not red (it is slightly 'less blue' than the adjacent surface). Its size, relative to Neptune, is about the same as that of the Great Red Spot relative to Jupiter. It has a rotation period of 18·296 hours, and drifts westward at about 30 m/s relative to the nearby clouds; it also rotates anti-clockwise, and shows more or less predictable changes in aspect ratio (the ratio of the long axis to the short axis) and orientation. Further south (latitude 42 deg) is a smaller, very variable feature, which has the quicker rotation period of 15·97 hours, and has been nicknamed the Scooter. Still further south (55 deg) is another dark spot, D2, which 'laps' the Great Dark Spot once in every five Earth days.

In June 1994 excellent images of Neptune were obtained with the Hubble Space Telescope. A bright cloud band near latitude 30 degrees north was shown, which apparently did not exist at the time of the Voyager 2 pass, and this was confirmed by further images taken in October 1994; it was very bright on 10 October, but had faded considerably only a week later, while two bright irregular patches appeared in the southern hemisphere. The real surprise was that the large dark spots were not found, and it is difficult to see how they could have been

Clouds on Neptune from Voyager 2, Aug. 1989; range 6 100 000 km. The Great Dark Spot is left of centre; the Scooter is below to the left, and the second dark spot with its bright core is below the Scooter.

missed – so that it may be that Neptune's surface features are much more variable than had been expected.

At Neptune's equator the winds blow westward (retrograde) at up to 450 metres per second; further south the winds slacken, and beyond latitude 50 deg they become eastward, reaching 300 metres per second, decreasing once more near the south pole. As the heat budget of Neptune is only one-twentieth that of Jupiter, it seems that the winds are strong because of the lack of turbulence.

Temperature measurements indicate that there is a cold mid-latitude region from 40 deg to 50 deg, with a warmer equator and pole. The surface temperature is about the same as that of Uranus, though Neptune is so much further from the Sun; this is because Neptune has a marked internal heat source, while Uranus has not.

The upper atmosphere consists of 85 per cent hydrogen, 13 per cent helium and 1 to 2 per cent methane. There are various cloud layers. At a level where the pressure is 3·3 bars there is a layer which seems to be of H_2S (hydrogen sulphide), and above this come layers of hydrocarbons such as ethylene and acetylene, with a methane layer and upper methane haze. Above the H_2S layer there are discrete clouds with diameters of the order of 100 km, and these cast perceptible shadows on the cloud-deck 50 to 75 km below, though the methane cirrus clouds above the Great Dark Spot and the D2 spot have not been observed to cast shadows.

Apparently there is a regular cycle of events. Ultra-violet radiation from the Sun destroys methane in the planet's upper atmosphere, converting it to hydrocarbons which descend into the cold lower stratosphere, where they evaporate and then condense. The hydrocarbon ice particles fall into the warmer troposphere, evaporate, and are converted back to methane; methane clouds then rise into the stratosphere – returning methane to it and keeping the overall amount of methane more or less constant.

THE RINGS OF NEPTUNE

Neptune's rings were studied from Voyager 2; previously it had been thought that there were only incomplete ring arcs. There are five rings, plus the 'plateau', or Lassell ring. The Arago ring moves in the same orbit as the small newly-discovered satellite Galatea.

Name	Distance from centre of Neptune, km	Width, km
Galle	41 900	50
Le Verrier	53 200	50
'Plateau' (Lassell)	53 200–59 100	4000
Arago	57 600	30
Adams	62 900	50

The Adams ring is the most pronounced, and is 'clumpy' with three brighter segments or arcs – named officially Liberté, Égalité and Fraternité, for reasons which are, to put it mildly, obscure! These arcs are thought to be due to the gravitational influence of Galatea. During the Voyager 2 pass, the Adams ring occulted the second-magnitude star σ Sagittarii, and proved to have a core only 17 km wide; the colour is reddish, not unlike that of Jupiter's ring and the Jovian satellite Amalthea. The Plateau is a diffuse band of material, containing the highest percentage of very small particles; there may also be 'dust' going all the way down to the planet, and there is a relatively empty region inside the Le Verrier ring. Probably the rings are relatively young, formed from the débris of small satellites.

Voyager 2 crossed the plane of the rings at a relative velocity of over 76 000 km/h. Impacts were recorded 40 minutes before the crossing, and reached a peak of 300 per second for 10 to 15 minutes either side of the actual crossing. To the relief of the Voyager team, the spacecraft emerged unscathed.

SATELLITES OF NEPTUNE

Only two of Neptune's satellites were known before the Voyager 2 encounter.

Triton was obviously unusual inasmuch as it has retrograde orbital motion – unique for a large planetary satellite. From Neptune the Sun would have an average apparent diameter of about 1 minute of arc, and Triton, with an apparent diameter of about 1 deg, could produce a total eclipse. Nereid was also exceptional; its eccentric orbit seems almost cometary, taking it from 1 345 500 km from Neptune out to as much as 9 688 500 km in a period of little under a year. The average apparent diameter as seen from Neptune would be only 19 seconds of arc. Nereid was not well imaged from Voyager 2; the minimum distance between the spacecraft and the satellite was 4 700 000 km. The albedo is 0·16, and the surface may well be icy, but no definite details could be made out. Previously suspected regular variations in brightness were not confirmed. It is unlikely that Nereid has a rotation period equal to its orbital period; more probably the rotation is chaotic.

The six inner satellites were all discovered from Voyager 2, though it is very probable that one of them, Larissa, had been accidentally noted in 1981 when it occulted a star. The occultation observation was made by a team led by J. Reitsema. Two of the satellites, Proteus and Larissa, were imaged; Larissa seems to contain craters from 30 to 50 km wide. Proteus is actually larger than Nereid, but is much more difficult to detect because of its closeness to Neptune, as well as having a much darker surface. It is somewhat squarish. On the Neptune-facing side surveyed from Voyager 2 there is one major feature, known provisionally as SH or the Southern Hemisphere Depression, which is almost 250 km in diameter and 10 km deep, with a rugged floor. There are also troughs and ridges, and several craters, one of which is 87 km across.

Triton proved to be the real surprise. It was smaller than expected, so that in size it ranks below all four of Jupiter's Galilean satellites, Titan in Saturn's system, and our own Moon; it is also more reflective, with an albedo rising to almost 0·8 in places, and it is very cold, with a temperature of –235°C (a mere 38 K above absolute zero). The mass is $2·14 \times 10^{22}$ kg. The density is fairly high (2·06

SATELLITES OF NEPTUNE

Name	Discoverer	Mean distance from Neptune, km	Sidereal period days	Orbital inclination degrees	Orbital eccentricity	Diameter, km	Albedo	Mag
Naiad	from Voyager 2, 1989	48 000	0·296	4·5	0	54	0·06	25
Thalassa	from Voyager 2, 1989	50 000	0·312	0	0	80	0·06	24
Despina	from Voyager 2, 1989	52 500	0·333	0	0	180	0·06	23
Galatea	from Voyager 2, 1989	62 000	0·429	0	0	150	0·054	23
Larissa	from Voyager 2, 1989	73 600	0·554	0	0	192	0·056	21
Proteus	from Voyager 2, 1989	117 600	1·121	0	0	416	0·050	20
Triton	Lassell, 1846	354 800	5·877	159·9	0·0002	2705	0·6–0·8	13·6
Nereid	Kuiper, 1949	5 514 000	360·16	27·2	0·749	240	0·16	18·7

times that of water) and the material is probably about ⅔ rock and ⅓ ice. The escape velocity is 1·44 km/s. This is enough for Triton to retain a very tenuous atmosphere, with a ground pressure of around 14 microbars (0·000014 bar), 100 000 times thinner than our own air. Nitrogen, in the form of N_2, makes up over 99 per cent of the atmosphere; most of the rest is methane, with a trace of carbon monoxide. There is considerable haze, seen from Voyager above the limb, which extends from 3 to 6 km above the surface (perhaps slightly more) and is probably composed of microscopic particles of ice crystals of methane or nitrogen. Winds in the atmosphere average around 5 metres per second westward, and there is a pronounced temperature inversion.

The surface of Triton is very varied. There is a general coating of ice, presumably water (H_2O) ice overlaid by nitrogen and methane ices. Water ice has not been detected spectroscopically, but it must exist, because nitrogen and methane ices are not hard enough to maintain surface relief over long periods – though in fact there is not a great deal of surface relief; it cannot amount to more than a few hundred metres on Triton, and there are no mountains on Triton. Normal craters are scarce; the largest is only 27 km in diameter. Extensive flows are seen, some of them 80 km wide, due probably to ammonia-water fluids. The most striking feature is the southern polar cap, which is pink, and is thought to be due to nitrogen snow and ice.

The Neptune-facing hemisphere, studied from Voyager 2, is divided into three regions: Bubembe Regio (western equatorial), Monad Regio (eastern equatorial) and Uhlanga Regio (polar). Uhlanga is covered by the bright cap, with some of the underlying geological units showing through in places. The edge of the cap, separating Uhlanga from Monad and Bubembe, is sharp and convoluted. The long Tritonian seasons

mean that the south pole has been in constant sunlight for over a century now, and along the borders there are signs of evaporation. The cap may eventually shift northward, so that the entire aspect of Triton may change according to the seasons.

In Uhlanga we find the nitrogen ice geysers which were so unexpected. According to the most plausible explanation, there is a layer of liquid nitrogen from 20 to 30 metres below the surface; here the pressure is high enough for the nitrogen to remain in a liquid state, but if for any reason it migrates toward the surface it will come to a region where the pressure is only about ¹/₁₀ that of the Earth's atmosphere at sea level. The nitrogen will then explode in a shower of ice and gas (about 80 per cent ice, 20 per cent vapour) and will travel up the nozzle of the geyser-like vent at up to 150 metres per second, fast enough to make it rise to many kilometres before falling back. The outrush sweeps dark débris along with it, and this débris is wafted downwind in the thin Tritonian atmosphere, producing plumes of dark material such as Viviane Macula and Namazu Macula – from 15 to 50 km wide and 50 to 75 km long.

North of the pink cap there is an 'edge' which looks darker and redder, perhaps because of the action of ultra-violet light upon methane. Running across this darker region, roughly parallel to the edge of the cap, is a slightly bluish layer which may be due to tiny crystals of methane ice, which scatter the incoming light to produce blueness.

Monad Regio is in part smooth, with knobbly and hummocky terrain; there are rimless or rimmed pits (paterae), with graben-like troughs (fossae) and strange, mushroom-like features sometimes called 'guttae' such as Zin and Akupara, whose origin is very uncertain. There are also walled plains or 'lakes' such as Tuonela and Ruach, edged

with terraces as though the original level has been changed several times by repeated melting, due to vulcanism, and refreezing. The interiors are flat and smooth. Most substances would flow like glaciers, but water ice would not, because Triton is so cold; we must therefore assume that water is the main material from which the lakes were formed. Undulating smooth planes cover many of the higher areas of Monad.

Bubembe Regio is characterized by the so-called cantaloupe terrain – a name given from a superficial resemblance to the skin of a melon! Fissures cross it, meeting in huge X or Y junctions (for example, where Slidr and Tano Sulci cross each other). Liquid material, presumably a mixture of water and ammonia, seems to have forced its way up some of these fissures, so that there are central ridges; material has even flowed out on to the plain before freezing there. The east part of Bubembe includes subdued circular pits with diameters of around 30 km. The cantaloupe areas are probably the oldest parts of Triton's surface; the paterae on the high smooth areas are probably explosive vents similar to calderas, while the pits could be collapse features.

Everything indicates that Triton was not originally a satellite of Neptune, but was captured well after its formation. Initially its orbit round Neptune would have been eccentric; over a period of perhaps 1000 million years it was forced into a circular form, and during this time there was great internal heating, coupled with surface activity. There may also have been a dense atmosphere.

Triton is unlike any other world in the Solar System. Unfortunately we cannot hope for another spacecraft there before southern midsummer on Neptune and Triton, which may be around the year 2006, though in view of Triton's wildly oscillating seasons this date is very uncertain.

Preliminary map of Proteus. It is heavily cratered, with a rugged surface; evidently it is a primitive body rather than an evolved one. The main feature is the Southern Hemisphere Depression (SHD), which dominates the southern part of the Neptune-facing hemisphere; basically circular, with a raised rim and a flat, rugged floor; 225 km across and 10 to 15 km deep. Proteus also shows linear streaks, which seem to be troughs 25 to 35 km wide, up to 100 km long, and several km deep; they form a global network and may be classed as graben of extensional origin. There are no cryovolcanic structures or coronæ.

PLUTO

DATA ....

Distance from the Sun:
mean: 5900 million km (39·5 a.u.)
max: 7375 million km (49·28 a.u.)
min: 4425 million km (29·65 a.u.)
Sidereal period: 247·7 years = 90 465 days
Synodic period: 366·7 days
Rotation period: 6 d 9 h 17 s
Mean orbital velocity: 4·7 km/s
Axial inclination: 122 deg ·46
Orbital inclination: 17 deg ·15

Orbital eccentricity: 0·248
Diameter: 2324 km
Density, water = 1: 2·0
Reciprocal mass, Sun = 1: 135 500 000
Mass, Earth = 1: 0·0022
Specific gravity: 2·03
Mean surface temperature: about −230°C
Escape velocity: 1·18 km/s
Surface gravity, Earth = 1: 0·06
Opposition magnitude at perihelion: 14

*T*he most recently discovered planet is Pluto, whose mean distance from the Sun is much greater than for any other planet. As yet our knowledge of it is very far from complete.

Pluto's comparatively eccentric orbit brings it more than 50 000 000 km closer to the Sun than Neptune can ever be, but since the orbit is also inclined by over 17 degrees there is no fear of collision; at the present epoch the distance between the two planets cannot be less than 348 000 000 km. The last aphelion fell on 12 July 1866, and the last perihelion on 12 September 1989; the next aphelion will be in 2113. Between 21 January 1979 and 14 March 1999 the radius vector of Pluto's orbit is less than that of Neptune. Pluto can depart from the conventional Zodiac; the 1995 opposition position is in Virgo.

OPPOSITION DATES, 1994–2005

		Declination at opposition deg
1994	May 17	−5
1995	May 20	−6
1996	May 22	−7
1997	May 25	−8
1998	May 28	−9
1999	May 31	−10
2000	June 1	−11
2001	June 4	−12
2002	June 7	−13
2003	June 9	−13
2004	June 11	−14
2005	June 14	−15

✳ **The first search for a trans-Neptunian planet** seems to have been made in 1877 by David Peck Todd, of the US Naval Observatory. From perturbations of Uranus, he predicted a planet at a distance of 52 astronomical units from the Sun, with a diameter of 80 000 km. He conducted a visual search, using the 66 cm USNO reflector with powers of 400 and 600, hoping to detect an object showing a sensible disk. He continued the hunt for 30 clear, moonless nights between 3 November 1877 and 5 March 1878, but

with negative results. A second investigation was conducted by the French astronomer Camille Flammarion in 1879. Flammarion based his suggestion upon the fact that several comets appeared to have their aphelia at approximately the same distance, well beyond the orbit of Neptune. Further comments were made in 1900 by G. Forbes, who assumed a planet larger than Jupiter, moving at a distance of about 100 a.u. in a period of 1000 years; in 1902 by T. Grigull of Münster, whose hypothetical planet was the size of Uranus, moving at 50 a.u. in a period of 360 years (Grigull even gave it a name – 'Hades'); by Gaillot in France – who believed that two planets existed – and by T. J. J. See in America, who increased this number to three.

✳ **The first systematic photographic and visual search for the planet** was undertaken by Percival Lowell, at Flagstaff, from 1905 to 1907. The results were negative. Lowell's

'Planet X' was believed to have a period of 282 years and a mass 7 times that of the Earth, with a rather eccentric orbit (0·202), and the date of next perihelion was given as 1991. A second search at Flagstaff, carried out by C. O. Lampland in 1914, was equally fruitless, and was given up on Lowell's death in 1916. Lowell's final calculations about his planet were published in 1915.

✳ **The second prediction** was made by W. H. Pickering in 1909; his planet had a mass twice that of the Earth and a period of 373·5 years. His method – unlike Lowell's – was essentially graphical, but the conclusions were much the same. On the basis of his results, Milton Humason, at Mount Wilson Observatory, undertook a photographic search in 1919, but again the results were negative.

✳ **The discovery of Pluto** was made at the Flagstaff Observatory by Clyde Tombaugh. Using a 13 in refractor, Tombaugh began work in 1929, again using photographic methods. Pluto was detected upon plates taken on 23 and 29 January, but the announcement was made on 13 March – 149 years after the discovery of Uranus and 78 years after Lowell's birth. It was later found that the planet had been recorded twice on the plates taken during the early searches, but had been missed because it was so unexpectedly faint. However, Humason's failure in 1919 was sheer bad luck. When his plates were re-examined, it was found that Pluto had been recorded twice; but once the image fell upon a flaw in the plate, and on the second occasion Pluto was masked by an inconveniently-placed bright star.

Pluto; Clyde Tombaugh's discovery plate, 1930. Pluto is arrowed. The bright overexposed star is Delta Geminorum. (Lowell Observatory)

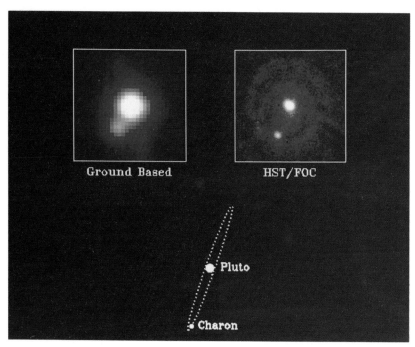

Pluto and Charon; ground-based, and from the HST (before the servicing mission). Below: the orbits, showing why in the late 1980s there was a series of mutual eclipses and occultations. (Image Select/NASA)

recorded the two bodies separately. The satellite has been named Charon.

DATA FOR CHARON
Distance from Pluto, km: 19 460
Orbital period: 6·387 days (= 6 d 9 h 17 m)
Orbital eccentricity: 0
Orbital inclination: 98 deg ·8
Diameter: 1270 km
Magnitude: 16·8
Specific gravity: 1·3
Escape velocity: 0·16 km/s
Mass (Pluto = 1): 0·08

The Pluto–Charon system is unique. Of the two, Pluto is much the more massive – $1·29 \times 10^{25}$ g, as against $1·77 \times 10^{24}$ g for Charon – but Charon has more than half the diameter of Pluto, so that it is probably inappropriate to regard it as a normal satellite. There is also serious doubt as to whether Pluto is worthy of planetary status. It is smaller than several planetary satellites, including Triton, and has only about 18 per cent the mass of our Moon. Moreover, Charon's orbital period is the same as Pluto's rotation period, so that the two are 'locked'. From one hemisphere of Pluto, Charon would remain fixed in the sky; from the other hemisphere it would never be visible.

Sunlight on Pluto is always more than 1500 times as brilliant as full moonlight on Earth. From Pluto, Charon would have an apparent magnitude of −9; its size would be one-fifth that of the Moon seen from Earth.

✳ **The first attempts to study the surface features of Pluto** originated with work by L. E. Andersson and D. Fix from 1973. They found that the amplitude of the light variations due to Pluto's rotation was increasing, but that the mean magnitude was becoming fainter. From this they inferred that there are bright poles, and that one of these poles, previously presented to the Earth, was turning slowly away, so that we are seeing more of the mid-latitude and equatorial zones where the ice is 'dirtier' and older.

✳ **The first detection of methane ice on Pluto** was made spectroscopically in 1976.

✳ **The first good infra-red observations of Pluto** were made from the IRAS satellite in 1983. From these it was found that methane ice cannot cover the entire surface; there is a band near the equator which is bright in infra-red but dark at visible wavelengths, so that it may be relatively frost-free. However, different results were obtained in 1993 by T. C. Owen and his colleagues, using UKIRT (the United Kingdom Infra-Red Telescope on Mauna Kea). They find that the surface of Pluto is dominated by nitrogen ice, not methane ice, with 98 per cent being in marble-sized fragments of fractured sheets. Methane ice exists, and a trace of frozen carbon monoxide, but no water ice or

Early diameter estimates indicated that Pluto is small by planetary standards – so small, indeed, that it could produce no measurable perturbations on Uranus or Neptune; yet it was by these alleged perturbations that the planet had been tracked down. In 1936 A. C. D. Crommelin, of Greenwich, suggested the theory of specular reflection. If Pluto were highly reflective, the bright image of the Sun might falsify the diameter estimates, so that Pluto could be much larger – and hence more massive – than it seemed.

✳ **The first suggestion that Pluto might be an escaped satellite of Neptune** was made by R. A. Lyttleton, of Cambridge, in 1936; it was supported by G. P. Kuiper in 1956. It is true that Pluto is much smaller than Triton, Neptune's large satellite, and Triton is also the only large satellite with a retrograde orbit, so that the idea of great disturbances in the remote past was logical enough. However, the theory has been weakened by the discovery of Pluto's satellite, Charon.

✳ **The first measurement of Pluto's diameter** was made by Kuiper, using the 82 in reflector at the McDonald Observatory in Texas, in 1949. The diameter was given as 10 200 km, with a mass 8/10 of that of the Earth. However, in 1950 Kuiper and Humason made new measurements with the Palomar 200 in reflector, and the diameter of Pluto was reduced to 5800 km – smaller than Mars, and possibly smaller

than Triton. *The first partial occultation of a star by Pluto* was observed in 1965. From this, it appeared that the diameter could not be more than 5800 km.

✳ *The first measurement of the rotation period of Pluto* was made in 1955 by M. Walker and R. Hardie. They found that the brightness of the planet varied by 20 per cent in 6·39 days, and this was presumably due to axial rotation. There are thus over 14 100 Plutonian 'days' in each Plutonian 'year'!

✳ *The first report of a satellite of Pluto* was made on 22 June 1978 by James W. Christy at the US Naval Observatory at Flagstaff, Arizona. Pluto had been repeatedly photographed with the 1·55 metre reflector with a view to predicting possible occultations of stars by the planet. Pictures taken in April and May showed that Pluto's image was elongated. Plates taken earlier (1965, 1970, 1971) were then checked, and the same effects were noted. Intensive studies were carried out at Flagstaff, and confirmation was obtained on 6 July 1978 by J. Graham, using the 401 cm reflector at the Cerro Tololo Observatory in Chile. Subsequently, the existence of the satellite as a separate body was confirmed by D. Bonneau and P. Foy, using the 3·6 metre Canada–France–Hawaii Telescope on Mauna Kea. Using the technique of speckle interferometry, they carbon dioxide ice. The polar caps presumably vary in size over a Plutonian year, and may have a maximum spread down to latitude 45 deg. The caps are three or four times brighter than the equator.

✳ *The first mutual transits and occultations of the Pluto/Charon system* were observed from 1985 to 1990, when the orbits were suitably positioned (a situation which will not recur for 120 years). The first observation, when the edge of Charon passed in front of Pluto, was made by R. Binzel on 17 February 1985. Total events began in 1987 and ended in 1988, while the whole series of phenomena ended in September 1990. When Charon passed behind Pluto it was completely hidden, and Pluto's spectrum could be seen alone; when Charon passed in front of Pluto the two spectra were seen together, but that of Pluto could be subtracted. Charon showed indications of water frost, with no methane ice. If both bodies had a common origin, it may be that Charon's original surface methane sublimated and escaped, possibly to be captured by Pluto. When both bodies are seen together, Pluto contributes 80 per cent of the total light of the system, partly because it is larger and also because it is more reflective. The separation between the two, as seen from Earth, is only 0 sec ·8. Charon shows no polar caps, but there may be a darkish band in one hemisphere and a brighter band in the other.

✳ *The best images of Pluto and Charon* have been taken by the Hubble Space Telescope. The two are shown well separated, and vague darker and lighter patches can be made out on Pluto.

✳ *The atmosphere of Pluto* was discovered in 1980. The density is very low, with a pressure about 100 000 times less than that of the Earth's air at sea-level. The atmosphere is, however, surprisingly extensive. Results of the observations of a 12th-magnitude star by Pluto, on 9 June 1988, showed that it must be at least as deep as the planet's diameter. The star began to fade at a distance of 1500 km from Pluto's centre, and it seems that there is an upper transparent layer about 300 km deep, with haze below – still not opaque enough to hide the surface.

The main constituent is thought to be methane, but the atmosphere is too extensive to be pure methane, and there may be come carbon monoxide, nitrogen and even argon. When Pluto is in the far part of its orbit, the temperature is so low that methane may be

expected to freeze out, so that it is possible that for part of its 'year' there is no appreciable atmosphere. (Remember, however, that Triton's atmosphere proved to be mainly nitrogen, and in many ways it seems that Triton and Pluto are not dissimilar, so that we may have to revise our ideas about the Plutonian atmosphere.) Charon, with its lower escape velocity, shows no trace of atmosphere.

✳ *The internal constitution of Pluto* is very uncertain. The density of 2·02 times that of water means that there is more rock than in the icy satellites of the giant planets – perhaps around 75 per cent – but it seems that the gravitational pressure cannot significantly increase the rock density near the core. The core rock must be hydrated, so presumably it accreted from silicates and warm water ice, and radioactive decay was inadequate to heat it up to the temperature needed for dehydration. Either Pluto is undifferentiated, or else it has a silicate core surrounded by a mantle of ice probably from 200 to 300 km thick.

✳ *The nature of Pluto* is still a matter for debate. Its mass is so small (even combined with that of Charon) that it could not possibly perturb the motions of giant planets such as Uranus and Neptune to any measurable extent, so that it was not the planet for which Lowell was searching. It does not seem to be worthy of true planetary status, and it cannot have been produced in the inner part of the Solar System, because it contains too much ice. There is good reason to regard it and Charon as Kuiper Belt objects – in fact, planetesimals – and the same may be true of Chiron and of Triton, which were presumably captured by Neptune long ago. Further information will be hard to obtain without using a spacecraft, and although probes to Pluto have been seriously discussed for the early years of the 21st century there are no definite commitments as yet.

At any rate, either Lowell's reasonably correct prediction was sheer luck, or else another planet exists, moving at a much greater distance. This hypothetical body is generally known as 'Planet X'.

One suggestion came from Pickering after Pluto had been found, and also in 1950 by K. H. Schütte of Munich on the basis of cometary aphelia. In 1962 A. G. W. Cameron wrote that 'it is difficult to escape the conclusion that there must be a tremendous mass of small solid material on the outskirts

of the Solar System'. In 1972 K. A. Brady (USA) investigated the motion of Halley's Comet, and suggested that there were perturbations due to a planet moving round the Sun in a retrograde orbit at about 8000 million km in a period of 512 years; its mass would be greater than that of Saturn. Brady gave a position for his 'planet' in Cassiopeia; searches were made, but it now seems that Brady's calculations were unsound.

It does seem that there are slight, unexplained irregularities in the movements of Uranus and Neptune which were more evident a century ago than they are now. In 1987 J. Anderson, of NASA, proposed that the planet responsible moves in a very eccentric and highly inclined orbit; the mean distance from the Sun was given as around 80 astronomical units (12 000 million km), with a period of from 700 to 1000 years and a mass five times that of the Earth. If this planet exists, and is at present within what may be called 'reasonable range', it could possibly be tracked down by observations of the movements of the probes now on their way out of the Solar System, though the planet would have to be in just the right place at the right time.

In 1988 R. Harrington, of the US Naval Observatory in Washington, undertook new studies of the perturbations of Uranus, and gave a different orbit for Planet X; the period is 600 years and the present distance about 9·6 thousand million kilometres. The mass is taken to be from two to five times that of the Earth. The predicted position is in the region of Scorpius and Sagittarius. Searches are planned, and, of course, there is always the chance that the movements of the Pioneers and Voyagers will provide clues before they pass out of range. At the moment, however, we have to admit that we are unsure about the existence of Planet X, and there have even been suggestions that a slight error in the accepted mass of Neptune may explain the residuals without any need for a more remote planet.

There is, of course, no theoretical reason why there should not be several planets orbiting the Sun at great distances. It has even been suggested that there may be a connection with the clouds of comets which are believed to exist; the perturbing influence of a remote planet might well disturb comets in these clouds, and send them into orbits which bring them into the inner part of the Solar System. However, this is no more than conjecture at the present time.

COMETS

Comets are members of the Solar System, but are made up of small particles together with extremely tenuous gas, so that they are very different from the planets and are of slight mass. Generally their orbits are highly eccentric.

For all practical purposes, comets may be divided into two classes: (1) Periodical comets, which are seen regularly and which move around the Sun in periods ranging from 3·3 years to over 150 years; and (2) Non-periodical comets, which cannot be predicted because their periods are too long. (There are also comets which may well travel in parabolic orbits, in which case they will never return to the neighbourhood of the Sun; but it is almost impossible to distinguish between a section of a parabola and of an extremely eccentric ellipse.) *The first telescopic observation of a comet* was made in 1618 by J. Cysat.

For many centuries comets were regarded as atmospheric phenomena – though Anaxagoras, around 500 BC, attributed them to clusters of faint stars. The first man to prove that they were extraterrestrial was Tycho Brahe, who found that the comet of 1577 showed no diurnal parallax, and must therefore be at least six times as remote as the Moon (actually, of course, it was much further away than that). There have been occasional panics due to comets – as in France in 1773, when a mathematical paper by J. J. de Lalande was misinterpreted, and people believed that a comet would collide with the Earth on 20 or 21 May. (Seats in Paradise were sold by members of the clergy at high prices.) Another alarm occurred in 1832, when it was suggested – wrongly – that there might be a near-encounter between the Earth and the now-lost Biela's Comet. In 1910, when Halley's Comet was in view, a manufacturer in America made a large sum of money by selling anti-comet pills, and in 1970 some Arabs mistook Bennett's Comet for an Israeli war weapon. Since 1950 there have appeared several books by an eccentric psychiatrist, Dr Immanuel Velikovsky, who confused planets with comets, and believed Venus to have been a comet only a few thousands of years ago!

✳ *The brightest periodical comet* is Halley's, with a period of 76 years. Most comets are named in honour of their discoverers, but this is not the case for Halley's Comet, which has been recorded regularly since well before the time of Christ. In 1682 it was observed by Edmond Halley (the actual discovery was made on 15 August of that year by G. Dörffel), and subsequently Halley decided that it must be identical with comets previously seen in 1607 and in 1531. He predicted a return for 1758. On Christmas Night 1758 the comet was duly found, by a German amateur, J. Palitzsch, and it passed perihelion in 1759. This was the *first predicted cometary return*, and it was appropriate that the comet concerned should be known as Halley's.

✳ *The first definitely recorded return of Halley's Comet* is that of 240 BC, though it seems that a comet observed by the Chinese in 1059 BC may well have been Halley's. Since 240 BC the comet has been seen at every return:

240 BC	141	530	912	1301	1682
164 BC	218	607	989	1378	1759
87 BC	295	684	1066	1456	1835
12 BC	374	760	1145	1531	1910
AD 66	451	837	1222	1607	1986

✳ *The brightest return* was that of 837, when the comet passed within 0·04 astronomical units of the Earth (11 April). The magnitude was about –3·5, and the tail had a length of 93 degrees. The return of 1986 was the worst for 2000 years, though the comet was visible with the naked eye. The next return, that of 2061, will also be unfavourable.

✳ *The only comet to have been referred to by a Roman emperor* was Halley's. Comets were once nicknamed 'hairy stars' and were regarded as evil. (In AD 79 Vespasian commented that a comet seen in that year – not Halley's – 'menaced rather the King of the Parthians; for he is hairy, while I am bald' – however, Vespasian died in the same year.) Another Roman emperor, Macrinus, died at the time of Halley's return in 218.

✳ *The first surviving drawing of Halley's Comet* refers to the return of 684. It comes from the *Nürnberg Chronicle*.

✳ *The only comet recorded in the Bayeux Tapestry* is Halley's. The comet was seen in spring 1066, as the Normans were preparing to invade England, and in the Tapestry it appears clearly, with King Harold toppling on his throne and the Saxon courtiers showing signs of great alarm. (Some scholars consider that the Tapestry was woven by William I's wife.)

✳ *The only comet to have been officially referred to by a Pope* is again Halley's. This was at the 1456 return, when Pope Calixtus III regarded it as an agent of the Devil. (Some scholars doubt this story, but it may well be true!)

✳ *The first observer to see Halley's Comet in 1835* was Dumouchel, from Rome; the last to see it in 1836 (May) was Sir John Herschel, from the Cape. The comet was recovered on 12 September 1909 by Max Wolf in Germany, and remained under observation until June 1911, when its distance from the Sun was more than 800 000 000 km. Aphelion was reached in 1948.

The last perihelion occurred on 9 February 1986. The comet was recovered on 16 October 1982 by a team of astronomers at Palomar (Jewitt, Danielson and Dressler) who used the Hale reflector to detect the comet as a tiny blur of magnitude 24·3 – making it one of the faintest objects ever recorded. It was a mere 8 seconds of arc away from its predicted position. The discovery was confirmed shortly afterwards at Kitt Peak. At the time of its recovery, the comet was still moving between the orbits of Saturn and Uranus.

✳ *The first attempt to observe a cometary transit across the face of the Sun* was made on 18–19 May 1910, with Halley's Comet. The American astronomer F. Ellerman went to Hawaii to observe under the best possible conditions, but could see no trace of the comet.

✳ *The first comet to be studied from close range* by a number of space-probes was Halley's, in 1986. There were five probes in all: two Japanese (Suisei and Sakigake), two Russian (Vega 1 and Vega 2), and one European (Giotto). Of these only Giotto was scheduled to pass into the comet's inner coma, and image the nucleus, but prior information sent back from the earlier vehicles was of immense value. Vega 1 made its closest approach on 6 March 1986 and Vega 2 on 9 March; Giotto passed within 605 km of the comet's nucleus on the night of 13–14 March.

It was generally believed that the 'dirty iceball' theory proposed by F. L. Whipple in 1950 was correct. On this theory a comet's nucleus was composed of rocky fragments held together with ices such as frozen methane, ammonia, carbon dioxide and water; when the comet approached perihelion these ices would start to evaporate, producing a coma and tail (or tails). Tails were of two kinds: ion and dust. Both would point more or less away from the Sun, as the ions would be repelled by sunlight and the dust by the solar wind, and when receding from the Sun a comet would move tail-first – though in general an ion tail would be straight and a dust-tail curved. (Smaller comets, of course, would never develop tails of either type.)

All the probes to Halley's Comet were successful. The camera on Giotto (the HMC, or

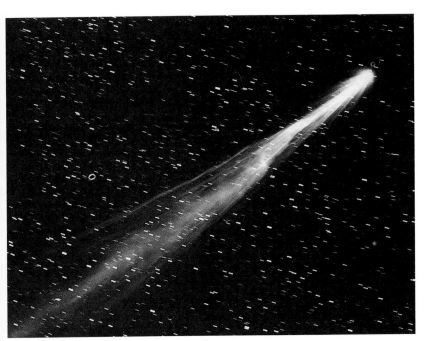

Comet Morehouse, 1908 III; photograph by E. E. Barnard, 16 Nov. 1908. This was a very active comet, showing rapid changes in its tail.

Halley Multicolour Camera) functioned until 14 seconds before the closest approach, when the spacecraft was made to gyrate by the impact of a 'dust' particle about the size of a grain of rice, and communications were temporarily interrupted; in fact the camera did not produce any further results, and the closest image was obtained at 1675 km from the nucleus. The nucleus itself was found to be $15 \times 8 \times 8$ km in dimension, with a total volume of over 500 cubic km, and a mass of from 50 000 million to 100 000 million tons and a density of 0·1 to 0·2 grams per cubic centimetre. (It would take 60 000 million comets of this mass to equal the mass of the Earth.) To the surprise of most people, the nucleus proved to be very dark, with an albedo from 2 to 4 per cent.

Water ice appeared to be the main constituent of the nucleus (84 per cent) followed by formaldehyde and carbon dioxide (each around 3 per cent) and smaller amounts of other volatiles, including nitrogen and carbon monoxide. The shape of the terminator showed that the central region was smoother than the ends; a bright patch 1·5 km in diameter was taken to be a hill, and there were features which appeared to be craters, around 1 km across. Dust-jets were active, though only from a small area of the nucleus on the sunward side. The sunward side was found to have a temperature of 47°C, far higher than expected, and from

this it was inferred that the icy nucleus was covered with a layer of warm, dark dust. The nucleus is eroded at around 1 cm per day near perihelion, and at each return the comet must lose about 300 000 000 tons of material. The rotation period was found to be 53 hours with respect to the long axis of the nucleus, with a 7·3-day rotational period around the axis; the nucleus is in fact 'precessing' rather in the manner of a toppling gyroscope.

✳ *The only major outburst to have been seen in Halley's Comet* was detected on 12 February 1991 by C. Hainaut and A. Smette, using the 1·54-metre Danish telescope at La Silla (European Southern Observatory). As the comet receded from the Sun it had been followed by R. West and others, with the Danish telescope, and had behaved in a conventional manner. In April–May 1988 and January 1989 the images were still diffuse, indicating some residual activity or possibly a cloud of dust, but by February 1990, when the distance from the Sun was 12·5 astronomical units, the image appeared stellar, with a magnitude of 24·3. Therefore the outburst seen by Hainaut and Smette was quite unexpected. The magnitude was 18·9! The distance from the Sun was then 14·5 a.u.

On 22 February, Smette used the New Technology Telescope at La Silla to obtain a spectrum. The coma showed a solar-type spectrum, with no emission features, which indicated a dust composition. It was subsequently found that structures within the coma varied with time, while the central region faded by about 1 magnitude per

month. It seems that a fan-like structure, in the approximate direction of the Sun, reached a radius of 61 000 km on 13 February, expanding to 142 000 km by 12 April. If the expansion of the coma material were about 14·5 m/s, the actual outburst would have occurred on 17 December 1990, lasting for three months or so. A short explosive event is not possible – it would have involved higher velocities for the dust than were observed.

The cause of the outburst is uncertain. A collision with a wandering body is possible, but seems unlikely; possibly a pocket of volatile carbon monoxide ice was exposed to sunlight, and the vaporizing gases carried the dust particles away from the nucleus, but this also seems improbable in view of the comet's distance from the Sun. We may have to await the 2061 return before solving the problem.

By June 1994 Halley's Comet had reached the half-way point between perihelion and aphelion, and was still under observation with the Danish telescope on Mauna Kea. It will reach aphelion in 2024.

The Giotto probe was put into 'hibernation' in April 1986, and re-activated on 19 February 1990. On 2 July it flew past Earth at 22 730 km, and used the gravity-assist technique to put it into a path to rendezvous with the periodical comet Grigg-Skjellerup. After a further hibernation period, Giotto was again reactivated on 4 May 1992, when it was 219 000 000 km from Earth, and on 10 July 1992 it encountered Comet Grigg-Skjellerup, passing less than 200 km from the nucleus. Most of the instruments on the spacecraft were still working (apart from the camera), and valuable results were obtained. Comet Grigg-Skjellerup is a much older comet than Halley's, and seldom produces a tail; however, the density of the gas near its nucleus was greater than expected, and there was a good deal of fine 'dust'. It was found that the gas coma extends to at least 50 000 km beyond the visible boundary. Giotto suffered no damage, though it was hit by a particle about 3 mm across. Giotto is still in solar orbit, though it does not retain sufficient gas to send it on to another comet, as had originally been hoped.

✳ *The first comet known to hit the Sun* was 1979 XI (Howard-Kooman-Michels), on 31 August of that year. Several other cases have since been established, and it may well be that cometary suicides are much more common than has been previously supposed.

✳ *The first comet observed to collide with a planet* was Shoemaker-Levy 9, which impacted Jupiter in July 1994.

✳ *The first short-period comet to be predicted* was Encke's, again now named in honour of the mathematician who computed the orbit. The comet was first seen on 17 January 1786 by P. Méchain, from France. It was again recorded on 7 November 1795 by Caroline Herschel; the next returns to be seen were those of 1805 (discovered by Thulis at Marseilles, 19 October) and 1818 (Jean Pons, 26 November). Encke, at Berlin, decided that these comets must be identical, and predicted a return for 1822. He was correct, and since then the comet has been seen at every return except that of 1944 (when it was badly placed, and most astronomers were otherwise engaged).

✳ *The comet with the shortest known period* is also Encke's: 3·3 years. The period has shortened slightly since its discovery; the decrease amounted to over a day between 1822 and 1858. Evaporation processes taking place inside the comet are responsible – not, as Encke himself supposed, some resisting medium in the inner part of the Solar System.

✳ *The comet which has presented most observed returns* is Encke's. The 1994 return was the 56th to be seen. Encke's Comet is therefore far ahead of its nearest rivals: Halley (30 returns, last perihelion passage 1986) and Pons-Winnecke (20 returns, last perihelion passage 1988). Modern instruments are now capable of recording Encke's Comet even when it is near aphelion.

The following data relate to comets with periods of between 3·3 and 20 years, and which have been observed at more than one return.

Greatest orbital eccentricity: Encke (0·85)
Smallest orbital eccentricity: Schwassmann-Wachmann 1 (0·11)
Greatest orbital inclination: Tuttle (54 deg ·5)
Smallest orbital inclination: Du Toit-Neujmin-Delporte (2 deg ·9)
Greatest perihelion distance: Schwassmann-Wachmann 1 (5·45 a.u.)
Smallest perihelion distance: Encke (0·34 a.u.)
Greatest aphelion distance: Neujmin I (12·16 a.u.)
Smallest aphelion distance: Encke (4·09 a.u.)

A few comets, with orbits of exceptionally low eccentricity, are visible throughout their orbits. The first-discovered of these was Schwassmann-Wachmann 1, first seen in 1925. Its orbit lies entirely between those of Jupiter and Saturn. Normally it is a very faint object, but it can show sudden, unpredictable outbursts which bring it within the range of small telescopes. A recent occasion was in the autumn of 1976, when the magnitude rose to above 12. Other comets which may be kept in view throughout their orbits are Gunn and Smirnova-Chernykh. Oterma's Comet used to have a very low

eccentricity and a period of 7·9 years, but a close approach to Jupiter in 1963 altered the period to 19·3 years.

✳ *The most numerous comet 'family' is Jupiter's.* The mean distance of Jupiter from the Sun is 5·2 a.u., and a glance at the table on page 129 shows that many short-period comets have their aphelia at about this distance. 'Comet families' of Saturn and Uranus are very questionable. However, the Neptune 'family' is rather more definite. The mean distance of Neptune from the Sun is approximately 30 a.u., which corresponds to the aphelia of at least four comets, including Halley's. It must however be added that some authorities are highly sceptical about the significance of these alleged comet families, and attempts to predict the existence of a still more remote planet by means of cometary aphelia are very speculative.

✳ *The comet with the longest confirmed period* is Herschel-Rigollet (1788 and 1939; 156 years). This and Swift-Tuttle (period 133 years) are the only two comets definitely seen at more than one return whose periods are longer than that of Halley's Comet. It has been suggested that Comet McNaught-Russell of 1993 may be identical with a comet seen in China in 574, and if so, then the period would be 1420 years, but obviously this is very uncertain.

✳ *The most famous 'lost' periodical comet* is Biela's. At the 1846 return it separated into two; the twins returned in 1852, but have never been seen since, though the comet's orbit is associated with a meteor shower. Westphal's Comet was seen in 1852 and again in 1913, but it evidently failed to survive the perihelion passage of 1913, and did not appear on schedule in 1976. Brorsen's Comet was seen at five returns between 1846 and 1879, but has not appeared since, and has evidently disintegrated. However, one must beware of jumping to conclusions. Comet Di Vico-Swift was 'lost' for 38 years after 1897, but was recovered in 1965; Holmes' 'Comet' 'went missing' for 58 years prior to its recovery in 1964. Both these latter successes were due to the mathematical work of the British astronomer B. Marsden.

✳ *The closest cometary approach to the Earth ever recorded* was that of Lexell's Comet of 1770 (discovered by C. Messier on 14 June of that year; A. Lexell of St Petersburg computed the orbit). The distance was reduced to 2·2 million km, and the comet was visible with the naked eye. It has never been seen again, since an approach to Jupiter in 1779 has altered the orbit, and the comet now comes nowhere near the Earth.

✳ *The closest known cometary approach to Jupiter* was that of Comet Brooks 2, in March

1886 (the approach was not, of course, actually observed). Evidently the comet passed within the orbit of Io, the closest to Jupiter of the four large satellites. The comet suffered partial disruption, and at the 1889 return was seen to be attended by four minor companions which were classified as 'splinters', but which did not last for long. In 1994 Comet Shoemaker-Levy 9 impacted Jupiter and was of course destroyed.

✳ *The largest comet ever recorded* seems to have been that of 1811; the diameter of the coma was approximately 2 000 000 km, larger than the Sun. The maximum length of the tail was over 160 000 000 km.

✳ *The longest tail ever recorded* was that of the Great Comet of 1843: 330 000 000 km – considerably greater than the distance between the Sun and the orbit of Mars.

✳ *The brightest comet of modern times* was probably that of 1843. According to the famous astronomer Sir Thomas Maclear, it was much more brilliant than the comet of 1811; and Maclear saw both.

✳ *The most beautiful comet of modern times* is said to have been Donati's of 1858, with its wonderfully curved main tail and its two shorter ones. It was discovered by G. Donati, from Florence, on 2 June 1858, and was last seen on 4 March 1859. During October 1858 the length of the tail was as much as 80 000 000 km. The period is unknown; 2000 years has been suggested, but with no certainty.

✳ *The most-tailed comet on record* was that of 1744, discovered on 9 December 1743 by Klinkenberg in Holland (and independently on 13 December by De Chéseaux in Switzerland; rather unfairly, it is usually called De Chéseaux' Comet). There were at least six bright, broad tails, but records are sparse, as the comet remained brilliant for only a few nights in March 1744.

✳ *The comet with the longest computed period* is Finsler's Comet of 1937: 13 600 000 years. This is naturally very arbitrary, and all we can really say is that the period is very long indeed.

✳ *The only comet to be associated with port wine* was that of 1811. In that year the vintage in Portugal was unusually good; for years afterwards 'Comet Wine' appeared in the price lists of wine-merchants! It would be interesting to know whether any of it remains to be drunk.

✳ *The first time that the Earth was known to pass through the tail of a comet* was late June 1861. The comet concerned – discovered by J. Tebbutt from Australia – was brilliant, but the nucleus never came within 18 million km of the Earth. No effects of the comet's tail were noticed with certainty.

✳ *The first comet to be discovered during a total eclipse of the Sun* was Tewfik's Comet of 1882; the eclipse was photographed from Egypt, and the comet was named after the then ruler of Egypt. It was not seen again, so that this is our only record of it. The bright comet of 1948 was also discovered during a total solar eclipse.

✳ *The first known attempt to photograph a comet* was made by the British pioneer W. de la Rue, with the great Comet of 1861. However, he was unsuccessful.

✳ *The first really good comet photograph* was taken in 1882 at the Cape by Sir David Gill. The comet was very brilliant, and was well shown. Many stars were also shown, and it was this picture which made Gill appreciate the endless potentialities of stellar photography.

✳ *The first cometary spectrum* was studied in 1864 by G. Donati. The comet concerned was reasonably bright, and Donati saw that the spectrum was not merely a reflected solar spectrum; there were lines which could be due only to materials in the comet itself.

✳ *The greatest distance at which a good cometary spectrum has been obtained* is 2140 million km (on 22 February 1991, when A. Smette, using the New Technology Telescope at La Silla, obtained a spectrum of Halley's Comet).

✳ *The first comet to be detected by radar was Encke's*, in 1980. Then came Grigg-Skjellerup (1982), Sugano-Saigusa-Fujikawa (1983) and Halley (1985).

✳ *The brightest comet so far seen during the 20th century* is probably the so-called Daylight Comet of 1910, discovered by some diamond miners in the Transvaal on 12 January. It was certainly brighter than Halley's, which was on view later in the year; and people who claim to remember Halley's Comet are warned that the comet they saw was probably the Daylight! The orbit was virtually parabolic.

✳ *The greenest comet of modern times* was Jurlov-Achmarov-Hassell (1939). The green colour was evident even in a small telescope.

✳ *The most famous 'spiked' comet* was Arend-Roland, which was easily visible with the naked eye in April 1957. There was an apparent tail directed sunward – though the appearance was actually due to thinly-spread matter in the comet's orbit, catching the sunlight at a suitable angle. (The Skylab astronauts also reported a 'spike' with Kohoutek's Comet of 1973.)

✳ *The first discovery of a hydrogen cloud associated with a comet* was made in 1969, when Comet Tago-Sato-Kosaka was studied from the space-vehicle OAO2 (Orbiting Astronomical Observatory 2) and was found to be surrounded by a hydrogen cloud approximately 1·6 million km in diameter. Similar clouds were associated later with Bennett's Comet of 1970 and Kohoutek's of 1973. Kohoutek's Comet was also found to contain large quantities of the ionized water molecule, H_2O^+.

✳ *The comet richest in large dust-grains* is P/Tempel 2. This was established from observations by IRAS (the Infra-Red Astronomical Satellite) in 1983. The particles range in size from pebbles about 6 cm across down to tiny grains only 0·006 mm across.

✳ *The greatest cometary disappointment of modern times* was Kohoutek's Comet of 1973, which was found on 7 March by L. Kohoutek at Hamburg when it was still almost 700 000 000 km from the Sun. Few comets are detectable as far away as this, and the comet was expected to become a magnificent object in the winter of 1973–4, but it failed to come up to expectations even though it was visible with the naked eye. It was, however, scientifically interesting, and was carefully studied by the astronauts then aboard the US space-station Skylab (Carr, Gibson and Pogue). Perhaps the comet will do better when it next returns to the Sun, in approximately 75 000 years' time!

✳ *The last brilliant comet* (to the spring of 1995) was West's, first recorded on a photograph taken at La Silla, in Chile, on 10 August 1975. During March 1976 it was a prominent naked-eye object with a major tail, but after perihelion, and as it receded, it showed signs of disintegration. The estimated period is 300 000 years.

✳ *The only comets to have been followed out to 2000 million km* have been Cernis (1983 XII), Bowell (1982 I) and Halley (1994).

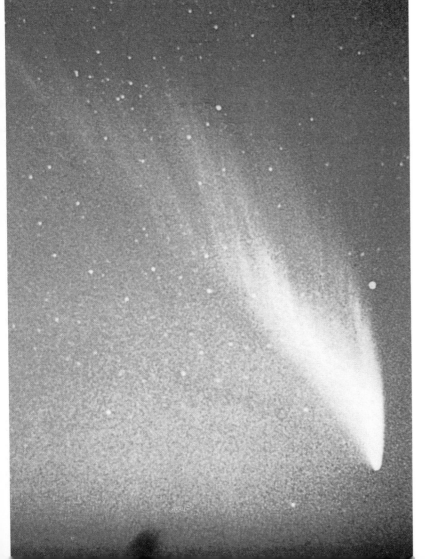

West's Comet of 1976; photograph by Richard West. The tail has a length of almost 50 000 000 km

127

Halley's Comet, 1910 – much brighter than it was in 1986: 12 and 15 May, as seen from Honolulu. (Palomar Obs.)

Bowell's Comet was under observation for more than four years, which is a record for a non-periodical comet; if it had come closer it would have been brilliant; it had an extensive coma and a peculiar narrow tail. Halley's Comet was photographed in February 1991 at a distance of 14·5 astronomical units.

✳ *The greatest perihelion distance for any comet* is that of Schuster's Comet of 1976, discovered by H. Schuster from Chile on 25 February. The perihelion distance was just over one thousand million km, not far from midway between the orbits of Jupiter and Saturn. (The previous record-holder had been Van den Bergh's Comet of 1974: 900 000 000 km.)

✳ *The record for comet discoveries* is held by J. L. Pons, who discovered a total of 37. C. Messier discovered 13. *The record year* for discoveries of new comets or recoveries of previously known ones was 34 in 1991.

✳ *The most successful 20th-century comet-hunter* is the Australian amateur Bill Bradfield, with 14 discoveries to his credit (as of 1995).

✳ *The first comet to be encountered by a space-probe* was Giacobini-Zinner, in September 1985. The probe was ICE (International Cometary Explorer), formerly known as ISEE (International Sun-Earth Explorer) which had been launched in 1978 mainly to study the solar wind. On 10 June 1982 a complicated series of manœuvres began, involving five passes of the Moon and leading to a rendezvous with the comet – though since ICE carried no camera, there were no pictures returned.

✳ *The origin of comets* is still a matter for debate. Most astronomers (though not all) believe that they are bona-fide members of the Solar System, and they are certainly very ancient. In 1950 the Dutch astronomer J. H. Oort suggested that they come from a cloud of bodies lying between 50 000 and 150 000 astronomical units from the Sun – that is to say, around one light-year (a light-year is equal to 63 240 astronomical units). The total population of the cloud could be of the order of 200 000 million, with a combined mass of up to 100 times that of the Earth. If one of these primordial, ice-rich bodies – never incorporated into a planet – were perturbed for some reason, perhaps by the pull of a passing star, it would start to fall inward towards the Sun. It might then swing round the Sun and return to the Oort Cloud; it might be perturbed by the pull of a planet and expelled from the Solar System altogether; it might be destroyed by collision with the Sun or a planet, or it might be forced into a short-period orbit.

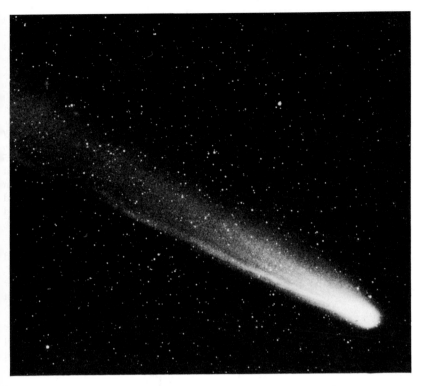

Kohoutek's Comet of 1973–4; Catalina Observatory, 15 Jan. 1974, with the 42 cm Schmidt telescope, 10 minute exposure. (by R. B. Minton, NASA)

This is still the favoured explanation for the long-period comets, but it now seems likely that the short-period comets come from the Kuiper Cloud, at a distance of between 40 and 100 astronomical units from the Sun – not so very far beyond the orbit of Neptune (30 astronomical units). The Kuiper Cloud may contain up to 1000 million bodies, with a total mass from 0·02 to 1 that of the Earth. Short-period comets have limited lifetimes, since they lose material by evaporation at every perihelion passage, and so there must presumably be some source of replenishment to maintain the numbers of short-period comets actually observed.

✳ *The first confirmed case of a comet becoming asteroidal* was that of Comet Wilson-Harrington, discovered in 1949; it had a short tail. The period was given as 2·3 years, but it was not seen again until 1979, when it was recovered as an asteroid by Eleanor Helin and given an asteroidal number, 4015. This lends support to the idea that at least some near-Earth asteroids are ex-comets which have lost all their volatiles, and moreover two periodical comets, Arend-Rigaux and Neujmin 1, now appear sensibly stellar. In 1982 a theory was proposed by two astronomers of the Royal Observatory Edinburgh, V. Clube and W. Napier, that the Apollo-type asteroid Hephaistos (diameter 10 km), discovered in 1978, was a comet in near-historical times, and was brilliant enough to give rise to some of the old myths about comets and disasters. They also maintain that every time the Sun passes through a spiral arm of the Galaxy, the Oort cloud is replenished, so that for a definite period after the Sun emerges from the arm brilliant comets and, later, collisions from Apollo or Aten type asteroids are more frequent than at other times. They also believe that such an impact occurred some 65 000 000 years ago, and altered conditions on Earth so markedly that dinosaurs became extinct. The theory is admittedly highly speculative!

COMET NOMENCLATURE

*N*ormally, a comet is named after its discoverer (e.g. Comet Lovas of 1974), after independent discoverers (Arend-Roland of 1957) or after the mathematicians who computed the orbits (Lexell, 1770). According to

PERIODICAL COMETS

Name	Period years	q (au) (perihelion)	Q (au) (aphelion)	Ecc	Incl. deg	Abs Mag
Encke	3·31	0·341	4·10	0·846	11·9	11
Grigg-Skjellerup	5·10	0·995	2·96	0·664	6·6	12
du Toit-Martley	5·21	1·197	3·01	0·602	2·9	
Maccholz 2	5·23	0·752	3·01	0·750	12·8	16
Tempel 2	5·29	1·195	4·81	0·545	12·4	10
Honda-Mrkós-Pajdusáková	5·30	0·542	5·54	0·822	4·2	11
Schwassmann-Wachmann 3	5·34	0·941	5·19	0·693	11·4	11
Neujmin 2	5·43	1·338	4·84	0·567	10·6	11
Tuttle-Giacobini-Kresak	5·46	1·065	3·10	0·656	9·2	
Tempel 1	5·49	1·491	4·73	0·521	10·6	9
Wirtanen	5·50	1·083	3·115	0·652	11·7	16
Clark	5·51	1·552	3·12	0·502	9·5	12
Tempel-Swift	5·68	1·153	5·22	0·638	5·4	13
Howell	5·94	1·615	4·94	0·508	4·9	
Russell 1	6·10	1·612	5·06	0·517	22·7	15
West-Kohoutek-Ikemura	6·12	1·401	5·29	0·581	30·1	10
Wild 2	6·17	1·494	5·24	0·556	3·3	6
Forbes	6·27	1·479	5·32	0·565	4·7	10
Pons-Winnecke	6·36	1·254	5·61	0·635	22·3	14
du Toit-Neujmin-Delporte	6·37	1·708	5·16	0·503	2·9	14
Spitaler	6·37	1·818	5·06	0·471	12·8	16
Kopff	6·44	1·576	5·35	0·545	4·7	13
Schwassmann-Wachmann 2	6·50	2·135	4·83	0·387	3·7	11
D'Arrest	6·51	1·346	3·49	0·614	19·5	6
Bus	6·52	2·183	3·49	0·374	2·6	
Wolf-Harrington	6·57	1·608	3·49	0·539	18·4	13
Giacobini-Zinner	6·61	1·034	3·52	0·706	31·8	10
Churyumov-Gerasimenko	6·61	1·306	5·74	0·629	7·1	10
Kohoutek	6·67	1·785	3·54	0·496	5·9	
Tsuchinshan 1	6·67	1·508	5·58	0·575	10·5	14
Perrine-Mrkos	6·77	1·293	3·58	0·639	17·8	18
Reinmuth 2	6·74	1·946	5·19	0·455	7·0	10
Finlay	6·76	1·036	3·58	0·710	3·7	13
Harrington	6·78	1·572	3·58	0·561	8·6	15
Gunn	6·82	2·459	4·74	0·316	10·4	10
Arend-Rigaux	6·82	1·438	3·60	0·600	17·9	9
Tsuchinshan 2	6·82	1·782	3·60	0·504	6·7	14
Borrelly	6·88	1·365	3·62	0·623	30·2	13
Brooks 2	6·89	1·843	3·62	0·490	5·5	13
Giclas	6·96	1·847	3·64	0·493	7·3	13
Wild 3	6·91	2·299	3·63	0·366	15·4	

Name	Period years	q (au) (perihelion)	Q (au) (aphelion)	Ecc.	Incl. deg	Abs. Mag.
Johnson	6·94	2·302	4·98	0·367	13·7	10
Longmore	6·98	2·399	3·65	0·343	24·4	
Taylor	6·99	1·961	5·35	0·464	20·5	12
Holmes	7·06	2·160	5·20	0·413	19·2	13
Daniel	7·06	1·649	3·68	0·551	20·1	11
Reinmuth I	7·31	1·874	3·77	0·562	8·2	14
de Vico-Swift	7·32	2·145	3·77	0·431	6·1	
Faye	7·34	1·593	3·78	0·578	9·0	8
Russell 2	7·38	2·276	3·79	0·400	12·0	
Shajn-Schaldach	7·46	2·331	5·31	0·390	6·1	12
Ashbrook-Jackson	7·47	2·307	5·34	0·396	12·5	7
Skiff-Kosai	7·54	2·846	3·85	0·260	3·2	
Harrington-Abell	7·59	1·774	3·86	0·540	10·2	16
Russell II	7·62	2·175	3·87	0·438	17·7	
Oterma	7·88	3·388	4·53	0·144	4·5	9
Kojima	7·89	2·414	5·51	0·391		
Gehrels 2	7·98	2·362	5·62	0·409	0·9	
Arend	7·99	1·850	4·00	0·537	19·9	14
Peters-Hartley	8·12	1·627	6·46	0·598	29·8	8
Gehrels 3	8·14	3·442	4·65	0·149	4·7	9
Jackson-Neujmin	8·24	1·381	4·08	0·661	13·5	17
Wolf	8·25	2·428	4·08	0·406	27·5	13
Schaumasse	8·26	1·213	6·96	0·703	11·8	11
Whipple	8·53	3·093	4·17	0·259	9·9	
Smirnova-Chernykh	8·57	3·572	4·19	0·147	6·6	8
Comas Solà	8·94	1·870	6·74	0·566	13·0	8
Kwerns-Kwee	8·99	2·224	6·42	0·486	9·0	11
Denning-Fujikawa	9·01	0·779	7·88	0·820	8·7	11
Swift-Gehrels	9·21	1·355	4·39	0·691	9·2	15
Neujmin 3	10·6	1·976	7·66	0·590	3·9	14
Väisälä I	10·9	1·800	8·02	0·633	11·6	13
Klemola	10·9	1·766	8·09	0·642	10·6	10
Boethin	11·2	1·114	8·91	0·778	5·8	10
Gale	11·3	1·214	5·02	0·758	10·7	10
Slaughter-Burnham	11·6	2·544	7·71	0·504	8·2	14
Van Biesbroeck	12·4	2·401	5·37	0·552	6·6	7
Machholz 2	12·8	0·752	3·01	0·750	12·8	
Wild I	13·3	1·981	9·24	0·647	9·2	14
Tuttle	13·5	0·998	5·67	0·824	54·6	8
Gehrels I	14·5	2·935	8·97	0·507	9·6	11
du Toit	15·0	1·294	10·9	0·787	18·7	16
Schwassmann-Wachmann I	15·0	5·448	6·73	0·105	6·7	5
Neujmin I	18·2	1·553	12·3	0·776	12·3	10
Crommelin	27·4	0·735	17·4	0·919	29·1	11
Tempel-Tuttle	32·9	0·982	19·6	0·982	162·7	13
Stephan-Oterma	37·7	1·574	20·9	0·860	18·0	5
Olbers	69·6	1·178	32·6	0·930	44·6	5
Pons-Brooks	70·9	0·774	33·5	0·955	74·2	6
Brorsen-Metcalf	72·0	0·485	34·1	0·972	19·2	9
Halley	76·0	0·587	35·3	0·967	162·2	4
Swift-Tuttle	133	0·958	51·7	0·964	113·4	4
Herschel-Rigollet	155	0·748	56·9	0·974	64·2	8

LOST PERIODICAL COMETS

Name	Period years	q	Q	Ecc.	Incl. deg	Returns	Last seen
Brorsen	5·5	0·59	5·61	0·81	29·4	5	1879
Biela	6·6	0·86	6·19	0·76	12·6	6	1852
Westphal	61·9	1·25	30·03	0·92	40·9	2	1913
Shoemaker-Levy 9	–	–	–	–	–	1	1994

It seems certain that of these comets, the first three have disintegrated. Shoemaker-Levy 9 was discovered when already in orbit round Jupiter, and impacted the planet in July 1994.

thus the comet formerly called Pons-Coggia-Winnecke-Forbes is now known as P/ Crommelin, since the identity of the four observed comets was established by A. C. D. Crommelin of Greenwich Observatory.

However, at the General Assembly of the International Astronomical Union held in 1994, it was agreed to abandon the old system and introduce a new one – partly because there can be some difficulty in deciding whether an object is cometary or asteroidal. The prefix P/ will be retained for periodical comets, and D/ for a comet which has been destroyed. In the new system, comets will be recorded by the half-month. Thus the third comet reported as discovered during the second half of February 1996 would be designated 1996 D3. In a process analogous to the permanent numbering of asteroids, sequential numbers will be assigned to comets whose periodicity has been well established, and in such cases these numbers immediately precede the P/ notation.

It must be remembered that these elements alter considerably for each revolution. The values given here apply to the beginning of 1992. Du Toit 1 (Comet 1944 III) was reported again in 1974, but the observations cannot be regarded as definite.

Orbits with inclinations over 90 deg indicate retrograde motion. There are two comets in the list with retrograde orbits, Tempel-Tuttle and Halley.

✳ *The greatest number of observed returns for a periodical comet* (to 1995) is 56 (Encke's). Comets which have been observed at 10 or more returns are:

Comet	Discovery	Returns
Encke	1786	56
Grigg-Skjellerup	1902	19
Tempel 2	1873	20
Pons-Winnecke	1819	20
D'Arrest	1851	16
Kopff	1906	13
Schwassmann-Wachmann	1929	11
Giacobini-Zinner	1900	11
Borrelly	1905	12
Brooks 2	1889	14
Finlay	1886	12
Faye	1843	18
Wolf I	1884	13
Tuttle	1790	10
Halley	240 BC	30

Periodical comets seen at only one return

There are various reasons for 'losing' comets. A few, listed separately, seem definitely to have disintegrated. In other cases the orbit has been so perturbed that we have

the old system, a comet is given a letter to correspond with its discovery; thus the first comet to be discovered in 1987 was 1987a, the next 1987b and so on. This is a provisional designation, and a permanent designation depends upon the time of perihelion; thus the first comet to reach perihelion in 1987 was 1987 I, the next 1987

II, and so on. These two systems may disagree with respect to the year; thus Comet 1995m may well become 1996 III!

The periodical comets are identified by P/ (e.g. P/Halley, P/Encke). They may be named according to their discoverers at different returns (P/Pons-Winnecke, P/Di Vico-Swift). Occasionally a name is altered;

lost contact; the classic case is that of Lexell's Comet of 1770. This list is restricted to comets whose sole recorded appearance took place before 1970, so that their recovery must be considered doubtful – though by no means impossible!

Brilliant Comets seen since 1700

1744 DeChéseaux' six-tailed comet (actually discovered by Klinkenberg).
1811 Great Comet: discovered by H. Flaugergues.
1843 Great Comet, superior to that of 1811
1858 Donati's Comet, with its curved tails – possibly the most beautiful comet ever seen.
1861 Brilliant comet, discovered by Tebbutt.
1862 Bright comet, discovered by Swift.
1874 Coggia's Comet; bright naked-eye object.
1882 Great Comet, photographed by Gill from South Africa.
1901 Bright southern comet, discovered by Paysandu.
1910 Daylight Comet; brightest of the century (so far).
1927 Skjellerup's Comet. Bright for a brief period.
1947 Bright southern comet; brief visibility.
1965 Comet Ikeya-Seki; brilliant as seen from the southern hemisphere for a brief period (also well seen from the USA).
1970 Bennett's Comet; fairly bright.
1976 West's Comet; bright naked-eye object during early mornings in March.
To this list must be added Halley's Comet at its returns of 1759, 1835 and 1910.

PERIODICAL COMETS SEEN AT ONLY ONE RETURN

Comet	Year	Period years	q (au)	Q (au)	Ecc.	Inclin.
Helfenzrieder	1766	4·35	0·406	4·92	0·848	7·9
Blanpain	1819	5·10	0·892	5·03	0·699	9·1
Barnard 1	1884	5·38	1·279	4·86	0·583	5·5
Brooks 1	1886	5·44	1·325	4·86	0·571	12·7
Lexell	1770	5·60	0·674	5·63	0·786	1·6
Pigott	1783	5·89	1·459	5·06	0·552	45·1
Harrington-Wilson	1951	6·36	1·664	5·10	0·515	16·4
Barnard 3	1891	6·52	1·432	5·55	0·590	31·3
Giacobini	1896	6·65	1·455	5·62	0·588	11·4
Schorr	1918	6·67	1·884	5·21	0·469	5·6
Swift	1895	7·20	1·298	6·16	0·652	3·0
Denning	1894	7·42	1·147	6·01	0·698	5·5
Metcalf	1906	7·78	1·631	6·22	0·584	14·6
Van Houten	1961	15·6	3·957	8·54	0·367	6·7
Pons-Gambart	1827	57·5	0·807	29·0	0·946	136·5
Dubiago	1921	62·3	0·929	30·3	0·929	22·3
de Vico	1846	76·3	0·664	35·3	0·963	85·1
Väisälä	1942	85·4	1·287	37·5	0·934	38·0
Barnard 2	1889	145	1·105	54·2	0·960	31·2
Mellish	1917	145	0·198	55·1	0·993	32·7
Wilk	1937	187	0·619	64·9	0·981	26·0

Comet Wilson-Harrington (1949) was recovered in 1979 as an asteroid, and given an asteroid number (4015)

Comets of very long period

Obviously, the periods are very uncertain!

Comet	Year	Period years	q (au)	Ecc.	Inclin.
Great Comet	1861 II	409	0·822	0·985	85
Great Comet	1843 I	517	0·0055	0·99991	14
Great Comet	1882 II	759	0·0077	0·99991	14
Ikeya-Seki	1965 VIII	880	0·008	0·9999	14
Pereyra	1963 V	903	0·0051	0·99995	14
Bennett	1970 II	1678	0·538	0·996	90
Donati	1858 VI	1951	0·578	0·996	11
Flaugergues	1811 I	3096	1·035	0·995	10
1680 Comet	1680	8917	0·006	0·9999	6

Among comets with hyperbolic orbits are Morehouse (1908 III), Arend-Roland (1957 III) and Kohoutek (1973 XII).

METEORS

Meteors may be regarded as the débris of the Solar System. They are small and friable, and cannot reach ground level intact. Sporadic meteors may appear from any direction at any moment, but many meteors are members of definite showers.

THE PRINCIPAL METEOR SHOWERS

The showers listed in the table occur annually, though not all of them are consistent: the Perseids are the most reliable, whereas the Leonids produce major showers only

Perseid meteor, 12 Aug. 1993. (M. de F. Maunder)

METEORS

Shower	Begins	Max	Ends	Max ZHR	Position of radiant R.A. h m	Dec deg	Associated Comet	Notes
Quadrantids	1 Jan.	4 Jan.	6 Jan.	60	15 28	+50	–	Radiant in Boötes. Short, sharp maximum. Blue meteors, with splendid trains.
Virginids	7 Apr.	12 Apr.	18 Apr.	5	13 36	–11	–	Slow, long paths. Several radiants in Virgo during March and April.
Lyrids	19 Apr.	21 Apr.	25 Apr.	10	18 08	+32	1861 I (Thatcher)	Occasional very rich displays, as in 1922 and 1982.
η Aquarids	24 Apr.	5 May	20 May	35	22 20	–01	P/Halley	Multiple radiant and broad maximum.
α Scorpiids	20 Apr.	13 May	19 May	5	16 04	–24	–	Several weak radiants from April to July.
Ophiuchids	19 May	9 June	July	5	17 20	–20	–	Feeble activity from several radiants.
α Cygnids	July	21 July, 21 Aug.	Aug.	5	21 00	+48	–	Several radiants. Less rich than formerly?
Capricornids	July	Mid-July	Aug.	5	21 00	–15	–	Bright meteors. Several maxima; multiple radiant.
δ Aquarids	15 July	29 July, 6 Aug.	20 Aug.	20	22 36	–17	–	Double radiant. Rich, but meteors faint.
Piscis Australids	15 July	31 July	20 Aug.	5	22 40	–30	–	Possibly two maxima. Not well studied as yet.
α Capricornids	15 July	2 Aug.	25 Aug.	5	20 36	–10	–	Slow, yellow fireballs. Triple maximum.
ι Aquarids	July	6 Aug.	Aug.	8	22 10	–15	–	Double radiant. Rich in faint meteors.
Perseids	23 July	12 Aug.	20 Aug.	75	03 04	+58	P/Swift-Tuttle	Rich annual shower. Consistent.
Piscids	Sept.	8 Sept.	Oct.	10	00 36	+07	–	Several radiants in Pisces and Aries.
Orionids	16 Oct.	22 Oct.	27 Oct.	25	06 24	+15	P/Halley	Swift, with fine trains.
Draconids	10 Oct.	10 Oct.	10 Oct.	var.	18 00	+54	P/Giacobini-Zinner	Usually feeble, but very rare great storms.
Taurids	20 Oct.	3 Nov.	30 Nov.	10	03 44	+14	P/Encke	Slow meteors. Fine display in 1988.
Leonids	15 Nov.	17 Nov.	20 Nov.	var.	10 08	+22	P/Tempel-Tuttle	Occasional major storms, as in 1966.
Andromedids	15 Nov.	20 Nov.	6 Dec.	V. low	00 50	+55	P/Biela	Now almost extinct. (Comet is defunct.)
Puppids-Velids	27 Nov.	9 Dec., 26 Dec.	Jan.	15	09 00	–48	–	Several radiants in Puppis, Carina and Vela from November to January.
Geminids	7 Dec.	13 Dec.	16 Dec.	75	07 28	+32	Phæthon (asteroid)	Rich annual shower.
Ursids	17 Dec.	23 Dec.	25 Dec.	5	14 28	+78	P/Tuttle	Can be rich, as in 1945 and 1986.

Permanent daytime showers include the Arietids (29 March to 17 June), the ξ Perseids (1 to 15 June) and the β Taurids (23 June to 7 July). The β Taurids seem to be associated with Encke's Comet.

occasionally (the last being in 1966). The ZHR or zenithal hourly rate is the probable hourly rate of meteors for an observer who has the radiant at his zenith, and is observing under ideal conditions. In practice, of course, these conditions are never fulfilled, so that the ZHR is the theoretical maximum only.

✳ **The only meteor shower named after a discarded constellation** is that of the Quadrantids. The old constellation, Quadrans Muralis, has been long since rejected.

✳ **The maximum velocity of meteors entering the atmosphere** is around 72 km/s. Most meteors vaporize above altitudes of 80 km. Micrometeorites, about 0·1 mm in diameter, are too small to cause luminous effects.

✳ **The total number of meteors entering the atmosphere daily** has been estimated at 75 000 000 for meteors of magnitude 5 or brighter. An observer under ideal conditions could normally be expected to see about 10 naked-eye meteors per hour (except during a shower, when the number would naturally be higher).

✳ **Meteor showers were first postulated** by Olmsted and Twining in 1834. H. Olbers suggested a period of 33 years for the Leonid shower. H. A. Newton came independently to the same conclusion, and traced the Leonids back to AD 902; he predicted another major shower for 1866, and it duly occurred.

✳ **The Perseid shower** was identified by Quetelet in 1836. The shower has been nicknamed 'the tears of St Lawrence'.

✳ **The richest showers** on record are probably been the Leonids, which on 17 November 1966 reached a rate of 60 000 per hour over a period of 40 minutes. There were major showers also in 1799, 1833 and 1866, but not in 1899 or 1933, because of perturbations by Jupiter and Saturn.

✳ **The first radiant point to be measured** was that of the Leonids, on 12–13 November 1833.

✳ **The first suggestion of an association between meteors and comets** was made by D. Kirkwood in 1861; he believed meteors to be the débris of disintegrated comets. In 1862 G. V. Schiaparelli showed the association of the Perseids with comet P/Swift-

Tuttle. The November Andromedids were subsequently shown to be associated with the defunct comet P/Biela. On 27 November 1872 the hourly rate of the Andromedids (sometimes termed Bieliids) reached 100 per hour, but in recent years the shower has become very feeble.

✳ **The greatest 'cometary' shower of meteors** in the present century (excluding the 1966 Leonids) was that of 9 October 1933, associated with Comet P/Giacobini-Zinner. The rate reached 350 meteors per minute for a brief period.

✳ **The first measures of meteor heights by the triangulation method** (i.e. observing the same meteors from widely separated points – at least two observers, at two different stations) were made by two German students, Brandes and Benzenberg, in 1798. They concluded that most meteors penetrate down to an altitude of 80 km above the ground.

✳ **The earliest meteor photograph** (of an Andromedid) was taken on 27 November 1885 by L. Weinek, at Prague.

✳ **The first radar measurements of meteors**

The greatest meteor shower ever recorded, the Leonids, over Arizona, USA on 17 Nov. 1966.

were made during the last war; pioneer work was done by the British radio astronomers Hey and Stewart, who in 1945 measured the Delta Aquarids by radar. Nowadays, radar has largely superseded the old method of visual observation.

✳ *The most reliable annual shower* is that of the August Perseids, which never fail to produce a spectacular display. It is noteworthy that so far as meteor observation is concerned, the northern hemisphere has the best of matters!

✳ *Noctilucent clouds*, which occur at heights of from 80 to 100 km, are thought to be due to ice crystals forming along meteor trails, though upon this point there is still no universal agreement among astronomers and meteorologists.

METEORITES

Meteorites reach ground level without being destroyed, and are not simply large meteors. They do not belong to showers, and any association with comets is dubious. Meteorites may in fact be more closely associated with asteroids, and it has been claimed that there is no distinction between a small asteroid and a large meteorite!

One important sub-group consists of the carbonaceous chondrites, which contain carbon compounds and organic matter (such as hydrocarbons). One famous example is the Orgueil Meteorite, which fell in France on 14 May 1864. In 1964 B. Nagy, at Columbia University together with his colleague

Claus, suggested that this meteorite contained 'organized elements' – leading to the suggestion that there might be evidence of past living material. However, others maintain that these substances are due to terrestrial contamination. (In 1908 Svante Arrhenius suggested that life was brought to Earth by a meteorite; this 'panspermia' theory never met with much support, but has recently been revived, in modified form, by F. Hoyle and C. Wickramasinghe.)

✳ *The largest meteorite widely assumed to have come from the Moon* consists of two stones, MAC 88104 and 88105, found in the MacAlpine Hills region of Antarctica in

1990. The combined mass is 724 grams; the stones are breccias (fused collections of rock fragments). It has been estimated that they took about 100 000 years to reach the Earth, and have been awaiting discovery for at least 30 000 years. Other lunar meteorites have been identified, and also some from Mars, such as the Alban Hills 84001 meteorite, again found in Antarctica. It may have fallen over 12 000 years ago. The evidence in favour of a lunar or Martian origin for some meteorites is strong, though it is not accepted by all authorities.

✳ *The earliest reports of meteoritic phenomena* are recorded on Egyptian papyrus, around 2000 BC. Early meteorite falls are, naturally, poorly documented, but it has been suggested that a meteorite fell in Crete in 1478 BC; stones near Orchomenos in Boetia in 1200 BC, and an iron on Mount Ida in Crete in 1168 BC. According to Livy, 'stones' fell on Alban Hill in 634 BC, and there is evidence that in 416 BC a meteorite fell at Ægospotamos in Greece. The Sacred Stone in Mecca is certainly a meteorite. *The oldest meteorite which can be positively dated* is the Ensisheim Meteorite, which fell on 16 November 1492 and is now on show in Ensisheim Church (Switzerland).

In India, it is said that the Emperor Jahangir ordered two sword-blades, a dagger and knife to be made from the Jalandhar Meteorite of 10 April 1621. A sword was made from a meteorite which fell in Mongolia in 1670, and in the 19th century part of a South African meteorite was used to make a sword for the Emperor Alexander of Russia. *Meteorite ages* are usually estimated at between 4 and 4·6 thousand million years.

formation in Northern Territory, Australia, is almost certainly an ancient impact structure.

✳ *The youngest known impact crater* may be the Merna crater, a circular depression in a Central Nebraska cornfield, now just over 20 m deep. W. Dort suggested that it was formed between 500 and 1000 years ago, and could have been observed by the Pawnee Indians; the original depth may have been over 150 m. However, the evidence is far from conclusive.

✳ *The greatest meteoritic impact* may have been that of 65 000 000 years ago – according to some authorities! It has been claimed that this impact caused such a climatic change that it led to the extinction of the dinosaurs. It has also been claimed that the Chicxulub formation in the northern Yucatán region of

✳ *The largest tektite known* was found in 1932 at Muong Nong, Laos; it weighs 3·2 kg. It was once thought that tektites were meteoric (in 1897 R. O. M. Verbeek even suggested that they might have come from the Moon), but nowadays it is more generally believed that they are terrestrial in origin, though they remain decidedly enigmatical.

✳ *The first suggestion that objects could 'fall from the sky'* was made by E. F. Chladni in 1794, but his idea was met with considerable scepticism. (As recently as 1807 Thomas Jefferson, President of the United States, is quoted as saying 'I could more easily believe that two Yankee professors would lie than that stones would fall from heaven.')

✳ *The first scientific proof of meteorite falls* was obtained by the French scientist J. B. Biot, who investigated the meteoritic shower at L'Aigle on 26 April 1803.

✳ *The classification of meteorites* is, broadly speaking, into three types: stones (aerolites), stony-irons (siderolites) and irons (siderites)

Meteorite Craters

Name	Diameter metres (largest crater)	Date of discovery	Notes
Meteor Crater, Arizona	1265	1891	Many metallic meteorites found.
Wolf Creek, Australia	850	1947	
Henbury, Australia	200 × 110	1931	Thirteen craters.
Boxhole, Australia	175	1937	
Odessa, Texas	170	1921	
Tswainga, S. Africa	132	1991	'Saltpan'. Depth 120 m.
Waqar, Arabia	100	1932	Two craters.
Oesel, Estonia	100	1927	Six craters.
Campo del Cielo, Argentina	75	1933	Many craters.
Dalgaranga, Australia	70	1928	
Sikhote-Alin, Siberia	28	1947	Total of 106 craters. (Fall observed.)

✳ *The most famous terrestrial meteorite crater* is in Arizona (lat 35 deg ·3 N., long 111 deg ·2 W.), 1265 metres in diameter and 175 metres deep. It is well-preserved, and there is no reasonable doubt about its origin – even though G. K. Gilbert, who was strongly of the opinion that the main lunar craters were due to impact, regarded the Arizona crater as volcanic! The crater is certainly more than 20 000 years old, and is a well-known tourist attraction. Other craters which are certainly meteoritic include those at Wolf Creek (Australia), Waqar (Arabia) and Tswainga (South Africa).

Various other craters have been attributed to meteoric impact. In some cases this may be true, but there are serious doubts about, for instance, the New Quebec crater in Canada (discovered 1950; diameter 3400 metres) and others in the Canadian shield, which appear to be associated with geological features. The Vredefort Ring near Pretoria, in South Africa, is often listed as an impact crater, but geologists who have studied it are unanimous in stating that it is of volcanic origin. The huge Gosse's Bluff

Mexico is the result of this impact. The structure is buried below a thick layer of sedimentary rock, and studies of the gravity and magnetic fields indicate that the hidden crater is about 180 km in diameter; there are three major ring structures round its rim, and the suggestion is that the whole multi-ring structure has a diameter of at least 300 km.

Not all investigators agree with these conclusions, or even that the disappearance of the dinosaurs was due to an impact event. However, the theory is certainly worthy of serious consideration.

✳ *Tektites* are small glassy objects, found only in a few regions. They seem to have been heated twice, and are usually aerodynamically shaped. Four main groups are known:

IRONS (SIDERITES)

Hexahedrites contain between 4 and 6 per cent of nickel, and consist of the cubic mineral kamacite. With increased nickel another cubic mineral, taenite, is formed. These two minerals orient themselves parallel to octahedral planes, and when the surface is etched with acid after polishing the characteristic Widmanstätten patterns are revealed (*octahedrites*). The *nickel-rich ataxites*, with more than 12 per cent nickel, do not show the Widmanstätten patterns, and consist of plessite.

STONY IRONS (SIDEROLITES)

Pallasites consist of a network of nickel-iron enclosing crystals of olivine. *Mesosiderites* are heterogeneous aggregates of silicate minerals and a nickel-iron alloy. There are two further groups, *siderophytes* and *lodranites*, but only one example of each is known.

STONES (AEROLITES)

Chondrites contain spherical particles called chondrules. These chondrules are fragments of minerals, and show a radiating structure. Chondrites account for 85 per cent of known specimens. In *achondrites*, chondrules are

Region	Name	Geological age
Australasia	Australites	Middle/Late Pleistocene
Ivory Coast	Ivory Coast tektites	Lower Pleistocene
Czechoslovakia	Moldavites	Miocene
United States	Bediasites (Texas)	
	Georgiaites (Georgia)	Oligocene

Meteor crater, Arizona, 1994.

absent; the achondrites are of coarser structure, with little free iron. They may be calcium-rich or calcium-poor.

The early systems of classification were due to G. Rose (1863), G. Tschermak (1883) and A. Brezina (1904); they were extended by G. Prior (1920) and latterly by G. J. H. McCall (1973). Stones are more commonly found than irons in the ratio of 96 per cent to 4 per cent, but this is misleading, as irons are much more durable and are more likely to survive. Over 10 000 meteorites have been located, though few have actually been seen to fall. Among famous falls which have resulted in meteorite recovery are those of the Příbram fireball (Czechoslovakia, which was recorded as magnitude −19!) on 7 April 1959, the Lost City meteorite (Oklahoma) in 1970, the Barwell Meteorite (Leicestershire) of 24 December 1965, and the Sikhote-Alin fall in Siberia on 12 February 1947 – the greatest of the century, apart from that of 1908 – which was an iron, and produced many craters, since it broke up during its descent.

Of special interest are eight meteorites known as the *SNC Meteorites* after the regions in which they were found (Shergotty in India, Nakhla in Egypt and Chassigny in France). They seem to have crystallized only 1·3 thousand million years ago, and their composition and texture indicates that they formed on or in a planet which had a strong gravitational field. They have concentration of volatile elements, and glassy incursions which were permanently formed in the extreme heat of whichever process ejected them from a parent body. These glassy incursions have trapped gases such as Ar, Kr, Xe and N. It has been suggested that they are of Martian origin, though this is of course highly speculative.

* *There is no record of any human death* due to a meteorite. Reports that a monk was killed at Cremona in 1511, and another monk at Milan in 1650, are unsubstantiated. However, in 1954 a woman in Alabama, USA, had a narrow escape when a meteorite fell through the roof of her house, and she suffered a minor injury to her arm. The only casualty has been a dog, which was in the wrong place at the wrong time when a meteorite landed in Egypt.

* *The narrowest escape of recent times* was that of José Martin, on 21 June 1994. He was driving his car from Madrid to Marbella, in Spain, when a 1·4 kg meteorite crashed through the windscreen, ricocheted off the dashboard and injured the driver's finger – fortunately not seriously. More than 50 fragments were later found within 200 m of the impact. There have been several more instances. Two boys in Noblesville, Indiana, were lucky on 31 August 1991; at 7 p.m. local time a meteorite landed 3·5 m away from them, making a crater 9 cm wide and 4 cm deep. The boys – Brodie Spaulding and Brian Kinzie – found the 'small black stone', which was still warm; it proved to be an unusual sort of chondrite. On 9 October 1992 a 12 kg meteorite landed on the bonnet of an unoccupied car at Peeksville, New York, belonging to Michelle Knapp; and on 10 December 1992 a house in Japan, belonging to Masaru and Maiko Matsumoto, was struck by a 6·5 kg chondrite.

Meteorites are not always easy to identify on sight, but when etched with acid irons will show the characteristic *Widmanstätten patterns*, not found elsewhere.

* *The largest known meteorite* is still lying where it fell, in prehistoric times, at Hoba West, near Grootfontein in South-West Africa; it weighs at least 60 tons. All the known meteorites weighing more than 10 tons are irons. They are:

Meteorite	Weight tonnes
Hoba West, Grootfontein, SW Africa	60
Ahnighito (The Tent), Cape York, W Greenland	30·4
Bacuberito, Mexico	27
Mbosi, Tanganyika	26
Agpalik, Cape York, W Greenland	20·1
Armanty, Outer Mongolia	20 (est.)
Willamette, Oregon, USA	14
Chupaderos, Mexico	14
Campo del Cielo, Argentina	13
Mundrabilla, Western Australia	12
Morito, Mexico	11

There is a report of an iron found by W. A. Cassidy in Argentina, in September 1969, which is said to have a weight of about 18 tonnes.

* *The largest stone meteorite* fell in Kirin Province, Manchuria, on 8 March 1976. It weighs 1766 kg.

* *The largest meteorite in a museum* is the Ahnighito, found by Robert Peary in Greenland in 1897; it is now in the Hayden Planetarium, New York, along with two other meteorites found at the same time and on the same site (known as The Woman and The Dog). Apparently the local Eskimos were rather reluctant to let them go! The Willamette meteorite is also in the Hayden Planetarium. (This meteorite was the subject of a lawsuit. It was found in 1902 on property belonging to the Oregon Iron and Steel Company. The discoverer moved it to his own property, and exhibited it; the Company sued him for possession, but the Court ruled in favour of the discoverer.)

* *The most famous fall of recent times* was that of 30 June 1908, in the Tunguska region of Siberia. As seen from Kansk, 600 km away, the descending object was said to outshine the Sun, and detonations were heard 1000 km away; reindeer were killed and pine-trees blown flat over a large area. The first expedition to the site was not until 1927, led by L. Kulik. No fragments have been found, and it has been suggested that the object was the nucleus of a small comet, or else a fragment from Encke's Comet – which, if icy in nature, would presumably evaporate during the descent and landing. (Inevitably, flying saucer enthusiasts have suggested that it was a space-ship in trouble!)

A second major fall in Siberia occurred on 12 February 1947, in the Sikhote-Alin area. The fall was observed, and more than 100 craters were located, of which thirty were of appreciable size. It is fortunate that both these Siberian falls occurred in uninhabited territory. If a meteorite of this size had hit a city, the death-toll would inevitably have been high.

* *Twenty-three meteorite falls in the British Isles* are known:

1623 January 10	Stretchleigh, Devon. 12 kg.	
1628 April 9	Hatford, Berkshire. Three stones; about 33 kg.	
1779 –	Pettiswood, West Meath.	
1795 December 13	Wold Cottage, Yorkshire. 25·4 kg.	
1804 April 5	High Possil, Strathclyde (Lanarkshire). 4·5 kg.	
1810 August ?	Mooresfort, Tipperary. 3·2 kg.	
1813 September 10	Limerick. 48 kg (shower).	
1830 February 15	Launton, Oxfordshire. 0·9 kg.	
1830 May 17	Perth. About 11 kg.	
1835 August 4	Aldsworth, Gloucestershire. Small shower, over 0·5 kg.	
1844 April 29	Killeter, Tyrone. Small shower; small amount preserved.	
1865 August 12	Dundrum, Tipperary. 1·8 kg.	
1876 April 20	Rowton, Shropshire. 3·2 kg. (Iron meteorite.)	
1881 March 14	Middlesbrough, Cleveland (Yorkshire). 1·4 kg.	
1902 September 13	Crumlin, Antrim. 4·1 kg.	
1914 October 13	Appley Bridge, Lancashire. 33 kg.	
1917 December 3	Strathmore, Tayside (Perthshire). 13 kg (four stones).	
1923 March 9	Ashdon, Essex. 0·9 kg.	
1931 April 14	Pontlyfni, Gwynedd (Caernarvon). 120 g.	
1949 September 21	Beddgelert, Gwynedd (Caernarvon). 723 g.	
1965 December 24	Barwell, Leicestershire. 46 kg (total).	
1969 April 25	Bovedy, N. Ireland. Main mass presumably fell in the sea.	
1991 May 5	Glatton, Cambridgeshire. 767 g.	

The Barwell fall was well observed. Many fragments of the meteorite were found; one was detected some time later nestling in a vase of artificial flowers on the windowsill of a house in Barwell village. Though it broke up during its descent, the stone is the largest known to have fallen over the British Isles.

The Bovedy meteorite was also well observed during descent, but the main mass has not been recovered.

The most recent British meteorite fell at Glatton, Cambridgeshire. The fall was witnessed by A. Pettifor, who heard a loud whining noise and the crash of the stone into a conifer hedge some 20 metres away from him. He found the meteorite, which had made a shallow depression 2 cm deep. It was warm, not hot, when picked up. It has a granular structure, indicating that soon after its formation as part of an asteroid it had been hot, not melted, causing the mineral grains to grow and interlock. It is an ordinary chondrite of the low-iron group (L-group); 23 per cent by weight iron, about 5 per cent of which is in nickel-iron metal, with 18 per cent of stony minerals – mainly pyroxene and olivine, which are common components of terrestrial basaltic lavas.

✳ *The largest meteorites to have been found in different regions* are:

Region	Meteorite	Weight
Africa	Hoba West, Grootfontein	60 tonnes
USA	Willamette, Oregon	14 tonnes
Asia	Armanty, Outer Mongolia	20 tonnes
South America	Campo del Cielo, Argentina	13 tonnes
Australia	Mundrabilla	12 tonnes

Region	Meteorite	Weight
Europe	Magura, Czechoslovakia	1·5 tonnes
Ireland	Limerick	48 kg
England	Barwell, Leicestershire	±46 kg (total)
Scotland	Strathmore, Tayside (Perthshire)	10·1 kg
Wales	Beddgelert, Gwynedd (Caernarvon)	723 g

✳ *The brightest fireball ever seen* may have been that of 1 February 1994. It passed over the Western Pacific at 22.38 GMT; the magnitude was about −25. Presumably it was a rocky object; if moving at 15 km/s, it would have been about 7 m across, weighing 400 tons.

✳ *The only known case of a meteorite entering the Earth's atmosphere and leaving it again* is that of the object of 10 August 1972. It seems to have approached the Earth 'from behind' at a relative velocity of 10 km/s, which increased to 15 km/s as the Earth's gravity accelerated it. The object entered the atmosphere at a slight angle, becoming detectable at a height of 76 km above Utah and reaching its closest point to the ground at 58 km above Montana. It then began to move outward, and became undetectable at just over 100 km above Alberta after a period of visibility of 1 minute 41 seconds; the magnitude was estimated by eye-witnesses to be at least −15, and the diameter of the object may have been as much as 80 m. After emerging from the Earth's atmosphere it re-entered a solar orbit, admittedly somewhat modified by its encounter, and presumably it is still orbiting the Sun.

GLOWS AND ATMOSPHERIC EFFECTS

AURORÆ

Auroræ, or polar lights (Aurora Borealis in the northern hemisphere, Aurora Australis in the southern) must have been observed in ancient times, since they may often become extremely brilliant. There is excellent evidence that the Roman emperor Tiberius, who reigned from AD 14 to 37, once dispatched his fire-fighters to the port of Ostia on account of a red glow in the sky, seen from Rome, which proved to be auroral.

✳ *The first use of the term 'aurora'* was due to the French astronomer P. Gassendi, following the brilliant display of 12 November 1621. However, *the first really good description of a display of Aurora Borealis* was given by K. Gesner of Zürich for the aurora of 27 December 1560. *The first Antarctic aurora* was described by Captain Cook on 20 February 1773.

✳ *The first major book dealing with auroræ* was the *Traité physique et historique de l'aurora boréale*, written by J. J. de Mairan in 1731.

✳ *The first suggestion of an association between auroræ and electrical discharges* may have been due to Halley in 1716, when he linked auroral displays with discharges associated with the Earth's magnetic field. (It must be remembered that Halley saw his first aurora in this year, following the end of the solar Maunder Minimum.)

✳ *The first really good photograph of an aurora* was obtained in 1885 by S. Tromholt, of Denmark.

✳ *The first suggestion of a connection between auroræ and magnetic effects* was made by Hjorter in 1741.

✳ *The first suggestion of a connection between auroræ and sunspot phenomena* was made by E. Loomis, of Yale, in 1870. The solar association was defined in more detail by the Italian astronomer G. Donati in 1872, following the great display of 4–5 February. In 1896 Birkeland reproduced 'miniature auroræ' by using a spherical electromagnet and a beam of cathode rays. In 1929 S. Chapman and V. Ferraro suggested that the cause of auroræ was solar plasma. It is now known that auroral displays are indeed due to electrified particles from space colliding with atoms and molecules in the upper atmosphere. Auroral activity is more or less permanent at high latitudes around the so-called auroral ovals, which are 'rings' asymmetrically displaced around the geomagnetic poles (at present the north magnetic pole lies in Nares Strait, at latitude 79 deg N longitude 71 deg W, between Greenland and Ellesmere Island). When there are major disturbances in the Sun, producing high-speed solar wind particles, the ovals broaden and expand, bringing auroral displays further from the main regions. *The maximum activity of auroræ* is seen along geomagnetic latitude 68 deg N or S. However, observers in North America are more favoured than observers at equivalent latitudes in Europe, because they are closer to the geomagnetic pole.

On average, auroræ of some kind or other are seen on 240 nights per year in North Alaska, North Canada, Iceland, North Norway and Novaya Zemlya; 25 nights per year along the Canada / United States border and in Central Scotland; and one night per year in Central France. Obviously, however, this varies according to the state of the solar cycle, and in general auroræ are most common two years after a spot-maximum.

✳ *Exceptionally brilliant auroræ* occurred on 24–5 October 1870, 4–5 February 1872, 25–6 January 1938 (spectacular in England, but not seen from Tromsø in Norway!), 13 March 1989 and 8–9 November 1991. In 1909 an aurora was seen from Singapore (lat. 1 deg 25 min), but auroræ are very rare in such low latitudes.

✳ *The heights of auroræ* vary. In general, the sharp lower boundary is at about 98 km above ground level, the maximum region of activity at about 110 km, and the normal upper boundary 300 km. In the cases of sunlit activity the upper limit may reach 700 km, or even 1000 km in extreme cases. Extremely low auroræ have been reported now and then, but with no certainty. There have also been many reports of noise accompanying auroral displays – either hissing or crackling. Many cases were listed by S. Tromholt in Norway in 1885 (reported in *Nature*, vol. 32, p. 499); it is extremely hard to explain the noise, but there does seem to be considerable weight of evidence that it occurs, even though final proof is still lacking.

Auroræ may be seen in various forms. There are glows; arcs, with or without rays; bands, more diffuse and irregular; draperies, or curtains made up of very long rays; individual rays or streamers; coronæ, or radi-ating systems of rays converging at the zenith; 'surfaces', which are diffuse patches, sometimes pulsating; flaming auroræ, composed of quickly-moving sheets of light; and ghost arcs, which may persist long after the main display has ended.

THE ZODIACAL LIGHT

*T*he Zodiacal Light may be seen as a faint cone of light rising from the horizon after sunset or before sunrise. It extends away from the Sun, and is generally observable for a fairly short period only after the Sun has disappeared or before it rises. On a clear, moonless night, under ideal conditions it contributes about one-third of the total sky light, and may be brighter than the average Milky Way region. It is due to particles scattered in the Solar System along and near the main plane of the system. The diameters of the particles are of the order of 0·1 to 0·2 micron. (One micron is equal to one-millionth of a metre.) Since the Zodiacal Light extends along the ecliptic, it is best seen when the ecliptic is most nearly vertical to the horizon – i.e. February to March and again in September to October.

✳ *The first scientific account* of the Zodiacal Light was given by G. D. Cassini in 1683; he correctly suggested that it was due to sunlight reflected from tiny particles orbiting the Sun near the plane of the ecliptic. However, it had been observed much earlier by the Persians and the Arabs, and had been clearly described by Joshua Childrey in the *Britannia Baconica* published in London in 1661.

THE GEGENSCHEIN

*T*he Gegenschein or Counterglow is seen as a faint patch of radiance in the position in the sky exactly opposite to the Sun. It is extremely elusive, and is generally visible only under near-perfect conditions. The best opportunities come when the anti-sun position is well away from the Milky Way (i.e. in February / April, and September / November) and when the anti-sun position is high, at local mid-night.* It has generally an oval shape, measuring 10 by 20 degrees, so that its maximum diameter is roughly 40 times that of the full moon.

✳ *The discovery of the Gegenschein* seems to have been due to Esprit Pézénas, who reported it to the Paris Academy in 1731. *It was named* by Humboldt, who saw it on 16 March 1803. It was described in more detail

---

* From England I have seen it only once – in March 1942, when the whole country was blacked out as a precaution against German air-raids.

---

by the Danish astronomer Theodor Brorsen, who saw it in 1854 and wrote about it in 1863, but (contrary to statements in many books) Brorsen did not claim the discovery for himself, as he was familiar with Humboldt's description.

✳ *The Zodiacal Band* is a very faint, parallel-sided band of radiance, which may extend to ether side of the Gegenschein, or be prolonged from the apex of the Zodiacal Light cone, or join the Zodiacal Light with the Gegenschein. It may be from 5 to 10 degrees wide, and is extremely faint. Like the Zodiacal Light and the Gegenschein, it is due to sunlight being reflected from interplanetary particles in the main plane of the Solar System.

OTHER GLOWS AND ATMOSPHERIC EFFECTS

*T*he Kordylewski Clouds were reported by the Polish astronomer K. Kordylewski in 1961. He believed them to be due to collections of interplanetary débris lying at the 'Lagrangian points' of the Moon's orbit – that is to say, moving in the same path as the Moon, but at distances of 60 degrees ahead and 60 degrees behind respectively. (The Trojan asteroids behave in this way with respect to Jupiter.) According to J. Rosch, they were recorded from the OSO6 satellite in 1976, and M. Winiarski claimed to have photographed them in 1991, describing them as redder than the Gegenschein; but their existence has yet to be properly confirmed.

✳ *The Airglow* (dayglow and nightglow) was named by O. Struve in 1950, who suggested the name in correspondence with C. T. Elvey. Nightglow prevents the sky from being completely dark at any time, quite apart from the diffusion of starlight. There are various causes for it; chemical reactions of the neutral constituents of the upper atmosphere, with light emissions; reactions from ionized constituents, again producing light emission; the influx of incoming particles from space, and so on. There also seems to be a sort of 'geo-corona' in the exosphere (1000 to 10000 km altitude), due to emissions from hydrogen and helium atoms in this highly rarefied region of the atmosphere.

✳ *Diffused starlight* is quite appreciable – coming as it does from the thousand million stars above magnitude 20, of which approximately one million are above magni-

Aurora Borealis, 13 Mar. 1989; photographed by Paul Doherty from Stoke-on-Trent, England. The rayed band and curtains are shown against the background of stars in Draco and Hercules.

tude 11·2, 4850 above magnitude 6, and 1620 above magnitude 5. Over a full visible hemisphere of the sky (20626 square degrees) the total starlight is approximately equal to 51 stars the brilliance of Sirius or 4 planets the brilliance of Venus at maximum.

It is, however, a tiny fraction of the light sent by the Moon.

✳ *The Green Flash* (or Green Ray) is an atmospheric effect. As the Sun sinks below the horizon, the last segment of it may flash brilliant green for an instant. There are vague references to this in Egyptian and Celtic folklore, but *the first scientific reference to it* was made by W. Swan, who saw it on 13 September 1865. However, Swan's account was not published until 1883. *The first published reference* to the phenomenon was due to J. P. Joule, in 1869. Much more recently, some superb photographs of it have

been taken by D. J. K. O'Connell at the Vatican Observatory; these were published in book form in 1958.

The Green Flash has also been seen with the planet Venus. The first published reference to it seems to be due to Admiral Murray, who saw it from HMS *Cornwall*, off Colombo, at 13.50 on 28 November 1939. Venus was setting over a sea horizon; Admiral Murray was using binoculars, and described the flash as 'emerald'.

✳ *The first green flash seen with Jupiter* was observed by G. Verschuur on 8 June 1978.

THE STARS

THE STARS: DISTANCES AND MOVEMENTS

*T*he number of stars visible with the naked eye is c. 5780, but this value is bound to be arbitrary. This means that it is seldom possible to see more than 2500 stars with the naked eye at any one time.

✳ It has long been known that the stars are suns, and are very remote, but it was not until 1838 that the first star-distance was measured. This was achieved by F.W. Bessel, using the method of trigonometrical parallax; the star concerned was 61 Cygni. At about the same time, also using the parallax method, T. Henderson measured the distance of Alpha Centauri, and F. Struve measured the distance of Vega.

✳ *The nearest star* (excluding the Sun) is Proxima Centauri, at 4·25 light-years; it is a member of the α Centauri system. The two bright components of α Centauri are only

slightly further away. Stars within 13 light-years of the Sun are as listed in the table below.

In 1976 O.J. Eggen reported the discovery of a red star, magnitude 10·8, in Sculptor (near τ Sculptoris) which was regarded as very close; later work showed that it is well beyond the 13 light-year limit of the table below.

If placed at the standard distance of 10 parsecs (32·6 light-years) only α Centauri, Sirius, Procyon and τ Ceti would be visible with the naked eye; it is interesting to note that if our Sun could be observed from Sirius it would be a second-magnitude star at R.A., 18 h 44 m, declination –16 deg 4 min. Most of the stars in the table are faint red dwarfs; thus the very feeble Wolf 359 has only 0·000 02 of the luminosity of the Sun.

The companions of Sirius and Procyon are white dwarfs. There are four flare stars; UV Ceti B (Luyten L 726–8 B), Proxima, Krüger 60 B and Ross 154. Lalande 21185 and Groombridge 34 A are excessively close binaries (too close to be separated visually) and there is a third body, of low mass, in the 61 Cygni system.

Radial velocities (the towards or away motion of the star relative to the Sun) are given as – for a velocity of approach, + for a velocity of recession.

✳ *The star with the greatest known proper motion* is Barnard's Star, discovered in June 1916 by E.E. Barnard at the Yerkes Observatory. The annual proper motion is 10 sec·31, so that in 180 years it crosses the sky by a distance equal to the apparent diameter of the Full Moon. It is approaching us at 108 km/s, so that its distance is decreasing at the rate of 0·036 light-years per century. The proper motion is increasing by about 0 sec·0013 per year, and will reach 25 sec by AD 11800, when the star will be at its closest to us – 3·85 light-years, closer than Proxima; the parallax will be 0 sec·87. The apparent magnitude will then have become 8·5. Subsequently the star will begin to recede.

THE NEAREST STARS

Star	R.A. h m (1950)		Dec. deg min		Spec	Apparent Mag	Absolute Mag	Luminosity Sun = 1	Parallax sec	Distance lt-yrs	Radial vel. km/s
Proxima	14	26·3	–62	28	dM5e	11·0	15·45	0·00006	0·772	4·249	–16
α Centauri A	14	36·2	–60	38	G2	–0·01	4·37	1·6	0·750	4·4	–22
α Centauri B	14	36·2	–60	38	K0	1·33	5·71	0·45	0·750	4·4	–22
Barnard's Star	17	55·4	+04	33	M5	9·54	13·22	0·00045	0·545	6·0	–108
Wolf 359	10	54·1	+07	20	dM8e	13·53	16·65	0·00002	0·421	7·8	+13
Lalande 21185	11	00·7	+36	18	M2	7·50	10·50	0·0055	0·397	8·2	–84
UV Ceti A	01	36·4	–18	13	dM6e	12·52	15·52	0·00006	0·387	8·4	+29
UV Ceti B	01	36·4	–18	13	dM6e	13·02v	15·96	0·00004	0·387	8·4	+32
Sirius A	06	42·9	–16	39	A1	–1·46	1·42	26	0·377	8·7	–8
Sirius B	06	42·9	–16	39	DA	8·5	11·2	0·003	0·377	8·7	–8
Ross 154	18	46·1	–23	53	dM5e	10·45	13·13	0·00048	0·345	9·5	–4
Ross 248	23	39·5	+43	56	dM6e	12·29	14·78	0·00011	0·314	10·4	–81
ε Eridani	03	30·6	–09	38	K2	3·73	6·14	0·30	0·303	10·8	+16
Ross 128	11	43·5	+01	06	dM5	11·10	13·47	0·00036	0·298	10·9	–13
61 Cygni A	21	04·7	+38	30	K5	5·22	7·56	0·082	0·294	11·1	–64
61 Cygni B	21	04·7	+38	30	K7	6·03	8·37	0·039	0·294	11·1	–64
ε Indi	21	59·6	–57	00	K5	4·68	7·00	0·14	0·291	11·2	–40
BD +43°44A	00	15·5	+43	44	M1	8·08	10·39	0·0061	0·290	11·3	+13
BD +43°44B	00	15·5	+43	44	M6	11·06	13·37	0·00039	0·290	11·3	+20
Luyten 789–6	22	35·7	–15	36	dM7e	12·18	14·49	0·00014	0·290	11·3	–60
Procyon A	07	36·7	+05	21	F5	0·38	2·64	7	0·285	11·4	–3
Procyon B	07	36·7	+05	21	DF	10·7	13·0	0·00055	0·285	11·4	–3
BD +59°1915A	18	42·2	+59	33	dM4	8·90	11·15	0·0030	0·282	11·6	0
BD +59°1915B	18	42·2	+59	33	dM5	9·69	11·94	0·0015	0·282	11·6	+10
Lacaille 9352	23	02·6	–36	09	M2	7·35	9·58	0·013	0·279	11·7	+10
G 51–15	08	26·8	+26	57	M	14·81	17·03	0·00001	0·278	11·7	
τ Ceti	01	41·7	–16	12	G8	3·50	5·72	0·45	0·277	11·8	–16
Luyten's Star	07	24·7	+05	29	dM5	9·82	11·94	0·0015	0·266	12·3	+26
L 725–32	01	09·8	–17	16	dM5e	12·04	14·12	0·00020	0·261	12·5	+28
Lacaille 8760	21	14·3	–36	09	M0	6·66	8·74	0·028	0·260	12·6	+21
Kapteyn's Star	05	09·7	–45	00	sdM0pec	8·84	10·88	0·0039	0·256	12·7	+245
Krüger 60A	22	26·3	+57	27	dM3	9·85	11·87	0·0016	0·253	12·9	–26
Krüger 60B	22	26·3	+57	27	dM5e	11·3	13·3	0·0004	0·253	12·9	–26

Barnard's Star is also known as Munich 15040; Lalande 21185 as BD +36°2147; Lacaille 9352 as CD –36°15693; and Lacaille 8760 as CD –39°14192. UV Ceti B is sometimes known as Luytens' Flare Star.

Star	Proper motion per year, seconds of arc
Barnard's Star	10·27
Lalande 21185	4·75
61 Cygni	4·12
ε Indi	3·93
Wolf 359	3·84
Proxima	3·75
α Centauri	3·61
Lalande 8760	3·53
Van Maanen's Star	3·01
Groombridge 34	2·87
τ Ceti	1·72
Sirius	1·21
Procyon	1·03
ε Eridani	0·97
Krüger 60	0·80

✳ *The only naked-eye star to have changed constellations in modern times* is ρ Aquilæ; in 1992 its proper motion carried it across the boundary of Delphinus. The list of 'migrations' up to the year 7000 is as follows, according to R.W. Sinnott:

Star	Mag	Enters	Year
ρ Aquilæ	5·0	Delphinus	1992
γ Cæli	4·6	Columba	2400
ε Indi	4·7	Tucana	2640
ε Sculptoris	5·3	Fornax	2920
λ Hydri	5·1	Tucana	3200
μ¹ Cygni	4·8	Pegasus	4500
χ Pegasi	4·8	Pisces	5200
μ Cassiopeiæ	5·1	Perseus	5200
η Sagittarii	3·1	Corona Australis	6300
ζ Doradûs	4·7	Pictor	6400

✳ *The brightest star within 13 light-years*, both apparently and absolutely, is of course Sirius A, with 26 times the luminosity of the Sun. The least massive star is UV Ceti B, with a mass of only 0·035 that of the Sun. Apart from the two white dwarfs, all the stars within 13 light-years belong to the Main Sequence; there are no giants.

✳ *The first predictions of the positions of invisible stars* were made by F.W. Bessel in 1844, when he forecast the positions of the companions to Sirius and Procyon; he based his calculations upon irregularities in the proper motions of the primaries. In 1862 Clark discovered Sirius B in almost the position predicted by Bessel; the orbital period of the system is 50 years. The companion to Procyon was found in 1898 by Schaeberle, again almost in Bessel's position; the orbital period of the system is 39 years.

✳ *The first star suspected to have a very low-mass companion* was the nearby Barnard's Star. At the Sproule Observatory in Pennsylvania, P. van de Kamp made a long series of observations of the star, beginning in 1937, and believed that the irregularities in the proper motion indicated the existence of two companions each rather less massive than Jupiter – so that they would be planetary rather than stellar. It seems, however, that instrumental errors make these results highly suspect. One of

the *smallest and least massive stars known* is Gliese 623B, 25 light-years away in Hercules. In June 1994 it was imaged by the Hubble Space Telescope. It is the smaller component of a binary system; the separation between the two members of the pair is only 300 million km. The star is a red dwarf, with 1/10 the mass of the Sun and only 1/60,000 of the Sun's luminosity. Originally it was detected by its gravitational effects on its companion, but only the Hubble telescope is capable of showing it visually. The orbital period is 4 years.

✳ *The nearest stars which are at all like the Sun* are τ Ceti and ε Eridani. From 4 April 1960 to March 1961 astronomers at the Green Bank Radio Astronomy Observatory in West Virginia, led by Dr Frank Drake, carried out a 'listening operation' at a wavelength of 21 cm, using the 80 ft radio telescope, in the hope of picking up rhythmical signals which could be interpreted as artificial. The project was known officially as Project Ozma, though unofficially as Operation Little Green Men. Not surprisingly, the results were negative, but since 1969 operations of rather similar type have been carried out in Russia, and a major programme was started in 1992 in the USA. This SETI programme (Search for Extra-Terrestrial Intelligence) was ambitious, but unfortunately the US Congress cut off all funds after a year, and the programme is now being continued on a less elaborate scale. It is, however, significant that at the General Assembly of the International Astronomical Union held in Buenos Aires in 1991 a whole day was given over to deciding upon the procedure to be followed in the event of ETI being detected!

✳ *The star remotest from the galactic plane* is PC 0832/676, at 60 000 light-years. It is a B-type star in Ursa Major; magnitude 14.

✳ *The remotest star seen in our Galaxy* is an RR Lyræ variable in the galactic halo. Its distance is at least 155 000 light-years, and possibly as much as 180 000 light-years. Its visual magnitude is 19.3. It lies in front of the giant elliptical galaxy M.49, in Virgo.

✳ *The most distant individual star ever recorded* was a supernova in the galaxy cluster AC118. It was discovered on 9 August 1988 by H. U. Nørgaard-Nielsen with the 1·5 m Danish telescope at La Silla. The distance was 5 000 000 000 light-years. By 6 September it had become too faint to be observable.

CLASSIFI-CATION

*T*he stars are of many different types, and show tremendous range in size and lumi-

nosity, though rather less in mass. The Sun is a normal star, and the only one sufficiently close to be studied in great detail.

✳ *The first attempt to classify the stars according to their spectra* was made by the Italian Jesuit astronomer, Angelo Secchi, in 1863–7. He divided the stars into four types:

I. White or bluish stars, with broad, dark spectral lines of hydrogen but obscure metallic lines. Example: Sirius.

II. Yellow stars; hydrogen less prominent, metals more so. Examples: Capella, the Sun.

III. Orange stars: complicated banded spectra. Examples: Betelgeux, Mira. The class included many long-period variables.

IV. Red stars, with prominent carbon lines: all below magnitude 5. Example: R Cygni. This class also included many variables.

Secchi's catalogue included over 500 stars. His system was, however, superseded by that worked out at Harvard in the United States.

✳ *The Harvard system was introduced* by E.C. Pickering in 1890, and was extended by two famous women astronomers, Miss A. Cannon and Mrs W. Fleming. It has been further modified into what is now called the MKK (after Morgan, Keenan and Kellerman of the USA) or Yerkes system.

Most stars lie in the sequence from B to M. The very hot W stars are (as noted) very rare; O stars are also rare. Types R, N and S consist entirely of giants, and are regarded as no 'later' than Type M. R and N are now often combined into a simple type, C.

Originally it had been planned to make the sequence alphabetical, beginning at A, representing the spectra with the strongest hydrogen lines. When it was realized that the spectra represented star temperatures, major modifications had to be introduced, and the final result was alphabetically chaotic. A good and famous mnemonic is 'Wow! Oh Be A Fine Girl Kiss Me Right Now Sweetie' (or, if you prefer it, 'Smack'!)

In addition, P is used for gaseous nebulæ, and Q for novæ.

In what is termed the *instability strip*, virtually all stars pulsate as Cepheid variables (for example the prototype star Delta Cephei itself; spectral type F, mean absolute magnitude –4.6).

The following additional letters are used: e (emission lines present), n (nebulous lines, i.e. stars rotating rapidly), s (sharp lines), k (interstellar lines), m (metallic lines), v (variable), and p or pec (peculiar).

✳ *The most luminous star known* is probably η Carinæ, which is unique in many

Type	Spectrum	Surface temperature	Examples	Notes
W	Many bright lines: divided into WN (nitrogen sequence) and WC (carbon sequence)	Up to 80 000°C	Rare: About 150 have been found in our Galaxy, and 50 in the Large Magellanic Cloud. 16 cases are known as planetary nebulæ in which the central star is a Wolf-Rayet; one of these is NGC 7009, the 'Saturn Nebula' in Aquarius. ζ Puppis O5.8	Known as Wolf-Rayet stars. Have expanding shells, moving outwards at up to 3000 km/s. All very remote, and appear faint despite their considerable luminosity
O	Both bright and dark lines	40 000°C–35 000°C	γ Velorum, WC7	Represent a transition between W and B stars, though this does not imply any evolutionary sequence
B	Bluish-white (B0) to white (B9). No emission lines, but dominant absorption lines of hydrogen and (particularly) helium	Over 25 000°C for B0 12 000°C for B9	Rigel B8 (i.e. 8/10 of way from B0 to A0) Spica. B1.	Rigel is particularly luminous, with peculiarities in its spectrum
A	White stars: spectra dominated by hydrogen lines	10 000°C–8000°C	Sirius, Vega, Altair	
F	Yellowish hue. Calcium very conspicuous, with less prominent hydrogen lines	7500°C–6000°C	Procyon, Polaris	Yellow hue so elusive that to the naked eye most F-stars will be regarded as white
G	Yellow; weaker hydrogen lines, numerous conspicuous metallic lines	5500°C to 4200°C (giants) 6000°C to 5000°C (dwarfs)	Capella (giant) Sun (dwarf)	Beginning of division into dwarf or Main Sequence stars and the giants
K	Orange; weak hydrogen lines, strong metallic lines	4000°C to 3000°C (giants) 5000°C to 4000°C (dwarfs)	Arcturus, Aldebaran, Pollux (giants) ε Eridani, τ Ceti (dwarfs)	K stars are more numerous than any other type
M	Orange-red; very complicated spectra with many bands due to molecules	3400°C (giants) 3000°C (dwarfs)	Mira Ceti (variable), Betelgeux, Antares (giants), Proxima Centauri (dwarf)	Strong differences between giants and Main Sequence dwarfs. Proxima Centauri closest star beyond the Solar System
R*	Reddish	2600°C	V Arietis, T Lyræ	Remote, and appear faint
N*	Reddish; strong carbon lines	2500°C	R Leporis, V Aquitæ	Remote, with many variable examples
S	Reddish; prominent bands of titanium oxide and zirconium oxide	2600°C	χ Cygni, R Cygni	Examples given are long period variables

* Types R and N are now often combined as Type C. A separate class Q, is used for novæ.

ways. Telescopically it has been described as looking more like a 'red blob' than a normal star, and it is associated with nebulosity; the distance is estimated at 6400 light-years, and the maximum luminosity may amount to several million times that of the Sun (one estimate gives 6 000 000 Suns). For some years in the 1830s and early 1840s it was the brightest star in the sky apart from Sirius, but for a century now it has been below naked-eye visibility, though it emits strongly in the infra-red. Yet there has been no decline in real luminosity. The star has become surrounded by dust, so that its light is absorbed by the dust and then re-radiated in infra-red. There have been speculative suggestions that it may suffer a supernova outburst in the foreseeable future (that is to say, within the next few tens of thousands of

years). At infra-red wavelangths, η Carinæ is the brightest object in the sky outside the Solar System, and it also produces the strongest known stellar wind.

Excluding η Carinæ, the most luminous star known is S Doradûs, in the Large Magellanic Cloud; the absolute magnitude is −8·9, so that it is about a million times as powerful as the Sun (it is somewhat variable). From Earth, it cannot be seen with the naked eye.

✳ *The least luminous star known* is MH18, identified in 1990 by M. Hawkins (Royal Observatory Edinburgh) from plates taken with the UKS telescope in Australia. It has 1/20 000 the luminosity of the Sun, and is 30 per cent fainter than any other known star. It is presumably a brown dwarf, as it has

only 5 per cent the mass of the Sun. Its distance is 68 light-years.

✳ *The most massive star known* is Plaskett's Star, HD 47129 in Monoceros, identified in 1922 by K. Plaskett when he discovered its binary nature. The combined mass of the two supergiants is 90 times that of the sun, the secondary (spectral type O6) is 18 per cent more massive than the primary (type O7.5). The orbital period is 14.4 days.

✳ *The hottest known star* is the central star of the planetary nebula NGC 2240: surface temperature 220 000 degrees K.

✳ *The first measures of stellar diameters using an intensity interferometer* have been made by J. Hanbury Brown, from Australia. Some of the values found are as follows:

Star	Diameter, km
Canopus	115 000 000
β Carinæ	12 400 000
Vega	7 700 000
Regulus	6 600 000
α Gruis	6 100 000
Procyon	5 700 000
Sirius	4 800 000
Altair	4 700 000
Fomalhaut	4 600 000
ζ Puppis	4 400 000
Spica	2 200 000

✳ **The largest star known** was formerly believed to be the invisible secondary of the eclipsing binary ε Aurigæ, with a diameter of about 5 700 000 000 km. Eclipses occur every 27 years; the latest eclipse began in 982. However, it is now thought more likely that the secondary is a relatively small, hot star surrounded by an immense shell of gas, as was originally proposed by M. Hack. The system of α Herculis, made up of a red variable supergiant with a double companion, is enveloped in a huge cloud of gas which could have a diameter of 250 000 000 000 km. Some red supergiants, such as Betelgeux, have diameters exceeding 400 000 000 km.

✳ **The star with the largest apparent diameter** is Betelgeux (α Orionis), 50 milliarc seconds. This is equal to the apparent diameter of a metre ruler placed on the Moon and observed from Earth.

✳ **The first surface detail to be recorded on a star** (other than the Sun) were detected on Betelgeux by J. Baldwin and P. Warner, using a complex technique with the William Herschel Telescope at la Palma, in 1990. It had been suggested that the outer layer of a red supergiant should consist of a few vast convection cells, where energy bubbles up from below. With Betelgeux, Baldwin and Warner found a bright patch toward one edge of the tiny disk – presumably a convection cell of this type.

✳ **The first solar-type oscillations in another star** were detected in 1994 by Danish and European Southern Observatory astronomers, using the 2·4 m Nordic telescope at Los Muchachos in the Canary Islands. It is known that the Sun oscillates in a period of about 5 minutes; the surface moves up and down by less than 25 m (the driving force is convective motion in the interior; convection cells rise toward the surface and jostle the Sun from inside). Similar oscillations were found for the star η Boötis, which is a G-type star 32 light-years away and 6·5 times as luminous as the Sun. The temperature varies periodically by a few hundredths of a degree; the mean surface temperature is rather less than 6000 °C. The star oscillates in at least 10 modes simultaneously, with periods of around 20 minutes.

EVOLUTION

Stellar populations were first noted by W. Baade in the early 1950s. Population I stars are metal-rich, with about 2 per cent of their mass being made up of elements heavier than helium. The most brilliant Population I stars are of types W, O and B. The disk and arms of the Galaxy (and other spirals) are mainly Population I.

Population II stars are metal-poor, and are clearly older, so that their most brilliant members are red giants which have already left the Main Sequence. They are found in the halo and nucleus of the Galaxy (and other galaxies) and in globular clusters. It must, however, be stressed that there is no hard and fast boundary between Population I and Population II regions.

✳ **Stellar Evolution.** It was originally believed that a star began its career as a red giant, condensing out of nebular material (type M; giant), shrank and heated up to become a Main Sequence star (type B or A), and then cooled while continuing to shrink, ending its career as a red dwarf of type M. This was – it was thought – shown by the Hertzsprung-Russell or H/R diagram.

✳ **The H/R diagram was originally due** to Ejnar Hertzsprung, a Danish astronomer, in 1911; it was proposed independently in 1913 by H.N. Russell of the United States. In an H/R diagram, a star is plotted according to its luminosity (or equivalent, such as absolute magnitude) against spectral type (or surface temperature). The Main Sequence, running diagonally from hot O and B stars at the top left, down to faint M-type dwarfs at the lower right, is very marked; so are the giant and supergiant areas, to the upper right. White dwarfs (unrecognized in 1913) are to the lower left.

✳ **The key to stellar energy was discovered in 1939** by H. Bethe (and, at about the same time, by G. Gamow). Bethe actually worked it out during a train journey from Washington to Cornell University! The stars – or, at least, the normal ones – shine by means of nuclear transformations. Thus inside the Sun, hydrogen is being converted into helium. It takes four hydrogen nuclei to make one helium nucleus; each time this happens, a little mass is lost and a little energy is released. Each second, the Sun converts 600 million tonnes of hydrogen into helium – and loses 4 million tonnes in mass. This may sound a great deal, but the Sun is very massive and has a great deal of fuel, which lasts for a long time: it is certainly older than the Earth, and will continue in virtually its present state for 5000 million years in the future – perhaps rather longer.

The career of a star depends upon its initial mass, and the H/R diagram itself is not indicative of a definite evolutionary sequence. In every case the star begins by condensing out of nebular material; this star formation is certainly going on in M.42, the Orion Nebula. We can also observe fairly small dark 'globules' against some bright nebulæ, known as Bok globules in honour of their discoverer, Bart J. Bok. Some of them are over 6 light-years in diameter, though others are much smaller. They may well indicate regions of star formation by gravitational contraction, though it is unlikely that all stars begin as Bok globules. There are many known cases in which many stars of similar type, and presumably similar age, are concentrated in a limited area; these are *stellar associations.*

Moreover, it is logical to assume that the stars in any cluster (such as the Pleiades) had a common origin in the same nebular cloud. Not all open clusters are similar in age: for instance the Pleiades cluster is relatively young, while M.67 in Cancer is comparatively old.

✳ **The first brown dwarf** to be definitely identified was detected by B. Zuckerman (University of California) and E. Becklin (University of Hawaii) in 1987. They found that the spectrum of a white dwarf star, Giclas 29–38, shows an infra-red excess in the region between 2 and 5 micrometres. This is unusual, and indicates the presence of an object with a diameter 15 per cent that of the Sun and a temperature of about 900 °C, which is below the lower limit for a star capable of triggering off nuclear reactions and starting to shine in the same way as the Sun. A brown dwarf is (just) hot enough to shine detectably, but is not the end product of an evolved star.

The evolution of a star depends on its inertial mass – so let us consider the different categories:

a) *Stars of less than 0·1 solar masses:* Condensation out of nebular material causes a rise in temperature, but the cores never become hot enough for nuclear reactions to begin (about 10 000 000 degrees is the 'starting point'). The star therefore turns into a dim red dwarf of type M, and then simply fades until it has lost all its energy.

b) *Stars of between 0·1 and 1·4 solar masses:* The star begins by condensing gravitationally out of nebular material. The 'protostar' will be large and red, though by no means the same as the red giants such as Betelgeux. As contraction continues, the surface temperature remains the same, so that the star becomes less luminous. On the H/R diagram, it passes along what are known as the Hayashi and Henyey tracks

Planetary nebula: NGC 7293, the Helix Nebula in Aquarius. Photographed with a 200in, reflector, Palomar.

Planetary nebula: M.27, the Dumbbell Nebula in Vulpecula. Palomar, 200in photograph.

(in honour of the astronomers who first described them) and then joins the Main Sequence. The T Tauri stars, named after the most famous member of the class, are still contracting toward the Main Sequence, and are irregularly variable. Condensation to the Main Sequence takes several million years, and a young star will also have to 'blow away' a surrounding cocoon of dust. It will then brighten, and cases have been found; thus the star V1057 Cygni brightened up from magnitude 16 to above magnitude 11, in 1969, over a period of only about 250 days, though it has subsequently faded. Not surprisingly, stars of this type are strong emitters of infra-red radiation.

The youngest star may be VLA 1623 in the ρ Ophiuchi cloud. Its probable age is below 10 000 years. The mass is 0·6 that of the Sun. It is still in the pre-Main Seqence stage.

The star joins the Main Sequence at a point known as the ZAMS (Zero Age Main Sequence); just where it joins depends upon its mass, and this also regulates the subsequent course of events. For example, a star such as the Sun will spend about 10 000 million years on the Main Sequence (so that our Sun is now about half-way through this stage in its career), while a star with half the Sun's mass will stay on the Main Sequence for up to 200 000 million years, while a star one and a half times as massive as the Sun will remain there for a mere 3000 million years.

When the core temperature reaches 10 000 000°C, nuclear reactions begin; at this stage the Sun was only 70 per cent as luminous as it is now, but as the hydrogen-into-helium conversion progressed, the core temperature rose to around 15 000 000°C. The conversion process – known, rather misleadingly, as hydrogen 'burning' – is accomplished by a rather roundabout process termed the proton–proton reaction; with hotter stars, what is termed the carbon–nitrogen cycle is dominant, so that these two elements are used as catalysts, but the end result is much the same. Four hydrogen nuclei combine to make one nucleus of helium, with release of energy and loss of mass.

This process can maintain the flow of radiation for a long time, but not for ever; helium is built up, in the core, and the supply of available hydrogen 'fuel' begins to run low. This means that less energy is generated, and the core has difficulty in supporting itself against the weight of the outer layers. The core is compressed, and this makes it hotter once more; hydrogen burning continues in a shell round the now helium-rich core. As the central temperature rises, the outer layers expand and cool, so that the star becomes a red giant. The luminosity has increased perhaps a hundredfold, and the bloated globe is now so large that in the case of the Sun it will engulf the inner planets. Certainly, the Earth cannot hope to survive as a habitable world.

When the core temperature reaches 100 000 000°C, helium suddenly reacts to form carbon and oxygen (the 'helium flash'). There follows a period of instability, during which the outer layers of the star are puffed off. This process is quite gentle; subsequently the expelled material becomes visible, excited by radiation from the star, to produce what is inappropriately termed a *planetary nebula*. This may dispose of 10 to 20 per cent of the star's mass. The expelled material will not fall back, but will expand and dissipate in space; planetary nebulæ, as such, are therefore relatively short-lived, and are unlikely to last for as long as 100 000 years.

What, then, of the core, in which nuclear reactions have ceased? The temperature is not high enough to make carbon and oxygen react, and what remains of the old star (that is to say, the burnt-out core) settles into a very small, super-dense object known as a white dwarf. The dim Companion of Sirius is of this type. Its escape velocity is 3400 km/s, and its surface gravity is 900 times that of the Sun.

In a white dwarf, the atoms are broken up and packed tightly together with little waste space. This leads to amazing density values, up to a million times the density of water in extreme cases; a tablespoonful of white dwarf material might well weigh several thousand tonnes. For instance, the Companion of Sirius is as massive as the Sun, but smaller than planets such as Uranus and Neptune; we even know of white dwarfs which are smaller than the Moon. They have been described as 'bankrupt stars', though their surface temperatures are still high (over 100 000°C in some cases). We must assume that eventually a white dwarf will fade away into a cold, dead 'black dwarf' sending out no energy at all but no white dwarfs are known with surface temperatures below 3000°C; our Galaxy is not yet old enough for the most ancient white dwarf to have cooled below this temperature.

Luminosity and absolute magnitude
Many people prefer to visualize a star's luminosity in terms of the Sun as unity rather than as absolute magnitude. The following conversion table is approximate, but good enough for most purposes.

CONVERSION OF ABSOLUTE MAGNITUDE (M) TO SOLAR LUMINOSITIES (L)

M	L		
−20	8 300 000 000	+0·5	52·5
−19·5	5 250 000 000	+1	33·1
−19	3 300 000 000	+1·5	20·9
−18·5	2 000 000 000	+2	13·2
−18	1 300 000 000	+2·5	8·3
−17·5	832 000 000	+3	5·2
−17	525 000 000	+3·5	3·3
−16·5	330 000 000	+4	2·1
−16	209 000 000	+4·5	1·3
−15·5	132 000 000	+5	0·8
−15	83 200 000	+5·5	0·5
−14·5	52 500 000	+6	0·3
−14	33 100 000	+6·5	0·2
−13·5	20 900 000	+7	0·1
−13	13 200 000	+7·5	0·08
−12·5	8 320 000	+8	0·05
−12	5 250 000	+8·5	0·03
−11·5	3 310 000	+9	0·02
−11	2 090 000	+9·5	0·01
−10·5	1 320 000	+10	0·008
−10	832 000	+10·5	0·005
−9·5	525 000	+11	0·003
−9	331 000	+11·5	0·002
−8·5	209 000	+12	0·001
−8	132 000	+12·5	0·0008
−7·5	83 200	+13	0·0005
−7	52 500	+13·5	0·0003
−6·5	33 100	+14	0·0022
−6	20 900	+14·5	0·000 1
−5·5	13 200	+15	0·000 08
−5	8 320	+15·5	0·000 05
−4·5	5 250	+16	0·000 03
−4	3 310	+16·5	0·000 02
−3·5	2 090	+17	0·000 01
−3	1 320	+17·5	0·000 008
−2·5	832	+18	0·000 005
−2	525	+18·5	0·000 003
−1·5	331	+19	0·000 002
−1	209	+19·5	0·000 001
−0·5	132	+20	0·000 000 8
0	83·2		

The most accurate 'cosmic clock' ever observed at visible wavelengths is the pulsation period of a 16th-magnitude white dwarf in Leo Minor, G117–B15A. It pulsates simultaneously in six different modes, but the light-curve is dominated by a single variation with an amplitude of 0·022 magnitude and a period of 215·2 seconds. The period is lengthening by no more than a second or so every few million years.

The Indian astronomer S. Chandrasekhar has shown that a star more than 1·4 times as massive as the Sun cannot become a white dwarf unless it sheds some of its mass – which happens, of course, during the unstable and planetary nebula stages.

The hottest white dwarf with an atmosphere of pure hydrogen is RE 1738+665, discovered from ROSAT observations in 1994 by M. Bairstow of Leicester University. The surface temperature is about 90 000°C. It seems to form a link between cooler, hydrogen-rich white dwarfs and the very hot hydrogen-rich stars at the centres of planetary nebulæ.

c) *Stars more than 1·4 times as massive as the Sun:* Here, everything happens at an accelerated rate; for instance, if a proto-star is 25 times as massive as the Sun it may reach the Main Sequence in a few thousand years. At the end of the helium burning, the temperature at the core is at least 200 million °C, and this is enough to make carbon react; as the temperatures rise still higher, heavier and heavier elements are built up – until iron is produced. Iron will not react, and so energy production stops. There is an 'implosion', as the outer layers of the star fall on to the core; the core rebounds, and a shock-wave spreads through the star, disrupting it in a Type II supernova outburst. The violence of the outburst is incredible; the peak luminosity may reach 1000 million times that of the Sun! The end result is a cloud of expanding material, while the remnant of the original star's core becomes a *neutron star*. Even the protons and electrons are forced to combine to make neutrons, which of course have no electrical charge. A rough timetable for a star 25 times as massive as the Sun is as follows:

Hydrogen burning:	7 000 000 years
Helium burning:	500 000 years
Carbon burning:	600 years
Oxygen burning:	6 months
Silicon burning:	1 day
Core collapse:	c. 0·1 second
Core bounce:	a few milliseconds
Explosive burning:	c. 10 seconds
Surface blows away:	c. 1 hour

The concept of neutron stars was proposed in 1932 by the Russian physicist Landau, and again in 1934 by F. Zwicky and W. Baade at Caltech (USA) but the first neutron star was not detected until 1967 – not by visible light, but by its radio emissions. A neutron star is indeed an amazing object. The diameter is of the order of 10 km, and the density perhaps a thousand million million tonnes per cubic metre, or 1000 million million times that of water. A pin's head of neutron star material would 'weigh' more than the liner *QE2*.

According to theory (which may or may not be correct!) the outer crust of a neutron star is crystalline and iron-rich, composed of what we may term 'normal' matter. Cracks occur in it, producing 'starquakes' which affect the object's behaviour. Below the crust comes the neutron-rich liquid mantle; below again, a superfluid core composed mainly of neutrons; and at the centre, material made up of 'hyperons', about which we can only speculate. If one could stand on the surface of a neutron star, one's weight would be 10 000 million times greater than on Earth. There are strong magnetic fields – perhaps a hundred million tesla, as against 30 millionths of a tesla for the Earth.

The first neutron star was discovered in 1967 by Miss Jocelyn Bell (now Dr Jocelyn Bell-Burnell) working with the Cambridge team led by Professor A. Hewish. During radio surveys of the sky, she found a weak source which was fluctuating quickly as though 'ticking'. The radio pulses were very regular and rapid, so that the object was called a *pulsar* (CP 1919). The official announcement was made on 29 February 1968. More pulsars were soon found, and today over 300 are known. Their periods have a limited range; that of CP 1919 is 1·3373 seconds.

The slowest pulsar is PSR 1845–19 (period 4·3 seconds). In general all pulsars seem to be slowing down by very small amounts; thus the period of CP 1919 is lengthening by a thousand millionth of a second each month, so that in 3000 years' time the period will be 1·3374 second instead of the present-day 1·3373 second. However, some pulsars show irregular changes in period, or glitches; thus on 1 March 1969 the pulsar PSR 0833–45 speeded up a full quarter of a millionth of a second!

It was some time before astronomers decided upon the nature of pulsars. There was even the short-lived LGM or 'Little Green Men' theory – that the radio signals were artificial! Pulsating or rotating white dwarfs were also suggested, but it became apparent that a pulsar must be smaller than even a white dwarf, and the idea of pulsars as neutron stars is now universally accepted.

The pulsar is spinning rapidly, and the more normal surface protons and electrons are accelerated outward at the north and south magnetic poles of the star in rather the manner of searchlight beams; every time the Earth passes through the beam, we receive a pulse. Glitches are due to the disturbances in rotational period caused by starquakes.

The first pulsar to be seen visually was NP 0532 in the Crab Nebula. This was achieved in January 1969 by a team at the Steward Observatory in Arizona, using a 36 in reflector. They identified a faint, flashing object whose mean magnitude was about 17, and whose period was the same as that of the pulsar. Its period is 0·0339 second, so that it pulses 30 times a second. Later in 1969 the pulsar was photographed from Kitt Peak (Arizona). It is undoubtedly the remnant of the supernova of 1054. (R. Minkowski and W. Baade had observed it in 1942, and had suspected that it was the centre of activity in the Crab.) The second visual identification was that of the pulsar PSR 0833–45 in the southern constellation of Vela; the pulsar lies in the Gum Nebula (named after its discoverer, the late Dr C.S. Gum), and it may be assumed that the nebula is a supernova remnant. The Vela pulsar itself was found in 1968 at the Molonglo Radio Observatory in Australia. The period is 0·089 second – the third shortest known. In 1977 a team working with the 3·9 m Anglo-Australian telescope at Siding Spring (Australia) identified the pulsar as a faint flashing object with a mean magnitude of 24·2.

The closest known pulsar is JO108–1431 in Cetus. It seems to be 280 light-years away; the probable age is about 160 000 000 years. It is a thousand times weaker than any other known pulsar.

A very close neutron star is known as Geminga, in Gemini; distance 300 light-years. Its optical counterpart is a 25th-magnitude star which has moved by 1·8 seconds of arc in 8 years. Geminga emits light, X-rays and gamma-rays in all directions, but presumably its radio beams never sweep across our line of sight.

The fastest-moving star is near the pulsar OSR 2224+65 in the Guitar Nebula, in Cepheus, 6000 light-years away. Its speed is from 800 to 1000 km/s, presumably as the result of an asymmetrical supernova blast. It spins 1·47 times per second.

The first-known millisecond pulsar was PSR 1937+215 in Vulpecula. It pulses at a record 642 flashes per second, regularly spaced at intervals of 1·557 milliseconds. It was identified in October 1982 by a team at Berkeley, California, led by D. Backer. The distance is about 7000 light-years. It still has the shortest-known period, though many other millisecond pulsars have been found.

Apparently a millisecond pulsar is an old neutron star which has only a weak magnetic field, and has been 'spun up'. It has been a member of a binary system, now disrupted; the other member was a red giant, whose outer layers were held only weakly by gravitation. The neutron star pulled material away from its companion, and the gases pulled toward it were heated sufficiently to produce X-rays. As the material struck the surface of the neutron star tangentially, it increased the rotation speed. Subsequently the two stars separated, perhaps in a collision, perhaps in a supernova outburst, leaving the millisecond pulsar isolated.

* *The first millisecond pulsar to be found in a globular cluster* (M.28, in Sagittarius) was identified in 1987 by A. Lyne and his team at Jodrell Bank. Other millisecond pulsars in globular clusters have since been found – in the cluster 47 Tucunæ there are no fewer than 11. However, not all these are members of binary systems. Apparently the companion star will be literally evaporated by the pulsar over a sufficient period of time, and only the millisecond pulsar will be left. Proof of this theory was obtained in 1988 by J. Taylor and A. Fruchter (Princeton University) who found that the pulsar PSR 1957+20 in Sagitta – spin rate 622 per second – is evaporating its companion, which is now left with only $\frac{1}{50}$ the mass of the Sun. Not inappropriately, the pulsar is known as the Black Widow! A second case, PSR 1744–24a, has been found in the globular cluster Terzan 5, where the pulsar seems to be engulfed in the clouds of material which it is pulling away from its doomed companion.

Pulsar planets were reported in 1994 by A. Wolszczan, using the radio telescope at Arecibo in Puerto Rico. The pulsar concerned is 1257+2 in Virgo; it is at least 1500 light-years away (perhaps as much as 3000), and has a period of 6·2 milliseconds, so that it 'beats' 161 times per second. From timing the pulses, Wolszczan came to the conclusion that the pulsar was orbiting with three invisible bodies. If the pulsar's mass is 1·4 times that of the Sun, the planets reported by Wolszczan have respectively 3·4, 2·8 and 0·014 the mass of the Earth, so that the least massive of them is comparable with the Moon.

If these bodies exist, we have to decide how they were produced. No planet could survive the supernova outburst from which the pulsar was presumably formed, so either the planets were captured from another system or they condensed from débris in some way. However, all this is highly speculative, and it is indeed difficult to believe that any body of planetary nature could survive in the immediate region of a supernova remnant.

d) *Stars too massive initially to form neutron stars:* Here we are by no means certain of the course of events, but it is suggested that once the final collapse starts nothing can stop it. Eventually the star has become so small and so dense that its escape velocity becomes greater than the velocity of light. Nothing can therefore escape, and to all intents and purposes the collapsed star or *collapsar* is cut off from the rest of the universe.

For obvious reasons, these **black holes** can be detected only by their effects upon objects which emit detectable radiation. One candidate is Cygnus X-1, so called because it is an X-ray source. The system consists of a B0 type supergiant, HDE 226868, with about 30 times the mass of the Sun and a diameter 23 times that of the Sun (18 000 000 km), together with an invisible secondary with 14 times the mass of the Sun. The orbital period is 5·6 days, the distance 6500 light-years, and the magnitude of the primary star 9 (it lies at RA 19h 56m 29s, dec. + 35 deg 03 min 55 sec, near η Cygni). The secondary would certainly be visible if it were a normal star, and it may be a black hole.

The best black hole candidate is probably V404 Cygni, which (having presumably been created in a much larger supernova explosion in the remote past) flared up as a nova in 1938 and rose from magnitude 19 to 12 before fading back to its normal brightness. In 1989 observations from the Japanese satelite Ginga showed that X-rays were being emitted, and studies by P. Charles, using the William Herschel Telescope at La Palma, indicate that the system consists of a visible star 70 per cent as massive as the Sun together with a black hole of 12 solar masses. The period of revolution is 6·5 days.

The size of a black hole will depend upon the mass of the collapsed star. The critical radius of a non-rotating black hole is called the Schwarzschild radius, after the German astronomer K. Schwarzschild, who investigated the problem mathematically in 1916; the boundary around the mass having this radius is the event horizon. For a body the mass of the Sun, the Schwarzschild radius would be about 3 km; for the Earth, less than 1 cm.

Conditions inside the event horizon are so different from our everyday experience that we cannot visualize them; it may even be that the old star finally crushes itself out of existence. However, it is important to add that the whole black hole concept is questioned by some astronomers – and we have as yet no decisive proof that they exist.

The basic principle was suggested as long ago as 1783 by the English natural philosopher John Michell, and again by the great French mathematician Laplace in 1798. Like Newton, Laplace believed light to consist of a stream of particles, and commented that if a body were sufficiently small and dense it would be invisible, since the light-particles would not be able to travel fast enough to escape from it.

Meanwhile, all we can really say is that black holes are theoretical possibilities, and that the concept has been widely accepted. If valid, then black holes represent the end products of the most massive stars in the universe.

The first star found to have a marked infrared excess was Vega. The discovery was made in 1983 by H. Aumann and F. Gillett, from results obtained from IRAS (the Infra-Red Astronomical Satellite). The infra-red excess was found when the IRAS instruments were being calibrated. It was presumably due to solid particles coming from an extended region round Vega, stretching out some 80 astronomical units from the star. The temperature of the material was about –185°C, and it was suggested that the material might be planet-forming. Other similar cases followed, notably with Fomalhaut.

The first probable proto-planetary system to be detected optically was that of β Pictoris. This is an A-type star, 78 light-years away and 68 times as luminous as the Sun. It was one of the stars found by IRAS to have a large infra-red excess, and the material round it was subsequently detected optically by R. Terrile and B. Smith at the Las Campanas Observatory in Chile. It was then said that β Pictoris is surrounded by a disk of material extending to nearly 80 000 000 000 km from the star; we see it almost edge-on, and it may be no more than a few hundred million years old. It was also found that the cloud of material showed a 'depleted region' in an area extending from 20 to 30 astronomical units from the star itself. It was suggested that this rarefied part of the cloud could be produced by the presence of planets which would 'sweep up' much of the dust.

Observations were also made with the IUE (International Ultra-Violet Explorer) satellite. Then, in May 1991, came results from the Goddard High Resolution Spectrograph on the Hubble Space Telescope which have led to a better understanding of the system. It is now thought that the solid particles in the β Pictoris disk which were discovered by IRAS are distributed in a disk with a diameter of about 100 astronomical units; the main gas-cloud is smaller, and extends to a few astronomical units from the star. The gas consists of several main components: first a diffuse disk circling β Pictoris in a stable manner; second, an inner gas disk which is slowly drifting in toward the star, and third, isolated clumps of gas which are spiralling in toward β Pictoris at velocities up to 200 m/s. As the gas clumps pass in front of β Pictoris, the star's ultra-violet spectrum shows marked changes. All three gaseous components seem to be embedded in the more rarefied circumstellar gas-cloud which is diffusing outward, possibly because of the radiation pressure of β Pictoris.

The origin of the circumstellar gas-cloud is uncertain. It may be due to the slow decomposition of the solid particles in the disk discovered by IRAS, or it may have evaporated from cool objects in the suspected proto-planetary system.

In 1994 new studies carried out by P. Lagage and E. Pantin, from La Silla in Chile, found that there is a decrease in particle density at about 30 astronomical units from the star. They suggest that there is a planet of about 20 Earth masses orbiting at a distance of around 20 astronomical units, and that this planet has swept the inner zone of the disk relatively free of dust.

DOUBLE STARS

*T*he term 'double star' was first used by Ptolemy, who wrote that η Sagittarii was 'διπλους'. There are, of course, several doubles which may be separated with the naked eye, so that they have presumably been known from antiquity; the most celebrated is ζ Ursæ Majoris (Mizar), which makes a naked-eye pair with 80 Ursæ Majoris (Alcor). The Arabs gave full descriptions of it – though it is true that they apparently regarded Alcor as rather a difficult object. (This is not true today, but it is unlikely that there has been any real change.)

✳ *The first double star discovered* telescopically was Mizar itself, which is made up of two components 14 sec·5 apart. The discovery was made by Riccioli in 1651. Alcor is 700 sec from the main pair, which is rather too wide for a recognized 'double' as entered in the official catalogues. The duplicity of γ Arietis was discovered by Robert Hooke in 1665, while he was searching telescopically for a comet.

Disk of material round ß Pictoris; R. Terrile and B. Smith, Las Campanas Observatory, Chile.

✳ *The first southern double star to be discovered* was α Crucis, by Father Guy Tachard in 1685. Tachard was on his way to Siam, by sea, and stopped off at the Cape of Good Hope, where he was warmly welcomed by the Dutch settlers and set up a temporary observatory, mainly for navigational purposes. He recorded that 'the foot of the Crozier marked in Bayer is a Double Star, that is to say, consisting of two bright stars distant from one another about their own Diameter, only much like to the most northern of the Twins; not to speak of a third much less, which is also to be seen, but further from these two.' The Crozier is the Southern Cross; 'Bayer' refers to J. Bayer's famous star catalogue of 1603, and the 'northern of the Twins' is Castor, already known to be double. Like most of his contemporaries, Tachard believed the stars to show definite apparent diameters rather than being virtual point sources.

Other doubles discovered at an early stage included α Centauri (1689), γ Virginis, and the 'Trapezium', θ Orionis, in the Orion Nebula, which is a multiple system.

✳ *The first true list of doubles* was published in 1871 by C. Mayer of Mannheim. His list included γ Andromedæ, ζ Cancri, α Herculis and β Cygni. Mayer used an 8 ft mural quadrant, with magnifications of 60 and 80.

✳ *The first comment upon possible physically-associated or binary pairs* was made by the Rev. John Michell in 1767, who wrote: 'It is highly probable in particular, and next to a certainty in general, that such double stars as appear to consist of two or more stars placed very near together, do really consist of stars placed near together, and under the influence of some general law.' Michell repeated this view in 1784. How-

ever, in 1782 William Herschel had commented that it was 'much too soon to form any theories of small stars revolving round large ones'.

✳ *The first proof of the existence of binary pairs* was given by Herschel in 1802. From 1779 he had been attempting to measure the parallaxes of stars, and he had concentrated upon the doubles, since if one member of the pair were more remote than the other it followed that the closer member should show an annual parallax relative to the more distant member. He failed, because his equipment was not sufficiently sensitive, but he made the fortuitous discovery that some of the pairs under study (such as Castor) showed orbital motion, and by 1802 he was confident enough to publish his findings. His classic paper actually appeared in the *Philosophical Transactions* on 9 June 1803.

✳ *The first successful measurement of a star-distance* was announced in 1838 by F.W. Bessel of Königsberg; the star concerned was 61 Cygni, which Bessel had selected because it had relatively large proper motion and was a wide binary – presumably indicating that by stellar standards it was comparatively close. At about the same time T. Henderson announced a parallax value for α Centauri, based upon measurements which he had made earlier (1831–33) at the Cape, while Struve in Russia gave a less accurate value for the parallax of Vega.

Rather surprisingly, binaries are more common than optical pairs, in which the components are not genuinely associated. It has even been estimated that more than 50 per cent of all stars are members of binary systems, with a mean separation from 10 to 20 astronomical units, though this may be too high a ratio. A visual binary will have an orbital period of at least two years. It is important to note that both components of a binary move round their common centre of gravity, and the orbits are not in general so dissimilar as might be thought, since in mass there is a far lower spread among the stars than there is in luminosity and in size.

✳ *The first reliable orbit for a binary pair* (ξ Ursæ Majoris) was worked out by the French astronomer Felix Savary in 1830. (The period is 60 years.) Such calculations are of great importance, since they lead to a determination of the combined masses of the components – something which is much more difficult to calculate for a single star. In fact, our knowledge of stellar masses depends very largely upon the orbital movements of binaries.

Many important catalogues of double stars have been published since Mayer's. Among them are the catalogues of F.G.W. Struve (1822, with later additions), E. Dembowski (Naples, 1852; over 20 000 measures), S.W. Burnham (1870, and again in 1906, listing 13 665 pairs – he was personally responsible for the discovery of 1340 of them), R. Aitken (1932; 17 180 pairs) and the Lick Index Catalogue or IDS (1963; 65 000 pairs, of which 40 000 are binaries). Work of the greatest importance was carried out at the Republic Observatory, Johannesburg (formerly the Union Observatory) between 1917 and 1965, under the successive directorships of R.T.A. Innes (1917–27), H.E. Wood (1927–41), W.H. van den Bos (1941–56) and W.S. Finsen (1957–65). The telescope used for most of the work was the 27 in refractor. It is a matter of great regret that the virtual closing of the Observatory, following the concentration of South African astronomy at Sutherland in Cape Province, has brought this work to an end – temporarily, one must hope.

✳ *The first spectroscopic binary* (Mizar A) was discovered in 1889 by E.C. Pickering at Harvard; another identification (β Aurigæ) soon followed. Spectroscopic binaries have too small a separation for the components to be seen individually, but the binary nature of the system betrays itself because of the Doppler shifts in the spectra. If both spectra are visible, the absorption lines will be periodically doubled; if one spectrum is too faint to be seen, the lines due to the primary will oscillate about a mean position. There are some 'borderline' cases; thus Capella was long known to be a spectroscopic binary, but only the world's largest telescopes can just indicate that it is not a single star. The orbital period is 100 days.

✳ *The first astrometric binaries to be studied* were Sirius and Procyon, by F.W. Bessel in 1844. In an astrometric binary, the presence of an invisible companion is inferred from slight 'wobblings' of the primary. In both these first cases, the companions were subsequently discovered, but with other astrometric binaries the secondary stars remain unseen.

✳ *The first White Dwarf binary companion to be discovered* (in fact, the first white dwarf of any kind) was Sirius B, or the Companion of Sirius. It was first seen in 1862 by the young son of the telescope maker Clark, at Washington. The orbital period is 50 years, and the maximum separation is 11 sec·5 (as in 1975). For many years after its discovery the Companion was assumed to be faint because it was cool and red, but in 1915 W.S. Adams, at Mount Wilson, studied its spectrum and found that it was white; the surface temperature was at least 8000°C.

Since the luminosity was only 1/10 000 that of Sirius itself, the Companion had to be very small; smaller, in fact, than Uranus or Neptune, and with a diameter only 3 times that of the Earth. The resulting density turned out to be 70 000 times that of water. It is worth noting that in ancient times Sirius was described as being red. There is no chance that the primary has changed in colour, and suggestions that the Companion was then going through its red giant stage, before collapsing into the white dwarf condition, seem to be untenable. We must therefore assume that there was some error in observation or interpretation in the old records, though an element of mystery remains.

Binary stars are of many different kinds; sometimes the components are dissimilar (as with Sirius), sometimes they are identical twins – as with γ Virginis, which has a period of 171 years. Several decades ago it was wide and easy, but is now closing, and by 2016 will be single except in giant telescopes. This does not, of course, indicate any actual closing; everything depends upon the angle from which we see the pair. Another binary, formerly easy but now much less so, is Castor. In fact both components of the bright pair are spectroscopic binaries, and also associated with the system is Castor C or YY Geminorum, made up of two red dwarfs; it is an eclipsing variable. Castor therefore consists of six stars, four luminous and two dim.

✳ *The first eclipsing binary to be discovered* was Algol, by G. Montanari in 1669 (it is not now thought that the 'winking' was known in ancient times, even though Algol was nicknamed the Demon Star). The cause of the variability was explained by J. Goodricke, the deaf-mute astronomer, in 1782.

✳ *The most extreme eclipsing binary* is probably ε Aurigæ. The variability was discovered by Fritsch in 1821, but was then thought to be irregular. In fact the system is made up of a powerful supergiant together with an invisible secondary, which may be a hot star surrounded by an immense shell of dust and gas. Close to it in the sky is another extreme eclipsing binary ζ Aurigæ, with a K-type supergiant primary and a B-type secondary; when the B-star passes behind the supergiant, its light comes to us for a period via the outer layers of the primary, and the spectral effects are both complicated and highly informative. Another interesting system is β Lyræ. Here, and also with other stars of the same type, the two components are almost in contact; they are tidally distorted into elliptical shapes, and the whole system is enveloped in gas.

✳ *The rarest stellar eclipse* is that of the binary Gamma Pegasi; the period is 14⅔ years, and the amplitude 0·3 magnitude. The next eclipse will be in April 2005, but the star will then be too close to the Sun in the sky to be observed, so that the next observable eclipse will be delayed until November 2019.

✳ *The binary with the shortest known period* is X-1820–303, an X-ray star in the globular cluster NGC 6624, distant 30 000 light-years. Its period is 685 seconds, or 11 minutes. It was discovered in 1987 by the aptly-named Luigi Stella and collaborators with the satellite Exosat. The shortest period visual binary is Wolf 630 Ophiuchi, 1·725 years

✳ *It is impossible to say which is the binary with the longest known period.* With very large separations the period may amount to millions of years, and it is better merely to say that very widely separated components share a common binary motion through space.

✳ *The first binary White Dwarf* was found by W. Luyten and P. Higgins in 1973. Its position is R.A. 9h 42m, dec. +23 deg 41 min. The separation is 13 sec, the position angle 052, and the period 12 000 years; at present the two components are 600 astronomical units apart.

✳ *The first binary neutron star* was discovered in July 1974. The orbital period is 8 hours. The discovery was made by means of radio observations, and the object (pulsar PSR 1913 + 16) has not been seen optically, but it is assumed to be a binary made up of two neutron-star components. Presumably there are binary black holes somewhere in the universe, but they are virtually impossible to identify except perhaps by their gravitational waves. Until we have found out how to detect gravitational waves, binary black holes will, it is expected, remain hidden.

✳ *The first X-ray binaries* were reported following the launch of the first X-ray astronomical satellite, UHURU, in 1970. X-ray binaries are white dwarfs or neutron stars in orbit around another star – a B star, K star or another white dwarf. The X-ray star is pulling material off the companion, perhaps at 100 million million tonnes per second, and X-rays are being emitted.

✳ *The origin of binary systems* is not known with certainty. The old theory – that a binary formed as a result of the fission of a single star – has been rejected by most astronomers. It has been suggested that binaries may be due to the mutual capture of the components, but in most cases it is believed that the components were formed from the same cloud of interstellar material in the same region of space. If the components are identical or near-identical twins, the spectra are generally the same. When there is a definite difference in brightness between the components, the spectra also differ. If both stars belong to the Main Sequence, the primary is usually of earlier type than the secondary, while if the primary is a giant the secondary is either a giant of earlier type or else a dwarf of similar spectral type. Novæ are binary systems (DQ Herculis 1934 is a good example; it is an eclipsing system with a period of 4 h 39 m made up of the old nova itself, probably so dense that it resembles a white dwarf, together with an invisible companion which is probably a red dwarf of type M). We also have the SS Cygni or U Geminorum stars, all of which seem to be binaries. In a star of this kind the primary is a white dwarf, the secondary a Main Sequence star, presumably a red dwarf. The components are very close together, and the white dwarf is pulling material away from the less massive secondary; gas moves across to the primary in a narrow stream, producing a disk or shell of gas round the white dwarf. Where the stream hits the disk there is a bright spot, and irregularities in the stream make the light we receive flicker rapidly with small amplitude; this flickering is detectable only with very sensitive photoelectric equipment. The larger outbursts in the SS Cygni stars originate in the white dwarf component of the system. It is easy to see why these variables are sometimes termed dwarf novæ.

✳ **The most spectacular multiple stars in the sky** are probably ε Lyræ and θ Orionis. ε Lyræ has two main components, making up a naked-eye pair; each component is again double, and all four may be distinguished with a small telescope. θ Orionis, in the great nebula M.42, is nicknamed the 'Trapezium' for obvious reasons. It lies on the outskirts of the nebula, and is responsible for making the nebulosity luminous.

✳ **The position angle** of a visual double star (either a binary or an optical pair) is measured according to the angular direction of the secondary (B) from the primary (A), reckoned from 000 at north round by east (090), south (180) and west (270) back to north. With rapid binaries the separations and position angles alter quickly; a good example is the fine binary ζ Herculis, where the magnitudes are 3 and 5·6, and the period is only 34 years. Some of the published catalogues are already out of date, and need revising. There is scope here for the skilful and well-equipped amateur as well as for the professional, and double-star research is a fascinating branch of astronomy.

VARIABLE STARS

The greatest number of variable stars discovered by one man is 9646, by C. Hoffmeister (Sonneberg Observatory). The Macho project in Australia, now looking for gravitational lens events in star-fields, is expected to produce the largest number of variable stars from one survey, but the enormous volume of data will take a very long time to analyse!

THE CLASSIFICATION OF VARIABLE STARS

*V*ariable stars are of many kinds. Elaborate systems of classifying them have been proposed, and the notes given below are not intended to be more than a general guide.

Eclipsing variables

Better termed 'eclipsing binaries', since there is no intrinsic variation. The light-changes are caused by one component of the binary system passing in front of the other.

✳ *Algol type* (EA). A principal minimum, and a secondary minimum which may be very slight (as with Algol itself) or quite deep. The star remains at the maximum brightness for most of the time.

✳ *Beta Lyræ type* (EB). Prototype, β Lyræ itself. Usually the components are semi-detached, so that the brighter component is distorted roughly into the shape of an egg, while the fainter component need not be noticeably distorted. (In Algol systems, the brighter component is nearly spherical, while the fainter one may be spherical, ellipsoidal or egg-shaped.)

✳ *W Ursæ Majoris* type (EW). Dwarf pairs, with ellipsoidal components which are generally in contact. The periods are less than one day.

Pulsating variables

The variations are intrinsic; the star expands and contracts.

✳ *Mira variables* (M), often called long-period variables. Named after Mira Ceti, the brightest and first-discovered member of the class. They are late-type giants of

types M, R, N or S, with periods of from about 80 to almost 1000 days, though neither the periods nor the amplitudes are constant. The median absolute magnitudes are from +2 to –2, and the range may be very great – 11 magnitudes in the case of χ Cygni. The spectra show emission lines, and many of the stars may be binary systems.

✳ *Semi-regular variables.* These are also of late spectral type. The periods are often so ill-defined as to be virtually unrecognizable, and the amplitudes are less than with the Mira stars. The periods range from around 30 to around 1000 days. There are various subdivisions. SRa: late-type supergiants with relatively stable cycles. SRb: also late-type, but less regular. SRc: young semi-regular supergiants. SRd: giants of types F, G or K.

✳ *Cepheids.* Named after the prototype star, δ Cephei. They are yellow supergiants, and show the celebrated period-luminosity law which gives a key to their distances. The period of pulsation is the time required for a vibration to travel from the surface of the star to the centre and back again, so that large stars have longer periods than smaller ones. *Classical Cepheids* (Cδ) are F, G or K supergiants, and are highly evolved, so that they have exhausted their available hydrogen and helium; the periods range from a few days to 70 days, with amplitudes of from 0·1 to 2 magnitudes. Either the rise to maximum is sharper than the subsequent decline, or else the light curves are symmetrical (or almost so). W Virginis stars or Population II Cepheids have lower masses, and are about 2 magnitudes fainter than classical Cepheids of equal periods. They are also less regular. The brightest example is the southern κ Pavonis.

The best-known ex-variable star is Polaris. In 1899 it was found to be a Cepheid variable, with a very low amplitude 1·92 to 2·07, period 3·969778 days (according to the Cambridge catalogue). But the amplitude has been steadily decreasing, and by 1992 was down to 0·010 of a magnitude. It is expected to cease fluctuating entirely in the near future, though it remains well inside what is termed the 'instability strip' in the HR Diagram where every star ought to pulsate as a Cepheid.

✳ *RR Lyræ variables.* Formerly known as cluster-Cepheids, though not all are found in clusters; RR Lyræ itself is not. The spectra are of type A or F. They are old, with masses lower than that of the Sun, but radii 4 to 5 times greater, and all are of about the same luminosity. The range is about 1 magnitude, and the periods are from 0·2 to 1·2 days. There are various subdivisions. RRAB stars (formerly divided into two sub-classes,

SUMMARY OF VARIABLE STAR CLASSIFICATION

ECLIPSING VARIABLES (or Eclipsing Binaries)

EA	Algol	Period 0·2 day to over 27 years. Almost spherical components. Maximum for most of the time.
EB	β Lyræ	Periods over 1 day; spectra B to A. Ellipsoidal components; magnitude continuously changing.
EW	W Ursæ Majoris	Dwarfs, periods usually less than 1 day. Almost or quite in contact. Primary and secondary minima almost equal.

PULSATING VARIABLES

M	Mira	Long-period late-type giants; spectra M-C-S, periods 80 to 1000 days. Amplitude may exceed 10 magnitudes. Periods and amplitudes vary from cycle to cycle.
SR	Semi-regular	Late-type giants (spectra M-C-S). Periods from 20 days to several years. SRA: persistent periodicity (Z Aquarii). SRB: rough periodicity (RR Coronæ Borealis). RRG: red supergiants (μ Cephei). SRD: F to K type giants and supergiants (SX Herculis).
PVTEL	PV Telescopii	Helium supergiants; periods from a few hours to a year; spectra Bp; small amplitudes (0·1 mag).
RR	RR Lyræ	Spectra A to F; periods 0·2 to 1·2 days; formerly called cluster-Cepheids. RRAV: steep rise to maximum. RRB: almost symmetrical light-curves. RRC: sinusoidal light-curves.
RV	RV Tauri	Supergiants, usually type F to K. Periods 30 to 150 days. RVA: stars with constant mean magnitude (AC Herculis). RVB: stars with variable mean magnitude.
L	Irregular	Types K, M, C, S. LB: slow giant variables (CO Cygni). LC: slow supergiant variables (TZ Cassiopeiæ).
CEP	Cepheids	Radial pulsating stars. Periods 1 to 135 days; spectra F to K.
CW	W Virginis	Population II Cepheids. CWA: longer period (W Virginis). CWB: shorter period (BL Herculis).
DSCT	δ Scuti	Types A to F; periods less than 1 day; small amplitude.
SXPHE	SX Phoenicis	Population II subdwarfs resembling δ Scuti stars. Types A to F, periods less than 1 day, amplitude up to 0·7 magnitude.
ACYG	α Cygni	Types B to A; supergiants, with low amplitudes. Most have short periods.
BCEP	β Cephei (or β Canis Majoris)	B-type subgiants. Low amplitude and short period.
ZZ	ZZ Ceti	Non-radially pulsating white dwarfs; amplitude up to 0·2 magnitude; periods 30 seconds to 1500 seconds. ZZA: hydrogen absorption lines only (spectrum DA). ZZB: helium absorption lines only (spectrum DV). ZZO: very hot stars (spectrum DO).

ERUPTIVE VARIABLES

FU	Fuors	Prototype, FU Orionis. Types A to G; slow rise to maximum (many years) and slower decline.
GCAS	γ Cassiopeiæ	Shell stars; type B; rapid rotaters. Amplitudes usually below 2 magnitudes.
I	Irregular	Types O to M. IA: early spectral type. IB: intermediate and late spectral type.
IN	Orion	INA: early type (T Orionis). INB: later type (spectra F to M).
IT	T Tauri	Similar to Orion variables: types F to M (RW Aurigæ). INT: types F to M; contained in diffuse nebulæ (T Tauri itself). IS: rapidly-varying stars, amplitudes up to 1 magnitude; ISA (early type, spectra B to A). ISB (later type, spectra F to M).
RCB	R Coronæ Borealis	Types B to R. Occasional deep minima; large amplitude (over 9 magnitudes in some cases).
RS	RS Canum Venaticorum	Small amplitude. Close binaries with active chromospheres.
SDOR	S Doradus	Types Bp to Fp. Very luminous supergiants with expanding shells, often in diffuse nebulæ.
UV	UV Ceti	Dwarfs of types K to M. Flare stars; amplitudes may be as much as 6 magnitudes (UV Ceti itself).
W	Wolf-Rayet	Very low amplitudes; spectra of type W. Non-stable mass outflow.

CATACLYSMIC VARIABLES

AM	AM Herculis	Polars; close binaries with one compact component. Amplitude up to 5 magnitudes.
UG	U Geminorum	Dwarf novæ. Periods from 10 to 1000 days, amplitudes from 2 to 9 magnitudes. SS Cygni stars have outbursts lasting for several days.
UGSU	SU Ursæ Majoris	Dwarf novæ with occasional supermaxima, brighter and longer than normal maxima.
UGZ	Z Camelopardalis	Dwarf novæ which have occasional 'standstills' when the normal cycle of variation is suspended.
ZAND	Z Andromedæ	Symbiotic stars; close binaries.
N	Novæ	Thermonuclear outburst on the white dwarf component of a binary system. NA: fast, fading by 3 magnitudes in 100 days or less (GK Persei 1901). NAB: fading at intermediate speed. NB: slow fading, no more than 3 magnitudes in 150 days (RR Pictoris 1925). NC: very slow novæ with maxima which may last for years, as with RR Telescopii.
NL	Nova-like	NL: poorly-studied stars with nova-like outbursts (V Sagittæ). NR: recurrent novæ (such as T Coronæ Borealis) which flared in 1866 and again in 1946.
SN	Supernovæ	SNI: explosion and destruction of the white dwarf component of a binary system. SNII: collapse of a very massive star, often leaving a neutron star or pulsar.

ROTATING VARIABLES

ACV	α² Canum Venaticorum	Types B to A; strong magnetic fields; spectra rich in silicon, strontium, chromium and rare earth lines. Periods from 12 hours up to 160 days.
BY	BY Draconis	Types G to M; periods up to 20 days. Rotating dwarfs with starspots and active chromospheres.
ELL	Ellipsoidal	Close binaries with no eclipses, but changing visible area. Low amplitudes (up to 0·1 magnitude).
FKCOM	FK Com	Types G to K. Periods up to several days. Amplitudes up to half a magnitude. Rapidly rotating giants with non-uniform surface brightness.
SXARI	SX Arietis	Helium stars (type B); amplitude around 0·1 magnitude and periods of around 1 day. High-temperature versions of α² Canum Venaticorum stars.

A few variables, such as VY Canis Majoris, do not seem to fit into any class, and there are also very exceptional stars such as η Carinæ. Pulsars and X-ray binaries are often included in variable star classification lists.

RRa and RRb) have amplitudes of from 0·3 to 1 magnitude, and periods of from about 0·5 to 0·7 day. RRc variables have almost sinusoidal light-curves, with amplitude around 0·5 magnitude and period around 0·3 day. Some RR Lyræ stars, such as AR Herculis, have several pulsation periods, and this results in a continuous deformation of the light-curve (Blazhko effect).

* RV Tauri variables. These are pulsating supergiants, usually of types F to K, but occasionally M. There are alternate deep and shallow minima; the interval between successive primary minima may be from 30 to 150 days. There are several subdivisions. RVa: the period is not constant, but is at least reasonably consistent; the brightest member of the class, R Scuti, is of this type. RVb: several superposed cycles, with rough periods from 30 to 1000 days.

* Dwarf Cepheids, formerly classed with the RR Lyræ stars as RRs; the prototype is AI Velorum. They are of types A to F, with absolute magnitudes of from +1 to +5 and periods from 0·05 to 0·25 day.

* δ Scuti variables (δS), once confused with dwarf Cepheids. They have smaller amplitudes, often below 0m·1, and periods from 0·02 to 0·25 day. They are young, with spectra of types A to F, and many are spectroscopic binaries. Sub-dwarfs showing the same characteristics are termed SX Phœnicis stars.

* β Canis Majoris or β Cephei variables. B0 to B3 giants or subgiants, with periods of from 0·1 to 0·3 day, and amplitudes from 0·1 to 0·3 magnitude. They are relatively massive stars which have almost exhausted their core hydrogen. Ultra-short periods of a few hundredths of a day are suffixed s; such a star is χ Centauri.

* α Cygni variables. Pulsating supergiants of types B or A, with short periods and amplitudes below 0m·1.

* ZZ Ceti variables. Pulsating white dwarfs with periods which may be as short as 30 seconds and amplitudes below 0m·2. The longest periods are around 23 minutes. Flares of about a magnitude, possibly due to activity in an otherwise undetected UV Ceti-type companion, are sometimes seen.

* Irregular variables. Slowly-varying stars denoted by L. Lb variables are of late spectral type, and Lc stars are late-type supergiants.

ERUPTIVE VARIABLES

Novæ. Novæ can be spectacular in the extreme. A formerly faint star suddenly flares up, and becomes thousands of times brighter than its usual brilliance, remaining at maximum for a brief period before declining again. The rise may take only a few hours – as with the bright nova V1500 Cygni of 1975, which reached an absolute magnitude of –10 and an apparent magnitude 1·8, but within a week or so had fallen below naked-eye visibility. This was a fast nova; slow novæ, such as HR Delphini of 1967, are much more gradual in their decline.

A nova is a binary system, consisting of a low-density red star with a white dwarf companion. The white dwarf pulls material away from the red star, and this material produces a ring or 'accretion disk' round the white dwarf. Eventually there is a nuclear outburst in the atmosphere of the white dwarf, and gas is ejected at high velocity; at the end of the outburst the system returns to its former state. Some stars, such as T Coronæ, have shown more than one outburst; these are the *recurrent novæ*.

The least luminous old nova known is CK Vulpeculæ (Anthelm's Star) of 1760. It is around 2000 light-years away, and has only about 1/100 the luminosity of the Sun.

* Dwarf novæ, otherwise known as U Geminorum or SS Cygni stars, show minor outbursts at roughly regular intervals. Of the two components, one is a K or M red dwarf and the other is a white dwarf. The amplitude is of the order of 3 to 5 magnitudes, and the interval between outbursts may be from 10 days to several years. The basic cause of variation is the same as with true novæ, but on a much reduced scale. *SU Ursæ Majoris* stars have both normal maxima and 'supermaxima' of greater amplitude. *Z Camelopardalis* stars show outbursts from 2 to 5 magnitudes every 10 to 40 days, but with unpredictable 'standstills' when variation is virtually suspended.

* Nova-like variables. There are various types. γ *Cassiopeiæ* variables (γ C) are rapidly-rotating B-type stars with emission lines in their spectra; they are losing mass from their equatorial regions as they evolve off the Main Sequence. They have small amplitudes, usually less than 2 magnitudes. Stars of the *Z Andromedæ* or *Symbiotic* type are close binaries, often involved in nebulosity. One component is a cool red star, while the other is a hot star; the variations are caused by pulsations in the red star together with interactions between the two. *RR Telescopii* stars show slow increases which may be of very long duration, and it is suggested that they may be in the process of becoming planetary nebulæ. *S Doradûs variables* are very luminous, with absolute magnitudes of around –10 and spectra of B to F; they show variations from 1 to 3 magnitudes

which may be either cyclic or irregular.

* P Cygni variables, also very luminous produce expanding gas-shells.

* Nebular variables are very young. In: Orionis stars, with spectra of from type B t K; they are of low mass, and fluctuate ir regularly. IT stars, such as RW Aurigæ, ar similar, but are not associated with nebula They have amplitudes of from 0·5 to 1 mag nitude, and vary over periods from a few hours to several days. T Tauri variables ar extremely young, and show irregular fluc tuations of small amplitude. FU Orionis star have spectra of from type A to F; they ma brighten up by several magnitudes over period of several months, and stay at max mum for decades. At maximum, emissio lines appear in their spectra.

* Flare stars (UV Ceti type) are dwarfs o type K or M. They show sudden outburst of from 1 to 6 magnitudes, lasting for sev eral minutes. 'Flash' variables (UVa) are o earlier spectral type, and are more lum nous; they are associated with nebulosity.

* R Coronæ Borealis variables are deficien in hydrogen but rich in carbon. They ar highly luminous, and of spectral type F, C K or R. They remain at maximum for mos of the time, but undergo sudden, unpredic able drops to minimum, taking from severa weeks to several months to recover. Th amplitudes are large – at least 10 magnitude in the case of R Coronæ itself. The fading are due to clouds of 'soot' accumulating i the star's atmosphere. R Coronæ variable are rare.

ROTATING VARIABLES

These have small amplitudes. BY Dracon stars are young, rapidly-rotating K or M dwarfs with emission lines in their spectra they are quasi-periodic, with amplitudes o no more than 0·3 magnitude, and period from a few hours to 120 days. The variation are due to their non-uniform surface brigh ness. FK Comæ stars are of type G; they hav fast rotation and great chromospheric activ ity. The rapid spin of FK Comæ itself may b due to the presence of a close, low-mas companion in a decaying orbit, leading pos sibly to eventual coalescence. RS Canur Venaticorum stars are binaries, of which on component is a rapidly-rotating F to K giar or subgiant; the periods range from a fev hours up to several months. The variation are due to 'starspots' and, in some case eclipses. Ellipsoidal variables are also bina ries, which are close together an gravitationally distorted; there are n eclipses, and the fluctuations are due to th differing amount of surface which is presented to us.

✳ *α² Canum Venaticorum* stars are magnetic or spectrum variables, usually of type A, but with enhanced lines of some elements such as silicon, strontium, chromium or europium. The changes in visual light are small, usually below 0m·1. *SX Arietis* stars are helium variables of type B, with intense helium and silicon lines in their spectra, and strong, variable magnetic fields.

✳ *Secular variables* are stars which may have brightened or faded in historical times. Thus Ptolemy and others ranked β Leonis and θ Eridani as being of the first magnitude, whereas today they are respectively of magnitude 2 and below 3. δ Ursæ Majoris was ranked as equal to the other stars of the Plough, but is now at least a magnitude fainter; on the other hand α Ophiuchi was ranked as of magnitude 3, and is now 2. These variations must be regarded as suspect, since it would be unwise to trust the old estimates too far.

The first variable star to be positively identified as such as Mira (o Ceti), in 1638. It had been previously recorded by Fabricius (1596) and by Bayer (1603) and had even been allotted a Greek letter, so that it is surprising that its fluctuations were not tracked down earlier. In the latter part of the 17th century two more variables were identified – Algol, now known to be an eclipsing binary, and the long-period χ Cygni.

The following variables were identified between 1638 and 1850 (novæ are not included):

Star	Discoverer	Date of variability
Mira (o Ceti)	Holwarda	1638
Algol (β Persei)	Montanari	1669
χ Cygni	Kirch	1686
R Hydræ	Maraldi	1704
Rasalgethi (α Herculis)	W. Herschel	1759
μ Cephei	W. Herschel	1782
R Leonis	Koch	1782
δ Cephei	Goodricke	1784
β Lyræ	Goodricke	1784
η Aquilæ	Pigott	1784
R Scuti	Pigott	1795
R Coronæ Borealis	Pigott	1795
R Virginis	Harding	1809
R Aquarii	Harding	1811
ε Aurigæ	Fritsch	1821
R Serpentis	Harding	1826
η Carinæ	Burchell	1827
S Serpentis	Harding	1828
U Virginis	Harding	1831
δ Orionis	J. Herschel	1834
S Vulpeculæ	Rogerson	1837
Betelgeux (α Orionis)	J. Herschel	1840
β Pegasi	Schmidt	1847
λ Tauri	Baxendell	1848
R Orionis	Hind	1848
R Pegasi	Hind	1848
R Capricorni	Hind	1848
S Hydræ	Hind	1848
S Cancri	Hind	1848

Star	Discoverer	Date of variability
S Geminorum	Hind	1848
R Geminorum	Hind	1848
T Geminorum	Hind	1848
R Tauri	Hind	1849
T Virginis	Boguslawsky	1849
T Cancri	Hind	1850
R Piscium	Hind	1850

Sir John Herschel regarded Alphard (α Hydræ) as a variable, but this has not been confirmed. It is now known that δ Orionis has a very small range, so that Herschel's discovery may not have been valid.

A CATALOGUE OF BRIGHT VARIABLE STARS

*T*he following list includes variable stars with a maximum of magnitude 6 or brighter, and a range of at least 0m·4.

MIRA VARIABLES

Star	Max.	Min.	Period, d.	Spectrum
R And	5·8	14·9	409	S
R Aql	5·5	12·0	284	M
R Car	3·9	10·5	309	M
S Car	4·5	9·9	149	K–M
R Cas	4·7	13·5	430	M
R Cen	5·3	11·8	546	M
o Cet	1·7	10·1	332	M Mira
S CrB	5·8	14·1	360	M
χ Cyg	3·3	14·2	407	S
U Cyg	5·9	12·1	462	N
R Gem	6·0	14·0	370	S
S Gru	6·0	15·0	401	M
R Hor	4·7	14·3	404	M
R Hya	4·0	10·0	390	M
R Leo	4·4	11·3	312	M
R Lep	5·5	11·7	432	N
V Mon	6·0	13·7	334	M
X Oph	5·9	9·2	334	M + K
U Ori	4·8	12·6	372	M
RU Sgr	6·0	13·8	240	M
RT Sgr	6·0	14·1	305	M
RR Sco	5·0	12·4	279	M
S Sci	5·5	13·6	365	M
R Ser	5·1	14·4	356	M
R Tri	5·4	12·6	266	M
SS Vir	6·0	9·6	355	N
R Vir	6·0	12·1	146	M

SEMI-REGULAR VARIABLES

Star	Max.	Min.	Period, d.	Spectrum
UU Aur	5·1	6·8	234	N
W Boö	4·7	5·4	450	M
VZ Cam	4·7	5·2	24	M
X Cnc	5·6	7·5	195	N
TU CVn	5·6	6·6	50	M
S Cen	6·0	7·0	65	N
T Cen	5·5	9·0	90	K–M
T Cet	5·0	6·9	159	M
FS Com	5·3	6·1	58	M
W Cyg	5·0	7·6	126	M
U Del	5·7	7·6	110	M
EU Del	5·8	6·9	59	M
R Dor	4·8	6·6	338	M

Star	Max.	Min.	Period, d.	Spectrum
UX Dra	5·9	7·1	168	N
RY Dra	5·6	8·0	173	N
η Gem	3·2	3·9	233	M Propus
π¹ Gru	5·4	6·7	150	S
g Her	5·7	7·2	70	M
α Her	3	4	±100	M Rasalgethi
R Lyr	3·9	5·0	46	M
ε Oct	4·9	5·4	55	M
α Ori	0·1	0·9	2110	M Betelgeux
W Ori	5·9	7·7	212	N
CK Ori	5·9	7·1	120	K
Y Pav	5·7	8·5	233	N
SX Pav	5·4	6·0	50	M
β Peg	2·3	2·8	38	M Scheat
ρ Per	3	4	33–55	M
TV Psc	4·6	5·4	70	M
L² Pup	2·6	6·2	140	M
R Sci	5·8	7·7	370	N
RR UMi	6·0	6·5	40?	M

CEPHEIDS

Star	Max.	Min.	Period, d.	Spectrum
η Aql	3·5	4·4	7·2	F–G
RT Aur	5·0	5·8	3·7	F–G
ZZ Car	3·3	4·2	36·5	F–K
U Car	5·7	7·0	38·8	F–G
SU Cas	5·7	6·2	1·9	F
δ Cep	3·5	4·4	5·4	F–G
AX Cir	5·6	6·1	5·3	F–G
X Cyg	5·9	6·9	16·4	F–G
β Dor	3·7	4·1	9·8	F–G
ζ Gem	3·7	4·1	10·1	F–G
T Mon	6·0	6·6	27·0	F–K
S Mus	5·9	6·4	9·7	F
R Mus	5·9	6·7	7·5	F
Y Oph	5·9	6·4	17·1	F–G
κ Pav	3·9	4·7	9·1	F (W Virginis type)
S Sge	5·3	6·0	8·4	F–G
X Sgr	4·2	4·8	7·0	F
W Sgr	4·3	5·1	7·6	F–G
Y Sgr	5·4	6·1	5·8	F
AH Vel	5·5	5·9	4·2	F
T Vul	5·4	6·1	4·4	F–G

RV TAURI VARIABLE

Star	Max.	Min.	Period, d.	Spectrum
R Sct	4·4	8·2	140	G–K

SYMBIOTIC VARIABLES

Star	Max.	Min.	Period, d.	Spectrum
R Aqr	5·8	12·4	387	M + P
AG Peg	6·0	9·4	830	WN + M

RECURRENT NOVÆ

Star	Max.	Min.	Spectrum	Outbursts
T CrB	2·0	10·8	M + Q	1866, 1946 Blaze Star
RS Oph	5·3	12·3	O + M	1901, 1933, 1958, 1967

R CORONÆ VARIABLES

Star	Max.	Min.	Spectrum
R CrB	5·7	15	Fp
RY Sgr	6·0	15	Gp

ERUPTIVE VARIABLES

Star	Max.	Min.	Spectrum
U Ant	5·7	6·8	N
η Car	−0·8	7·9	Pec
ρ Cas	4·1	6·2	F–K (Occasional fades)
α Cas	2·1?	2·5?	K (Suspected variable)
γ Cas	1·6	3·3	B
μ Cen	2·9	3·5	B

	Max.	Min.	Spectrum
μ Cep	3·4	5·1	M
θ Cir	5·0	5·4	B
P Cyg	3	4	Bp
T Cyg	5·0	5·5	K
BU Gem	5·7	7·5	M
RX Lep	5·0	7·0	M
S Mon	4	5	O7
B0 Mus	6·0	7·7	M
χ Oph	4·2	5·0	B
λ Pav	3·4	4·3	B
X Per	6·0	7·0	O9·5 X-ray star
d Ser	4·9	5·9	G + A
VY UMa	5·9	6·5	N
BU Tau	4·8	5·5	Bp Pleione

ECLIPSING BINARIES
(Algol type)

	Max.	Min.	Period, d.	Spectrum
R Ara	6·0	6·9	4·4	B
WW Aur	5·8	6·5	2·5	A + A
R CMa	5·7	6·3	1·1	F
RS Cha	6·0	6·7	0·1	A + F
δ Lib	4·9	5·9	2·3	B
U Oph	5·9	6·6	1·7	B + B
β Per	2·2	3·4	2·9	B + G Algol
ζ Phe	3·9	4·4	1·7	B + B
RS Sgr	6·0	6·9	2·4	B + B
λ Tau	3·3	3·8	3·9	B + A
HU Tau	5·9	6·7	2·1	A

(Beta Lyræ type)

UW CMa	4·0	5·3	4·3	O7
u Her	4·6	5·3	2·0	B + B
GG Lup	5·4	6·0	2·1	B + A
β Lyr	3·3	4·3	12·9	B + A Sheliak
V Pup	4·7	5·2	1·4	B + B

(Long period)

ε Aur	2·9	3·8	9892	F
ζ Aur	3·7	4·1	972	K + B
VV Cep	4·8	5·4	7430	M + B

BRIGHT NOVÆ

*T*he following list includes all novæ since 1600 to have attained magnitude 6·0 or brighter.

Nova	Year	Max. mag.	Discoverer
CK Vulpeculæ	1670	3	Anthelm
WY Sagittæ	1783	6	D'Agelet
V. 841 Ophiuchi	1848	4	Hind
Q Cygni	1876	3	Schmidt
T Aurigæ	1891	4·2	Anderson
V. 1059 Sagittarii	1898	4·9	Fleming
GK Persei	1901	0·0	Anderson
DM Geminorum	1903	5·0	Turner
OY Aræ	1910	6·0	Fleming
DI Lacertæ	1910	4·6	Espin
DN Geminorum	1912	3·3	Enebo
V. 603 Aquilæ	1918	−1·1	Bower
GI Monocerotis	1918	5·7	Wolf
V. 476 Cygni	1920	2·0	Denning
RR Pictoris	1925	1·1	Watson
XX Tauri	1927	6·0	Schwassmann and Wachmann
DQ Herculis	1934	1·2	Prentice
V. 368 Aquilæ	1936	5·0	Tamm
CP Lacertæ	1936	1·9	Gomi
V. 630 Sagittarii	1936	4·5	Okabayasi
BT Monocerotis	1939	4·3	Whipple and Wachmann
CP Puppis	1942	0·4	Dawson

Nova	Year	Max. mag.	Discoverer
DK Lacertæ	1950	6·0	Bertaud
RW Ursæ Minoris	1956	6·0	Satyvaldiev
V. 446 Herculis	1960	5·0	Hassell
V. 533 Herculis	1963	3·2	Dahlgren and Peltier
HR Delphini	1967	3·7	Alcock
LV Vulpeculæ	1968	4·9	Alcock
FH Serpentis	1970	4·4	Honda
V. 1500 Cygni	1975	1·8	Honda
NQ Vulpeculæ	1976	6·0	Alcock
V. 1370 Aquilæ	1982	6	Honda
QU Vulpeculæ	1984	5·6	Collins
V. 842 Centauri	1986	4·6	McNaught
Nova Herculis	1991	5·0	Alcock
Nova Cygni	1992	4·3	Collins
Nova Cassiopeiæ	1993	5·4	Kanatsu

SUPERNOVÆ

*S*upernovæ are the most colossal outbursts known in nature. Many have been seen in external galaxies, but the last supernova to be seen in our Galaxy was observed as long ago as 1604 – before the invention of the telescope. However, the 1987 supernova in the Large Cloud of Magellan attained naked-eye visibility.

Supernovæ are of two types, quite different in origin.

Type I. Here we are dealing with a binary system, of which one component (A) is initially more massive than its companion (B). A therefore evolves more quickly into the red giant stage. Material from it is pulled across to B, so that B grows in mass while A declines; eventually B becomes the more massive of the two. Meanwhile, A is reduced to a very small, dense core, made up chiefly of carbon. The situation is then reversed; B evolves in its turn, swells out, and starts to lose material back to the shrunken A. The white dwarf builds up a gaseous layer, composed largely of hydrogen. But when the mass of the white dwarf exceeds the Chandrasekhar limit (around 1·4 times the mass of the Sun) the carbon detonates, and in a matter of a few seconds the white dwarf is completely destroyed. The resulting outburst takes a long time to die down. When the carbon white dwarf explodes, it creates other elements including neon, oxygen and silicon, ending up as nickel. Nickel decays to cobalt and then to iron, so that these elements can be detected spectroscopically. The energy produced is remarkable; the absolute magnitude of a Type I supernova at its peak can reach −19, or approximately 4 000 000 000 times the luminosity of the Sun.

Type II. A supernova outburst of this type is due to the collapse of a very massive star – at least 8 times as massive as the Sun – which has used up its main nuclear fuel, and has

produced a nickel-iron core with a mass about 1·5 times that of the Sun. The structure of the star prior to the outburst has been likened to that of an onion! Outside the iron core is a zone of silicon and sulphur; next comes a layer of neon and magnesium; then a layer of carbon, neon and oxygen; then a layer of helium, and finally an outer layer of hydrogen. Eventually the core can no longer support the pressure of the outer layers, and collapses to form a neutron star of immense density (one could pack over 2500 million tons of neutron star into a matchbox). The temperature now is of the order of 300 000 million degrees C, and vast numbers of neutrinos are produced, most of which pass straight through the star into space. The creation of the neutron star causes the core to 'bounce', and a shock-wave moves outward – colliding with the material which is falling inward. The result is a catastrophic explosion. (An alternative explanation for Type I supernovæ which is finding increasing favour is that they are produced when two white dwarfs in a binary system merge to form a single white dwarf with mass above the Chandrasekhar limit. The outcome is the same.) Most of the star's material is blown away, leaving only a small, incredibly dense remnant which may be a neutron star or – in extreme cases – a black hole. The peak luminosity is not so great as with a Type I supernova, but the absolute magnitude may reach a value of around −17, which is more than 1000 million times as luminous as the Sun. A supernova remnant (SNR) may be detectable as a pulsar, the classic case being that of the Crab Nebula – known to be the remnant of the supernova observed in the year 1054.

The first supernova since 1604 clearly visible with the naked eye was SN 1987A, in the Large Cloud of Magellan. It was discovered on 24 February 1987 by Ian Shelton, at the Las Campanas Observatory in Chile, on a routine photograph he had taken, and almost at the same time by O. Duhalde, also at Las Campanas, with the naked eye. It had been photographed earlier by R. McNaught, in Australia, but McNaught had not checked his observations!

When discovered, it was of magnitude 4·5, but brightened slowly to magnitude 2·3 before fading. The progenitor star was known: it was Sanduleak −69°202, which was a blue supergiant, not a red star. It was thought to be about 20 000 000 years old, with a mass 20 times that of the Sun. It had been previously assumed that supernovæ were produced by old red supergiants. The peak luminosity was low by supernova standards – only about 250 000 000 times that of the Sun. Its distance was 169 000 light-years.

Maximum brightness was reached in mid-May 1987, and then came a slow decline. Various phenomena were seen, including 'light rings' due to dust in the Large Magellanic Cloud which was now being illuminated by the outburst. The supernova was observed at many wavelengths, including X-rays (by the Japanese satellite Ginga) and ultra-violet (the IUE satellite). It seems that the progenitor star was originally hot and massive. As it exhausted its available hydrogen, it expanded to become a red supergiant. It then shed its outer layers, and became smaller and hotter. It was then, with the star a blue supergiant, that the outburst happened.

✳ *The first report of a pulsar in the remnant of the supernova 1987A* came on 18 January 1989, when a team at the Cerro Tololo Observatory in Chile, led by J. Middleditch, claimed to have observed a pulsar spinning at a rate of 1968·63 turns per second. Various explanations were offered, but it was finally found that the observational effects were due to the mechanism of the telescope.

GALACTIC SUPERNOVÆ

*T*he following supernovae have been reported in our Galaxy in historic times:

Date	Constellation	Max magnitude	Time visible with naked eye, months	Remnant name	Remnant diameter light-years	Estimated distance light-years	Max absolute magnitude	Type
185	Centaurus	−8	20	RCW 86	115	9800	−19±2	I ?
386	Sagittarius	+1·5	3	–	20?	16 000		II
393	Scorpius	0	8	CTB 37A–B	80	33 900		II
1006	Lupus	−9·5	24	PKS 1459–41	30	3260	−19·8±1	I
1054	Taurus	−5	22	Crab, 3C 144	9	6500		II
1181	Cassiopeia	0	6	3C 58	17	8500		II
1572	Cassiopeia	−4	16	3C 10	18	7500	−17·6±0·5	I Tycho's Star
1604 Ophiuchus Star		−3	12	3C 358	12	14 200	−19·6±0·5	I Kepler's Star

185 December Discovered 7 December, near α and β Centauri; Chinese sources. The optical object RCW86 (G.315·4–2·3) is an X-ray emitter. However, it has recently been suggested that there may have been an error, so that the position was near α and ξ Centauri and the object was a comet.

386 Chinese sources. Several radio sources lie in the general area (near λ Sagittarii). Probably a supernova.

393 February/March Chinese sources. Near the Scorpion's tail (λ Scorpii). Probably a supernova.

1006 3 April (discovery) Chinese and Korean sources. Unfortunately it is not very well documented.

1054 July Chinese, Japanese and Korean sources. It was seen for 23 days in daylight and 22 months in all. This is the Crab Nebula progenitor.

1181 August Between ε and ι Cassiopeiæ; Chinese and Japanese sources. Probably a supernova, though the evidence is not absolutely conclusive.

1572 Tycho's Star, B Cassiopeiæ. Seen by W. Schüler of Wittenberg on 6 November; Tycho saw it on 11 November; it may have been seen as early as 3 November. Fortunately Tycho Brahe left a full account of it. The remnant is an X-ray emitter. Near ξ.

1606 October Kepler's Star, near λ Ophiuchi. Discovered on 9 October; Kepler saw it on 11 October. Korean astronomers followed it for several months.

There must have been a supernova in Cassiopeia around 1700, which has left the remnant known as the radio source Cassiopeia A, but it was not certainly observed, as it was too heavily obscured by interstellar material near the main plane of the Galaxy.

Apart from SN1987A in the Large Cloud of Magellan, only one extragalactic supernova has approached naked-eye visibility: this was S Andromedæ (1885) in M.31.

Using the 4 m telescope at Kitt Peak on 4 November 1988, R.A. Feisen identified the remnant of S Andromedæ; the iron-rich remnant shows up as a dark spot against the background.

✳ *The only supernova to have been discovered by a Hungarian baroness* was S Andromedæ. On 22 August 1885 the Baroness de Podmaniczky was holding a house party, and for the entertainment of her guests had set up a small telescope on the lawn. Using this, a small star was seen in M.31. True, the significance of the observation was not recognized, and the supernova had been seen on 20 August by Hartwig at the Dorpat Observatory in Estonia; however, the observation by the Baroness was certainly independent!

The brightest of these galactic supernovæ was certainly the star of 1006.

Many supernovae have been seen in external galaxies. The most powerful supernova at radio wavelengths is 1986J, in the galaxy NGC 891. Its luminosity is 4000 times that of Cassiopeia A. The outburst probably occurred in 1982; the progenitor star seems to have been from 20 to 30 times as massive as the Sun.

✳ *The brightest supernova in the northern hemisphere of the sky in telescopic times,* SN 1993J, flared up in the galaxy M.81 on 28 March 1993. It was discovered by F. Garcia from Madrid, and rose to magnitude 10·5. It was unusual in being a Type II supernova which later showed some characteristics of a Type I outburst. The distance is 11 000 000 light-years.

✳ *The only case of two supernovæ appearing simultaneously in the same galaxy* relates to the spiral NGC 10–24–007 in Draco, in April 1992. The supernovæ appeared within a month of each other.

STAR CLUSTERS AND NEBULÆ

*C*lusters and nebulæ are among the most striking of stellar objects. Several are easily visible with the naked eye. Few people can fail to recognize the lovely star-cluster of the Pleiades or Seven Sisters, which has been known since prehistoric times and about which there are many old legends. The nebula in the Sword of Orion, the Sword-Handle in Perseus, and Præsepe in Cancer are other objects visible without optical aid; keen-sighted people have little difficulty in locating the great Andromeda Spiral and the globular cluster in Hercules, while in the far south there are the two Clouds of Magellan, which cannot possibly be overlooked, as well as the bright globulars ω Centauri and 47 Tucanæ.

The most famous of all catalogues of nebulous objects was compiled by the French astronomer Charles Messier, and published in 1781. Ironically, Messier was not interested in the objects he listed; he was a comet-hunter, and merely wanted a quick means of identifying misty patches which were non-cometary in nature! In 1888 J. L. E. Dreyer, Danish by birth (though he spent much of his life in Ireland, and ended it in England), published his New General Catalogue (NGC), augmented in 1895 and again in 1908 by his Index Catalogue (I. or I.C.).

✳ *The brightest nebular object* is the Large Cloud of Magellan, which may be seen even in moonlight. Novæ have been seen in it, and one supernova (1987). A pulsar in it has been detected by J. Ables (Parkes, NSW). Next comes the Small Cloud of Magellan. The brightest cluster visible from England is the cluster of the Pleiades. (This claim may be disputed; the Hyades are also bright, but are somewhat drowned by the brilliant orange-red light of Aldebaran, which is not a bona-fide member of the cluster, but merely happens to lie roughly half-way between the Hyades and ourselves.)

Nebular objects may be divided into various types, all of which are represented in Messier's Catalogue and the NGC:

✳ *Open or Loose Clusters.* Aggregations of stars, with no particular shape. The Pleiades, the Hyades and Præsepe are good exam-

Supernova 1987 A in the large cloud of Magellan.

ples. These three (and many others!) are contained in our Galaxy, but other galaxies also include open clusters.

✳ *The nearest star cluster* is that of the Hyades; distance 147 light-years, ±10 light-years.

✳ *Globular Clusters.* Symmetrical systems, which may contain up to a million stars, strongly concentrated towards the centre. In our Galaxy, the globulars lie around the edge of the main system, so that all are very remote. Leading examples are ω Centauri, 47 Tucanæ and M.13 Herculis. Again, other galaxies have globular clusters of their own.

Globular clusters are very old, so that their brightest stars tend to be red giants well advanced in their evolution. However, in 1953 A. Sandage, at Palomar, found some blue giants also. They ought not to be there, so they were nicknamed Blue Stragglers.

They can hardly be young, as the globular clusters have long since used up all their star-forming material; neither is it likely that they have mixed their hydrogen 'fuel' so thoroughly that they could keep on shining for much longer than normal – they would have to increase their efficiency by 100 per cent. In 1964 F. Hoyle and W.H. McCrea pointed out that in the core of a globular cluster the star density is very high, and stars could pass each other at only a few thousands of kilometres per second, so that stars might capture each other to form binary systems. The less massive star would draw hydrogen-rich material from its more evolved, less dense companion, and would heat up, becoming blue again. Spectroscopic results with the Hubble Space Telescope indicate that at least some blue stragglers are indeed doubles, and in a direct collision, the two stars could even merge. The resulting star would be of greater than average mass, and such stars would collect near the core of the cluster – which is just where we find the Blue Stragglers. The cluster with the greatest known number of Blue Stragglers is 47 Tucanæ, with 21 – eleven of which were found with the Hubble Space Telescope in 1991. However, the Hubble results did not support the suggestion that a globular cluster may have a central black hole.

✳ *The most luminous outlying globular cluster of the Milky Way* is NGC 2419, in Lynx.

✳ *The first measurements of the proper motion of a globular cluster* (ω Centauri) were published in the 1960s by R.v.d.R. R. Woolley and M.P. Candy, but more accurate measurements were made in 1993 for another globular, M.3 by Scholz. Between 1200 and 2200 background galaxies provided the reference frame. The annual proper motion of M.3 is only 0·005 of a second of arc per year. This is the rate at which the slowly growing trunk of an oak tree, as seen from a distance of 32 km adds tree rings!

✳ *Gaseous nebulæ*, also termed *galactic nebulæ*. These may be described as 'stellar birth-places', since fresh stars are condensing out of the nebular gas and dust. The best known example is M.42, the Sword of Orion. They are immensely rarefied; D. A. Allen has pointed out that if it were possible to take a 1-in core sample right through the Orion Nebula, which is of the order of 30 light-years in diameter, the total 'weight' of material collected would weigh not more than one new British penny. Some nebulæ (such as that in the Pleiades) shine by reflecting the light of stars in or very near them; others (such as the Orion Nebula)

shine also because the very hot stars embedded in them ionize the hydrogen, and cause it to emit light on its own account. For this reason, bright emission nebulæ are termed H.II regions. If a nebula has no suitable stars in or near it to cause luminosity, it remains dark, and is detectable only because it blots out the light of stars beyond.

In some nebulæ we find what are termed *Herbig-Haro Objects*, associated with newly forming stars; the first three were discovered in 1946–7 by G. Herbig and G. Haro, in nebulosity in Orion. The best-known dark nebula is the Coal Sack in Crux. The first exhaustive catalogue of dark nebulæ was published by E.E. Barnard in 1919. There is no basic difference between a bright nebula and a dark one; and indeed a nebula which appears bright to us could well seem dark if viewed from elsewhere in the Galaxy. We cannot see the objects deep inside nebulæ, but they can be detected by their infra-red radiations; thus the Orion nebula contains what is known as the Becklin–Neugebauer (BN) object, now known to be a very luminous star permanently hidden.

Protoplanetary disks have been found with stars in the Orion Nebula by C. R. O. Day and Zheng Wen. Of 110 stars surveyed, using the Hubble Space Telescope, 56 were found to have pancake-shaped disks; one of these has a diameter 7·5 times that of the Solar System.

Supernova remnants should not be classed with gaseous nebulæ of the Orion type. The classic example is M.1 Tauri (the Crab Nebula), identified with the supernova of 1054. This was first seen by Messier in August 1758; he had not known of its existence, though it had been recorded by J. Bevis in 1731 – and the observation led directly on to

The Omega Nebula in Sagittarius. (California Institute of Technology and Carnegie Institution of Washington)

Messier's resolve to draw up his famous Catalogue.

✳ *Planetary nebulæ* are neither planets nor nebulæ, but small, hot stars which are at an advanced stage of evolution, and are surrounded by shells of tenuous gas. The most famous example (though not actually the brightest) is M. 57, the Ring Nebula in Lyra. The brightest planetary is NGC 7293, in Aquarius, which was not included in Messier's list. The faintest planetary in Messier's catalogue is M.76 Persei, magnitude 12·2. (This is in fact the faintest of all Messier objects.)

The central stars of planetary nebulæ are among the hottest known, with surface temperatures of from 50 000 to 200 000 degrees. They are usually smaller than the Sun, and are still contracting, because they are slowly collapsing to the white dwarf stage.

Planetary nebulæ are transient; they cannot last, as such, for as much as 100 000 years, and 30 000 to 40 000 years is a more likely estimate. Their material is extremely rare-

fied. If you take the amount of gas in an average village hall and spread it out over an area equivalent to the volume of the Earth, the resulting density will be about the same as that of the gas in a planetary nebula.

✳ *The only star observed to change into a planetary nebula* is Hen 1357, in Ara (the 1357th object in the catalogue compiled by R. Henize). It is 17 000 light-years away, and has now assumed the characteristics of a planetary nebula. Although the central star may have started to expel gas a few thousand years ago, only recently has there been enough radiation from the star to make the gas glow.

✳ *The first planetary nebula to be detected in the Large Cloud of Magellan* was N.66, on 26 June 1991. It is 1.9 light-years in diameter.

GALAXIES

✳ *Galaxies* are external systems, beyond the limits of the Galaxy in which we live. Our Galaxy is often termed the Milky Way, though this name is better restricted to denoting the luminous band seen stretching across the night sky. Galaxies are of various types. In 1925 the American astronomer E. E. Hubble drew up a simple system of classification by shape, which is still widely used even though many more complex classifications have been proposed. Hubble's classes are as follows:

✳ *Spirals*, resembling Catherine-wheels. Sa: conspicuous, often tightly-wound arms issuing from a well-defined nucleus. Sb: arms looser, nucleus less condensed. Sc: incon-

spicuous nucleus, loose arms. Our Galaxy belongs to type Sb.

✳ *Barred spirals*, in which the spiral arms issue from the ends of a kind of 'bar' through the nucleus. They are divided into types SBa, SBb and SBc in the same way as for the normal spirals.

✳ *Ellipticals*, with no indications of spirality; they range from E7 (highly flattened) down to E0 (virtually spherical, looking superficially like globular clusters even though there is a great difference in mass).

✳ *Irregular galaxies*, with no definite form. It is thought that among the galaxies approximately 30 per cent are spiral, 60 per

cent elliptical and 10 per cent irregular, though these values are very approximate.

✳ *The first spiral form* was identified in 1845 by the third Earl of Rosse. Using his newly-completed 72 in reflector at Birr Castle in Ireland, he saw the spiral shape of M.51 in Canes Venatici, now often nicknamed the Whirlpool. Others were subsequently found, though for some years the 72 in was the only telescope of aperture large enough to show the spiral forms clearly. Our Galaxy is an Sb spiral. It has an overall diameter of about 100 000 light-years (though some astronomers believe this to be an over-estimate), and contains approximately 100 000 million stars. The central bulge surrounding the galactic nucleus has a thickness of about 20 000 light-years.

✳ *The first determination of the position of the centre of the Galaxy* was derived from radio observations; the source is known as Sagittarius A* (pronounced Sagittarius A-star), as it lies in the direction of the Sagittarius star-clouds.

✳ *The first optical image* of the central object was reported in October 1990 by H. Zinnecker, M. Rosa and A. Moneti, using the New Technology Telescope at La Silla. They found the object, GZ-A, 0·3 arc-second from the position of Sagittarius A*, well within the limits of uncertainty, and concluded that the blue GZ-A really marked the actual centre. The distance is 28 000 light-years.

The Sun lies on the inner edge of one of the spiral arms of the Galaxy, known as the Carina-Centaurus arm. The Galaxy is in rotation; the Sun takes about 225 000 000 years to complete one circuit – a period sometimes termed the 'cosmic year', moving at about 220 km/s.

Other terms are also in use. *Seyfert galaxies*, first identified by Carl Seyfert in 1943, have bright, almost stellar nuclei which show variations in brightness; a good example is M.77 Ceti, the most massive galaxy in Messier's list. Many Seyferts are thought to be highly active. Some emit radio waves, infrared and even X-rays which are detectable from Earth. *The weakest Seyfert galaxy* is NGC 4395, a dwarf spiral 8 500 000 light-years away; it was identified in 1994 by A. Filippenko and W. Sargent with the Hale reflector at Palomar. It is less luminous than some supergiant stars, but shows the usual characteristics of a Seyfert system. The so-called N-galaxies also have bright nuclei, and may be closely associated with the Seyferts.

It is not now generally thought that spiral galaxies turn into ellipticals or vice versa, so that Hubble's classification does not indicate an evolutionary sequence. Ellipticals consist mainly of Population II objects, so that the most brilliant stars are old red giants, while Population I objects are dominant in the spirals, though the populations are always to some extent mixed, and no hard and fast boundaries can be drawn. Certainly the ellipticals contain relatively little interstellar matter – whereas there is plenty of it in the spirals, where star formation is still going on.

✳ *The first positive proof of interstellar matter in our Galaxy* was obtained by

Hartmann in 1904, when he was studying the spectrum of the star δ Orionis. δ Orionis is a spectroscopic binary, so that its lines show Doppler shifts corresponding to the orbital motion. However, some of the lines remained stationary, so that clearly they were associated not with the star itself, but with material between δ Orionis and ourselves. Further indications were obtained by J. Trümpler in 1930, because some of the Milky Way clusters appeared fainter than logically they should have done, and were presumably being dimmed. The gas between the stars is mainly made up of hydrogen and helium, with smaller amounts of carbon, nitrogen, oxygen and neon. In the 1930s astronomers at Mount Wilson found the first indications of interstellar molecules, but firm proof was postponed until 1963, when the hydroxyl radical, OH, was identified.

In 1968 C. Townes and his colleagues at Berkeley detected interstellar ammonia (NH_3) at a wavelength of 1·26 cm. The first organic molecules (that is to say, molecules containing carbon) were identified in 1969 by L. Snyder, D. Buhl, B. Zuckerman and P.

Palmer, who detected formaldehyde (H_2CO). Others have since been found, including carbon monoxide, hydrogen cyanide and methanoic acid. Ethyl alcohol (CH_3CH_2OH) has also been detected. One 'cloud' contains enough of it to fill the entire globe of the Earth with alcohol, or make 10^{28} bottles of whisky!

Of supreme importance was the detection, in 1951, of cold hydrogen clouds emitting radiation at a wavelength of 21·1 cm. This had been predicted in 1945 by the Dutch astronomer H.C. van de Hulst; the 1951 identification was made by Ewen and Purcell in America. Studies of the 21-cm line have enabled astronomers to map the spiral arms of the Galaxy, since these are the regions in which the hydrogen clouds lie.

However, the average density of interstellar gas in our Galaxy is very low – about 500 000 hydrogen atoms per cubic metre, which is much lower than any laboratory vacuum. There are also dust grains, about 0·1 micron (10^{-7} metres) in radius, which are chiefly responsible for the reddening and dimming of distant objects. We cannot see through to the galactic centre, though as

Lord Rosse's drawing of the Whirlpool Galaxy made in 1845, the first recognition of the spiral structure of a galaxy. The photograph reproduced in the frontispiece demonstrates the accuracy of Rosse's observation.

long ago as 1927 J.H. Oort established that the nucleus lies beyond the glorious star-clouds in Sagittarius. Dust is thought to make up about 1 per cent of the total mass of the Galaxy, though this figure is very arbitrary.

More molecules are being detected yearly. Also, in 1974 M. Cohen, who was engaged in studying the T Tauri variable HL Tauri, detected associated ice grains. HL Tauri is though to be a very young star – perhaps no more than 300 000 years old.

✳ *The first intergalactic cloud of molecular hydrogen* was found in 1993 by N. Brouillet, A. Baudry and C. Henkel. It is 14 minutes of arc east of M.81, and contains between one and ten million solar masses of cold gas. Another such cloud may be in the Antennæ, the colliding galaxies NGC 4038 and 4039.

✳ *The first proof that the 'resolvable nebulæ' are independent galaxies* was obtained by Hubble in 1923, when he was able to measure the periods of Cepheids in the Andromeda Spiral, M.31, and hence to find their distances. The idea was not new. It had been suggested as early as 1755 by Immanuel Kant, and had at one time been supported by Sir William Herschel. However, it had been very much of an open question, and was the subject of a great debate in 1920 between two American astronomers, Harlow Shapley and Heber D. Curtis. By studying short-period variables in the globular clusters, Shapley, in 1918, had been able to give the first reasonably accurate measurement of the size of our Galaxy, but he regarded the spirals as being contained in the Galaxy, whereas Curtis believed them to be external. Hubble's work, carried out with the Mount Wilson 100 in reflector (then the only telescope in the world of sufficient power) proved that Curtis had been right. It was thought that the Andromeda Spiral must be 900 000 light-years away, a figure

subsequently modified to 750 000 light-years. In 1952 W. Baade found that there had been an error in the Cepheid scale; it was found that there were two kinds of Cepheids, one more luminous than the other, and those in the Spiral had been wrongly identified. They were more powerful than had been believed, and hence more remote. The distance of the Andromeda Spiral is now known to be 2 200 000 light-years. It is *the most remote object clearly visible with the naked eye*. It has been reported that the Triangulum Spiral, M.33, has been seen without optical aid, but this needs exceptional eyesight. The only external systems brighter than these two spirals are, of course, the Magellanic Clouds.

✳ *The first indication of an expanding universe* was given by V.M. Slipher, who from 1912 to 1925 found that almost all the galaxies showed red shifts in their spectra, indicating recession. However, the significance of this was not immediately realized, since when Slipher began his work – and for more than a decade after – it was by no means certain that the 'resolvable nebulæ' were external galaxies, and men of the calibre of Shapley believed otherwise. Subsequently, Hubble showed a relationship between distance and velocity of recession; the further away the galaxy, the greater the recessional velocity.

Assuming that the spectral red shifts are true Doppler effects, the only galaxies not receding from us are the members of the *Local Group*. This consists of three or four large systems and more than two dozen dwarf galaxies. The whole Local Group is at least 5 000 000 light-years across.

✳ *The nearest galaxy* is a dwarf spheroidal system in Sagittarius, discovered by Cambridge astronomers R. Ibata, M. Irwin and G. Gilmore on 5 April 1994. It is 80 000 light-years from the Sun, and 50 000 light-years

from the centre of our Galaxy. It is receding at 153 km/s, but is probably making its first circuit of our Galaxy, in a period of 15 000 million years; it will not survive, and during the next few hundred million years it will be gradually disrupted. It already has a length of about 10 000 light-years, and is elongated in a direction pointing to the galactic centre. It probably contains about a million stars, many of which are old red giants. This was the 11th satellite galaxy found. Later in 1994 a smaller companion galaxy was found near it.

✳ *The largest galaxy in the Local Group is M.31*, the Andromeda Spiral, which is 1·5 times the size of our Galaxy. It contains over 300 globular clusters, and objects of all kinds (one supernova has been seen – S Andromedæ of 1885 – and numerous normal novæ). In 1993 results from the Hubble Space Telescope showed that M.31 has a double nucleus. The brighter component may be the core of a galaxy which has been swallowed up by M.31, though it is also possible that we are seeing a single nucleus divided by a dark dust-lane. Second in size comes our Galaxy – possibly excluding Maffei 1, to be described below. Third is the Triangulum Spiral, M.33, half the size of our Galaxy and with 10 000 million stars. The Large Magellanic Cloud is a quarter the size of our Galaxy, and the Small Cloud one-sixth the size; both have been classified as irregular, though it has now been shown that the Large Cloud shows traces of spirality.

In 1968 the Italian astronomer Maffei discovered two galaxies, only about half a degree from the plane of our Galaxy, and therefore heavily obscured by interstellar matter. It now seems that the first of these, Maffei 1, is about 3 000 000 light-years away, and is a member of our Local Group; it appears to be a giant elliptical, but since we receive only about one-hundredth of its light our information about it is scanty. Maffei 2 was originally though to be in the Local Group, but recent work indicates that it is well beyond, at a distance of about 15 000 000 light-years.

The other Local Group members are dwarfs, mainly less than 5000 light-years in diameter. Some of them, such as the Ursa Minor, Draco, Sculptor, Fornax, and Leo I and II systems are well within a million light-years of us, but are so sparse that they are inconspicuous. No doubt many more of these

Messier 13, the globular cluster in Hercules, the brightest globular cluster in the northern hemisphere of the sky. The photograph was taken with a 200 in reflector at Palomar.

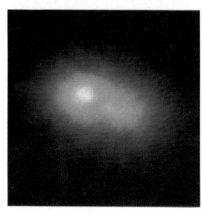

The Andromeda galaxy, M.31. Top: ground-based view. Centre: ground-based view of the centre of the galaxy; individual points of light are globular clusters. Bottom: view of the core from the Hubble Space Telescope, showing the 'double' nucleus; the true centre of the galaxy is the fainter component. (NASA)

dwarfs exist, concealed by interstellar matter in the region of the galactic plane. *The smallest known galaxy* is GR-8, a dwarf irregular, a mere 1000 light-years in diameter. *The least massive Messier galaxy* is M.32.

It is worth noting that one globular cluster, NGC 5694, seems to be moving in a way that indicates its escape from our Galaxy. If so, it will become what is termed an 'intergalactic tramp'!

The following are the nearest members of the Local Group:

Name	Type	Abs. Mag.	Dist thousands of light-years	Diam thousands of light-years
Sagittarius dwarf	dE	−11	8	10
The Galaxy	Sb	−20·5	−	100
Large Magellanic Cloud	SBm	−18·5	170	30
Small Magellanic Cloud	SBm	−16·8	190	16
Ursa Minor dwarf	dE6	−8·8	250	2
Draco dwarf	dE3	−8·6	250	3
Sculptor dwarf	dE3	−11·7	280	5
Fornax dwarf	dE3	−13·6	420	7
Carina dwarf	dE	−	550	−
Leo I	dE3	−11·0	750	2
Leo II	dE0	−9·4	750	3
NGC 6822 (Barnard's Galaxy)	Ir	−15·7	1700	5
M.31 (Andromeda Galaxy)	Sb	−21·1	2200	130
M.32	E2	−16·4	2200	6
NGC 205	E6	−16·4	2200	12
NGC 185	dE0	−11	2200	8
NGC 147	dE4	−14·9	2200	2
Andromeda I	dE0	−11	2200	2
Andromeda II	dE0	−11	2200	−
Andromeda III	dE2	−11	2200	−
NGC 1613	Ir	−14·8	2400	8
M.33 (Triangulum Galaxy)	Sc	−18·9	2900	52

Other clusters of galaxies are known, some of them much more populous than the Local Group. The most famous is the Virgo Cluster, at a mean distance of 50 000 light-years with over 1000 members – notably the giant elliptical M.87, a strong radio source. According to E. Holmberg, its mass is as great as 790 000 million Suns, and the diameter is almost 40 kiloparsecs. Even more populous is the Coma cluster, 300 000 000 light-years away; the diameter is at least 10 million light-years, and there are two huge elliptical systems considerably larger than our Galaxy. Their combined absolute magnitude is about −24, as against −21 for our Galaxy.

✶ *The nearest colliding galaxies* are NGC 4038 and 4039, known as the Antennæ, at 80 million light-years. Two huge tails of material make it clear that interactions are going on, and that the two systems will eventually merge.

✶ *The remotest known 'normal' galaxy* may be 8C 1435+63, discovered in 1995 by a team led by Hyron Spinrad, using the Keck telescope on Mauna Kea; the red shift is 4.25. It seems to be at least 50 000 light-years in diameter.

✶ *The most luminous object known* is IRAS F 10214+4724, which has 300 million times the luminosity of the Sun (30 000 times that of our Galaxy). It was discovered in 1991 by M. Rowan-Robinson and his team, using the William Herschel Telescope at La Palma, and was confirmed by the VLA (Very Large Array) in New Mexico. Only 1 per cent of the emitted energy is in the optical range; 99 per cent is in the infra-red. It seems that there is a power source shrouded in dust, which absorbs optical and ultra-violet radiation and re-emits it as infra-red. There is a huge dust-cloud 100 000 light-years across, inside which there is either a hidden quasar or else a region where stars are being formed; 1000 million massive stars would be needed. If this latter explanation is correct, we may be seeing the process of formation of a massive galaxy.

The faintest galaxies ever recorded are in Sextans, and were imaged with CCD equipment on the New Technology Telescope (La Silla) in 1991. Their magnitudes are about 29.

✶ *The largest known galaxy* is the central galaxy in Abell 2029 (in the Virgo cluster). Its diameter is 8 000 000 light-years, and its luminosity is 2×10^{12} times that of the Sun.

✶ *The most massive galaxy* is Malin 1 – 1000 times more massive than our Galaxy.

✶ *The faintest known galaxy* is the Sextans dwarf, which has only 100 000 times the luminosity of the Sun. Its mass is 10 000 000 times that of the Sun. Its distance is 300 000 light-years.

✶ *The first known large-scale structure in galaxy distribution* is the so-called Great Wall, a huge sheet of galaxies 300 000 000 light-years across which joins the Coma cluster to the Hercules cluster. An even larger structure, 500 000 000 light-years across, joins the Virgo, Hydra and Centaurus clusters, and these merge into the Pavo cluster on the other side of the Milky Way. It is now thought that the overall pattern of the distribution of galaxies is cellular, with vast 'voids' containing few or no galaxies.

QUASARS

*T*he word 'Quasar' is really an abbreviation of Quasi-Stellar Radio Source, since it was originally thought that all members of the class were strong radio emitters. In fact this is most certainly not the case, and the term QSO or Quasi-Stellar Object is favoured by many astronomers.

A quasar is now known to be the core of a very active galaxy, probably powered by a central black hole. It must, however, be added that although almost all astronomers believe that the red shifts in the spectra of quasars are 'cosmological', i.e. Doppler effects, there are some dissentients. H. Arp claims that he has identified alignments of quasars and galaxies which are presumably associated, but which have different red shifts; F. Hoyle regards the quasars as reasonably local to our Galaxy. However, if the majority view is accepted, we may be observing quasars not so very far from the boundary of the observable universe, since an object receding at the full velocity of light would naturally be beyond our range.

✳ *The first quasar to be identified* was 3C–273, in Virgo. (The prefix 3C indicates the third Cambridge catalogue of radio sources, published in 1962.) 3C–273 was known to be a strong radio emitter, but identifying it with a visual object proved to be difficult. On 5 August 1962, however, radio astronomers in Australia, working at the Parkes observatory, followed an occultation of the radio source by the Moon, and were able to fix its position very exactly. From this result, the source was identified with what seemed to be a faint bluish star of magnitude 12·8. In 1963 M. Schmidt, at Palomar in California, obtained an optical spectrum, and found that the object was not a star at all; its spectrum was quite different, and showed hydrogen lines which were tremendously red-shifted, indicating a high recessional velocity and therefore immense distance and luminosity. Other quasars were quickly found, together with quasar-like objects which were not radio emitters.

✳ *Among the reddest known galaxies* are two objects near the quasar PC1643+4631A. Their nature is very uncertain; they may be infant galaxies, or perhaps clusters of galaxies consisting largely of red stars.

✳ *The brightest quasar* is 3C–273. No others exceed magnitude 16.

✳ *The most distant quasar known* is PC 1247+3406, where z = 4·897. (z is a measure of the red shift. $z = \frac{L-L_0}{L_0}$, where L_0 is the

laboratory wavelength and L is the observed wavelength.)

✳ *The most luminous quasar known* was discovered by observers using the William Herschel Telescope, at La Palma, in 1991. It has z = 4·73. It is 5 times more luminous than the remotest known quasar, PC 1247+3406.

✳ *The first quasar to be detected in infrared* was 3C–345, by P. Harvey at Texas in May 1982.

✳ *The first giant cloud of cold dust near a quasar* was detected around NR1202–0727, a quasar 12 000 million light-years away (red shift 4.69), on 30 March 1994 with the 30 m radio telescope at Sierra Nevada in Spain; it was subsequently confirmed with the James Clerk Maxwell Telescope on Mauna Kea. The temperature is –193°C.

✳ *The first gravitational lens system* (Q0956 + 561) was discovered in 1979. If a galaxy lies on or near the line of sight of a more remote quasar, the result will be that the quasar will show multiple images, simply because the intervening galaxy acts as a lens. Many examples of this are now known, notably the 'Clover Leaf' H 1413 + 117 – discovered from La Silla in 1988 – where there are four images of the distant quasar. The distance of the quasar is about 13 000 million light-years.

✳ *The most distant gravitational lens known* is Q1208 + 1011, discovered from La Silla in 1991. The object lies in Virgo, and is the most remote quasar whose image has been found to be double. The luminosity is perhaps 100 million million times that of the Sun. It has a high red shift: z = 3·8.

✳ *The first systematic studies of microlensing* were carried out in 1993 by astronomers using the Melbourne telescope at Mount Stromlo Observatory in Australia. (When an image is gravitationally lensed by a star rather than a galaxy, it is termed microlensing.) In 1986 B. Paczynski proposed that if what is termed a MACHO – a Massive Astronomical Compact Halo Object – passes in front of a star, it will act in the manner of a lens and cause a temporary increase in the star's apparent brightness. Effects of this kind have been reported with fair certainty, though the exact nature of the

MACNOs remains very uncertain indeed; brown dwarfs, black holes, neutron stars and other faint stars have all been suggested as candidates.

✳ *The first 'Einstein ring'* was found by G. Soucail, Y. Mellier, B. Fort, G. Mather and M. Cailloux in 1985, from the CFH telescope on Mauna Kea. They detected a giant luminous arc round the cluster of galaxies Abell 370, 500 000 light-years long and 25 000 light-years wide. The effect is due to the light from a background galaxy being bent in the field of the Abell 370 cluster.

✳ *The first true binary quasar* was identified in 1987 by S. Djorgevski, G. Meylan, R. Perley and P. McCarthy. The binary quasar (QQ 1145–071) is about 12 000 million light-years away. The spectra of the two components are not identical, so that we are dealing with two separate objects rather than a gravitational lens effect.

BL LACERTÆ OBJECTS

*T*he first BL Lacertæ object to be studied was BL Lacertæ itself, so that it has given its name to the whole class. In 1941 it was noted as a variable star, but in 1968 it was identified as a radio source (by Schmidt, at Palomar) and was found to have a featureless spectrum. Other similar objects, such as AP Libræ and W Comæ, were then found. In 1973 Palomar astronomers found a fuzzy surround which did show spectral lines, and deduced a recessional velocity of 20 000 km/s. In 1976 J. Wampler, at the Lick Observatory, found some lines in the spectrum of the BL Lac object 0548–322, and was able to measure the red shift.

It now seems that BL Lacertæ objects and quasars are the same types of objects, but with a quasar we are seeing it at a full angle – even broadside on – whereas with a BL Lacertæ object we are looking 'straight down' a jet. This means that the jet appears as a bright point, overpowering the dim galaxy surrounding it. By now over 100 BL Lacertæ objects are known.

THE EVOLUTION OF THE UNIVERSE

*M*ost astronomers believe that the universe was created at one set moment in what is usually called the Big Bang (a term originally introduced scornfully by Sir Fred Hoyle, who does not believe in it!). Space,

time and matter came into existence simultaneously. It is impossible to discuss what happened before that, because there was no 'before'. Neither can we say just where the Big Bang happened, because if it included

the whole universe it happened 'everywhere'.

The concept was originally described in 1927 by the Belgian abbé Georges Lemaître, and has been elaborated since. Theory can take us back to 10^{-43} second after the Big Bang. (This is a convenient mathematical shorthand; 10^{-43} is equal to 1 divided by a 1 followed by 43 zeros.) The temperature at this time was of the order of 10^{32} degrees, but cooling was rapid. Between 10^{-36} and 10^{-32} second there was an 'inflationary' period when the universe expanded rapidly. Subsequent expansion was less rapid. By 10^{-6} seconds, the temperature had dropped to about 10^{13} K and protons and neutrons had formed. At the end of 3 minutes, the temperature had fallen to a thousand million degrees and the first atomic nuclei – hydrogen, helium and lithium – had formed but it was only after perhaps 300 000 years, when the temperature had dropped to 3000 degrees, that the first complete atoms formed and the universe became transparent to radiation; this is known as the decoupling stage. Galaxies began to form 200 million years after the Big Bang, and expansion continued.

All this sounds highly speculative, and indeed it is; but supporting evidence comes in the form of weak microwave radiation, coming in from all directions all the time, indicating an overall temperature of 3 degrees Kelvin – that is to say, 3 degrees above absolute zero, the coldest temperature there can possibly be ($-273°C$). The microwave radiation was detected in 1965 by A. Penzias and R. Wilson in the United States, using a special radio antenna which had been built for a completely different investigation; but the radiation had been predicted earlier by R. Dicke, who calculated that the temperature of the universe should by now have fallen to precisely this value: 3°K.

One initial problem was that the background radiation seemed to be quite uniform, and it was not easy to see how a 'lumpy' universe could have been formed from a very smooth expansion; how could galaxies begin to condense? However, to the relief of theorists, measurements carried out in 1992 from an artificial satellite, COBE (the Cosmic Background Explorer) showed that there were tiny irregularities in the background radiation, which masked the seedcorn condensations for galaxies. These results were confirmed in 1993 by S.S. Meyer and his team using a balloon-borne radiometer, and by a Jodrell Bank team using a radiometer on Teide in Tenerife. All this seemed to give the *coup de grâce* to a theory proposed in 1947 by H. Bondi and T. Gold, who believed in a 'steady state' universe which had no beginning and will never end.

There is, however, another pressing problem. We know that the universe is expanding; but how long has this expansion been going on? This depends upon what is termed the Hubble Constant (abbreviation H_0), which is defined as the rate at which the groups of galaxies at various distances are racing away from each other. The previously accepted value has been about 50 km/s per megaparsec. (One parsec is equal to 3·26 light-years; a megaparsec is a million parsecs.) This agreed with an age of the universe of between 15 000 million and 20 000 million years. However, in 1994 the Hubble Space Telescope was used to make very accurate measurements of the distance of the spiral galaxy M.100, in the Virgo Cluster, and found it to be 51 000 000 light-years or 17 megaparsecs away. Matching this with its recessional velocity, as determined by the red shift in its spectral lines, it follows that the value of the Hubble Constant is at least 80 km/s/Mpc. Interpreted in a simplified formula this means that the expansion cannot have been going on for as long as had been believed, and that the age of the universe can be no more than 12 000 million years at most.

Yet we have excellent evidence that many stars are older than this. All the elements apart from the three lightest (hydrogen, helium and lithium) have been built up inside stars, and later-generation stars have been formed from material which has been sent out from their precursors which have exploded as supernovæ. Our Sun contains large quantities of heavier elements, so that it cannot be a first-generation star; its age is certainly of the order of 5000 million years. The star CS 22876–32 consists almost entirely of hydrogen and helium, and its content of heavier elements is only 1/30 000 that of the Sun – it has been referred to as a stellar coelocanth! Studies of the percentage of elements in stars shows that the oldest stars must date back well over 12 000 million years. Since we cannot have stars which are older than the universe, there is a major error somewhere or other. Presumably the universe is more complicated than the simple formula implies.

No doubt the problem will be resolved before long, but we must also consider the work of Halton Arp, who has produced images of galaxies and quasars which are clearly joined by luminous 'bridges', and are therefore presumably associated, but which have completely different red shifts. Arp believes that the shifts are not pure Doppler effects, but there is a marked non-velocity component, so that all our distance-measures beyond the Local Group are unreliable; he even things it possible that quasars are minor features shot out of galaxies! This is a most unpopular view among cosmologists, and it may be significant that Arp was refused further observing time on the large American telescopes because he was obtaining results which were too embarrassing to official theory (subsequently he went to Germany, and is now working at the Max Planck Institute in Munich). If Arp is right, then many of our cherished theories will have to be drastically modified, or even abandoned.

Adopting the conventional view rather than Arp's, we have to decide whether or not the expansion of the universe will continue indefinitely. Everything depends upon the amount of matter present – that is to say, the total mass of the universe. If the mean density is less than a critical value of about one hydrogen atom per cubic metre, then expansion will never stop, and will continue until all the groups of galaxies lose contact with each other. If the density is higher, then the expansion will be succeeded by a period of contraction until all the galaxies rush together in what may be termed a Big Crunch. This might happen in around 90 000 million years, and could be followed by a new period of expansion. The amount of visible matter is not nearly enough to halt the expansion, but studies of the movements of stars inside galaxies, and of groups of galaxies themselves, indicate that there is a tremendous amount of 'dark matter' which we cannot see at all; indeed, this could account for well over 90 per cent of the total mass of the universe. What this invisible matter may be – if, indeed, it exists at all – is a complete mystery at the present time.

All in all, it is only too clear that our present knowledge of the evolution of the universe is very incomplete. Moreover, no theory can explain how matter came into existence in the first place, so that we are not genuinely discussing the origin of the universe, we are dealing with its development, which is a very different thing.

THE FIRST KNOWN CLUSTERS AND NEBULÆ

*I*t seems that a few nebulous objects have been known since prehistoric times. This certainly applies to the Pleiades, which can hardly be overlooked. In his great work the Almagest, Ptolemy (circa AD 120–180) records the Pleiades, and also the Sword-Handle in Perseus (NGC 869 and 884), M.44 (Præsepe) and, with almost certain identification, the open cluster M.7 in Scorpius; probably Ptolemy himself discovered the last of these. We may also assume that ancient men of the southern hemisphere knew the Magellanic Clouds, though they did not come to the notice of European astronomers until about 1520.

Clusters	Nebulae	Discoverer
M.45	Pleiades	(Prehistoric)
NGC 869/884	Sword-Handle in Perseus	(Listed by Ptolemy)
M.44	Præsepe	(Listed by Ptolemy)
M.7	Open cluster in Scorpio	Ptolemy, c. AD 140
M.31	Andromeda Spiral	Al-Sûfi, c. 964
o Velorum	Open cluster	Al-Sûfi, c. 964
Large Cloud of Magellan		1519
Small Cloud of Magellan		1519
M.42	Great Nebula in Orion	N. Pieresc, 1610
M.22	Globular in Sagittarius	A. Ihle, 1665
NGC 5139	ω Centauri	Halley, 1677
M.8	Lagoon Nebula	Flamsteed, 1680
M.11	Wild Duck Cluster	G. Kirch, 1681
NGC 2244	12 Monocerotis	Flamsteed, 1690
M.41	Open cluster in Canis Major	Flamsteed, 1702
M.5	Globular in Serpens	G. Kirch, 1702
M.50	Open clusters in Monoceros	G. D. Cassini, c. 1711
M.13	Hercules Globular	Halley, 1714
M.43	Part of Orion Nebula	de Mairan, c. 1731
M.1	Crab Nebula	J. Bevis, 1731

From the year 1745 more and more objects were found, mainly by astronomers such as De Chéseaux, Legentil, Lacaille and, of course, the two great Frenchmen, Messier and Méchain. By 1781, when Messier published his Catalogue, 138 nebulous objects were known. The above list includes all the objects found before 1745.

This makes in all 21 objects. Many others had been previously listed, but subsequently found to be mere groups of stars rather than true clusters or nebulæ. One remarkable fact relates to the Great Spiral in Andromeda, which was recorded by the Persian astronomer Al-Sûfi. It was not noted again until 1612, when Simon Marius described it. Amazingly, it was completely overlooked by the greatest observer of pre-telescopic times, Tycho Brahe.

MESSIER'S CATALOGUE OF NEBULOUS OBJECTS

Messier's original catalogue ends with M.103. The later numbers are objects which, apart from the last, were discovered by Méchain (M.110 may have been discovered by Messier himself, but it is still more generally known as NGC 205, one of the companions to the Great Spiral in Andromeda). M.104 was added to the Messier catalogue by Camille Flammarion in 1921, on the basis of finding a handwritten note about it in Messier's own copy of his 1781 catalogue. M.105 to 107 were listed by H. S. Hogg in the 1947 edition of the catalogue; M.108 and 109 by Owen Gingerich in 1960. The naming of NGC 205 as M.110 was suggested by K. G. Jones in 1968, but does not seem to have received general acceptance.

M.	NGC	Constellation	Type	
1	1952	Taurus	Supernova remnant	Crab Nebula
2	7089	Aquarius	Globular	
3	5272	Canes Venatici	Globular	
4	6121	Scorpius	Globular	
5	5904	Serpens	Globular	
6	6405	Scorpius	Open cluster	
7	6475	Scorpius	Open cluster	
8	6523	Sagittarius	Nebula	Lagoon Nebula
9	6333	Ophiuchus	Globular	
10	6254	Ophiuchus	Globular	
11	6705	Scutum	Open cluster	Wild Duck cluster
12	6218	Ophiuchus	Globular	
13	6205	Hercules	Globular	Great Hercules cluster
14	6402	Ophiuchus	Globular	
15	7078	Pegasus	Globular	
16	6611	Serpens	Nebula and embedded cluster	Eagle Nebula
17	6618	Sagittarius	Nebula	Omega or Horseshoe Nebula
18	6613	Sagittarius	Open cluster	
19	6273	Ophiuchus	Globular	
20	6514	Sagittarius	Nebula	Trifid Nebula
21	6531	Sagittarius	Open cluster	
22	6656	Sagittarius	Globular	
23	6494	Sagittarius	Open cluster	
24	6603	Sagittarius	Star-cloud in Milky Way	
25	I. 4725	Sagittarius	Open cluster	
26	6694	Scutum	Open cluster	
27	6853	Vulpecula	Planetary	Dumbbell Nebula
28	6626	Sagittarius	Globular	
29	6913	Cygnus	Open cluster	
30	7099	Capricornus	Globular	
31	224	Andromeda	Spiral galaxy	Great Spiral
32	221	Andromeda	Elliptical galaxy	Companion to M.31
33	598	Triangulum	Spiral galaxy	Triangulum Spiral
34	1039	Perseus	Open cluster	
35	2168	Gemini	Open cluster	
36	1960	Auriga	Open cluster	
37	2099	Auriga	Open cluster	
38	1912	Auriga	Open cluster	
39	7092	Cygnus	Open cluster	
40	–	–	Missing. Possibly a comet?	
41	2287	Canis Major	Open cluster	
42	1976	Orion	Nebula	Great Nebula
43	1982	Orion	Nebula	Part of Orion Nebula
44	2632	Cancer	Open cluster	Præsepe
45	–	Taurus	Open cluster	Pleiades
46	2437	Puppis	Open cluster	
47*	2422	Puppis	Open cluster	
48*	2548	Hydra	Open cluster	
49	4472	Virgo	Elliptical galaxy	
50	2323	Monoceros	Open cluster	
51	5194	Canes Venatici	Spiral galaxy	Whirlpool Galaxy
52	7654	Cassiopeia	Open cluster	
53	5024	Coma Berenices	Globular	
54	6715	Sagittarius	Globular	
55	6809	Sagittarius	Globular	
56	6779	Lyra	Globular	

M.	NGC	Constellation	Type	
57	6720	Lyra	Planetary	Ring Nebula
58	4579	Virgo	Spiral galaxy	
59	4621	Virgo	Elliptical galaxy	
60	4649	Virgo	Elliptical galaxy	
61	4303	Virgo	Spiral galaxy	
62	6266	Ophiuchus	Globular	
63	5055	Canes Venatici	Spiral galaxy	Sunflower Galaxy
64	4826	Coma Berenices	Spiral galaxy	Black-eye Galaxy
65	3623	Leo	Spiral galaxy	
66	3627	Leo	Spiral galaxy	
67	2682	Cancer	Open cluster	Famous old cluster
68	4590	Hydra	Globular	
69	6637	Sagittarius	Globular	
70	6681	Sagittarius	Globular	
71	6838	Sagitta	Globular	
72	6981	Aquarius	Globular	
73	6994	Aquarius	Four faint stars (Not a cluster)	
74	628	Pisces	Spiral galaxy	
75	6864	Sagittarius	Globular	
76	650	Perseus	Planetary	
77	1068	Cetus	Spiral galaxy	
78	2068	Orion	Nebula	
79	1904	Lepus	Globular	
80	6093	Scorpius	Globular	
81	3031	Ursa Major	Spiral galaxy	Bode's Nebula
82	3034	Ursa Major	Irregular galaxy	Companion to M.31
83	5236	Hydra	Spiral galaxy	
84	4374	Virgo	Spiral galaxy	
85	4382	Coma Berenices	Spiral galaxy	
86	4406	Virgo	Elliptical galaxy	
87	4486	Virgo	Elliptical galaxy	Seyfert galaxy
88	4501	Coma Berenices	Spiral galaxy	
89	4552	Virgo	Elliptical galaxy	
90	4569	Virgo	Spiral galaxy	
91	–	–	Not identified. Possibly a comet?	
92	6341	Hercules	Globular	
93	2447	Puppis	Open cluster	
94	4736	Canes Venatici	Spiral galaxy	
95	3351	Leo	Barred spiral galaxy	
96	3368	Leo	Spiral galaxy	
97	3587	Ursa Major	Planetary	Owl Nebula
98	4192	Coma Berenices	Spiral galaxy	
99	4254	Coma Berenices	Spiral galaxy	
100	4321	Coma Berenices	Spiral galaxy	
101	5457	Ursa Major	Spiral galaxy	
102	–	–	Missing. Possibly a faint spiral in Draco, or else identical with M.101.	
103	581	Cassiopeia	Open cluster	
104	4594	Virgo	Spiral galaxy	Sombrero Hat Galaxy
105	3379	Leo	Elliptical galaxy	
106	4258	Ursa Major	Spiral galaxy	
107	6171	Ophiuchus	Globular	
108	3556	Ursa Major	Spiral galaxy	
109	3992	Ursa Major	Spiral galaxy	
110	205	Andromeda	Elliptical galaxy	Companion to M.31

SOME ASTRONOMICAL 'NICKNAMES'

*T*here are many 'unofficial' names for stellar objects in fairly common use. The following list is very far from complete, but includes some of the more familiar names.

Antennæ	Colliding galaxies NGC 4038 and 4039; in Corvus.
Barnard's Galaxy	NGC 6822; in the Local Group. RA 19h 45, dec. −14 deg 48 min.
Barnard's Loop	Extensive ring of nebulosity in Orion.
Barnard's Star	Munich 15040; the nearest star apart from the α Centauri group.
Bear Paw Galaxy	NGC 2537; galaxy in Lynx.
Becklin-Neugebauer Object	Infra-red source in M.42 (Orion Nebula).
Beehive Cluster	M.44, Præsepe.
Black-eye Galaxy	M.64; in Coma (spiral galaxy).
Blaze Star	The recurrent nova T Coronæ Borealis.
Blinking Nebula	NGC 6826; planetary nebula in Cygnus.
Blue Planetary	NGC 3918; planetary nebula in Centaurus.
Bode's Nebula	M.81; spiral galaxy in Ursa Major.
Bubble Nebula	NGC 7635; nebula in Cassiopeia.
Bug Nebula	NGC 6302; planetary nebula in Scorpius.
Burnham's Nebula	T Tauri nebula.
Butterfly Cluster	M.6; open cluster in Scorpius.
California Nebula	NGC 1499; nebula in Perseus.
Coal Sack	Dark nebula in Crux.
Coat-Hanger	Cluster in Vulpecula (also known as Brocchi's Cluster).
Cocoon Nebula	IC 5146; nebula in Cygnus.
Cone Nebula	IC 2264 (S Monocerotis); nebula in Monoceros.
Crab Nebula	M.1; supernova remnant in Taurus; NGC 1952.
Crescent Nebula	NGC 6888; nebula in Cygnus.

Crimson Star	The Mira variable R Leporis.
Demon Star	Algol (β Persei).
Dog Star	Sirius (α Canis Majoris).
Dumbbell Nebula	M.27; planetary nebula in Vulpecula.
Eagle Nebula	M.16; nebula in Serpens Cauda.
Eskimo Nebula	NGC 2392; planetary nebula in Gemini (also known as the Clown Face Nebula).
Flaming Star Nebula	IC 405; nebula associated with AE Aurigæ.
Flying Star	61 Cygni (because of its large proper motion).
Garnet Star	μ Cephei.
Geminga	Gamma-ray source in Gemini.
Ghost of Jupiter	NGC 3242; planetary nebula in Hydra.
Gum Nebula	Nebula in Vela (supernova remnant).
Helix Nebula	NGC 7293; planetary nebula in Aquarius.
Hind's Variable Nebula	NGC 1555; nebula near T Tauri.
Homunculus Nebula	Core of η Carinæ Nebula.
Horse's Head Nebula	Barnard 33; dark nebula in Orion.
Hubble's Variable Nebula	NGC 2261; nebula round R Monocerotis.
Hyades	Mel. 25, open cluster in Taurus.
Innes' Star	Proxima Centauri.
Intergalactic Tramp	NGC 2419; globular cluster in Lynx.
Jewel Box	NGC 4755; open cluster (κ Crucis).
Kepler's Star	The supernova of 1604 (in Ophiuchus).
Keyhole Nebula	Dark nebula; in the η Carinæ nebulosity.
Lagoon Nebula	M.8; nebula in Sagittarius.
Little Dumbbell	M.76; planetary nebula in Perseus.
North America Nebula	NGC 7000; nebula in Cygnus.
Omega Nebula	M.17; nebula in Sagittarius (also known as the Horseshoe Nebula).
Owl Nebula	M.97; planetary nebula in Ursa Major.
Pelican Nebula	IC 5067/70; nebula in Cygnus.
Pipe Nebula	Barnard 59, 65-7, 76; dark nebulosity in Ophiuchus.
Plaskett's Star	Massive O-type binary star; HD 47129.
Pleiades	M.45; open cluster in Taurus.
Pup	Sirius B (Companion of Sirius, the Dog Star.)
Ring Nebula	M.57; planetary nebula in Lyra.
Rosette Nebula	NGC 2237-9; nebula in Monoceros.
Runaway Star	Another nickname for Barnard's Star.
Sagittarius A-*	Centre of the Galaxy.
Saturn Nebula	NGC 7009; planetary nebula in Aquarius.
Seven Sisters	The Pleiades.
Sidus Ludovicianum	Star near Mizar and Alcor.
Snake Nebula	Barnard 72; dark nebula near θ Ophiuchi.
Snickers	Nearby dwarf galaxy (RA 6 h 28 m, dec. +15 deg.)
Sombrero Hat Galaxy	M.104; galaxy in Virgo.
Southern Pleiades	The θ Carinæ cluster.
Spindle Galaxy	NGC 3115; galaxy in Sextans.
Stephan's Quintet	NGC 7317-20; galaxies in Pegasus.
Sunflower Galaxy	M.63; spiral galaxy in Canes Venatici.
Swan Nebula	Alternative name for the Omega Nebula.
Sword of Orion	Region of M.42; nebula in Orion.
Sword-Handle	NGC 869-884; double cluster in Perseus.
Tarantula Nebula	NGC 2070; 30 Doradûs; nebula in Large Magellanic Cloud.
Trapezium	The multiple star θ Orionis.
Trifid nebula	M.20; nebula in Sagittarius.
Tycho's Star	B Cassiopaiæ; the 1572 supernova.
Veil Nebula	NGC 6960-92-95; nebulosity in Cygnus.
Whirlpool Galaxy	M.51; spiral galaxy in Canes Venatici.
Wild Duck Cluster	M.11; cluster in Scutum.

THE CONSTELLATIONS

		1st mag	Stars to mag: 2.00	Stars to mag: 4.00	Stars to mag: 5.00	Area sq. deg.	Number of stars above mag. 5 per 100 square degrees (star density)
Andromeda	Andromeda	–	0	7	25	722	9·19
Antlia	The Airpump	–	0	0	4	239	1·67
Apus	The Bird of Paradise	–	0	2	6	206	2·91
Aquarius	The Water-bearer	–	0	7	31	980	3·16
Aquila	The Eagle	Altair	1	8	16	652	2·45
Ara	The Altar	–	0	7	10	237	4·21
Aries	The Ram	–	1	4	11	441	2·49
Auriga	The Charioteer	Capella	2	7	21	657	3·20
Boötes	The Herdsman	Arcturus	1	8	24	907	2·65
Cælum	The Graving Tool	–	0	0	2	125	1·60
Camelopardus	The Giraffe	–	0	0	11	757	1·45
Cancer	The Crab	–	0	1	6	506	1·19
Canes Venatici	The Hunting Dogs	–	0	1	7	465	1·51
Canis Major	The Great Dog	Sirius	4	10	26	380	6·84

		1st mag	Stars to mag: 2.00	4.00	5.00	Area sq. deg.	Number of stars above mag. 5 per 100 square degrees (star density)
Canis Minor	The Little Dog	Procyon	1	2	4	183	2·19
Capricornus	The Sea-Goat	–	0	5	16	414	3·86
Carina	The Keel	Canopus	3	14	40	494	8·10
Cassiopeia	Cassiopeia	–	0	7	23	598	3·85
Centaurus	The Centaur	Alpha Cen., Agena	2	14	49	1060	4·62
Cepheus	Cepheus	–	0	8	20	588	3·40
Cetus	The Whale	–	0	8	24	1232	1·95
Chamæleon	The Chameleon	–	0	0	6	132	4·55
Circinus	The Compasses	–	0	1	4	93	4·30
Columba	The Dove	–	0	4	9	270	3·33
Coma Berenices	Berenice's Hair	–	0	0	8	386	2·07
Corona Australis	The Southern Crown	–	0	0	7	128	5·47
Corona Borealis	The Northern Crown	–	0	3	10	179	5·59
Corvus	The Crow	–	0	5	6	184	3·26
Crater	The Cup	–	0	1	6	282	2·13
Crux Australis	The Southern Cross	Acrux, Beta Crucis	3	5	13	68	19·12
Cygnus	The Swan	Deneb	1	11	43	804	5·35
Delphinus	The Dolphin	–	0	4	6	189	3·17
Dorado	The Swordfish	–	0	1	8	179	4·47
Draco	The Dragon	–	0	11	26	1083	2·40
Equuleus	The Foal	–	0	0	3	72	4·17
Eridanus	The River	Achernar	1	12	43	1138	3·78
Fornax	The Furnace	–	0	1	5	398	1·26
Gemini	The Twins	Pollux	3	13	23	514	4·47
Grus	The Crane	–	1	4	13	366	3·55
Hercules	Hercules	–	0	15	37	1225	3·02
Horologium	The Clock	–	0	1	2	249	0·80
Hydra	The Watersnake	–	1	9	32	1303	2·46
Hydrus	The Little Snake	–	0	3	9	243	3·70
Indus	The Indian	–	0	2	7	294	2·38
Lacerta	The Lizard	–	0	1	11	201	5·47
Leo	The Lion	Regulus	2	10	26	947	2·75
Leo Minor	The Little Lion	–	0	1	6	232	2·59
Lepus	The Hare	–	0	7	14	290	4·83
Libra	The Balance	–	0	5	13	538	2·42
Lupus	The Wolf	–	0	8	32	334	9·58
Lynx	The Lynx	–	0	2	12	545	2·20
Lyra	The Lyre	Vega	1	4	11	286	3·85
Mensa	The Table	–	0	0	0	153	0·00
Microscopium	The Microscope	–	0	0	4	210	1·90
Monoceros	The Unicorn	–	0	2	13	482	2·70
Musca Australis	The Southern Fly	–	0	4	11	138	7·97
Norma	The Rule	–	0	0	6	165	3·64
Octans	The Octant	–	0	1	4	291	1·37
Ophiuchus	The Serpent-bearer	–	0	12	36	948	3·80
Orion	Orion	Rigel, Betelgeux	5	15	42	594	7·07
Pavo	The Peacock	–	1	4	14	378	3·70
Pegasus	The Flying Horse	–	0	9	29	1121	2·59
Perseus	Perseus	–	1	10	34	615	5·52
Phœnix	The Phoenix	–	0	7	17	469	3·62
Pictor	The Painter	–	0	2	5	247	2·02
Pisces	The Fishes	–	0	3	24	889	2·70
Piscis Australis	The Southern Fish	Fomalhaut	1	1	7	245	2·86
Puppis	The Poop	–	0	11	42	673	6·24
Pyxis	The Compass	–	0	1	7	221	3·17
Reticulum	The Net	–	0	2	7	114	6·14
Sagitta	The Arrow	–	0	2	5	80	6·25
Sagittarius	The Archer	–	1	14	33	867	3·80
Scorpio	The Scorpion	Antares	3	17	38	497	7·64
Sculptor	The Sculptor	–	0	0	6	475	1·26
Scutum	The Shield	–	0	0	6	109	5·50
Serpens	The Serpent	–	0	9	17	637	2·67
Sextans	The Sextant	–	0	0	2	314	0·63
Taurus	The Bull	Aldebaran	2	14	44	797	5·52
Telescopium	The Telescope	–	0	1	4	252	1·59
Triangulum	The Triangle	–	0	2	3	132	2·27
Triangulum Australe	The Southern Triangle	–	1	3	6	110	5·45
Tucana	The Toucan	–	0	2	7	295	2·37
Ursa Major	The Great Bear	–	3	19	35	1280	2·73
Ursa Minor	The Little Bear	–	1	3	9	256	3·51
Vela	The Sails	–	2	10	30	500	6·00
Virgo	The Virgin	Spica	1	8	26	1294	2·01
Volans	The Flying Fish	–	0	3	7	141	4·96
Vulpecula	The Fox	–	0	0	10	268	3·73
Totals		21	50	455	1417		(Average: 3·8)

These counts do not include variable stars which can rise above the fifth magnitude, but whose average magnitude is below this limit. For instance, Mira Ceti is excluded even though its brightest maxima exceed magnitude 2.

FORMING THE CONSTELLATIONS

Ptolemy gave a list of 48 constellations: 21 northern, 12 Zodiacal and 15 southern, as follows:

Northern

Ursa Minor
Ursa Major
Draco
Cepheus
Boötes
Corona Borealis
Hercules
Lyra
Cygnus
Cassiopeia
Perseus
Auriga
Ophiuchus
Serpens
Sagitta
Aquila
Delphinus
Equuleus
Pegasus
Andromeda
Triangulum

Zodiacal

Aries
Taurus
Gemini
Cancer
Leo
Virgo
Libra
Scorpio
Sagittarius
Capricornus
Aquarius
Pisces

Southern

Cetus
Orion
Eridanus
Lepus
Canis Major
Canis Minor
Argo Navis
Hydra
Crater
Corvus
Centaurus
Lupus
Ara
Corona Australis
Piscis Australis

All these are still to be found in modern maps, though the huge, unwieldy Argo Navis has been divided up into Carina, Vela and Puppis.

Various constellations have been added since. With those that survive, the names have often been shortened; thus Piscis Volans, the Flying Fish, has become simply Volans, while Mons Mensæ, the Table Mountain, has become Mensa. There was also some confusion over two of Bayer's constellations, Apis (the Bee) and Avis Indica (the Bird of Paradise); modern maps give it as Apus. Surviving constellations were added as follows:

By Tycho Brahe, c. 1590:
Coma Berenices

By Bayer, 1603:
Pavo
Tucana
Grus
Phœnix
Dorado
Volans (originally Piscis Volans)
Hydrus
Chamæleon
Apus (originally Avis Indica)
Triangulum Australe
Indus

By Royer, 1679:
Columba (originally Columba Noachi, Noah's Dove)
Crux Australis

By Hevelius, 1690:
Camelopardus
Canes Venatici
Vulpecula (originally Vulpecula et Anser, the Fox and Goose)
Lacerta
Leo Minor
Lynx
Scutum Sobieskii
Monoceros
Sextans (originally Sextans Uraniæ, Urania's Sextant)

By La Caille, 1752:
Sculptor (originally Apparatus Sculptoris, the Sculptor's Apparatus)
Fornax (originally Fornax Chemica, the Chemical Furnace)
Horologium
Reticulum (originally Reticulus Rhomboidalis, the Rhomboidal Net)
Cælum (originally Cæla Sculptoris, the Sculptor's Tools)
Pictor (originally Equuleus Pictoris, the Painter's Easel)
Pyxis (originally Pyxis Nautica, the Mariner's Compass)
Antlia (originally Antlia Pneumatica, the Airpump)

Octans
Circinus
Norma (or Quadra Euclidis, Euclid's Square)
Telescopium
Microscopium
Mensa (originally Mons Mensæ, the Table Mountain)

The list of rejected constellations is very long. The most notable are as follows:

Proposed by Tycho Brahe, c. 1559:
Antinoüs

Proposed by Royer, 1679:
Nubes Major (the Great Cloud)
Nubes Minor (the Little Cloud)
Lilium (the Lily)

Proposed by Halley, c. 1680:
Robur Caroli (Charles' Oak)

Proposed by Flamsteed, c. 1700:
Mons Mænalus (the Mountain Mænalus)
Cor Caroli (Charles' Heart)

Proposed by Hevelius, 1690:
Triangulum Minor (the Little Triangle)
Cerberus

Proposed by Le Monnier, 1776:
Tarandus (the Reindeer)
Solitarus (the Solitaire)

Proposed by Lalande, 1776:
Messier

Proposed by Poczobut, 1777:
Taurus Poniatowski (Poniatowski's Bull)

Proposed by Hell, c. 1780:
Psalterium Georgianum (George's Lute)

Proposed by Bode, c. 1775:
Honores Frederici (the Honours of Frederick)
Sceptrum Brandenburgicum (the Sceptre of Brandenburg)
Telescopium Herschelii (Herschel's Telescope)
Globus Ærostaticus (the Balloon)
Quadrans Muralis (the Mural Quadrant)
Lochium Funis (the Log Line)
Machina Electrica (the Electrical Machine)
Officiana Typographica (the Printing Press)
Felis (the Cat).

There seem to have been two Muscas, one formed by La Caille to replace Bayer's Apis and the other (rejected) formed by Bode out of stars near Aries. One discarded group, Quadrans, has at least given its name to the Quadrantid meteor shower of early January.

STARS OF MAGNITUDE 2·00 OR BRIGHTER

STARS OF MAGNITUDE 2·00 OR BRIGHTER

Star	Name	Apparent magnitude	Luminosity Sun=1	Spectrum	Distance light-years
1 α Canis Maj.	Sirius	−1·46	26	A1	8·8
2 α Carinæ	Canopus	−0·72	200 000	F0	1200
3 α Centauri		−0·27	1·7 + 0·45	K1 + G2	4·3
4 α Boötis	Arcturus	−0·04	115	K2	36
5 α Lyræ	Vega	0·03	52	A0	26
6 α Aurigæ	Capella	0·08	90 + 70	G8 + G1	43
7 β Orionis	Rigel	0·12v	60 000	B8	900
8 α Canis Min.	Procyon	0·38	7	F5	11·4
9 α Eridani	Achermar	0·46	400	B5	85
10 α Orionis	Betelgeux	var.	15 000	M2	310
11 β Centauri	Agena	0·61	10 500	B1	460
12 α Aquilæ	Altair	0·77	10	A7	16·6
13 α Crucis	Acrux	0·83	3200 + 2000	B1 + B3	360
14 α Tauri	Aldebaran	0·85	100	K5	68
15 α Scorpii	Antares	0·96v	7500	M1	330
16 α Virginis	Spica	0·98v	2100	B1	260
17 β Geminorum	Pollux	1·14	60	K0	36
18 α Pisc. Aust.	Fornhault	1·16	13	A3	22
19=α Cygni	Deneb	1·25	70 000	A2	1800
19=β Crucis		1·25v	8200	B0	425
21 α Leonis	Regulus	1·35	130	B7	85
22 ε Canis Maj.	Adhara	1·50	5000	B2	490
23 α Geminorum	Castor	1·58	45	A0	46
24=γ Crucis		1·63	160	M3	88
24=λ Scorpii	Shaula	1·63	1300	B2	275
26 γ Orionis	Bellatrix	1·64	2200	B2	360
27 β Tauri	Al Nath	1·65	400	B7	130
28 β Carinæ	Miaplacidus	1·68	130	A0	85
29 ε Orionis	Alnilam	1·70	23 000	B2	1200
30 α Gruis	Alnair	1·74	230	B5	69
31=ζ Orionis	Alnitak	1·77	19 000	O9·5	1100
31=ε Ursæ Maj.	Alioth	1·77	60	A0	62
33 γ Velorum	Regor	1·78	3800	WC7	520
34 α Ursæ Maj.	Dubhe	1·79	60	K0	75
35 α Persei	Mirphak	1·80	6000	F5	620
36 ε Sagittarii	Kaus Australis	1·85	110	B9	85
37=δ Canis maj.	Wezea	1·86	130 000	F8	3000
37=ε Carinæ	Avior	1·86	600	K0	200
37=η Ursæ Maj.	Alkaid	1·86	450	B3	108
40 θ Scorpii	Sargas	1·87	14 000	G0	900
41 β Aurigæ	Menkarlina	1·90	50	A2	46
42 α Tri. Aust.	Atria	1·92	96	K2	55
43 γ Geminorum	Alhema	1·93	82	A0	85
44 α Pavonis		1·94	700	B3	230
45 δ Velorum	Koo She	1·96	50	A0	68
46=α Hydræ	Alphard	1·98	105	K3	85
46=β Canis Maj.	Mirzam	1·98v	7200	B1	710
48=γ Leonis	Algieba	1·99	60	K0 + G7	90
48=α Ursæ Min.	Polaris	1·99	6000	F8	680
50 α Arietis	Hamal	2·00	96	K2	85

Star	Name	Apparent magnitude	Luminosity Sun = 1	Spectrum	Distance light-years
− o Ceti	Mira	var.	140v	Md	95
− γ Cassiopeiæ	Cih	2·2v	6000v	B0p	780

Stars of above magnitude 1·4 are conventionally termed 'first magnitude' stars – that is to say, down the list as far as Regulus.

Obviously the distances and luminosities of the stars, apart from the relatively close ones, are not known precisely, and in this list the values have been 'rounded off'. The distances are given in light-years, though in the main catalogue which follows it has seemed better to use parsecs. One parsec is equal to 3·2633 light-years.

Excluding the unique η Carinæ, there are two other stars which may exceed magnitude 2 though they are generally fainter. Mira Ceti attained 1·2 at the maximum of 1772, according to the Swedish astronomer Per Wargentin, though most maxima are well below 2; the official maximum magnitude is usually given as 1·7. γ Cassiopeiæ is an irregular variable; in 1936 it reached magnitude 1·6, but at present (1995) and for some years past its magnitude has been about 2·2.

According to a recent estimate, the most luminous star in the list is Canopus; next come Wezea, Deneb and Rigel. The difficulties of estimating very remote stars can be judged that in some other lists the luminosity of Canopus is reduced to only about 4000 times that of the Sun! The most remote star in the list is Wezea; the closest is, of course, α Centauri.

CONSTELLATIONS AND STARS

The largest constellation is Hydra, with an area of 1303 square degrees. (Formerly it was exceeded by Argo Navis, but Argo has now been unceremoniously chopped up into a keel, sails and a poop.)

✱ The smallest constellation is Crux Australis, with an area of 68 square degrees.

✱ The largest Zodiacal constellation is Virgo. The smallest is Aries.

✱ The constellation with the greatest number of stars above the second magnitude is Orion (5 stars).

✱ The constellation with the greatest number of stars above the fourth magnitude is Ursa Major (19 stars).

✱ The constellation with the greatest number of stars above the fifth magnitude is Centaurus (49 stars).

✱ The only constellation with no star as bright as the fifth magnitude is Mensa. Next in order of 'dimness' come Cælum, Horologium and Sextans, each with only 2 stars above the fifth magnitude.

✱ For stars above the fifth magnitude, the greatest 'star density' is Crux Australis, which averages out at 19·12 stars per 100 square degrees. Its nearest rival, Lupus, has a figure of only 9·58.

✱ The least star density, excluding Mensa, is for Sextans (0·63 star per 100 square degrees).

✱ The brightest star is Sirius (α Canis Majoris), magnitude −1·46.

✱ The brightest star in the northern hemisphere of the sky is Arcturus (α Boötis), magnitude −0·04.

✱ The only naked-eye star which is said to have a greenish tint is β Libræ – though most observers will certainly class it as white!

✱ The reddest naked-eye star is probably μ Cephei, called by Herschel 'the Garnet star', though optical aid is needed to bring out its colour well. The reddest first-magnitude stars are Antares, Betelgeux and Aldebaran. Among the reddest telescopic stars are R Leporis and U Cephei.

The following data apply to the stars above magnitude 5·00, listed in the Catalogue which follows:

✷ *Most powerful stars*, all of absolute magnitude –7·5 or brighter; Canopus (α Carinæ), s Carinæ, ε Aurigæ, ι'Scorpii, b Velorum, ρ Cassiopaiæ, β Doradûs, Deneb (α Cygni), ν Cephei and 3 Puppis.

✷ *Least powerful stars* are ε Indi (absolute magnitude 7·0), ε Eridani (6·1), o² Eridani (6·0) and σ Draconis (5·9).

THE STAR CATALOGUE

This catalogue has been compiled from various sources; positions of the brightest stars (down to magnitude 4·75) are given for epoch 2000·0. The published values for absolute magnitudes and distances differ somewhat, and are subject to uncertainty; therefore the values given here have been in general 'rounded off' except for those stars for which the data are known with true precision. I have followed the data given in the authoritative *Cambridge Sky Catalogues* (1987).

The list of double stars, variable stars, and nebular objects does not pretend to be complete, and could be extended almost *ad infinitum*, but I have included most variable stars whose maximum magnitudes are 8·0 or brighter and where the range is at least half a magnitude. For double stars, I have given pairs which are within the range of modest telescopes together with some which may be regarded as test objects. For binary stars of reasonably short period, the values of position angle and distance are for approximate date 1990.

Data given are:

Variable Stars Range, type, period in days, and spectral type.

Double Stars Position angle in degrees, separation in seconds of arc, and magnitudes of the components.

Open Clusters Diameter in minutes of arc, magnitude, and approximate number of stars (though in many cases this is subject to great uncertainty).

Globular Clusters Diameter in minutes of arc, and approximate total magnitude.

Planetary Nebulæ Dimensions in seconds of arc, total magnitude, and magnitude of the central star.

Nebulæ Dimensions in minutes of arc, and, where appropriate, the magnitude of the illuminating star.

Galaxies Magnitude; dimensions, in minutes of arc; and type.

These symbols identify the various features shown on the maps

☀	◉ ○	VARIABLE STARS
⊖	⊖	PLANETARY NEBULAE
⁝⁝⁝	⁝⁝	GLOBULAR CLUSTERS
◯	○	GALAXIES
◌	◌	OPEN CLUSTERS
🌿	🌿	GASEOUS NEBULA

●	Mag.1	**Star Magnitudes** These are given by different symbols, though of course there are many gradations; for example, the first symbol indicates stars between magnitudes –1. (Sirius) down to Regulus (+1.4).
●	Mag. 2	
●	Mag. 2.5	
●	Mag. 3	
●	Mag. 4	

ANDROMEDA

(Abbreviation: And).

A large and important northern constellation; one of Ptolemy's 'originals'. It contains the Great Spiral, M.31. One of the leading stars in the constellation – Alpheratz or α Andromedæ – is actually a member of the Square of Pegasus, and was formerly known, more logically, as δ Pegasi.

In mythology, Andromeda was the beautiful daughter of King Cepheus and Queen Cassiopeia. Cassiopeia offended the sea-god Neptune by her boasting about Andromeda's beauty, which, she claimed, was greater than that of any sea-nymph. Neptune thereupon sent a sea-monster to ravage the kingdom, and the Oracle stated that the only solution was to chain Andromeda to a rock by the shore where she would be devoured by the monster.

BRIGHTEST STARS

Star	R.A. h	m	s	Dec deg	min	sec	Mag.	Abs. mag.	Spectrum	Dist pc	
1 o	23	01	55·1	+42	19	34	3·6v	1·4	B6+A2	35	
7	23	12	32·9	+49	24	33	4·52	2·6	F0	23	
16 λ	23	37	33·7	+46	27	30	3·82v	1·8	G8	24	
17 ι	23	38	08·0	+43	16	05	4·29	–0·2	B8	79	
19 κ	23	40	24·4	+44	20	02	4·14	–0·2	B8	60	
21 α	00	08	23·2	+29	05	26	2·06	–0·1	A0p	22	Alpheratz
24 θ	00	17	05·4	+38	40	54	4·61	1·4	A2	44	
25 σ	00	18	19·6	+36	47	07	4·52	1·4	A2	42	
29 π	00	36	52·8	+33	43	09	4·36	–1·1	B5	120	
30 ε	00	38	33·3	+29	18	42	4·37	0·3	G8	41	
31 δ	00	39	19·6	+30	51	40	3·27	–0·2	K3	49	
34 ξ	00	47	20·3	+24	16	02	4·06v	–2·2	K1	48	Al Dhail
35 ν	00	49	48·8	+41	04	44	4·53	–1·1	B5	130	
37 μ	00	56	45·1	+38	29	58	3·87	2·1	A5	25	
38 η	00	57	12·4	+23	25	04	4·42	1·8	G8	35	
42 φ	01	09	30·1	+47	14	31	4·25	–1·2	B8	120	
43 β	01	09	43·8	+35	37	14	2·06	–0·4	M0	27	Mirach
50 υ	01	36	47·8	+41	24	20	4·09	4·0	F8	13	
51	01	37	59·5	+48	37	42	3·57	–0·2	K3	57	
57 γ	02	03	53·9	+42	19	47	2·18	–0·1	K2 }	37	Almaak
	02	03	54·7	+42	19	51	5·03		A0 }		
65	02	25	37·3	+50	16	43	4·71	–0·3	K4	82	

o is a β-Lyræ variable with a small range (3·5–3·7), period 1·6 days.
ζ is an ellipsoidal variable with a small range, 4·06–4·20.
λ is slightly variable (3·7–4·1) and is probably a spectroscopic binary.

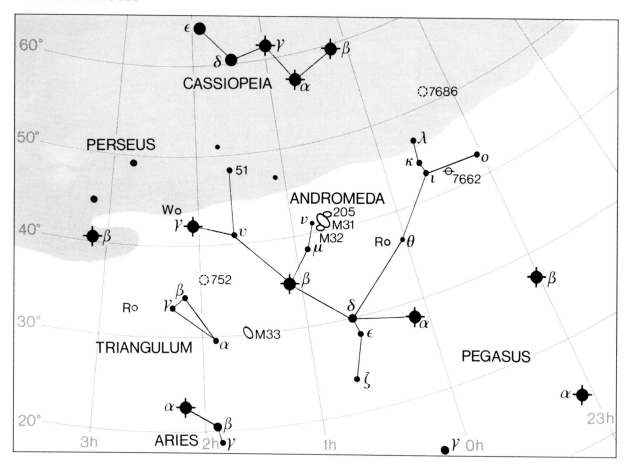

ANDROMEDA (continued)

This was duly done, but the situation was redeemed by the hero Perseus, who was on his way home after killing the Gorgon, Medusa. Mounted upon his winged sandals, Perseus arrived in the nick of time, turned the monster to stone by the simple expedient of showing it Medusa's head, and then, in the best story-book tradition, married Andromeda. Cepheus, Cassiopeia and of course Perseus are to be found in the sky; the Gorgon's head is marked by the 'Demon Star' Algol, and the sea-monster has sometimes been identified with the constellation of Cetus.

Andromeda has no first-magnitude star, but there are eight above magnitude 4:

	Mag.	Luminosity Sun = 1	Dist lt-yrs
β	2·06	115	88
α	2·06	96	72
γ	2·14	95	121
δ	3·27	105	160
51	3·57	105	186
o	3·6v	26	114
λ	3·82	16	78
μ	3·87	11	81

Also above mag. 5:

	Mag.	Abs. mag.	Spectrum	Dist	
3	4·65	0·2	K0	71	
5	5·70	3·4	F5	29	
8	4·85	−0·5	M2	100	
46 ξ	4·88	1·7	K0	35	Adhil
48 ω	4·83	2·0	F4	36	
52 χ	4·98	0·3	G8	86	
53 τ	4·94	−0·6	B8	130	
58	4·80	1·9	A4	38	
60	4·83	−0·3	K4	97	

VARIABLE STARS

	R.A. h	m	Dec. deg	min	Range	Type	Period, d.	Spectrum
Z	23	33·7	+48	49	8·0–12·4	Z And	–	M
ST	23	38·8	+35	46	7·7–11·8	Semi-reg.	328	R
SV	00	04·3	+40	07	7·7–14·3	Mira	325·2	M
KU	00	06·9	+43	05	6·5–10·5	Mira	750	M
VX	00	19·9	+44	43	7·8–9·3	Semi-reg.	369	N
T	00	22·4	+27	00	7·7–14·5	Mira	280·8	M
R	00	24·0	+38	35	5·8–14·9	Mira	409·3	S
TU	00	32·4	+26	02	7·8–13·1	Mira	316·8	M
RW	00	47·3	+32	41	7·9–15·7	Mira	430·3	M
W	02	17·6	+44	18	6·7–14·6	Mira	395·9	S

DOUBLE STARS

	R.A. h	m	Dec. deg	min	P.A. deg	Sep. sec	Mags.	
π	00	36·9	+33	43	173	35·9	4·4, 8·6	
γ	02	03·9	+42	20	063	9·8	2·3, 4·8	
γ²					106	0·5	5·5, 6·3	Binary, period 61 years

OPEN CLUSTERS

M	NGC	R.A. h	m	Dec. deg	min	Diameter min	Mag.	No. of stars
	7686	23	30·2	+49	08	15	5·6	20
	752	01	57·8	+37	41	50	5·7	60

ANDROMEDA (continued)

PLANETARY NEBULA

M	NGC	R.A.		Dec.		Dimensions	Mag.	Mag. of central star
		h	m	deg	min	sec		
	7662	23	25·9	+42	33	20 × 130	9·2	13·2

GALAXIES

M	NGC	R.A.		Dec.		Mag.	Dimensions	Type	
		h	m	deg	min		min		
	205	00	40·4	+41	41	8·0	17·4 × 9·8	E6	Companion to M.31
31	224	00	42·7	+41	16	3·5	178 × 63	Sb	
32	221	00	42·7	+40	52	8·2	7·6 × 5·8	E2	Companion to M.31
	891	02	22·6	+42	21	9·9	13·5 × 2·8	Sb	

ANTLIA

(Abbreviation: Ant).

A small southern constellation, covering 239 square degrees, added to the sky by Lacaille in 1752; its original name was Antlia Pneumatica. It contains no star brighter than magnitude 4·4, and no mythological legends are associated with it.

See chart for Carina.

BRIGHTEST STARS

Star	R.A.			Dec			Mag.	Abs.	Spectrum	Dist
	h	m	s	deg	min	sec		mag.		pc
ε	09	29	14·7	−35	57	05	4·51	−0·4	M0	96
α	10	27	09·1	−31	04	04	4·25	−0·4	M0	85
ι	10	56	43·0	−37	08	16	4·60	0·3	F2	79

Also above mag. 5:

	Mag.	Abs. mag.	Spectrum	Dist
θ	4·79	3·8	F7	14

VARIABLE STARS

	R.A.		Dec.		Range	Type	Period, d.	Spectrum
	h	m	deg	min				
S	09	32·3	−28	38	6·4–6·9	W Uma	0·65	A
U	10	35·2	−39	34	5·7–6·8	Irr.	–	N

DOUBLE STARS

	R.A.		Dec.		P.A.	Sep.	Mags.
	h	m	deg	min	deg	sec	
θ	09	44·2	−27	46	005	0·1	5·4, 5·6
δ	10	29·6	−30	36	226	11·0	5·6, 9·6

GALAXIES

M	NGC	R.A.		Dec.		Mag.	Dimensions	Type
		h	m	deg	min		min	
	2997	09	45·6	−31	11	10·6	8·1 × 6·5	Sc
	3223	10	21·6	−34	16	11·8	4·1 × 2·6	Sb
	3347	10	42·8	−36	22	12·5	4·4 × 2·6	SBb

APUS

(Abbreviation: Aps).

Originally Avis Indica; it was introduced by Bayer in 1603. It lies in the far south, and has only two stars brighter than magnitude 4.

	Mag.	Luminosity	Dist
	Sun = 1	lt-yrs	
α	3·83	110	218
γ	3·89	4·5	46

See chart for Musca

BRIGHTEST STARS

Star	R.A.			Dec			Mag.	Abs.	Spectrum	Dist
	h	m	s	deg	min	sec		mag.		pc
α	14	47	51·6	−79	02	41	3·83	−0·3	K5	67
δ	16	20	20·7	−78	41	44	4·68	−0·5	M4 + K5	110
γ	16	33	27·1	−78	53	49	3·89	3·2	K0	14
β	16	43	04·5	−77	31	02	4·24	0·2	K0	42
ζ	17	21	59·3	−67	46	13	4·78	−0·3	gK5	100

The magnitude of δ is given as the combination of δ^1 and δ^2, which are separated by 103 sec and make up a very wide physically connected pair.

Also above mag. 5:

	Mag.	Abs. mag.	Spectrum	Dist
η	4·91		A2p	

VARIABLE STARS

	R.A.		Dec.		Range	Type	Period, d.	Spectrum
	h	m	deg	min				
θ	14	05·3	−76	48	6·4–8·6	Semi-reg.	119	M
S	15	04·3	−71	53	9·5–15	R CrB	–	R
DW	17	23·5	−67	56	7·9–9·1	Algol	2·31	B

(R Apodis, magnitude 5·3, has been suspected of variability.)

DOUBLE STAR

	R.A.		Dec.		P.A.	Sep.	Mags.
	h	m	deg	min	deg	sec	
δ	16	20·3	−78	42	012	102·9	4·7, 5·1

GLOBULAR CLUSTERS

M	NGC	R.A.		Dec.		Diameter	Mag.
		h	m	deg	min	min	
	IC 4499	15	00·3	−82	13	7·6	10·6
	6101	16	25·8	−72	12	10·7	9·3

APUS (continued)

GALAXY

M	NGC	R.A.		Dec.		Mag.	Dimensions	Type
		h	m	deg	min		min	
	5967	15	48·1	−75	40	12·5	2·9 × 1·8	SBc

AQUARIUS

(Abbreviation: Aqr).

A Zodiacal constellation, and of course one of Ptolemy's 'originals'. Oddly enough there are no well-defined legends attached to it, though it has been associated with Ganymede, cup-bearer of the Olympian gods. There are 9 stars brighter than magnitude 4·00:

	Mag.	Luminosity Sun = 1	Dist lt-yrs
β	2·91	5250	980
α	2·96	5250	945
δ	3·27	105	98
ζ	3·6	50	98
c²	3·66	60	107
λ	3·74	120	256
ε	3·77	28	107
γ	3·84	50	91
b¹	3·97	60	160

BRIGHTEST STARS

Star		R.A.			Dec			Mag.	Abs. mag.	Spectrum	Dist pc	
		h	m	s	deg	min	sec					
2	ε	20	47	40·3	−09	29	45	3·77	1·2	A1	33	Albali
3	k	20	47	44·0	−05	01	40	4·42	−0·5	M3	91	
6	μ	20	52	39·0	−08	59	00	4·73	1·0	A8	30	
13	ν	21	09	35·4	−11	22	18	4·51	0·3	G8	70	
22	β	21	31	33·3	−05	34	16	2·91	−4·5	G0	300	Sadalsuud
23	ξ	21	37	44·9	−07	51	15	4·69	2·4	A7	36	Bunda
31	o	22	03	18·7	−02	09	19	4·69	−0·2	B8	95	
34	α	22	05	46·8	−00	19	11	2·96	−4·5	G2	290	Sadalmelik
33	ι	22	06	26·1	−13	52	11	4·27	−0·2	B8	78	
43	θ	22	16	49·9	−07	47	00	4·16	1·8	G8	26	Ancha
48	γ	22	21	39·2	−01	23	14	3·84	0·6	A0	28	Sadachiba
52	π	22	25	16·4	+01	22	39	4·66	−4·1	B0	450	Seat
55	ζ¹	22	28	49·5	−00	01	13	4·53	0·6	F2	30	
	ζ²	22	28	49·9	−00	01	12	4·31		F2		
62	η	22	35	21·2	−00	07	03	4·02	−0·2	B8	46	
71	τ	22	49	35·3	−13	35	33	4·01	−0·4	M0	74	
73	λ	22	52	36·6	−07	34	47	3·74	−0·5	M2	71	
76	δ	22	54	38·8	−15	49	15	3·27	−0·2	A2	30	Scheat
88	c²	23	09	26·6	−21	10	21	3·66	0·2	K0	33	
90	φ	23	14	19·2	−06	02	56	4·22	−0·5	M2	75	
91	ψ¹	23	15	53·4	−09	05	16	4·21	1·8	K0	30	
93	ψ²	23	27	54·1	−09	10	57	4·39	−1·1	B5	130	
98	b¹	23	22	58·0	−20	06	02	3·97	0·2	K0	49	
99	b²	23	26	02·5	−20	38	31	4·39	−0·3	K5	87	
101	b³	23	33	16·4	−20	54	32	4·71	—	A0	25	
105	ω²	23	42	43·2	−14	32	42	4·49	0·4	B9·5	43	

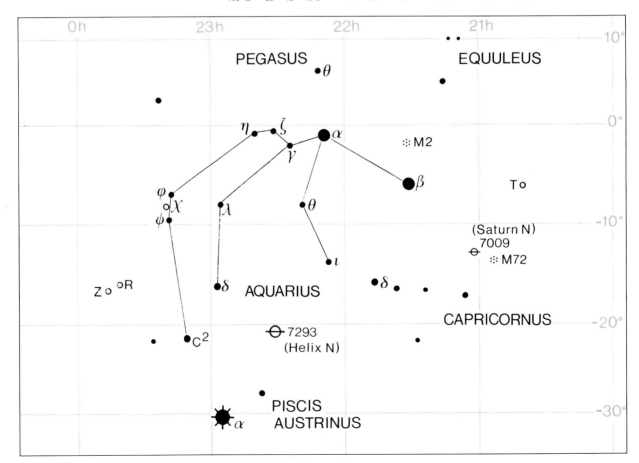

AQUARIUS (continued)

Also above mag. 5:

	Mag.	Abs. mag.	Spectrum	Dist
57 σ	4·82	0·0	A0	92
95 ψ³	4·98	0·6	A0	75
102 ω¹	5·00	—	A5	18
104 A²	4·82	−2·0	G0	220

VARIABLE STARS

	R.A.		Dec.		Range	Type	Period, d.	Spectrum
	h	m	deg	min				
W	20	46·4	−04	05	8·4–14·9	Mira	381·1	M
V	20	46·8	+02	26	7·6–9·4	Semi-reg.	244	M
T	20	49·9	−05	09	7·2–14·2	Mira	202·1	M
X	22	18·7	−20	54	7·5–14·8	Mira	311·6	S
S	22	57·1	−20	21	7·6–15·0	Mira	297·3	M
R	23	43·8	−15	17	5·8–12·4	Symb.	387	M+Pec

DOUBLE STARS

	R.A.		Dec.		P.A.	Sep.	Mags.	
	h	m	deg	min	deg	sec		
β	21	31·6	−05	34	AB 321	35·4	2·9, 10·8	
					AC 186	57·2	11·4	
41	22	14·3	−21	04	AB 114	5·0	5·6, 7·1	
					AC 043	212·1	9·0	
51	22	24·1	−04	50	AB 324	0·5	6·5, 6·5	
					AB+D 191	116·0	10·1	
					AC 342	54·4	10·2	
					AE 133	132·4	8·6	
ζ	22	28·8	−00	01	200	2·0	4·3, 4·5	Binary, p. 856 y
89	23	09·9	−22	27	007	0·4	5·1, 5·9	
107	23	46·0	−18	41	136	6·6	5·7, 6·7	

ASTERISM

M	NGC	R.A.		Dec.		
		h	m	deg	min	
73	6994	20	58·9	−12	38	Four stars; not a true cluster.

GLOBULAR CLUSTERS

M	NGC	R.A.		Dec.		Diameter	Mag.
		h	m	deg	min	min	
72	6981	20	53·5	−12	32	5·9	9·3
2	7089	21	33·5	−00	49	12·9	6·5

PLANETARY NEBULÆ

M	NGC	R.A.		Dec.		Dimensions	Mag.	Mag. of central star	
		h	m	deg	min	sec			
	7009	21	04·2	−11	22	2·5 × 100	8·3	11·5	Saturn Nebula
	7293	22	29·6	−20	48	770	6·5	13·5	Helix Nebula

GALAXIES

M	NGC	R.A.		Dec.		Mag.	Dimensions	Type
		h	m	deg	min		min	
	7184	22	02·7	−20	49	12·0	5·8 × 1·8	Sb
	7606	23	19·1	−08	29	10·8	5·8 × 2·6	Sb
	7723	23	38·9	−12	58	11·1	3·6 × 2·6	Sb
	7727	23	39·9	−12	18	10·7	4·2 × 3·4	SBap

AQUILA

(Abbreviation: Aql).

One of the most distinctive of all the northern constellations, and, of course, an 'original'. Mythologically it represents an eagle which was sent by Jupiter to collect a Phrygian shepherd-boy, Ganymede, who was destined to become cup-bearer of the Gods – following an unfortunate episode in which the former holder of the office, Hebe, tripped and fell during a particularly solemn ceremony.

BRIGHTEST STARS

Star	R.A.			Dec			Mag.	Abs. mag.	Spectrum	Dist pc	
	h	m	s	deg	min	sec					
13 ε	18	59	37·2	+15	04	06	4·02	−0·1	K2	65	
12 i	19	01	40·7	−05	44	20	4·02	0·0	K1	64	
17 ζ	19	05	24·4	+13	51	48	2·99	0·2	B9	32	Dheneb
16 λ	19	06	14·7	−04	52	57	3·44	0·0	B8·5	30	Althalimain
30 δ	19	25	29·7	+03	06	53	3·36	2·1	F0	16	
32 ν	19	26	30·9	+00	22	44	4·66	−4·6	F2	530	
38 μ	19	34	05·2	+07	22	44	4·45	−0·2	K3	39	
41 ι	19	36	43·1	−01	17	11	4·36	−2·2	B5	180	
50 γ	19	46	15·4	+10	36	48	2·72	−2·3	K3	87	Tarazed
53 α	19	50	46·8	+08	52	06	0·77	2·2	A7	5·1	Altair
55 η	19	52	28·1	+01	00	20	var.	−4·5v	G0v	440	
59 ξ	19	54	14·7	+08	27	41	4·71	0·2	K0	73	
60 β	19	55	18·5	+06	24	24	3·71	3·2	G8	11	Alshain
65 θ	20	11	18·1	−00	49	17	3·23	−0·8	B9	61	
71	20	38	20·1	−01	06	19	4·32	0·3	G8	64	

AQUILA (continued)

The leading star is Altair. Altogether there are eight stars above magnitude 4:

	Mag.	Luminosity Sun = 1	Dist lt-yrs
α	0·77	10	16·6
γ	2·72	700	186
ζ	2·99	60	104
θ	3·23	180	199
δ	3·36	11	52
λ	3·44	82	98
η	3·5 (max)	5000v	1400
β	3·71	4·5	36

Also above mag. 5:

	Mag.	Abs. mag.	Spectrum	Dist	
39 κ	4·95		B0·5		
67 ρ	4·95	1.4	A2	51	(Now in Delphinus)
69	4·91	−0·1	K2	100	
70	4·89	−2·3	K5	240	

VARIABLE STARS

	R.A. h	m	Dec. deg	min	Range	Type	Period, d.	Spectrum
V	19	04·4	−05	41	6·6–8·4	Semi-reg.	353	N
R	19	06·4	+08	14	5·5–12·0	Mira	284·2	M
TT	19	08·2	+01	18	6·4–7·7	Cepheid	13·75	F–G
W	19	15·4	−07	03	7·3–14·3	Mira	490·4	S
U	19	29·4	−07	03	6·1–6·9	Cepheid	7·02	F–G
X	19	51·5	+04	28	8·3–15·5	Mira	347·0	M
η	19	52·5	+01	00	3·5–4·4	Cepheid	7·18	F–G
RR	19	57·6	−01	53	7·8–14·5	Mira	394·8	M

DOUBLE STARS

	R.A. h	m	Dec. deg	min	P.A. deg	Sep. sec	Mags.
ε	18	59·6	+15	04	187	131·1	4·0, 9·9
23	19	18·5	+01	05	005	3·1	5·3, 9·3
31	19	25·0	+11	57	343	105·6	5·2, 8·7
δ	19	25·5	+03	07	271	108·9	3·4, 10·9
ν	19	26·5	+00	20	288	201·0	4·7, 8·9
U	19	29·4	−07	03	228	1·5	var., 11·7
χ	19	42·6	+11	50	077	0·5	5·6, 6·8
γ	19	46·3	+10	37	258	132·6	2·7, 10·7
π	19	48·7	+11	49	110	1·4	6·1, 6·9
α	19	50·8	+08	52	301	165·2	0·8, 9·5
57	19	54·6	−08	14	170	35·7	5·8, 6·5

OPEN CLUSTERS

M	NGC	R.A. h	m	Dec. deg	min	Diameter min	Mag.	No. of stars
	6709	18	51·5	+10	21	13	6·7	40
	6755	19	07·8	+04	14	15	7·5	100

PLANETARY NEBULÆ

M	NGC	R.A. h	m	Dec. deg	min	Dimensions sec	Mag.	Mag. of central star
	6741	19	02·6	−00	27	6	10·8	14·7
	6751	19	05·9	−06	00	20	12·5	13·9
	6790	19	23·2	+01	31	7	10·2	13·5
	6803	19	31·3	+10	03	6	11·3	15·2

ARA

(Abbreviation: Ara).

An original constellation, though apparently without any definite legends attached to it. There are 7 stars above magnitude 4·00:

	Mag.	Luminosity Sun = 1	Dist lt-yrs
β	2·85	5000	780
α	2·95	450	190
ζ	3·13	110	137
γ	3·34	5000	1075
δ	3·62	105	95
θ	3·66	9000	1570
η	3·76	110	190

BRIGHTEST STARS

Star	R.A. h	m	s	Dec deg	min	sec	Mag.	Abs. mag.	Spectrum	Dist pc	
η	16	49	47·0	−59	02	29	3·76	−0·3	K5	58	
ζ	16	58	37·1	−55	59	24	3·13	−0·3	K5	42	
ε¹	16	59	34·9	−53	09	38	4·06	−0·5	M1	82	
β	17	25	17·9	−55	31	47	2·85	−4·4	K3	240	
γ	17	25	23·5	−56	22	39	3·34	−4·4	B1	330	
δ	17	31	05·8	−60	41	01	3·62	−0·2	B8	29	
α	17	31	50·3	−49	52	34	2·95	−1·7	B3	58	Choo
σ	17	35	39·4	−46	30	20	4·59	—	A0	47	
θ	18	06	37·6	−50	05	30	3·66	−5·1	B1	480	

Also above mag. 5:

	Mag.	Abs. mag.	Spectrum	Dist
λ	4·77	3·4	dF5	19

VARIABLE STARS

	R.A. h	m	Dec. deg	min	Range	Type	Period, d.	Spectrum
X	16	36·4	−55	24	8·0–13·5	Mira	175·8	M
R	16	39·7	−57	00	6·0–6·9	Algol	4·42	B
U	17	53·6	−51	41	7·7–14·1	Mira	225·2	M

DOUBLE STARS

	R.A. h	m	Dec. deg	min	P.A. deg	Sep. sec	Mags.
γ	17	25·4	−56	23	AB 328	17·9	3·3, 10·3
					AC 066	41·6	11·8

OPEN CLUSTERS

M	NGC	R.A. h	m	Dec. deg	min	Diameter min	Mag.	No. of stars
	6193	16	41·3	−48	46	15	5·2	—
	6204	16	46·5	−47	01	5	8·2	45
	6208	16	49·5	−53	49	16	7·2	60
	6250	16	58·0	−45	48	8	5·9	60
	6253	16	59·1	−52	43	5	10·2	30
	H.13	17	05·4	−48	11	15	—	15
	IC 4651	17	24·7	−49	57	12	6·9	80

GLOBULAR CLUSTERS

M	NGC	R.A. h	m	Dec. deg	min	Diameter min	Mag.
	6352	17	25·5	−48	25	7·1	8·1
	6362	17	31·9	−67	03	10·7	8·3
	6397	17	40·7	−53	40	25·7	5·6

GALAXIES

M	NGC	R.A. h	m	Dec. deg	min	Mag.	Dimensions min	Type
	6215	16	51·1	−58	59	11·8	2·0 × 1·6	Sc
	6221	16	52·8	−59	13	11·5	3·2 × 2·3	SBc

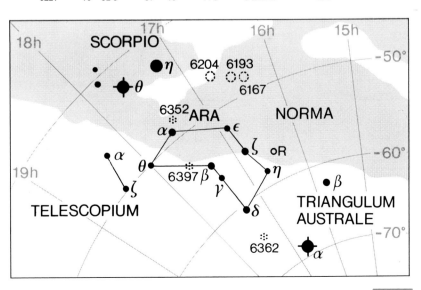

ARIES

(Abbreviation: Ari).

The first constellation of the Zodiac – though since the vernal equinox has shifted into Pisces, Aries should logically be classed as second! In mythology it represents a ram with a golden fleece, sent by the god Mercury (Hermes) to rescue the children of the king of Thebes from an assassination plan by their stepmother. The ram, which had the remarkable ability to fly, carried out its mission; unfortunately, the girl (Helle) lost her balance and fell to her death in that part of the sea now called the Hellespont, but the boy (Phryxus) arrived safely. After the ram's death, its golden fleece was hung in a sacred grove, from which it was later removed by the Argonauts commanded by Jason.

Aries is moderately conspicuous. There are four stars above the fourth magnitude:

	Mag.	Luminosity Sun = 1	Dist lt-yrs
α	2·00	96	85
β	2·64	11	46
c	3·63	105	117
γ	3·9	60 + 56	117

Of these, γ is an excellent double with two equal components.

See chart for Andromeda.

BRIGHTEST STARS

Star	R.A. h	m	s	Dec deg	min	sec	Mag.	Abs. mag.	Spectrum	Dist pc	
5 γ	01	53	31·7	+19	17	45	4·68	0·4	A	36 }	Mesartim
	01	53	31·8	+19	17	37	4·59	0·2	B9	36 }	
6 β	01	54	38·3	+20	48	29	2·64	2·1	A5	14	Sheratan
13 α	02	07	10·3	+23	27	45	2·00	−0·1	K2	26	Hamal
35	02	43	27·0	+27	42	26	4·66	−1·7	B3	180	
39	02	47	54·5	+29	14	50	4·51	0·0	K1	58	
41 c	02	49	58·9	+27	15	38	3·63	−0·2	B8	38	Nair al Butain
48 ε	02	59	12·6	+21	20	25	4·63	1·4	A2	48	
57 δ	03	11	37·7	+19	43	36	4·35	−0·1	K2	78	Botein

Also above mag. 5:

	Mag.	Abs. mag.	Spectrum	Dist
9 λ	4·79	1·7	F0	42
14	4·98	0·6	F2	75
58 ζ	4·89	0·0	A0	91

VARIABLE STARS

	R.A. h	m	Dec deg	min	Range	Type	Period, d.	Spectrum
R	02	16·1	+25	03	7·4–13·7	Mira	186·8	M
T	02	48·3	+17	31	7·5–11·3	Semi-reg.	317 (var.)	M
U	03	11·0	+14	48	7·2–15·2	Mira	371·1	M

DOUBLE STARS

	R.A. h	m	Dec deg	min	P.A. deg	Sep. sec	Mags.
γ	01	53·5	+19	18	000	7·8	4·8, 4·8
π	02	49·3	+17	28	AB 120	3·2	5·2, 8·7
					AC 110	25·2	10·8
ε	02	59·2	+21	20	191	1·5	5·2, 5·5

GALAXIES

M	NGC	R.A. h	m	Dec deg	min	Mag.	Dimensions min	Type	
	772	01	59·3	+19	01	10·3	7·1 × 4·5	Sb	Arp 78
	976	02	34·0	+20	59	12·4	1·7 × 1·5	Sb	

AURIGA

(Abbreviation: Aur).

One of the most brilliant northern constellations, with Capella outstanding. Mythologically it honours Erechthonius, son of Vulcan, the blacksmith of the gods, who became King of Athens and also invented the four-horse chariot.

There are eight stars brighter than the fourth magnitude:

	Mag.	Luminosity Sun = 1	Dist lt-yrs
α	0·08	90 + 70	43
β	1·90	50	46
θ	2·62	75	82
ι	2·69	700	267
ε	2·99v	200 000	4600
η	3·17	450	200
δ	3·72	60	163
ζ	3·75v	700	520

In addition, Al Nath used to be called γ Aurigæ, but has now been given a free transfer, and is included in Taurus as β Tauri.

BRIGHTEST STARS

Star	R.A. h	m	s	Dec deg	min	sec	Mag.	Abs. mag.	Spectrum	Dist pc	
3 ι	04	56	59·5	+33	09	58	2·69	−2·3	K3	82	Hassaleh
7 ε	05	01	58·1	+43	49	24	2·99v	−8·5	F0	1400	Almaaz
8 ζ	05	02	28·6	+41	04	33	3·75v	−2·3	K4	160	Sadatoni
10 η	05	06	30·8	+41	14	04	3·17	−1·7	B3	61	
13 α	05	16	41·3	+45	59	53	0·08	0·3	G8	13	V. close double Capella
15 λ	05	19	08·4	+40	05	57	4·71	4·4	G0	13	
29 τ	05	49	10·4	+39	10	52	4·52	0·3	G8	70	
31 υ	05	51	02·4	+37	18	20	4·74	−0·5	gM1	110	
32 ν	05	51	29·3	+39	08	55	3·97	0·2	K0	45	
33 δ	05	59	31·6	+54	17	05	3·72	0·2	K0	50	
34 β	05	59	31·7	+44	56	51	1·90	0·6	A2	22	Menkarlina
37 θ	05	59	43·2	+37	12	45	2·62	0·1	A0p	25	
35 π	05	59	56·1	+45	56	12	4·26	−2·4	M3	200	
44 κ	06	15	22·6	+29	29	53	4·35	0·3	G8	46	

Also above mag. 5:

	Mag.	Abs. mag.	Spectrum	Dist	
4 ω	4·94	0·6	A0	69	
2	4·78	−0·2	K3	190	
9	5·00	2·6	F0	30	
11 μ	4·86	1·8	A3	22	
21 σ	4·89	−0·3	K4	110	
25 χ	4·76	−6·3	B5	930	
30 ξ	4·99	0·8	A2		
46 ψ¹	4·91	−5·7	M0	600	Dolones

VARIABLE STARS

	R.A. h	m	Dec deg	min	Range	Type	Period, d.	Spectrum
RX	05	01·4	+39	58	7·3–8·0	Cepheid	11·62	F–G
ε	05	02·0	+43	49	2·9–3·8	Eclipsing	9892	F
ζ	05	02·5	+41	05	3·7–4·1	Eclipsing	972·1	K + B

AURIGA (continued)

	R.A.		Dec.		Range	Type	Period, d.	Spectrum
	h	m	deg	min				
R	05	17·3	+53	35	6·7–13·9	Mira	457·5	M
UV	05	21·8	+32	31	7·4–10·6	Mira	394·4	M
U	05	42·1	+32	02	7·5–15·5	Mira	408·1	M
X	06	12·2	+50	14	8·0–13·6	Mira	163·8	M
RT	06	28·6	+30	30	5·0–5·8	Cepheid	3·73	F – G
WW	06	32·5	+38	27	5·8–6·5	Algol	2·53	A + A
UU	06	36·5	+38	27	5·1–6·8	Semi-reg.	234	N

DOUBLE STARS

	R.A.		Dec.		P.A.	Sep.	Mags.
	h	m	deg	min	deg	sec	
ω	04	59·3	+37	53	359	5·4	5·0, 8·0
R	05	17·3	+53	35	339	47·5	var., 8·6
ν	05	51·5	+39	09	206	54·6	4·9, 9·3
δ	05	59·5	+54	17	AB 271	115·4	3·7, 9·5
					AC 067	197·1	9·5
θ	05	59·7	+37	13	AB 313	3·6	2·6, 7·1
					AC 297	50·0	10·6

OPEN CLUSTERS

M	NGC	R.A.		Dec.		Diameter	Mag.	No. of stars
		h	m	deg	min	min		
	1664	04	51·1	+43	42	18	7·6	–
	1778	05	08·1	+37	03	7	7·7	25
	1857	05	20·2	+39	21	6	7·0	40
	1893	05	22·7	+33	24	11	7·5	60
38	1912	05	28·7	+35	50	21	6·4	100
36	1960	05	36·1	+34	08	12	6·0	60
37	2099	05	52·4	+32	33	24	5·6	150
	2126	06	03·0	+49	54	6	10·2	40
	2281	06	49·3	+41	04	15	5·4	30

NEBULA

M	NGC	R.A.		Dec.		Dimensions	Mag. of illuminating star
		h	m	deg	min	min	
	IC 405	05	16·2	+34	16	30 × 19	6v AE Aurigæ; Flaming Star Nebula

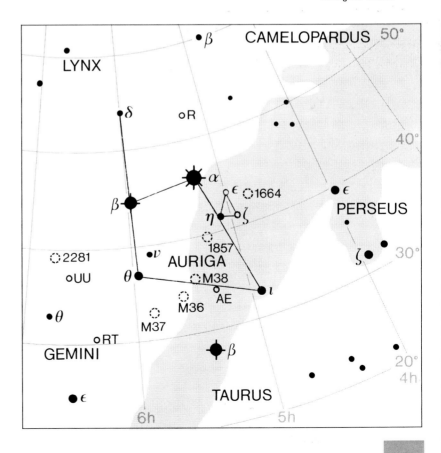

BOÖTES

(Abbreviation: Boö).

An original constellation, dominated by Arcturus. There are various myths attached to it, but none is very definite. According to one version, Boötes was a herdsman who invented the plough drawn by two oxen, for which service he was transferred to the sky.

There are eight stars above magnitude 4·00:

	Mag.	Luminosity Sun = 1	Dist lt-yrs
α	−0·04	115	36
ε	2·37	200	150
η	2·68	6·5	32
γ	3·03	53	104
δ	3·47	58	140
β	3·50	58	137
ρ	3·58	105	183
ζ	3·78	105	205

BRIGHTEST STARS

Star	R.A. h	m	s	Dec. deg	min	sec	Mag.	Abs. mag.	Spectrum	Dist pc	
4 τ	13	47	15·6	+17	27	24	4·50	3·8	F7	16	
5 υ	13	49	28·6	+15	47	52	4·06	−0·3	K5	72	
8 η	13	54	41·0	+18	23	51	2·68	2·7	G0	9·8	
17 κ	14	13	27·6	+51	47	15	4·40	1·5	A7	38	
16 α	14	15	39·6	+19	10	57	−0·04	−0·2	K2	11	Arcturus
21 ι	14	16	09·8	+51	22	02	4·75	2·4	A7	28	
19 λ	14	16	22·9	+46	05	18	4·18	1·8	A0p	29	
23 θ	14	25	11·7	+51	51	02	4·05	3·8	F7	13	
25 ρ	14	31	49·7	+30	22	17	3·58	−0·2	K3	56	
27 γ	14	32	04·6	+38	13	30	3·03	0·5	A7	32	Seginus
28 σ	14	34	40·7	+29	44	42	4·46	3·0	F2	17	
30 ζ	14	41	08·8	+13	43	42	3·78	−0·2	A2	63	
36 ε	14	44	59·1	+27	04	27	2·37	−0·9	K0	46	Izar
35 ο	14	45	14·4	+16	57	51	4·60	0·2	K0	29	
37 ξ	14	51	23·2	+19	06	04	4·55	5·5	G8	6·8	
42 β	15	01	56·6	+40	23	26	3·50	0·3	G8	42	Nekkar
43 ψ	15	04	26·6	+26	56	51	4·54	−0·1	K2	75	
49 δ	15	15	30·1	+33	18	53	3·47	0·3	G8	43	
51 μ¹	15	24	29·3	+37	22	38	4·31	2·6	F0	18	Alkalurops

Also above mag. 5:

	Mag.	Abs. mag.	Spectrum	Dist	
6	4·91	−0·3	K4	110	
12	4·83	2·4	F8	29	
20	4·86	−0·2	K3	100	
29 π¹	4·93	−0·3	B9	40	Alazal
31	4·86	0·3	G8	75	
34	4·81		gM0	30	
41 ω	4·81	−0·3	K4	91	
44	4·76	4·4	G0	12	
45	4·93	3·4	F5	19	
A	4·83	0·7	K1		

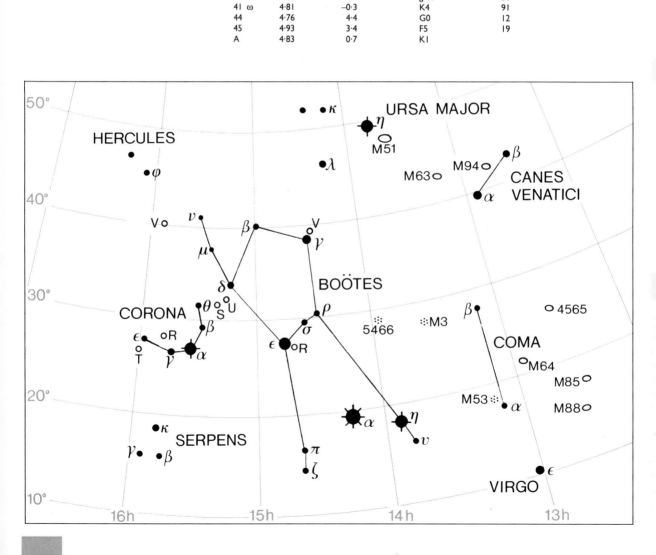

BOÖTES (continued)

VARIABLE STARS

	R.A. h	m	Dec. deg	min	Range	Type	Period, d.	Spectrum
S	14	22·9	+53	49	7·8–13·8	Mira	270·7	M
V	14	29·8	+38	52	7·0–12·0	Semi-reg.	258	M
R	14	37·2	+26	44	6·2–13·1	Mira	223·4	M
W	14	43·4	+26	32	4·7–5·4	Semi-reg.	450	M
i (44)	15	03·8	+47	39	6·5–7·1	W UMa	0·27	G + G

DOUBLE STARS

	R.A. h	m	Dec. deg	min	P.A. deg	Sep. sec	Mags.	
κ	14	13·5	+51	47	236	13·4	4·6, 6·6	
ι	14	16·2	+51	22	033	38·5	4·9, 7·5	
π	14	40·7	+16	25	108	5·6	4·9, 5·8	
ζ	14	41·1	+13	44	AB 303	1·0	4·5, 4·6	Binary, 123·3 y
					AC 259	99·3	10·9	
i (44)	15	03·8	+47	39	040	1·0	5·3v, 6·2	Binary, 225 y
μ	15	24·5	+37	23	171	108·3	4·3, 7·0	
ε	14	45·0	+27	04	339	2·8	2·5, 4·9	
ξ	14	51·4	+19	06	095	282	4·7, 6·8	Binary, 150 y

GLOBULAR CLUSTER

M	NGC	R.A. h	m	Dec. deg	min	Diameter sec	Mag.
	5466	14	05·5	+28	32	11	9·1

GALAXIES

M	NGC	R.A. h	m	Dec. deg	min	Mag.	Dimensions min	Type
	5248	13	37·5	+08	53	10·2	6·5 × 4·9	Sc
	5676	14	32·8	+49	28	10·9	3·9 × 2·0	Sc

CÆLUM

Abbreviation: Cae).

This entirely unremarkable constellation was introduced in 1752 by Lacaille, under the name of Cæla Sculptoris. It has no star brighter than the fourth magnitude, and only two which are brighter than the fifth.

See chart for Columba.

BRIGHTEST STARS

Star	R.A. h	m	s	Dec. deg	min	sec	Mag.	Abs. mag.	Spectrum	Dist pc
α	04	40	33·6	–41	51	50	4·45	3·0	F2	20
γ	05	04	24·3	–35	29	00	4·55	0·2	gK0	52

VARIABLE STARS

	R.A. h	m	Dec. deg	min	Range	Type	Period, d.	Spectrum
R	04	40·5	–38	14	6·7–13·7	Mira	390·9	M
T	04	47·3	–36	13	7·0–9·8	Semi-reg.	156	N

DOUBLE STARS

	R.A. h	m	Dec. deg	min	P.A. deg	Sep. sec	Mags.
α	04	40·6	–41	52	121	6·6	4·5, 12·5
γ	05	04·4	–35	29	308	2·9	4·6, 8·1

CAMELOPARDUS

Abbreviation: Cam).

A very barren northern constellation. It was introduced to the sky by Hevelius in 1690, and some historians have maintained that it represents the camel which carried Rebecca to Isaac! It is interesting to note that several of the apparently faint stars are in fact highly luminous and remote; for instance α Cam, which is below the fourth magnitude, is well over 20000 times more luminous than the Sun.

There are no stars in Camelopardus above the fourth magnitude.

See chart for Cassiopeia.

BRIGHTEST STARS

Star	R.A. h	m	s	Dec. deg	min	sec	Mag.	Abs. mag.	Spectrum	Dist pc
2 H	03	29	04·1	+59	56	25	4·21	–7·1	B9	1100
γ	03	50	21·5	+71	19	57	4·63	0·9	A3	56
9 α	04	54	03·0	+66	20	34	4·29	–6·2	O9·5	860
7	04	57	17·1	+53	45	08	4·47	1·2	A1	45
10 β	05	03	25·1	+60	26	32	4·03	–4·5	G0	460

Also above mag. 5:

	Mag.	Abs. mag.	Spectrum	Dist
22	4·80	0·6	A0	66

VARIABLE STARS

	R.A. h	m	Dec. deg	min	Range	Type	Period, d.	Spectrum
U	03	37·5	+62	29	7·7–8·9	Semi-reg.	400	N
RV	04	30·7	+57	25	7·1–8·2	Semi-reg.	101	M
T	04	40·1	+66	09	7·3–14·4	Mira	373·2	M
X	04	45·7	+75	06	7·4–14·2	Mira	143·6	K–M
S	05	41·0	+68	48	7·7–11·6	Semi-reg.	327	R
V	06	02·5	+74	30	7·7–16·0	Mira	522·4	M
VZ	07	31·1	+82	25	4·8–5·2	Semi-reg.	24	M
R	14	17·8	+83	50	7·0–14·4	Mira	270·2	S

CAMELOPARDUS
(continued)

OPEN CLUSTER

M	NGC	R.A.		Dec.		Diameter	Mag.	No. of stars
		h	m	deg	min	min		
	1502	04	07·7	+62	20	8	5·7	45

PLANETARY NEBULA

M	NGC	R.A.		Dec.		Diameter	Mag.	Mag. of central star
		h	m	deg	min	sec		
	IC 3568	12	32·9	+82	33	6	11·6	12·3

GALAXIES

M	NGC	R.A.		Dec.		Mag.	Dimensions	Type
		h	m	deg	min		min	
	IC 342	03	46·8	+68	06	9·2	17·8 × 17·4	SBc
	1961	05	42·1	+69	23	11·1	4·3 × 3·0	Sb
	2146	06	18·7	+78	21	10·5	6·0 × 3·8	SBb
	2366	07	28·9	+69	13	10·9	7·6 × 3·5	Irr.
	2403	07	36·9	+65	36	8·4	17·8 × 11·0	Sc
	2460	07	56·9	+60	21	11·7	2·9 × 2·2	Sb
	2655	08	55·6	+78	13	10·1	5·1 × 4·4	SBa
	2715	09	08·1	+78	05	11·4	5·0 × 1·9	Sc

CANCER

(Abbreviation: Cnc).

Cancer is an obscure constellation, redeemed only by the presence of two famous star-clusters, Præsepe and M.67. However, it lies in the Zodiac, and it has a legend attached to it. It represents a sea-crab which Juno, queen of Olympus, sent to the rescue of the multi-headed hydra which was doing battle with Hercules. Not surprisingly, Hercules trod on the crab, but as a reward for its efforts Juno placed it in the sky!

BRIGHTEST STARS

Star		R.A.			Dec.			Mag.	Abs.	Spectrum	Dist	
		h	m	s	deg	min	sec		mag.		pc	
16	ζ	08	12	12·6	+17	38	52	4·67	3·2, 4·1	F7, G2	16	Tegmine
17	β	08	16	30·9	+09	11	08	3·52	−0·3	K4	52	Altarf
43	γ	08	43	17·1	+21	28	06	4·66	1·2	A1	48	Asellus Borealis
47	δ	08	44	41·0	+18	09	15	3·94	0·2	K0	47	Asellus Australis
48	ι¹	08	46	41·8	+28	45	36	4·02	−2·1	G8	130	
65	α	08	58	29·2	+11	51	28	4·25	1·9	A3	23	Acubens

VARIABLE STARS

	R.A.		Dec.		Range	Type	Period, d.	Spectrum
	h	m	deg	min				
R	08	16·6	+11	44	6·1–11·8	Mira	361·6	M
V	08	21·7	+17	17	7·5–13·9	Mira	272·1	S
X	08	55·4	+17	14	5·6–7·5	Semi-reg.	195	N
T	08	56·7	+19	51	7·6–10·5	Semi-reg.	482	R–N

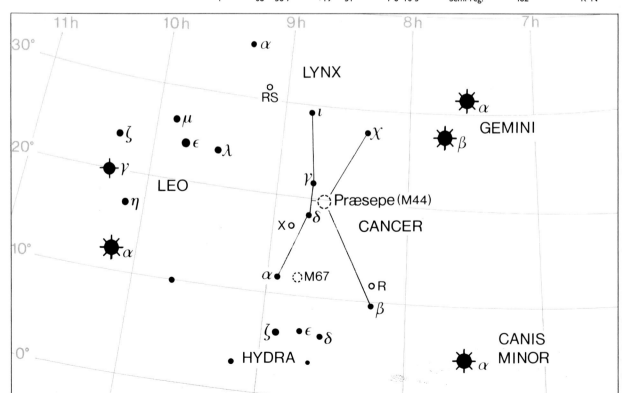

CANCER (continued)

There are only two stars above the fourth magnitude:

	Mag.	Luminosity Sun = 1	Dist lt-yrs
	3·52	110	170
	3·94	60	153

	R.A. h	m	Dec. deg	min	Range	Type	Period, d.	Spectrum
W	09	09·9	+25	15	7·4–14·4	Mira	393·2	M
RS	09	10·6	+30	58	6·2–7·7	Semi-reg.	120	M

DOUBLE STAR

	R.A. h	m	Dec. deg	min	P.A. deg	Sep. sec	Mags.	
ζ	08	12·2	+17	39	AB + C 088	5·7	5·0, 6·2	Binary, 1150 y
					AB 182	0·6	5·3, 6·0	Binary, 59·7 y
					AB + D 108	287·9	9·7	

OPEN CLUSTERS

M	NGC	R.A. h	m	Dec. deg	min	Diameter min	Mag.	No. of stars
44	2632	08	40·1	+19	59	95	3·1	50 Præsepe
67	2682	08	50·4	+11	49	30	6·9	200

GALAXY

M	NGC	R.A. h	m	Dec. deg	min	Mag.	Dimensions min	Type
	2775	09	10·3	+07	02	10·3	4·5 × 3·5	Sa

CANES VENATICI

(Abbreviation: CVn).

One of Hevelius' constellations, dating only from his maps of 1690, and evidently representing two hunting dogs (Asterion and Chara) which are being held by the herdsman Boötes – possibly to stop them from chasing the two Bears round and round the celestial pole. The name of Cor Caroli was given to α² CVn by Edmond Halley in honour of King Charles I. Cor Caroli (mag. 2·90) is the only star above the fourth magnitude. It is also the prototype 'magnetic variable'.

See chart for Ursa Major.

	Mag.	Luminosity Sun = 1	Dist lt-yrs
α²	2·90	75	65

BRIGHTEST STARS

Star	R.A. h	m	s	Dec. deg	min	sec	Mag.	Abs. mag.	Spectrum	Dist pc	
8 β	12	33	44·4	+41	21	26	4·26	4·5	G0	9·2	Chara
12 α²	12	56	01·6	+38	19	06	2·90	0·1	A0p	20	Cor Caroli
20	13	17	32·5	+40	34	21	4·73	−0·7	F0	52	
24	13	34	27·2	+49	00	57	4·70	1·9	A4	35	

Also above mag. 5:

	Mag.	Abs. mag.	Spectrum	Dist
5	4·80	0·3	G7	38
25	4·82	0·5	A7	73

VARIABLE STARS

	R.A. h	m	Dec. deg	min	Range	Type	Period, d.	Spectrum	
T	12	30·2	+31	30	7·6–12·6	Mira	290·1	M	
Y	12	45·1	+45	26	7·4–10·0	Semi-reg.	157	N	La Superba
U	12	47·3	+38	23	7·2–11·0	Mira	345·6	M	
TU	12	54·9	+47	12	5·6–6·6	Semi-reg.	50	M	
V	13	19·5	+45	32	6·5–8·6	Semi-reg.	192	M	
R	13	49·0	+39	33	6·5–12·9	Mira	328·5	M	

DOUBLE STAR

	R.A. h	m	Dec. deg	min	P.A. deg	Sep. sec	Mags.
α²	12	56·0	+38	19	22·9	19·4	2·9, 5·5

GLOBULAR CLUSTER

M	NGC	R.A. h	m	Dec. deg	min	Diameter min	Mag.
3	5272	13	42·2	+28	23	16·2	6·4

GALAXIES

M	NGC	R.A. h	m	Dec. deg	min	Mag.	Dimensions min	Type
	4111	12	07·1	+43	04	10·8	4·8 × 1·1	S0
	4138	12	09·5	+43	41	12·3	2·9 × 1·9	E4
	4145	12	10·0	+39	53	11·0	5·8 × 4·4	Sc
	4151	12	10·5	+39	24	10·4	5·9 × 4·4	Pec. (Strong UV source)
	4618	12	41·5	+41	09	10·8	4·4 × 3·8	Sc
	4214	12	15·6	+36	20	9·8	7·9 × 6·3	Irr.
	4217	12	15·8	+47	06	11·9	5·5 × 1·8	Sb
	4242	12	17·5	+45	37	11·0	4·8 × 3·8	S
106	4258	12	19·0	+47	18	8·3	18·2 × 7·9	Sb
	4395	12	25·8	+33	33	10·1	12·9 × 11·0	S
	4449	12	28·2	+44	06	9·4	5·1 × 3·7	Irr.
	4490	12	30·6	+41	38	9·8	5·9 × 3·1	Sc
	4631	12	42·1	+32	32	9·3	15·1 × 3·3	Sc
	4656–7	12	44·0	+32	10	10·4	13·8 × 3·3	Sc
94	4736	12	50·9	+41	07	8·2	11·0 × 9·1	Sb
	5005	13	10·9	+37	03	9·8	5·4 × 2·7	Sb
	5033	13	13·4	+36	36	10·1	10·5 × 5·6	Sb
63	5055	13	15·8	+42	02	8·6	12·3 × 7·6	Sb
	5112	13	21·9	+38	44	11·9	3·9 × 2·9	Sc

CANES VENATICI
(continued)

M	NGC	R.A.		Dec.		Mag.	Dimensions	Type	
		h	m	deg	min		min		
51	5194	13	29·9	+47	12	8·4	11·0 × 7·8	Sc	Whirlpool
	5195	13	30·0	+47	16	9·6	5·4 × 4·3	Pec.	Companion to M.51
	5371	13	55·7	+40	28	10·7	4·4 × 3·6	Sb	

CANIS MAJOR

(Abbreviation: CMa).

An original constellation, representing one of Orion's hunting dogs. Though dominated by Sirius, it contains several other bright stars, of which one (Adhara) is only just below the first magnitude. Altogether there are 11 stars brighter than the fourth magnitude:

	Mag.	Luminosity Sun = 1	Dist lt-yrs
α	−1·46	26	8·6
ε	1·50	5000	490
δ	1·86	132 000	3060
β	1·98v	7200	720
η	2·44	52 500	2500
ζ	3·02	450	290
o²	3·03	43 000	2800
o¹	3·86	21 000	1700
ω	3·86	700	550
ν²	3·95	82	235
κ	3·96	800	650

It is interesting to note that three of these stars – Wezea, Aludra and o² – are extremely luminous. Compared with them, Sirius is extremely feeble, as it has only 26 times the Sun's luminosity.

BRIGHTEST STARS

Star		R.A.			Dec.			Mag.	Abs. mag.	Spectrum	Dist pc	
		h	m	s	deg	min	sec					
1	ζ	06	20	18·7	−30	03	48	3·02	−1·7	B3	88	Phurad
2	β	06	22	41·9	−17	57	22	1·98v	−4·8	B1	220	Mirzam
	λ	06	28	10·1	−32	34	48	4·48	−0·5	B5	17	
4	ξ¹	06	31	51·2	−23	25	06	4·34	−3·9	B1	440	
5	ξ²	06	35	03·3	−22	57	53	4·54	0·6	A0	61	
7	ν²	06	36	41·0	−19	15	22	3·95	0·0	K1	72	
8	ν³	06	37	53·3	−18	14	15	4·43	0·0	K1	72	
9	α	06	45	08·9	−16	42	58	−1·46	1·4	A1	2·7	Sirius
13	κ	06	49	50·4	−32	30	31	3·96	−2·5	B2	200	
16	o¹	06	54	07·8	−24	11	02	3·86	−6·0	K3	520	
14	θ	06	54	11·3	−12	02	19	4·07	−0·3	K4	73	
19	π	06	55	37·3	−20	08	11	4·68	0·6	gF2	65	
20	ι	06	56	08·1	−17	03	14	4·38	−3·9	B3	410	
21	ε	06	58	37·5	−28	58	20	1·50	−4·4	B2	150	Adhara
22	σ	07	01	43·1	−27	56	06	3·46	−5·7	M0	460	
24	o²	07	03	01·4	−23	50	00	3·03	−6·8	B3	860	
23	γ	07	03	45·4	−15	38	00	4·11	−3·4	B8	320	Muliphen
25	δ	07	08	23·4	−26	23	36	1·86	−8·0	F8	940	Wezea
27		07	14	15·1	−26	21	09	4·66	−2·9	B3	320	
28	ω	07	14	48·6	−26	46	22	3·86	−2·3	B3	170	
29	UW	07	18	40·3	−24	33	32	4·98v		O7·8		
30	τ	07	18	41·4	−24	57	15	4·39	−6·0	O9	1100	
31	η	07	24	05·6	−29	18	11	2·44	−7·0	B5	760	Aludra

Sirius has a white dwarf companion, mag. 8; binary, period 50 years. Though the companion does not appear fai... it is difficult to see because of the overpowering glare of the primary. The separation ranges between 3 sec... 11 sec ·5. It was widest in 1975, so that it is now closing up. β (Mirzam) is variable over a very small range, and... also a spectrum variable. ξ¹ is also a spectrum variable, and optically variable over a very small range of less than 0... magnitude.

Also above mag. 5:

		Mag.	Abs. mag.	Spectrum	Dist
15	EY	4·82	−3·9	B1	560
29		4·98v		O7·8	

VARIABLE STARS

	R.A.		Dec.		Range	Type	Period, d.	Spectrum
	h	m	deg	min				
W	07	08·1	−11	55	6·4−7·9	Irr.	−	N
RY	07	16·6	−11	29	7·7−8·5	Cepheid	4·68	F−G
UW	07	18·4	−24	34	4·0−5·3	Beta Lyræ	4·39	O7
R	07	19·5	−16	24	5·7−6·3	Algol	1·14	F

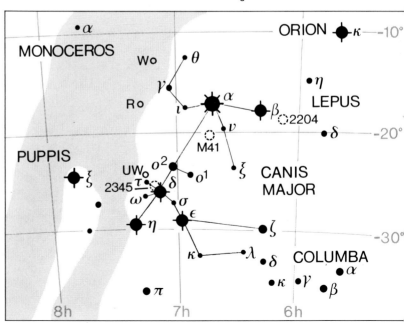

CANIS MAJOR
(continued)

DOUBLE STARS

	R.A.		Dec.		P.A.	Sep.	Mags.	
	h	m	deg	min	deg	sec		
ν¹	06	36·4	−18	40	262	17·5	5·8, 8·5	
α	06	45·1	−16	43	005	4·5	−1·5, 8·3	Binary, 50 y
17	06	55·0	−20	24	AB 147	44·4	5·8, 9·3	
					AC 184	50·5	9·0	
					AD 186	129·9	9·5	
π	06	55·6	−20	08	018	11·6	4·7, 9·7	
μ	06	56·1	−14	03	AB 340	3·0	5·3, 8·6	Isis
					AC 288	88·4	10·5	
					AD 061	101·3	10·7	
ε	06	58·6	−28	58	161	7·5	1·5, 7·4	

OPEN CLUSTERS

M	NGC	R.A.		Dec.		Diameter	Mag.	No. of stars	
		h	m	deg	min	min			
	2204	06	15·7	−18	39	13	8·6	80	
41	2287	06	47·0	−20	44	38	4·5	80	
	2345	07	08·3	−13	10	12	7·7	20	
	2360	07	17·8	−15	37	13	7·2	80	
	2362	07	17·8	−24	57	8	4·1	60	τ CMa

GALAXIES

M	NGC	R.A.		Dec.		Mag.	Dimensions	Type
		h	m	deg	min	min		
	2207	06	16·4	−21	22	10·7	4·3 × 2·9	Sc
	2217	06	21·7	−27	14	10·4	4·8 × 4·4	SBa
	2223	06	24·6	−22	50	11·4	3·3 × 3·0	SBb
	2280	06	44·8	−27	38	11·8	5·6 × 3·2	Sb

CANIS MINOR

(Abbreviation: CMi)

The second of Orion's two dogs. There are two stars above the fourth magnitude:

	Mag.	Luminosity Sun = 1	Dist lt-yrs
α	0·38	7	11·4
β	2·90	105	137

BRIGHTEST STARS

Star	R.A.			Dec.			Mag.	Abs. mag.	Spectrum	Dist pc	
	h	m	s	deg	min	sec					
3 β	07	27	09·0	+08	17	21	2·90	−0·2	B8	42	Gomeisa
4 γ	07	28	09·7	+08	55	33	4·32	−0·2	K3	65	
6	07	29	47·7	+12	00	24	4·54	−0·1	K2	70	
10 α	07	39	18·1	+05	13	30	0·38	2·6	F5	3·5	Procyon
HD 66141	08	02	15·8	+02	20	04	4·39	−0·1	K2	70	

Procyon has a white dwarf companion; magnitude 13, period 40 years. It is very difficult to observe because of the glare from the primary.

Also above mag. 5:

	Mag.	Abs. mag.	Spectrum	Dist
2 ε	4·99	0·3	G8	78

VARIABLE STARS

	R.A.		Dec.		Range	Type	Period, d.	Spectrum
	h	m	deg	min				
V	07	07·0	+08	53	7·4–15·1	Mira	366·1	M
R	07	08·7	+10	01	7·3–11·6	Mira	337·8	S
S	07	32·7	+08	19	6·6–13·2	Mira	332·9	M

DOUBLE STAR

	R.A.		Dec.		P.A.	Sep.	Mags.	
	h	m	deg	min	deg	sec		
α	07	39·3	+05	14	021	5·2	0·4, 12·9	Binary, 40·7 y

CAPRICORNUS

(Abbreviation: Cap).

A Zodiacal constellation, though by no means brilliant. It has been identified with the demigod Pan, but there seems to be no well-defined mythological associations. There are five stars above magnitude 4:

	Mag.	Luminosity Sun = 1	Dist lt-yrs
δ	2·87v	13	49
β	3·08	2	104
α²	3·57	60	117
γ	3·68	28	65
ζ	3·74	5200	1470

α¹ and α² make up a naked-eye pair, separated by 376 sec (position angle 291 deg), but the two components are not genuinely associated, and each is itself double; the fainter component of α² is also double!

BRIGHTEST STARS

Star	R.A. h	m	s	Dec. deg	min	sec	Mag.	Abs. mag.	Spectrum	Dist pc	
5 α¹	20	17	38·6	−12	30	30	4·24	−4·5	G3	490	Al Geidi
6 α²	20	18	03·1	−12	32	42	3·57	0·2	G9	36	
9 β	20	21	00·5	−14	46	53	3·08	4·0	F8	32	Dabih
16 ψ	20	46	05·5	−25	16	16	4·14	3·4	F5	12	Wei
18 ω	20	51	49·1	−26	55	09	4·11	−0·3	K5	64	
23 θ	21	05	56·6	−17	13	58	4·07	0·6	A0	49	
24 A	21	07	07·5	−25	00	21	4·50	−0·5	M1	95	
32 ι	21	22	14·6	−16	50	05	4·28	0·3	G8	63	
34 ζ	21	26	39·9	−22	24	41	3·74	−4·5	G4	450	Yen
36 b	21	28	43·2	−21	48	26	4·51	0·2	gG5	67	
39 ε	21	37	04·7	−19	27	58	4·68	−2·3	B3	250	
40 γ	21	40	05·2	−16	39	45	3·68	1·2	F0p	18	Nashira
43 κ	21	42	39·3	−18	51	59	4·73	0·3	G8	77	
49 δ	21	47	02·3	−16	07	38	2·87v	2·0	A5	15	Deneb al Giedi

Also above mag. 5:

	Mag.	Abs. mag.	Spectrum	Dist	
8 ν	4·76	−0·3	B9	99	Alshat
11 ρ	4·78	0·6	F2	32	
22 η	4·84	2·6	A4	12	Chow

VARIABLE STARS

	R.A. h	m	Dec. deg	min	Range	Type	Period, d.	Spectrum
RT	20	17·1	−21	19	6·5–8·1	Semi-reg.	393	N
RR	21	02·3	−27	05	7·8–15·5	Mira	277·5	M
RS	21	07·2	−16	25	7·0–9·0	Semi-reg.	340	M

DOUBLE STARS

	R.A. h	m	Dec. deg	min	P.A. deg	Sep. sec	Mags.	
α	20	18·1	−12	33	291	377·7	3·6, 4·2	(α¹–α²)
α¹	20	17·6	−12	30	AB 182	44·3	4·2, 13·7	
					AC 221	45·4	9·2	
α²	20	18·1	−12	33	AB 172	6·6	3·6, 11·0	
					AD 156	154·6	9·3	
					BC 240	1·2	11·3	
σ	20	19·6	−19	07	179	55·9	5·5, 9·0	
β	20	21·0	−14	47	267	205	3·1, 6·0	B is double
π	20	27·3	−18	13	148	3·2	5·3, 8·9	
ρ	20	28·9	−17	49	158	0·5	5·0, 10·0	
ε	21	37·1	−19	28	047	68·1	4·7, 9·5	
τ	20	39·3	−14	57	118	0·3	5·8, 6·3	Binary, 200 y

GLOBULAR CLUSTER

M	NGC	R.A. h	m	Dec. deg	min	Diameter min	Mag.
30	7099	21	40·4	−23	11	11·0	7·5

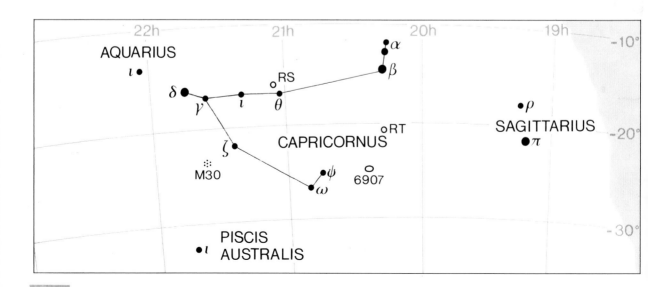

CAPRICORNUS
(continued)

GALAXY

M	NGC	R.A. h m	Dec. deg min	Mag.	Dimensions min	Type
	6907	20 25·1	−24 49	11·3	3·4 × 3·0	SBb

CARINA

Abbreviation: Car).

The keel of the ship Argo, in which Jason and his companions sailed upon their successful if somewhat unprincipled expedition to remove the Golden Fleece of the sacred ram (Aries) from its grove. Carina is the brightest part of Argo, and contains Canopus, which is second only to Sirius in brilliance. Canopus is a highly luminous star, but estimates of its power and distance vary considerably. A recent estimate makes it over 200 000 times as luminous as the Sun; other estimates are much lower. Also in this constellation is η Carinæ, which during part of the last century rivalled Sirius, but which is now well below naked-eye visibility. It is unique, and at its peak may have been the most luminous star in the Galaxy. Excluding η, there are 15 stars in Carina above the fourth magnitude:

	Mag.	Luminosity Sun = 1	Dist lt-yrs
α	−0·72	200 000	1200
β	1·68	130	85
ε	1·86	600	200
ι	2·25	7500	800
θ	2·76	3800	750
υ	2·97	520	320
ρ	3·32v	450	310
ω	3·32	200	230
q	3·40	5000	900
a	3·44	1300	620
χ	3·47	1300	590
u	3·78	17	85
R	3·8–10	var	800
c	3·84	200	300
x	3·91	132 000	4600

BRIGHTEST STARS

Star	R.A. h m s	Dec. deg min sec	Mag.	Abs. mag.	Spectrum	Dist pc	
α	06 23 57·1	−52 41 44	−0·72	−8·5	F0	360	Canopus
N	06 34 58·5	−52 58 32	4·39	−0·8	B9	100	
A	06 49 51·3	−53 37 20	4·40	0·4	gG3	52	
χ	07 56 46·7	−52 58 56	3·47	−3·0	B2	180	
d	08 40 37·0	−59 45 40	4·33	−3·6	B2	350	
ε	08 22 30·8	−59 30 34	1·86	−2·1	K0	62	Avior
f	08 46 42·6	−56 46 11	4·49	−2·5	B2	250	
c	08 55 02·8	−60 38 41	3·84	−1·0	B8	93	
G	09 05 38·4	−70 32 20	4·71	−2·5	B2	280	
a	09 10 57·9	−58 58 01	3·44	−3·0	B2	190	
i	09 11 16·7	−62 19 02	3·97	−2·3	B3	180	
β	09 13 12·2	−69 43 02	1·68	−0·6	A0	26	Miaplacidus
g	09 16 12·2	−57 32 28	4·34	−0·3	gK5	70	
ι	09 17 05·4	−59 16 31	2·25	−4·7	F0	250	Tureis (or Aspidiske)
R	09 32 14·7	−62 47 19	3·8–10	var.	M5	250	
h	09 34 26·6	−59 13 46	4·08	−3·7	B5	350	
m	09 39 20·9	−61 19 40	4·52	0·2	B9	73	
l(ZZ)	09 45 14·8	−62 30 28	4·1v	4·4	G0	16	
υ	09 47 06·1	−65 04 18	2·97	−2·0	A7	99	
ω	10 13 44·3	−70 02 16	3·32	−1·0	B7	70	
q	10 17 04·9	−61 19 56	3·40	−4·4	K5	280	
I	10 24 23·7	−74 01 54	4·00	0·6	F2	17	
s	10 27 24·4	−57 38 19	4·68	−8·5	F0	3600	
ρ	10 32 01·4	−61 41 07	3·32v	−1·7	B3	96	
r	10 35 35·2	−57 33 27	4·45	−0·4	gM0	92	
t²	10 38 45·0	−59 10 58	4·66	−5·9	cK	1300	
θ	10 42 57·4	−64 23 39	2·76	−4·1	B0	230	
W	10 43 32·1	−60 33 59	4·57	−0·3	K5	75	
u	10 53 29·6	−58 51 12	3·78	1·7	K0	26	
x	11 08 35·3	−58 58 30	3·91	−8·0	G0	1400	
y	11 12 36·0	−60 19 03	4·60	−8·5	F0	3300	

Also above mag. 5:

	Mag.	Abs. mag.	Spectrum
Q	4·92	−0·8	M0
D¹	4·96	−2·4	B3
B	4·80	3·5	F5
e²	4·80	0·2	G6
p	4·94		F1
k	4·87	1·5	G4
K	4·94	−0·5	A2
z¹	4·76	1·1	G5

VARIABLE STARS

	R.A. h m	Dec. deg min	Range	Type	Period, d.	Spectrum	
V	08 28·7	−60 07	7·1–7·8	Cepheid	6·70	F–H	
X	08 31·3	−59 14	7·9–8·6	Beta Lyræ	1·08	A + A	
R	09 32·2	−62 47	3·9–10·5	Mira	308·7	M	
ZZ	09 45·2	−62 30	3·3–4·2	Cepheid	35·53	F–K	(l Carinæ)
S	10 09·4	−61 33	4·5–9·9	Mira	149·5	K–M	
η	10 45·1	−59 41	−0·8–7·9	Irreg.	—	Pec.	
BO	10 45·8	−59 29	7·2–8·5	Irreg.	—	M	
U	10 57·8	−59 44	5·7–7·0	Cepheid	38·77	F–G	

DOUBLE STARS

	R.A. h m	Dec. deg min	P.A. deg	Sep. sec	Mags.
η	10 45·1	−59 41	195	0·2	var., 8·6
u	09 47·1	−65 04	127	5·0	3·1, 6·1

OPEN CLUSTERS

M	NGC	R.A. h m	Dec. deg min	Diameter min	Mag.	No. of stars	
	2516	07 58·3	−60 52	30	3·8	80	
	3114	10 02·7	−60 07	35	4·2	—	
	IC 2581	10 27·4	−57 38	8	4·3	25	
	IC 2602	10 43·2	−64 24	50	1·9	60	θ Carinæ cluster
	3532	11 06·4	−58 40	55	3·0	150	

CARINA (continued)

M	NGC	R.A. h	m	Dec. deg	min	Diameter min	Mag.	No. of stars
	Mel 101	10	42·1	−65	06	14	8·0	50
	3572	11	10·4	−60	14	7	6·6	35
	3590	11	12·9	−60	47	4	8·2	25
	Mel 105	11	19·5	−63	30	4	8·5	70
	IC 2714	11	17·9	−62	42	12	8·2	100
	3680	11	25·7	−43	15	12	7·6	30

GLOBULAR CLUSTER

M	NGC	R.A. h	m	Dec. deg	min	Diameter min	Mag
	2808	09	12·0	−64	52	13·8	6·3

NEBULA

M	NGC	R.A. h	m	Dec. deg	min	Mag.	Mag. of illuminating star
	3372	10	43·8	−59	52	6·2	var. η Carinæ nebula

PLANETARY NEBULÆ

M	NGC	R.A. h	m	Dec. deg	min	Diameter sec	Mag.	Mag. of central star
	IC 2448	09	07·1	−69	57	8	11·5	12·9
	2867	09	21·4	−58	19	11	9·7	13·6
	IC 2501	09	38·8	−60	05	25	11·3	−
	3211	10	17·8	−62	40	12	11·8	−
	IC 2621	11	00·3	−65	15	5	−	13·6

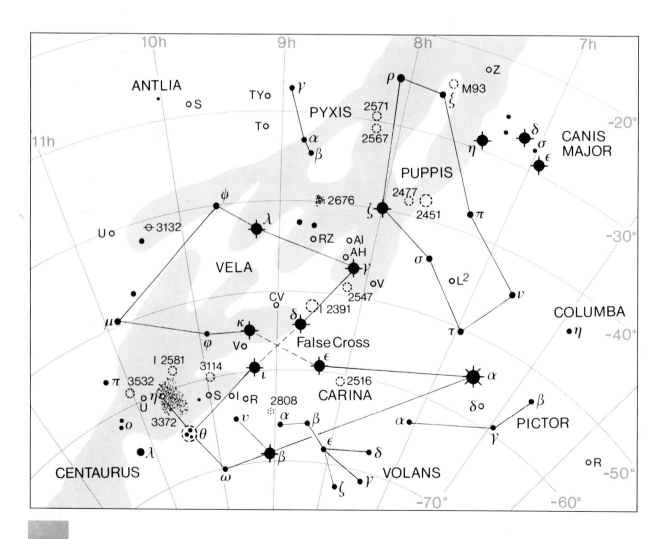

CASSIOPEIA

(Abbreviation: Cas).

An original constellation – one of the most distinctive in the sky. Mythologically, Cassiopeia was Andromeda's mother and wife of Cepheus; it was her boasting which led to the unfortunate contretemps with Neptune's sea-monster.

There are seven stars above magnitude 4. Of these, γ (Cih) is variable, and it is also suspected that α (Shedir) is variable over a small range – possibly 2·1 to 2·4, though opinions differ. κ, which looks obscure, is extremely luminous.

There are 8 stars brighter than magnitude 4:

	Mag.	Luminosity	Dist
		Sun = 1	lt-yrs
γ	2·2v	6000v	780
α	2·23v?	200	120
β	2·27	14	42
δ	2·68	11	62
ε	3·38	1200	520
η	3·44	1·2	19
ζ	3·67	830	550
ι	3·98	28	117

BRIGHTEST STARS

Star	R.A. h	m	s	Dec. deg	min	sec	Mag.	Abs. mag.	Spectrum	Dist pc	
7 ρ	23	54	22·9	+57	29	58	var.	−8·0	F8	1500	
11 β	00	09	10·6	+59	08	59	2·27	1·9	F2	13	Chaph
14 λ	00	31	46·3	+54	31	20	4·73	−0·9	B8	13	
15 κ	00	32	59·9	+62	55	55	4·16	−6·6	B1	930	
17 ζ	00	36	58·2	+53	53	49	3·67	−2·5	B2	170	
18 α	00	40	30·4	+56	32	15	2·23v?	−0·9	K0	37	Shedir
22 o	00	44	43·4	+48	17	04	4·54	−2·5	B2	210	
24 η	00	49	05·9	+57	48	58	3·44	4·6	G0	5·9	Achird
28 υ²	00	56	39·7	+59	10	52	4·63	1·8	G8	29	
27 γ	00	56	42·4	+60	43	00	2·2v	−4·6	B0p	240	Cih
33 θ	01	11	06·1	+55	09	00	4·33	2·4	A7	37	Marfak
37 δ	01	25	48·9	+60	14	07	2·68	2·1	A5	19	Ruchbah
36 ψ	01	25	55·9	+68	07	48	4·74	0·2	K0	74	
39 χ	01	33	55·8	+59	13	56	4·71	0·2	K0	80	
45 ε	01	54	23·6	+63	40	13	3·38	−2·9	B3	160	Segin
48	02	01	57·3	+70	54	26	4·48	1·9	A4	33	
50 ι	02	03	26·0	+72	25	17	3·98	1·2	A1	36	

Also above mag. 5:

	Mag.	Abs. mag.	Spectrum	Dist	
1	4·85	−4·4	B1	570	
5 τ	4·87	0·0	K1	92	
4	4·97	−0·3	K5	86	
8 σ	4·88	−3·5	B1	400	
10 ξ	4·80	−2·5	B2	260	
25 ν	4·89		B9		
26 υ¹	4·83	−0·1	K2	91	Castula
46 ω	4·99		B8		

VARIABLE STARS

	R.A. h	m	Dec. deg	min	Range	Type	Period, d.	Spectrum
V	23	11·7	+59	42	6·9–13·4	Mira	228·8	M
ρ	23	54·4	+58	30	4·1–6·2	?	–	F–K
R	23	58·4	+51	24	4·7–13·5	Mira	430·5	M
T	00	23·2	+55	48	6·9–13·0	Mira	444·8	M
TU	00	26·3	+51	17	6·9–8·1	Cepheid	2·14	F
α	00	40·5	+56	22	?2·1–2·5	Suspected	–	K
U	00	46·4	+48	15	8·0–15·7	Mira	277·2	S
RV	00	52·7	+47	25	7·3–16·1	Mira	331·7	M
W	00	54·9	+58	34	7·8–12·5	Mira	405·6	N
γ	00	56·7	+60	43	1·6–3·3	Irreg.	–	B
S	01	19·7	+72	37	7·9–16·1	Mira	612·4	S
SU	02	52·0	+68	53	5·7–6·2	Cepheid	1·95	F
RZ	02	48·9	+69	38	6·2–7·7	Algol	1·19	A

DOUBLE STARS

	R.A. h	m	Dec. deg	min	P.A. deg	Sep. sec	Mags.	
λ	00	31·8	+54	31	176	0·5	5·3, 5·6	
η	00	49·1	+57	49	293	12·2	3·4, 7·5	Binary, 480 y
ψ	01	25·9	+68	08	113	25·0	4·7, 9·6	
ι	02	29·1	+67	24	232	2·4	4·6, 6·9	Binary, 840 y
σ	23	59·0	+55	45	326	3·0	5·0, 7·1	

OPEN CLUSTERS

M	NGC	R.A. h	m	Dec. deg	min	Diameter min	Mag.	No. of stars	
52	7654	23	24·2	+61	35	13	6·9	100	
	7788	23	56·7	+61	24	9	9·4	20	
	7789	23	57·0	+56	44	16	6·7	300	
	H.21	23	54·1	+61	46	4	9·0	6	
	129	00	29·9	+60	14	21	6·5	35	Contains DL Cas.
	133	00	31·2	+63	22	7	9·4	5	
	146	00	33·1	+63	18	7	9·1	20	
	381	01	08·3	+61	35	6	9·3	50	
	436	01	15·6	+58	49	6	8·8	30	
	457	01	19·1	+58	20	13	6·4	80	φ Cas cluster.
	559	01	29·5	+63	18	4·4	9·5	60	
	IC 1805	02	32·7	+61	27	22	6·5	40	
103	581	01	33·2	+60	42	6	7·4	25	
	637	01	42·9	+64	00	3·5	8·2	20	
	1027	02	42·7	+61	33	20	6·7	40	
	654	01	44·1	+61	53	5	6·5	60	
	659	01	44·2	+60	42	5	7·9	40	
	663	01	46·0	+61	15	16	7·1	80	

CASSIOPEIA (continued)

NEBULÆ

M	NGC	R.A.		Dec.		Dimensions	Mag. of illuminating star
		h	m	deg	min	min	
	7635	23	20·7	+61	12	15 × 8	7 Bubble Nebula
	281	00	52·8	+56	36	35 × 30	8
	IC 1805	02	33·4	+61	26	60 × 60	–
	IC 1848	02	51·3	+60	25	60 × 30	–

GALAXIES

M	NGC	R.A.		Dec.		Mag.	Dimensions	Type
		h	m	deg	min		min	
	147	00	33·2	+48	30	9·3	12·9 × 8·1	dE4
	185	00	39·0	+48	20	9·2	11·5 × 9·8	dE0

CENTAURUS

(Abbreviation: Cen).

A brilliant southern constellation (one of Ptolemy's originals), containing many important objects, including the nearest of all the bright stars (α) and the superb globular cluster ω. There are 19 stars brighter than the fourth magnitude.

	Mag.	Luminosity	Dist
		Sun = 1	lt-yrs
α	−0·27	1·7+0·45	4·3
β	0·61	10 000	460
θ	2·06	17	46
γ	2·17	130	110
ε	2·30	2100	490
η	2·31v	1200	360
ζ	2·55	1300	360
δ	2·60v	830	325
ι	2·75	26	52
μ	3·04v	450	300
κ	3·13	830	420
λ	3·13	180	190
ν	3·41	2600	1000
φ	3·83	830	620
τ	3·86	26	98
υ¹	3·87	700	550
d	3·88	60	120
π	3·89	230	323
σ	3·91	450	425

BRIGHTEST STARS

Star	R.A.			Dec.			Mag.	Abs.	Spectrum	Dist	
	h	m	s	deg	min	sec		mag.		pc	
π	11	21	00·4	−54	29	27	3·89	−1·1	B5	99	
λ	11	35	46·8	−63	01	11	3·13	−0·8	B9	57	
65G	11	46	30·7	−61	10	12	4·11	−2·0	G0	140	
j	11	49	41·0	−45	10	25	4·32	−1·1	B5	120	
B	11	51	08·5	−45	43	05	4·46	−0·3	K4	90	
δ	12	08	21·5	−50	43	20	2·60v	−2·5	B2	100	
ρ	12	11	39·1	−52	22	07	3·96	−1·4	B4	120	
τ	12	37	42·1	−48	32	28	3·86	1·4	A2	30	
σ	12	28	02·4	−50	13	51	3·91	−1·7	B3	130	
γ	12	41	30·9	−48	57	34	2·17	−0·6	A0	34	Menkent
w	12	42	35·3	−48	48	47	4·66	0·0	gK1	86	
e	12	53	06·8	−48	56	35	4·33	−0·1	gK2	54	
n	12	53	26·1	−40	10	44	4·27	0·5	A7	57	
ξ²	13	06	54·5	−49	54	22	4·27	−3·0	B2	280	
ι	13	20	35·8	−36	42	44	2·75	1·4	A2	16	
J	13	22	37·8	−60	59	18	4·53	−1·1	B5	130	
m	13	24	0·05	−64	32	09	4·53	1·8	G5	35	
d	13	31	02·6	−39	24	27	3·88	0·3	G8	36	
ε	13	39	53·2	−53	27	58	2·30	−3·5	B1	150	
I	13	45	43·0	−33	02	30	4·23	0·8	F2	28	
M	13	46	39·3	−51	25	58	4·65	0·2	gK0	78	
2	13	49	26·6	−34	27	03	4·19	−0·5	M1	87	
ν	13	49	30·2	−41	41	16	3·41	−2·5	B2	150	
μ	13	49	36·9	−42	28	25	3·04	−1·7	B3	89	
3	13	51	50·0	−32	59	41	4·32	−1·6	B5	91	
4	13	53	12·4	−31	55	39	4·73	−1·6	B5	190	
ζ	13	55	32·3	−47	17	17	2·55	−3·0	B2	110	Al Nair al Kentaurus
294G	13	57	38·9	−63	41	11	4·71	−0·3	K4	100	
φ	13	58	16·2	−42	06	02	3·83	−2·5	B2	190	
υ¹	13	58	40·7	−44	48	13	3·87	−2·3	B3	170	
υ²	14	01	43·3	−45	36	12	4·34	−2·0	F5	160	
β	14	03	49·4	−60	22	22	0·61	−5·1	B1	140	Agena

CENTAURUS (continued)

Star	R.A. h	m	s	Dec. deg	min	sec	Mag.	Abs. mag.	Spectrum	Dist pc	
χ	14	16	02·7	−41	10	47	4·36	−2·5	B2	240	
5 θ	14	06	40·9	−36	22	12	2·06	1·7	K0	14	Haratan
ν	14	20	19·4	−56	23	12	4·33	−3·7	B5	310	
ψ	14	20	33·3	−37	53	07	4·05	0·0	A0	65	
a	14	23	02·1	−39	09	34	4·44	−1·9	B6	180	
η	14	35	30·3	−42	09	28	2·31v	−2·9	B3	110	
α²	14	39	35·4	−60	50	13	1·39	5·7	K1	1·3 }	
α¹	14	39	36·7	−60	50	02	0·00	4·4	G2	1·3 }	
b	14	41	57·5	−37	47	37	4·00	−1·7	B3	140	
c¹	14	43	39·3	−35	10	25	4·05	−0·3	K5	71	
κ	14	59	09·6	−42	06	15	3·13	−2·5	B2	130	Ke Kwan

Strangely, α has no official proper name. It has been called Al Rijil, and also Toliman, but astronomers in general prefer to call it simply α Centauri.

Also above mag. 5:

	Mag.	Abs. mag.	Spectrum	Dist
	4·85	0·6	A0	67
l	4·79	0·5	B8p	230
o¹	4·96	1·1	G4	
4 n	4·76	−1·3	B7	
f	4·96	−2·1	B3	

VARIABLE STARS

	R.A. h	m	Dec. deg	min	Range	Type	Period, d.	Spectrum
RS	11	20·5	−61	52	7·7–14·1	Mira	164·4	M
X	11	49·2	−41	45	7·0–13·8	Mira	315·1	M
W	11	55·0	−59	15	7·6–13·7	Mira	201·6	M
S	12	24·6	−49	26	6·0–7·0	Semi-reg.	65	N
U	12	33·5	−54	40	7·0–14·0	Mira	220·3	M
RV	13	37·5	−56	29	7·0–10·8	Mira	446·0	N
XX	13	40·3	−57	37	7·3–8·3	Cepheid	10·95	F–G
T	13	41·8	−33	36	5·5–9·0	Semi-reg.	60	K–M
μ	13	49·6	−42	28	2·9–3·5	Irr.	–	B
R	14	16·6	−59	55	5·3–11·8	Mira	546·2	M
V	14	32·5	−56	53	6·4–7·2	Cepheid	5·49	F–G

DOUBLE STARS

	R.A. h	m	Dec. deg	min	P.A. deg	Sep. sec	Mags.	
γ	12	41·5	−48	58	353	1·4	2·9, 2·9	Binary, 84·5 y
ε	13	39·9	−53	28	158	36·0	2·3, 12·7	
3	13	51·8	−33	00	108	7·9	4·5, 6·0	
4	13	53·2	−31	56	185	14·9	4·8, 8·4	
β	14	03·8	−60	22	251	1·3	0·7, 3·9	
η	14	35·5	−42	09	270	5·0	2·6, 13·5	
α	14	39·6	−60	50	215	19·7	0·0, 1·2	Binary, 79·9 y

OPEN CLUSTERS

M	NGC	R.A. h	m	Dec. deg	min	Diameter min	Mag.	No. of stars	
	3766	11	36·1	−61	37	12	5·3	100	
	IC 2944	11	36·6	−63	02	15	4·5	30	λ Cen cluster
	3960	11	50·9	−55	42	7	8·3	45	
	5138	13	27·3	−59	01	8	7·6	40	
	5281	13	46·6	−62	54	5	5·9	40	
	5316	13	53·9	−61	52	14	6·0	80	
	5460	14	07·6	−48	19	25	5·6	40	
	5617	14	29·8	−60	43	10	6·3	80	
	5662	14	35·2	−56	33	12	5·5	70	

GLOBULAR CLUSTERS

M	NGC	R.A. h	m	Dec. deg	min	Diameter min	Mag.	
	5139	13	26·8	−47	29	36·3	3·6	ω Centauri
	5286	13	46·4	−51	22	9·1	7·6	

PLANETARY NEBULA

M	NGC	R.A. h	m	Dec. deg	min	Dimensions sec	Mag.	Mag. of central star	
	3918	11	50·3	−57	11	12	8·4	10·9	Blue Planetary

NEBULA

M	NGC	R.A. h	m	Dec. deg	min	Dimensions min	
	5367	13	57·7	−39	59	4·3	Includes IC 4347. Double nucleus

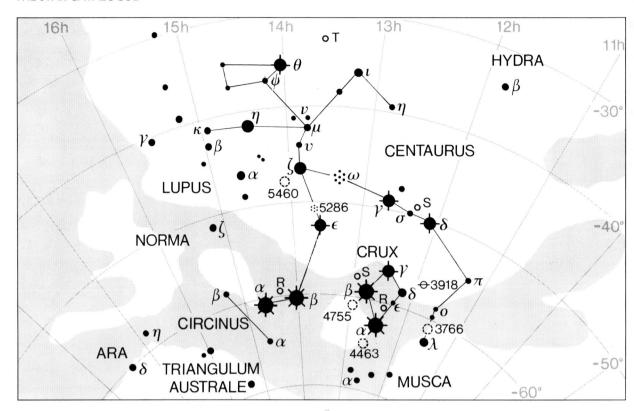

CENTAURUS (continued)

GALAXIES

M	NGC	R.A.		Dec.		Mag.	Dimensions	Type	
		h	m	deg	min		min		
	4603	12	40·9	−40	59	12·0	3·8 × 2·5	Sc	
	4696	12	48·8	−41	19	10·7	3·5 × 3·2	E1p	
	4945	13	05·4	−49	28	9·5	20·0 × 4·4	SBc	
	4976	13	08·6	−49	30	10·2	4·3 × 2·6	E4p	
	5128	13	25·5	−43	01	7·0	18·2 × 14·3	S0p	Centaurus A
	5253	13	39·9	−31	39	10·6	4·0 × 1·7	E5	
	5483	14	10·4	−43	19	12·0	3·1 × 2·8	Sc	
	3557	11	10·0	−37	32	10·4	4·0 × 2·7	E3	

CEPHEUS

(Abbreviation: Cep).

A rather undistinguished constellation, though it contains δ Cephei – the prototype Cepheid. Mythologically, Cepheus was Andromeda's father and Cassiopeia's husband. There are 8 stars above the 4th magnitude – though one of these, the 'Garnet Star' μ, is an irregular variable which is generally rather below this limit:

	Mag.	Luminosity Sun = 1	Dist lt-yrs
α	2·44	14	46
γ	3·21	10	59
β	3·23v	2200	750
ζ	3·35	5000	700
η	3·43	4·5	46
ι	3·52	82	130
μ	3·6v	53 000	1560
δ	3·7v	6000v	1340

The relatively nearby double Krüger 60 lies near δ; R.A. 22 h 26 m ·3, dec. +57 deg 27 min.

BRIGHTEST STARS

Star		R.A.			Dec.			Mag.	Abs. mag.	Spectrum	Dist pc	
		h	m	s	deg	min	sec					
1	κ	20	08	53·1	+77	42	41	4·39	−0·8	B9	100	
2	θ	20	29	34·7	+62	59	39	4·22	1·8	A0	34	
3	η	20	45	17·2	+61	50	20	3·43	3·2	K0	14	
5	α	21	18	34·6	+62	35	08	2·44	1·9	A7	14	Alderamin
8	β	21	28	39·4	+70	33	39	3·23v	−3·6	B2	230	Alphirk
9		21	37	55·0	+62	04	55	4·73	−5·7	B2	640	
11		21	41	55·1	+72	18	42	4·56	0·2	K0	60	
	μ	21	43	30·2	+58	46	48	var.	−7·0	M2	480	The 'Garnet Star'
10	ν	21	45	26·8	+61	07	15	4·29	−7·5	A2	1200	
17	ξ	22	03	45·7	+64	37	42	4·29	2·3	F7 + G	37	Kurdah
21	ζ	22	10	51·1	+58	12	05	3·35	−4·4	K1	220	
23	ε	22	15	01·8	+57	02	37	4·19	1·7	F0	30	
27	δ	22	29	10·1	+58	24	55	var.	−4·6v	F8	410	Prototype Cepheid
32	ι	22	49	40·6	+66	12	02	3·52	0·0	K1	39	
33	π	23	07	53·7	+75	23	16	4·41	0·4	G2	59	
35	γ	23	39	20·7	+77	37	57	3·21	2·2	K1	16	Alrai
43H		01	08	44·6	+86	15	26	4·25	−0·1	K2	68	

Also above mag. 5:

	Mag.	Abs. mag.	Spectrum	Dist
24	4·79	0·3	G8	360
34 o	4·90	1·2	G7	

CEPHEUS (continued)

The mean separation is between 2 and 3 seconds of arc, but the period is only 44·5 years, so that the separation and position angle alter quite quickly. The magnitudes are 9·8 and 11·4. The real separation is about the same as that between the Sun and Saturn. Krüger 60B, the fainter component, is a flare star, and in variable star catalogues is listed as DO Cephei.

VARIABLE STARS

	R.A. h	m	Dec. deg	min	Range	Type	Period, d.	Spectrum
T	21	09·5	+68	29	5·2–11·3	Mira	388·1	M
VV	21	56·7	+63	38	4·8–5·4	Eclipsing	7430	M + B
S	21	35·2	+78	37	7·4–12·9	Mira	486·8	N
μ	21	43·5	+58	47	3·4–5·1	Irreg.?	–	M
δ	22	29·2	+58	25	3·5–4·4	Cepheid	5·37	F–G
W	22	36·5	+58	26	7·0–9·2	Semi-reg.	Long	K–M
U	01	02·3	+81	53	6·7–9·2	Algol	2·49	B + G

DOUBLE STARS

	R.A. h	m	Dec. deg	min	P.A. deg	Sep. sec	Mags.	
κ	20	08·9	+72	43	122	7·4	4·4, 8·4	
β	21	28·7	+70	34	249	13·3	3·2, 7·9	
ξ	22	03·8	+64	38	277	7·7	4·4, 6·5	Binary, 3800 y
δ	22	29·2	+58	25	191	41·0	var., 7·5	
π	23	07·9	+75	23	346	1·2	4·6, 6·6	Slow binary
o	23	18·6	+68	07	220	2·9	4·9, 7·1	Binary, 796 y

OPEN CLUSTERS

M	NGC	R.A. h	m	Dec. deg	min	Diameter min	Mag.	No. of stars
	IC 1396	21	39·1	+57	30	50	3·5	50
	7160	21	53·7	+62	36	7	6·1	12
	7235	22	12·6	+57	17	4	7·7	30
	7261	22	20·4	+58	05	6	8·4	30
	7510	23	11·5	+60	34	4	7·9	60
	188	00	44·4	+85	20	14	8·1	120

PLANETARY NEBULA

M	NGC	R.A. h	m	Dec. deg	min	Dimensions sec	Mag.	Mag. of central star
	40	00	13·0	+72	32	37	10·7	11·6

NEBULÆ

M	NGC	R.A. h	m	Dec. deg	min	Dimensions min	Mag.	
	7023	21	01·8	+68	12	18 × 18	6·8	
	Sh2–155	22	56·8	+62	37	50 × 30	7·7	Cave Nebula

GALAXIES

M	NGC	R.A. h	m	Dec. deg	min	Mag.	Dimensions min	Type
	6946	20	34·8	+60	09	9·7	11·0 × 9·8	Sc
	6951	20	37·2	+66	06	12·2	3·8 × 3·3	Sbp

CETUS

(Abbreviation: Cet).

One of the largest of all constellations; sometimes associated with the sea-monster of the Perseus legend, at others relegated to the status of a harmless whale. It contains the prototype long-period variable Mira, which can occasionally rise to magnitude 1·7, but which spends most of its period below naked-eye visibility. Cetus abounds in faint galaxies.

Excluding Mira, there are 8 stars above the fourth magnitude:

	Mag.	Luminosity Sun = 1	Dist lt-yrs
β	2·04	60	68
α	2·53	120	130
η	3·45	96	117
γ	3·47	26	75
τ	3·50	0·45	11·8
ι	3·56	96	160
θ	3·60	60	114
ζ	3·73	96	190

BRIGHTEST STARS

Star		R.A. h	m	s	Dec. deg	min	sec	Mag.	Abs. mag.	Spectrum	Dist pc	
2		00	03	44·3	−17	20	10	4·55	−0·3	B9	91	
7		00	14	38·4	−18	55	58	4·44	−0·5	gM1	86	
8	ι	00	19	25·6	−08	49	26	3·56	−0·1	K2	50	Baten Kaitos Shemali
16	β	00	43	35·3	−17	59	12	2·04	0·2	K0	21	Diphda
17	φ¹	00	44	11·3	−10	36	34	4·75	0·2	K0	81	Al Nitham
31	η	01	08	35·3	−10	10	56	3·45	−0·1	K2	36	
45	θ	01	24	01·3	−08	11	01	3·60	0·2	K0	35	
53	χ	01	49	35·0	−10	41	11	4·67	1·9	F2	30	
52	τ	01	44	04·0	−15	56	15	3·50	5·7	G8	3·6	
55	ζ	01	51	27·6	−10	20	06	3·73	−0·1	K2	58	Baten Kaitos
59	υ	02	00	00·2	−21	04	40	4·00	−0·5	M1	79	
65	ξ¹	02	12	59·9	+08	50	48	4·37	−2·1	G8	200	
68	ο	02	19	20·6	−02	58	39	var.	−0·5v	Md	29	Mira
73	ξ²	02	28	09·5	+08	27	36	4·28	−0·8	B9	98	
76	σ	02	32	05·1	−15	14	41	4·75	2·1	F5	33	
82	δ	02	39	28·9	+00	19	43	4·07	−3·0	B2	260	
86	γ	02	43	18·0	+03	14	09	3·47	1·4	A2	23	Alkaffaljidhina
89	π	02	44	07·3	−13	51	32	4·25	−0·6	B7	93	
87	μ	02	44	56·4	+10	06	51	4·27	1·7	F0	30	
91	λ	02	59	42·8	+08	54	27	4·70	−2·2	B5	220	
92	α	03	02	16·7	+04	05	23	2·53	−0·5	M2	40	Menkar

Also above mag. 5:

		Mag.	Abs. mag.	Spectrum	Dist
6		4·89	3·7	F6	18
20		4·77	−0·4	M0	110
46		4·90	−0·2	K3	110
72	ρ	4·89	0·2	B9	85
78	ν	4·86	0·3	G8	82
83	ε	4·84	2·8	F5	22
96	κ	4·83	5·0	G5	9·3

VARIABLE STARS

	R.A. h	m	Dec. deg	min	Range	Type	Period, d.	Spectrum
W	00	02·1	−14	41	7·1−14·8	Mira	351·3	S
T	00	21·8	−20	03	5·0−6·9	Semi-reg.	159	M
S	00	24·1	−09	20	7·6−14·7	Mira	320·5	M
U	02	33·7	−13	09	6·8−13·4	Mira	234·8	M
UV	01	38·8	−17	58	6·8−13·0	Flare	−	dM
ο	02	19·3	−02	59	1·7−10·1	Mira	332·0	M

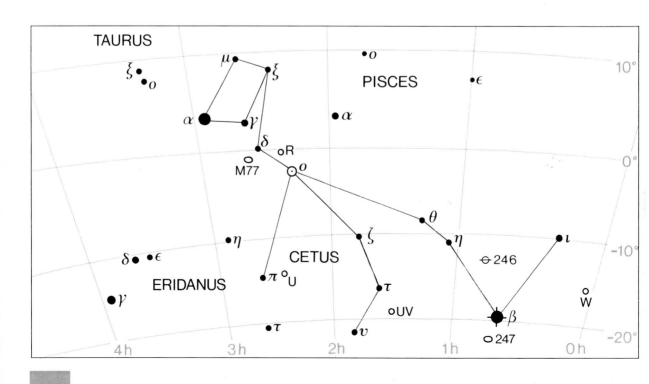

CETUS (continued)

DOUBLE STARS

	R.A. h	m	Dec. deg	min	P.A. deg	Sep. sec	Mags.	
37	01	14·4	−07	55	331	49·7	5·2, 8·7	
χ	01	49·6	−10	41	250	183·8	4·9, 6·9	
66	02	12·8	−02	24	AB 234	16·5	5·7, 7·5	
					AC 061	172·7	11·4	
o	02	19·3	−02	59	085	0·3	var., 9·5	(B is VZ Ceti)
ν	02	35·9	+05	36	081	8·1	4·9, 9·5	
ε	02	39·6	−11	52	039	0·1	5·8, 5·8	Binary, 2·7 y
γ	02	43·3	+03	14	294	2·8	3·5, 7·3	

PLANETARY NEBULA

M	GC	R.A. h	m	Dec. deg	min	Diameter sec	Mag.	Mag. of central star
	246	00	47·0	−11	53	225	8·0	11·9

GALAXIES

M	NGC	R.A. h	m	Dec. deg	min	Mag.	Dimensions min	Type
	45	00	14·1	−23	11	10·4	8·1 × 5·8	S
	247	00	47·1	−20	46	8·9	20·0 × 7·4	S
	428	01	12·9	+00	59	11·3	4·1 × 3·2	Scp
	578	01	30·5	−22	40	10·9	4·8 × 3·2	Sc
	584	01	31·3	−06	52	10·3	3·8 × 2·4	E4
	720	01	53·0	−13	44	10·2	4·4 × 2·8	E3
	864	02	15·5	+06	00	11·0	4·6 × 3·5	Sc
	895	02	21·6	−05	31	11·8	3·6 × 2·8	Sb
	908	02	23·1	−21	14	10·2	5·5 × 2·8	Sc
	936	02	27·6	−01	09	10·1	5·2 × 4·4	SBa
IC	1613	01	04·8	+02	07	9·9	12·0 × 11·2	Irr
	1042	02	40·4	−08	26	10·9	4·7 × 3·9	Sc
	1055	02	41·8	+00	26	10·6	7·6 × 3·0	Sb
77	1068	02	42·7	−00	01	8·8	6·9 × 5·9	SBp
	1073	02	43·7	+01	23	11·0	4·9 × 4·6	SBc
	1087	02	46·4	−00	30	11·0	3·5 × 2·3	Sc

CHAMÆLEON

(Abbreviation: Cha).

A small southern constellation of no particular note; there are no legends attached to it, and there are no stars above the fourth magnitude.

See chart for Musca.

BRIGHTEST STARS

Star	R.A. h	m	s	Dec. deg	min	sec	Mag.	Abs. mag.	Spectrum	Dist pc
α	08	18	31·7	−76	55	10	4·07	2·2	F6	24
θ	08	20	38·7	−77	29	04	4·35	1·7	K0	24
γ	10	35	28·1	−78	36	27	4·11	−0·4	M0	77
δ²	10	45	46·7	−80	32	24	4·45	−1·7	B3	170
β	12	18	20·7	−79	18	43	4·26	−0·9	B6	110

Also above mag. 5:

	Mag.	Abs. mag.	Spectrum	Dist
ε	4·91	0·2	B9	88

VARIABLE STARS

	R.A. h	m	Dec. deg	min	Range	Type	Period, d.	Spectrum
R	08	21·8	−76	21	7·5–14·2	Mira	334·6	M
RS	08	43·2	−79	04	6·0–6·7	Algol + δ Scuti	1·67	A–F

DOUBLE STARS

	R.A. h	m	Dec. deg	min	P.A. deg	Sep. sec	Mags.
δ	10	45·3	−80	28	076	0·6	6·1, 6·4
ε	11	59·6	−78	13	188	0·9	5·4, 6·0

CIRCINUS

(Abbreviation: Cir).

A very small southern constellation, in the area of α and β Centauri. It is associated with no legends. The only star above magnitude 4 is α.

See chart for Centaurus.

	Mag.	Luminosity Sun = 1	Dist lt-yrs
α	3·19	7	46

BRIGHTEST STARS

Star	R.A. h	m	s	Dec. deg	min	sec	Mag.	Abs. mag.	Spectrum	Dist pc
α	14	42	28·0	−64	58	43	3·19	2·6	F0	14
β	15	17	30·8	−58	48	04	4·07	1·7	A3	22
γ	15	23	22·6	−59	19	14	4·51	−1·0	B5	84

Also above mag. 5:

	Mag.	Abs. mag.	Spectrum	Dist
ε	4·86	0·3	K4	110

VARIABLE STARS

	R.A. h	m	Dec. deg	min	Range	Type	Period, d.	Spectrum
AX	14	52·6	−63	49	5·6–6·1	Cepheid	5·27	F–G
θ	14	56·7	−62	47	5·0–5·4	Irreg.	−	B

CIRCINUS (continued)

DOUBLE STARS

	R.A. h m	Dec. deg min	P.A. deg	Sep. sec	Mags.	
α	14 42·5	−64 59	232	15·7	3·2, 8·6	
δ	15 16·9	−60 57	270	50·0	5·1, 13·4	
γ	15 23·4	−59 19	033	0·6	5·1, 5·5	Binary, 180 y

OPEN CLUSTER

M	NGC	R.A. h m	Dec. deg min	Diameter min	Mag.	No. of stars
	5823	15 05·7	−55 36	10	7·9	100

COLUMBA

(Abbreviation: Col).

A 'modern' constellation dating from 1679, when it was introduced to the sky by Royer. Apparently it represents the dove which Noah released from the Ark; it was originally called Columba Noachi. There are five stars above magnitude 4:

	Mag.	Luminosity Sun = 1	Dist lt-yrs
α	2·64	105	120
β	3·12	96	143
δ	3·85	120	117
ε	3·87	60	137
η	3·96	60	143

δ Columbæ was formerly called 3 Canis Majoris.

BRIGHTEST STARS

Star	R.A. h m s	Dec. deg min sec	Mag.	Abs. mag.	Spectrum	Dist pc	
ε	05 31 12·7	−35 28 15	3·87	0·2	gK0	42	
α	05 39 38·9	−34 04 27	2·64	−0·2	B8	37	Phakt
β	05 50 57·5	−35 46 06	3·12	−0·1	K2	44	Wazn
γ	05 57 32·2	−35 17 00	4·36	−2·3	B3	210	
η	05 59 08·8	−42 48 55	3·96	0·2	K0	44	
κ	06 16 33·0	−35 08 26	4·37	0·3	G8	58	
δ	06 22 06·7	−33 26 11	3·85	0·5	gG1	36	

μ (mag. 5·16), position RA 05·44·1 dec. −32 deg 10 min, is one of the three 'runaway stars' which seem to be moving away from the nebulous region of Orion (the others are 53 Arietis and AE Aurigæ). μ Columbæ has an annual proper motion of 0 sec ·025. Its spectral type is O9.5.

Also above mag. 5:

	Mag.	Abs. mag.	Spectrum	Dist
ξ	4·97	0·0	K1	97
λ	4·87	−1·1	B5	160
o	4·83	3·2	K0	21

VARIABLE STARS

	R.A. h m	Dec. deg min	Range	Type	Period, d.	Spectrum
T	05 19·3	−33 42	6·6–12·7	Mira	225·9	M
R	05 50·5	−29 12	7·8–15·0	Mira	327·6	M

DOUBLE STARS

	R.A. h m	Dec. deg min	P.A. deg	Sep. sec	Mags.
α	05 39·6	−34 04	359	13·5	2·6, 12·3
γ	05 57·5	−35 17	110	33·8	4·4, 12·7
π²	06 07·9	−42 09	150	0·1	6·2, 6·3

GLOBULAR CLUSTER

M	NGC	R.A. h m	Dec. deg min	Diameter min	Mag.	
	1851	05 14·1	−40 03	11·0	7·3	X-ray source

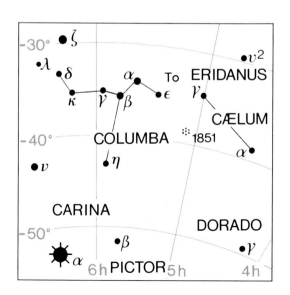

COLUMBA (continued)

GALAXIES

M	NGC	R.A.		Dec.		Mag.	Dimensions	Type
		h	m	deg	min		sec	
	1792	05	05·2	−37	59	10·2	4·0 × 2·1	Sb
	1808	05	07·7	−37	32	9·9	7·2 × 4·1	SBa
	2090	05	47·0	−34	14	11·7	4·5 × 2·3	Sc

COMA BERENICES

(Abbreviation: Com).

At first glance this constellation gives the impression of being a vast, dim cluster. Coma has no star above magnitude 4·3, but it abounds in faint ones, and there are many telescopic galaxies.

Though the constellation is not 'original', there is a legend attached to it. When Ptolemy Euergetes, King of Egypt, set out in an expedition against the Assyrians, his wife Berenice vowed that if he returned safely she would cut off her lovely hair and place it in the temple of Venus. The King returned; the Queen kept her vow, and Jupiter placed the shining tresses in the sky.

See chart for Boötes.

BRIGHTEST STARS

Star	R.A.			Dec.			Mag.	Abs.	Spectrum	Dist	
	h	m	s	deg	min	sec		mag.		pc	
15 γ	12	26	56·2	+28	16	06	4·35	−0·2	K1	31	
42 α	13	09	59·2	+17	31	45	4·32	3·4	F5	18	Diadem
43 β	13	11	52·3	+27	52	41	4·26	4·7	G0	8·3	

α is a very close binary, with a period of 26 years; the separation is never more than 0 sec ·3. The components are approximately equal.

Also above mag. 5:

	Mag.	Abs. mag.	Spectrum	Dist
12	4·79	0·7	F8	27
7	4·94	0·2	K0	89
11	4·74	0·3	G8	69
14	4·93		F0	24
16	4·99		A4	21
23	4·81	−0·6	A0	110
31	4·94	0·6	G0	74
36	4·78	−0·5	M1	110
37	4·90		K1	27
41	4·80	−0·3	K5	110

VARIABLE STARS

	R.A.		Dec.		Range	Type	Period, d.	Spectrum
	h	m	deg	min				
R	12	04·0	+18	49	7·1–14·6	Mira	362·8	M
FS	13	06·4	+22	37	5·3–6·1	Semi-reg.	58	M

GALAXIES

M	NGC	R.A.		Dec.		Mag.	Dimensions	Type	
		h	m	deg	min		min		
	4136	12	09·3	+29	56	11·7	4·1 × 3·9	Sc	
98	4192	12	13·8	+14	54	10·1	9·5 × 3·2	Sb	
	4251	12	18·1	+28	10	11·6	4·2 × 1·9	E7	
99	4254	12	18·8	+14	25	9·8	5·4 × 4·8	Sc	
	4278	12	20·1	+29	17	10·2	3·6 × 3·5	E1	
	4314	12	22·6	+29	53	10·5	4·8 × 4·3	SBa	
100	4321	12	22·9	+15	49	9·4	6·9 × 6·2	Sc	
	4448	12	28·2	+28	37	11·1	4·0 × 1·6	Sb	
	4450	12	28·5	+17	05	10·1	4·8 × 3·5	Sb	
	4459	12	29·0	+13	59	10·4	3·8 × 2·8	E2	
	4473	12	29·8	+13	26	10·2	4·5 × 2·6	E4	
	4477	12	30·0	+13	38	10·4	4·0 × 3·5	SBa	
88	4501	12	32·0	+14	25	9·5	6·9 × 3·9	SBb	
	4548	12	35·4	+14	30	10·2	5·4 × 4·4	SBb	
	4559	12	36·0	+27	58	9·8	10·5 × 4·9	Sc	
	4565	12	36·3	+25	59	9·6	16·2 × 2·8	Sb	
	4651	12	43·7	+16	24	10·7	3·8 × 2·7	Sop	
	4689	12	47·8	+13	46	10·9	4·0 × 3·5	Sb	
	4725	12	50·4	+25	30	9·2	11·0 × 7·9	SBb	
64	4826	12	56·7	+21	41	8·5	9·3 × 5·4	Sb	Black-Eye Galaxy

OPEN CLUSTER

M	NGC	R.A.		Dec.		Diameter	Mag.	No. of stars	
		h	m	deg	min	min			
	Mel 111	12	25	+26		275	4	80	Coma Berenices

GLOBULAR CLUSTER

M	NGC	R.A.		Dec.		Diameter	Mag.
		h	m	deg	min	min	
53	5024	13	12·9	+18	10	12·6	7·7

CORONA AUSTRALIS

(Abbreviation: CrA).

An original constellation. It has no bright stars, but is easy to recognize because of its distinctive shape.

See chart for Sagittarius.

BRIGHTEST STARS

Star	R.A.			Dec.			Mag.	Abs. mag.	Spectrum	Dist pc	
	h	m	s	deg	min	sec					
θ	18	33	29·9	−42	18	45	4·64	0·3	G5	60	
ζ	19	03	06·7	−42	05	43	4·75	0·6	A0	68	
γ	19	06	24·9	−37	03	48	4·21	4·0	F8	12	
δ	19	08	20·6	−40	29	48	4·59	0·2	gK0	63	
α	19	09	28·2	−37	54	16	4·11	1·6	A2	14	Meridiana
β	19	10	01·5	−39	20	27	4·11	0·3	gG5	34	

Also above mag. 5:

	Mag.	Abs. mag.	Spectrum	Dist
ε	4·8v	2·6	F0	28

DOUBLE STARS

	R.A.		Dec.		P.A. deg	Sep. sec	Mags.	
	h	m	deg	min				
κ	18	33·4	−38	44	359	21·6	5·9, 5·9	
λ	18	43·8	−38	19	214	29·2	5·1, 9·7	
γ	19	06·4	−37	04	109	1·3	4·8, 5·1	Binary, 120·4 y

GLOBULAR CLUSTER

M	NGC	R.A.		Dec.		Diameter min	Mag.
		h	m	deg	min		
	6541	18	08·0	−43	42	13·1	6·6

NEBULA

M	NGC	R.A.		Dec.		Dimensions min	Mag.	Mag. of illuminating star
		h	m	deg	min			
	6729	19	01·9	−36	57	1 (var.)	var.	9·7v (R Coronæ Australis)

PLANETARY NEBULA

M	NGC	R.A.		Dec.		Diameter sec	Mag.	Mag. of central star
		h	m	deg	min			
	IC 1297	19	17·4	−39	37	7	—	12·9v RU Coronæ Australis

CORONA BOREALIS

(Abbreviation: CrB).

A small but very distinctive constellation representing a crown given by Bacchus to Ariadne, the daughter of King Minos of Crete. There are three stars above magnitude 4; α is an eclipsing binary with a very small range:

	Mag.	Luminosity Sun = 1	Dist lt-yrs
α	2·23v	130	78
β	3·68	28	59
γ	3·84v	110	210

See chart for Boötes.

BRIGHTEST STARS

Star	R.A.			Dec.			Mag.	Abs. mag.	Spectrum	Dist pc	
	h	m	s	deg	min	sec					
3 β	15	27	49·7	+29	06	21	3·68	1·2	F0p	18	Nusakan
4 θ	15	32	55·7	+31	21	32	4·14	−1·1	B5	110	
5 α	15	34	41·2	+26	42	53	2·23v	0·6	A0	24	Alphekka
7 ζ²	15	39	22·6	+36	38	09	4·7	−0·6	B7	130	Wide binary with ζ¹, mag. 6·0
8 γ	15	42	44·5	+26	17	44	3·84v	−0·3	A0	64	
10 δ	15	49	35·6	+26	04	06	4·63	1·8	G5	38	
13 ε	15	57	35·2	+26	52	40	4·15	−0·2	K3	74	

α is an eclipsing binary with a very small range (0 m·1).
γ is a δ Scuti variable; range only 0·06 mag.
β is a magnetic spectrum variable.

Also above mag. 5:

	Mag.	Abs. mag.	Spectrum	Dist
2 η	4·98	4·4	G0	14
ζ	5·0 + 6·0	−0·6	B6 + B7	130
17 σ	5·22	4·4	G0	21
14 ι	4·99	−0·6	A0	130
16 τ	4·76	0·2	K0	52
19 ξ	4·85	0·2	K0	85
11 κ	4·82	1·7	K0	33

VARIABLE STARS

	R.A.		Dec.		Range	Type	Period, d.	Spectrum	
	h	m	deg	min					
U	15	18·2	+31	39	7·7–8·8	Algol	3·45	B + F	
S	15	21·4	+31	22	5·8–14·1	Mira	360·3	M	
R	15	48·6	+28	09	5·7–15	R CrB	—	F8p	
V	15	49·5	+39	34	6·9–12·6	Mira	357·6	N	
T	15	59·5	+25	55	2·0–10·8	Recurrent nova		M + Q	(1866, 1946)
W	16	15·4	+37	48	7·8–14·3	Mira	238·4	M	

DOUBLE STARS

	R.A.		Dec.		P.A. deg	Sep. sec	Mags.
	h	m	deg	min			
ο	15	20·1	+29	37	337	147·3	5·5, 9·4

CORONA BOREALIS
(continued)

	R.A. h m	Dec. deg min	P.A. deg	Sep. sec	Mags.	
η	15 23·2	+30 17	AB 030	1·0	5·6, 5·9	Binary, 41·6 y
			AC 012	57·7	12·5	
			AB + D 047	215·0	10·0	
ζ	15 39·4	+36 38	305	6·3	5·1, 6·0	
γ	15 42·7	+26 18	118	0·6	4·1, 5·5	Binary, 91 y
ε	15 57·6	+26 53	003	1·8	4·2, 12·6	
ρ	16 01·0	+33 18	071	89·8	5·5, 8·7	
σ	16 14·7	+33 52	234	7·0	5·6, 6·6	Binary, 1000 y

CORVUS

(Abbreviation: Crv).

An original group. When the god Apollo became enamoured of Coronis, mother of the great doctor Æsculapius, he sent a crow to watch her and report on her behaviour. To be candid, the crow's report was decidedly adverse; but Apollo rewarded the bird with a place in the sky!

Corvus is distinctive, since its leading stars form a quadrilateral. There are four stars above magnitude 4:

	Mag.	Luminosity Sun = 1	Dist lt-yrs
γ	2·59	250	185
β	2·65	600	290
δ	2·95	60	117
ε	3·00	96	104

Curiously, the star lettered α is more than a magnitude fainter than any of these.

See chart for Hydra.

BRIGHTEST STARS

Star	R.A. h m s	Dec. deg min sec	Mag.	Abs. mag.	Spectrum	Dist pc	
1 α	12 08 24·7	−24 43 44	4·02	1·9	F2	21	Alkhiba
2 ε	12 10 07·4	−22 37 11	3·00	−0·1	K2	32	
4 γ	12 15 48·3	−17 32 31	2·59	−1·2	B8	57	Minkar
7 δ	12 29 51·8	−16 30 55	2·95	0·2	B9	36	Algorel
8 η	12 32 04·1	−16 11 46	4·31	1·7	F0	29	
9 β	12 34 23·2	−23 23 48	2·65	−2·1	G5	89	Kraz

VARIABLE STARS

	R.A. h m	Dec. deg min	Range	Type	Period, d.	Spectrum
R	12 19·6	−19 15	6·7–14·4	Mira	317·0	M
SV	12 49·8	−15 05	6·8–7·6	Semi-reg.	70	M

DOUBLE STAR

	R.A. h m	Dec. deg min	P.A. deg	Sep. sec	Mags.
δ	12 29·9	−16 31	214	24·2	3·0, 9·2

PLANETARY NEBULA

M	NGC	R.A. h m	Dec. deg min	Dimensions sec	Mag.	Mag. of central star
	4631	12 24·5	−18 48	45 × 110	10·3	13·2

GALAXIES

M	NGC	R.A. h m	Dec. deg min	Mag.	Dimensions min	Type	
	4038	12 01·9	−18 52	11·3	2·6 × 1·8	Sc	Antennæ
	4039	12 01·9	−18 53	13	3·2 × 2·2	Smp	Antennæ

CRATER

(Abbreviation: Crt).

Like Corvus, a small constellation adjoining Hydra; it has been identified with the wine-goblet of Bacchus. The only star above the fourth magnitude is δ.

	Mag.	Luminosity Sun = 1	Dist lt-yrs
δ	3·56	16	72

See chart for Hydra.

BRIGHTEST STARS

Star	R.A. h m s	Dec. deg min sec	Mag.	Abs. mag.	Spectrum	Dist pc	
7 α	10 59 46·4	−18 17 56	4·08	0·2	K0	37	Alkes
11 β	11 11 39·4	−22 49 33	4·48	0·2	A2	72	Al Sharasif
12 δ	11 19 20·4	−14 46 43	3·56	1·8	G8	22	
15 γ	11 24 52·8	−17 41 02	4·08	2·1	A5	24	
21 θ	11 36 40·8	−09 48 08	4·70	0·2	B9	79	
27 ζ	11 44 45·7	−18 21 03	4·73	0·3	G8	74	

Also above mag. 5:

	Mag.	Abs. mag.	Spectrum	Dist
ε	4·83	−0·3	K5	96

DOUBLE STAR

	R.A. h m	Dec. deg min	P.A. deg	Sep. sec	Mags.
γ	11 24·9	−17 41	096	5·2	4·1, 9·6

GALAXIES

M	NGC	R.A. h m	Dec. deg min	Mag.	Dimensions min	Type
	3511	11 03·4	−23 05	11·6	5·4 × 2·2	Sc
	3513	11 03·8	−23 15	12·0	2·8 × 2·3	SBc
	3571	11 11·5	−18 17	12·8	3·3 × 1·3	Sa
	3672	11 25·0	−09 48	11·5	4·1 × 2·1	Sb
	3887	11 47·1	−16 51	11·0	3·3 × 2·7	Sc
	3981	11 56·1	−19 54	12·4	3·9 × 1·5	Sb

CRUX AUSTRALIS

(Abbreviation: Cru).

Though Crux is the smallest constellation in the entire sky, it is also one of the most famous. Before Royer introduced it, in 1679, it had been included in Centaurus. Strictly speaking it is more like a kite than a cross. As well as its brilliant stars it contains the glorious 'Jewel Box' cluster, and also the dark nebula known as the Coal Sack.

There are five stars above the fourth magnitude:

	Mag.	Luminosity Sun = 1	Dist lt-yrs
α	0·83	3200 + 2000	360
β	1·25v	8200	425
γ	1·63	120	88
δ	2·80v	1320	260
ε	3·59	80	59

Of the four main stars, any casual glance will show the difference between γ, a red giant, and the other three, which are hot and bluish-white.

See chart for Centaurus.

BRIGHTEST STARS

Star	R.A. h	m	s	Dec. deg	min	sec	Mag.	Abs. mag.	Spectrum	Dist pc	
θ¹	12	03	01·6	−63	18	46	4·33	−0·5	A0	17	
θ²	12	04	19·2	−63	09	56	4·72	−3·0	B2	290	
η	12	06	52·8	−64	36	49	4·15	0·6	F0	33	
δ	12	15	08·6	−58	44	55	2·80	−3·0	B2	79	
ζ	12	18	26·1	−64	00	11	4·04	−2·3	B3	150	
ε	12	21	21·5	−60	24	04	3·59	0·0	K2	18	
α	12	26	35·9	−63	05	56	1·41	−3·9	B1 }	110	Acrux
	12	26	36·5	−63	05	58	1·88	−3·4	B3 }		
γ	12	31	09·9	−57	06	47	1·63	−0·5	M3	27	
ι	12	45	37·8	−60	58	52	4·69	0·0	K1	87	
β	12	47	43·2	−59	41	19	1·25v	−5·0	B0	130	(Mimosa)
μ¹	12	54	35·6	−57	10	40	4·03	−2·3	B3	190	

β is variable over a small range (below 0m·1); β Canis Majoris type.

VARIABLE STARS

	R.A. h	m	Dec. deg	min	Range	Type	Period, d.	Spectrum
BH	12	16·3	−56	17	7·2–10·0	Mira	421	S
T	12	21·4	−62	17	6·3–6·8	Cepheid	6·73	F
R	12	23·6	−61	38	6·4–7·2	Cepheid	5·83	F–G
S	12	54·4	−58	26	6·2–6·9	Cepheid	4·69	F–G

DOUBLE STARS

	R.A. h	m	Dec. deg	min	P.A. deg	Sep. sec	Mags.
θ¹	12	03·0	−63	19	325	4·5	4·3, 13·6
η	12	06·9	−64	37	299	44·0	4·2, 11·7
α	12	26·6	−63	06	AB 115	4·4	1·4, 1·9
					AC 202	90·1	1·0, 4·9
γ	12	31·2	−57	07	AB 031	110·6	1·6, 6·7
					AC 082	155·2	9·5
ι	12	45·6	−60	59	022	26·9	4·7, 9·5
μ¹	12	54·6	−57	11	017	34·9	4·0, 5·2

OPEN CLUSTERS

M	NGC	R.A. h	m	Dec. deg	min	Diameter min	Mag.	No. of stars	
	4052	12	01·9	−63	12	8	8·8	80	
	4103	12	06·7	−61	15	7	7·4	45	
	4337	12	23·9	−58	08	3·5	8·9	−	
	4349	12	24·5	−61	54	16	7·4	30	
	H.5	12	29·0	−60	46	5	9·0	40	
	4439	12	28·4	−60	06	4	8·4	−	
	4609	12	42·3	−62	58	5	6·9	40	
	4755	12	53·6	−60	20	10	4·2	50+	Jewel Box (κ Crucis)

DARK NEBULA

M	NGC	R.A. h	m	Dec. deg	Dimensions min	Area, sq. deg	
		12	53	−63	400 × 300	26·2	Coal Sack

CYGNUS

(Abbreviation: Cyg).

Cygnus is one of the richest constellations in the sky; it is often nicknamed the Northern Cross – certainly it is much more nearly cruciform than is Crux. Various legends are associated with it. According to one, the group was placed in the sky to honour a swan into which Jupiter once transformed himself when on a visit to the wife of the King of Sparta!

There are 14 stars above the fourth magnitude. To these must be added the red variable χ, which can rise above magnitude 4 at times.

BRIGHTEST STARS

Star	R.A. h	m	s	Dec. deg	min	sec	Mag.	Abs. mag.	Spectrum	Dist pc	
1 κ	19	17	06·0	+53	22	07	3·77	0·2	K0	52	
10 ι	19	29	42·1	+51	43	47	3·79	2·1	A5	41	
6 β	19	30	43·1	+27	57	35	3·08	−2·3	K5	120	Albireo
8	19	31	46·1	+34	27	11	4·74	−2·3	B3	240	
13 θ	19	36	26·2	+50	13	16	4·48	2·1	F5	19	
12 φ	19	39	22·4	+30	09	12	4·69	1·8	G8	34	
18 δ	19	44	58·4	+45	07	51	2·87	−0·6	A0	49	
χ	19	50	33·7	+32	54	51	var.	0·2v	S	25	
21 η	19	56	18·2	+35	05	00	3·89	0·2	K0	52	
33	20	13	23·7	+56	34	04	4·30	1·3	A3	39	
31 o¹	20	13	37·1	+46	44	29	3·79	−2·2	K2	160	
32 o²	20	15	28·1	+47	42	51	3·98	−3·4	K3	280	
34 P	20	17	47·0	+38	01	59	var.	var.	B1p	1400	
37 γ	20	22	13·5	+40	15	24	2·20	−4·6	F8	230	Sadr
39	20	23	51·4	+32	11	25	4·43	−0·2	K3	78	
41	20	29	23·6	+30	22	07	4·01	−2·0	F5	160	
47	20	33	54·0	+35	15	03	4·61	−4·4	K2	570	
50 α	20	41	25·8	+45	16	49	1·25	−7·5	A2	560	Deneb
52	20	45	39·6	+30	43	11	4·22	0·2	K0	59	
53 ε	20	46	12·5	+33	58	13	2·46	0·2	K0	25	Gienah

CYGNUS (continued)

The very remote, unstable P Cygni is usually of about magnitude 4·8, but in 1600 it was seen as of magnitude 3 and was classed as a nova. At other times, as around the year 1660, it has fallen to below magnitude 6.

	Mag.	Luminosity Sun = 1	Dist lt-yrs
α	1·25	70 000	1800
γ	2·20	6000	750
ε	2·46	60	81
δ	2·87	130	160
β	3·08	700	390
	3·20	600	390
	3·72	5000	950
τ	3·72	17	68
κ	3·77	105	170
ι	3·79	11	134
ο¹	3·79	650	520
η	3·89	60	170
ν	3·94	50	147
ο²	3·98	1900	910

Star	R.A. h	m	s	Dec. deg	min	sec	Mag.	Abs. mag.	Spectrum	Dist pc	
54 λ	20	47	24·3	+36	29	27	4·53	−1·1	B5	130	
58 ν	20	57	10·2	+41	10	02	3·94	0·6	A0	45	
59 f¹	20	59	49·3	+47	31	16	4·74v	−3·9	B1	480	(V.832 Cyg)
62 ξ	21	04	55·7	+43	55	40	3·72	−4·4	K5	290	
63	21	06	35·9	+47	38	54	4·55	−2·3	K4	220	
64 ζ	21	12	56·0	+30	13	37	3·20	−2·1	G8	120	
65 τ	21	14	47·3	+38	02	44	3·72	1·7	F0	21	
67 σ	21	17	24·7	+39	23	41	4·23	−7·1	B9	1600	
66 υ	21	17	54·9	+34	53	48	4·43	−2·5	B2	240	
73 ρ	21	33	58·7	+45	35	30	4·02	0·3	G8	56	
80 π¹	21	42	05·5	+51	11	23	4·67	−1·7	B3	180	Azelfafage
78 μ²	21	44	08·2	+28	44	35	6·14	3·7	dF3 }	17	Combined
μ¹	21	44	08·4	+28	44	34	4·78		F6 }		mag. 4·4
81 π²	21	46	47·4	+49	18	35	4·23	−2·9	B3	250	

Also above mag. 5:

	Mag.	Abs. mag.	Spectrum	Dist	
29 b³	4·97	1·4	A2	33	
30	4·83	0·0	A3	92	
28 b²	4·93	−1·7	B3	210	
22	4·94	−1·9	B6	220	
24 ψ	4·92	1·3	A3	52	
17	4·99		F5		
15	4·89	0·3	G8	83	
2	4·97	−2·3	B3	260	
45 ω¹	4·95	−2·5	B2	270	Ruchba
55	4·84	−6·8	B3	1000	
57	4·78	−1·1	B5	150	
68	5·00		O8		
72	4·90	0·0	K1	96	

VARIABLE STARS

	R.A. h	m	Dec. deg	min	Range	Type	Period, d.	Spectrum
CH	19	24·5	+50	14	6·4–8·7	Z Andromedæ	±97	M + B
R	19	36·8	+50	12	6·1–14·2	Mira	426·4	M
RT	19	43·6	+48	47	6·4–12·7	Mira	190·2	M
SU	19	44·8	+29	16	6·5–7·2	Cepheid	3·84	F
χ	19	50·6	+32	55	3·3–14·2	Mira	406·9	S
Z	20	01·4	+50	03	7·4–14·7	Mira	263·7	M
RS	20	13·4	+38	44	6·5–9·3	Semi-reg.	417	N
P	20	17·8	+38	02	3–6	Irreg.	–	Bp
CN	20	17·9	+59	48	7·3–14·0	Mira	198·5	M
U	20	19·6	+47	54	5·9–12·1	Mira	462·4	N
V	20	41·3	+48	09	7·7–13·9	Mira	421·4	N
X	20	43·4	+35	35	5·9–6·9	Cepheid	16·39	F–G

CYGNUS (continued)

	R.A.		Dec.		Range	Type	Period, d.	Spectrum
	h	m	deg	min				
T	20	47·2	+34	22	5·0–5·5	Irreg.	–	K
W	21	36·0	+45	22	5·0–7·6	Semi-reg.	126	M
SS	21	42·7	+43	35	8·4–12·4	SS Cygni	±50	A–G
WY	21	48·7	+44	15	7·5–14·0	Mira	304·5	M

DOUBLE STARS

	R.A.		Dec.		P.A.	Sep.	Mags.	
	h	m	deg	min	deg	sec		
β	19	30·7	+27	58	054	34·4	3·1, 5·1	Yellow, blue
δ	19	45·0	+45	07	225	2·4	2·9, 6·3	Binary, 828 y
ψ	19	55·6	+52	26	178	3·2	4·9, 7·4	
γ	20	22·2	+40	15	196	41·2	2·2, 9·9	B is a close db
61	21	06·9	+38	45	148	39·9	5·2, 6·0	
τ	21	14·8	+38	03	015	0·5	3·8, 6·4	Binary, 50 y
μ	21	44·1	+28	45	300	1·6	4·8, 6·1	Binary, 507 y

OPEN CLUSTERS

M	NGC	R.A.		Dec.		Diameter	Mag.	No. of star
		h	m	deg	min	min		
	6811	19	38·2	+46	34	13	6·8	70
	6819	19	41·3	+40	11	5	7·3	–
	6834	19	52·2	+29	25	5	7·8	50
	6866	20	03·7	+44	00	7	7·6	80
	6871	20	05·9	+35	47	20	5·2	15
	6910	20	23·1	+40	47	8	7·4	50
29	6913	20	23·9	+38	32	7	6·6	50
	6939	20	31·4	+60	38	8	7·8	80
	7067	21	24·2	+48	01	3	9·7	20
39	7092	21	32·2	+48	26	32	4·6	30

PLANETARY NEBULÆ

M	NGC	R.A.		Dec.		Dimensions	Mag.	Mag. of central sta
		h	m	deg	min	sec		
	6826	19	44·8	+50	31	30 × 140	9·8	10·4 Blinking Nebula
	7048	21	14·2	+46	16	61	11·3	18

NEBULÆ

M	NGC	R.A.		Dec.		Dimensions	Mag. of illuminating star
		h	m	deg	min	min	
	6888	20	12·0	+38	21	20 × 10	7·4 Crescent Nebula
	6960	20	45·7	+30	43	70 × 6	– Filamentary Nebula, 52 Cygni
	IC 5067/70	20	50·8	+44	21	80 × 70	– Pelican Nebula
	6992/5	20	56·4	+31	43	60 × 8	– Veil Nebula: SNR
	7000	20	58·8	+44	20	120 × 100	6 North America Nebula
	IC 5146	21	53·5	+47	16	12 × 12	10 Cocoon Nebula, with sparse cluster

DELPHINUS

(Abbreviation: Del).

A small but compact constellation; one of Ptolemy's originals. It honours the dolphin which carried the great singer Arion to safety, after he had been thrown overboard by the crew of the ship carrying him home after winning all the prizes in a competition. The curious names of α and β were allotted by one Nicolaus Venator, for reasons which are obvious!

There are 3 stars above the fourth magnitude:

	Mag.	Luminosity Sun = 1	Dist lt-yrs
β	3·54	46	108
α	3·77	60	170
γ	3·9 (4·3 + 5·2)	4·5	75

See chart for Aquila.

BRIGHTEST STARS

Star	R.A.			Dec.			Mag.	Abs. mag.	Spectrum	Dist pc	
	h	m	s	deg	min	sec					
2 ε	20	33	12·6	+11	18	12	4·03	–1·9	B6	150	Deneb Dulfine
4 ζ	20	35	18·4	+14	40	27	4·68	1·7	A3	38	
6 β	20	37	32·8	+14	35	43	3·54	0·7	F5	33	Rotanev
9 α	20	39	38·1	+15	54	43	3·77	0·2	B9	52	Svalocin
11 δ	20	43	27·3	+15	04	28	4·43	0·5	A7	56	
12 γ	20	46	39·3	+16	07	27	3·9	3·2	G5+F8	23	

VARIABLE STARS

	R.A.		Dec.		Range	Type	Period, d.	Spectrum
	h	m	deg	min				
R	20	14·9	+09	05	7·6–13·8	Mira	284·9	M
EU	20	37·9	+18	16	5·8–6·9	Semi-reg.	59	M
HR	20	42·3	+19	10	3·7–12·7	Nova	–	Q
U	20	45·5	+18	05	5·7–7·6	Semi-reg.	110	M
S	20	43·1	+17	05	8·3–12·4	Mira	277·2	M
V	20	47·8	+19	20	8·1–16·0	Mira	533·5	M

DOUBLE STARS

	R.A.		Dec.		P.A.	Sep.	Mags.	
	h	m	deg	min	deg	sec		
I	20	30·3	+10	54	AB 346	0·9	6·1, 8·1	
					AC 349	16·8	14·1	
β	20	37·5	+14	36	167	0·3	4·0, 4·9	Binary, 26·7 y
α	20	39·6	+15	55	AB 224	29·5	3·8, 13·3	
					AC 272	43·4	11·8	

DELPHINUS (continued)

	R.A.		Dec.		P.A.	Sep.	Mags.
	h	m	deg	min	deg	sec	
κ	20	39·1	+10	05	286	28·8	5·1, 11·7
γ	20	46·7	+16	07	268	9·6	4·5, 5·5
13	20	47·8	+06	00	194	1·6	5·6, 9·2

GLOBULAR CLUSTERS

M	NGC	R.A.		Dec.		Diameter	Mag.
		h	m	deg	min	min	
	6394	20	34·2	+07	24	5·9	8·9
	7006	21	01·5	+16	11	2·8	10·6

PLANETARY NEBULA

M	NGC	R.A.		Dec.		Dimensions	Mag.	Mag. of central star
		h	m	deg	min	sec		
	6891	20	15·2	+12	42	12 × 74	11·7	12·4

DORADO

(Abbreviation: Dor).

A 'modern' southern constellation. The only star above the fourth magnitude is α, but Dorado contains part of the Large Magellanic Cloud (LMC), in which is the magnificent Tarantula looped nebula round 30 Doradûs.

	Mag.	Luminosity Sun = 1	Dist lt-yrs
α	3·27	130	192

See chart for Reticulum.

BRIGHTEST STARS

Star	R.A.			Dec.			Mag.	Abs. mag.	Spectrum	Dist pc
	h	m	s	deg	min	sec				
γ	04	16	01·6	−51	29	12	4·25	2·6	F0	45
α	04	33	59·8	−55	02	42	3·27	−0·6	A0	59
ζ	05	05	30·6	−57	28	22	4·72	4·0	F8	13
β	05	33	37·5	−62	29	24	var.	−8·0	F9v	2300
δ	05	44	46·5	−65	44	08	4·35	2·4	A7	450
HD 40409	05	54	06·1	−63	05	23	4·65	1·4	K0	13

Also above mag. 5:

	Mag.	Abs. mag.	Spectrum	Dist
θ	4·83	−0·1	K2	80

VARIABLE STARS

	R.A.		Dec.		Range	Type	Period, d.	Spectrum
	h	m	deg	min				
R	04	36·8	−62	05	4·8–6·6	Semi-reg.	338	M
β	05	33·6	−62	29	3·7–4·1	Cepheid	9·84	F–G

DOUBLE STAR

	R.A.		Dec.		P.A.	Sep.	Mags.
	h	m	deg	min	deg	sec	
α	04	34·0	−55	03	AB 182	0·2	3·8, 4·3
					AB + C 101	77·7	9·8

GALAXIES

M	NGC	R.A.		Dec.		Mag.	Dimensions	Type
		h	m	deg	min		min	
	1549	04	15·7	−55	36	9·9	3·7 × 3·2	E0
	1553	04	16·2	−55	47	9·5	4·1 × 2·8	S0
	1596	04	27·6	−55	02	11·0	3·9 × 1·2	S0
	1617	04	31·7	−54	36	10·4	4·7 × 2·4	SBa
	1672	04	45·7	−59	15	11·0	4·8 × 3·9	SBb
	LMC	05	23·6	−69	45	0·1	650 × 550	Large Cloud of Magellan. Contains 30 Doradûs and 3 planetary nebulæ, NGC 1714, 1722 and 1743
	1947	05	26·8	−63	46	10·8	3·0 × 1·6	S0p

NEBULA

M	NGC	R.A.		Dec.		Dimensions	
		h	m	deg	min	min	
	2070	05	38·7	−69	06	40 × 25	30 Doradûs. In the LMC; Tarantula Nebula

DRACO

(Abbreviation: Dra).

A long, sprawling northern group. In mythology it has been identified either with the dragon killed by Cadmus before the founding of the city of Bœotia, or with the dragon which guarded the golden apples in the Garden of the Hesperides. Thuban (α) was the pole star in ancient times.

BRIGHTEST STARS

Star	R.A.			Dec.			Mag.	Abs. mag.	Spectrum	Dist pc	
	h	m	s	deg	min	sec					
HD 81817	09	37	05·2	+81	19	35	4·29	−0·2	K3	57	
1 λ	11	31	24·2	+69	19	52	3·84	−0·4	M0	65	Giansar
5 κ	12	33	28·9	+69	47	17	3·87	0·2	B7	22	
10 i	13	51	25·8	+64	43	33	4·66	−0·5	gM3	110	
11 α	14	04	23·2	+64	22	33	3·65	−0·6	A0	71	Thuban
12 ι	15	24	55·6	+58	57	58	3·29	−0·1	K2	48	Edasich
13 θ	16	01	53·2	+58	33	55	4·01	3·2	F8	16	
14 η	16	23	59·3	+61	30	50	2·74	0·3	G8	25	Aldhibain
22 ζ	17	08	47·0	+65	42	53	3·17	−1·9	B6	97	Aldhibah
23 β	17	30	25·8	+52	18	05	2·79	−2·1	G2	82	Alwaid

DRACO (continued)

There are 12 stars above the fourth magnitude:

	Mag.	Luminosity Sun = 1	Dist lt-yrs
γ	2·23	110	101
η	2·74	110	81
β	2·79	600	270
δ	3·07	60	117
ζ	3·17	500	316
ι	3·29	96	156
χ	3·57	2	25
α	3·65	130	230
ξ	3·75	96	189
ε	3·83	58	166
λ	3·84	115	212
κ	3·87	60	72

Star	R.A. h	m	s	Dec. deg	min	sec	Mag.	Abs. mag.	Spectrum	Dist pc	
31 ψ	{17	41	56·1	+72	08	56	4·58	2·8	F5	23}	Dziban
	{17	41	57·7	+72	09	24	5·79	4·0	F0	23}	
32 ξ	17	53	31·5	+56	52	21	3·75	−0·1	K2	58	Juza
33 γ	17	56	36·2	+51	29	20	2·23	−0·3	K5	31	Eltamin
43 φ	18	20	45·2	+71	20	16	4·22	−0·6	A0p	33	
44 χ	18	21	03·0	+72	43	58	3·57	4·1	F7	7·8	
47 o	18	51	12·0	+59	23	18	4·66	−0·9	K0	67	
57 δ	19	12	33·1	+67	39	41	3·07	0·2	G9	36	Taïs
60 τ	19	15	32·8	+73	21	20	4·45	−0·2	K3	70	
58 π	19	20	39·9	+65	42	52	4·59	0·6	A2	63	
61 σ	19	32	21·5	+69	39	40	4·68	5·9	K0	5·7	Alrakis
63 ε	19	48	10·2	+70	16	04	3·83	0·3	G8	51	Tyl
67 ρ	20	02	48·9	+67	52	25	4·51	−0·2	K3	82	

Also above mag. 5:

	Mag.	Abs. mag.	Spectrum	Dist	
42	4·82	−0·1	K2	96	
39	4·98	1·2	A1	53	
28 ω	4·80	3·4	F5	22	Al Dhih
24 v¹	4·88		A8		}Kuma
25 v²	4·87		A4		
21 μ	4·92		F5		
4	4·95	−0·5	M4	120	
6	4·94	−0·1	K2	84	
18 g	4·83	−4·5	K1		
19	4·89	3·7	F6	17	
45	4·77	−4·6	F7	450	
52 υ	4·82	0·2	K0	64	
54	4·99	−0·1	K2	100	

VARIABLE STARS

	R.A. h	m	Dec. deg	min	Range	Type	Period, d.	Spectrum
RY	12	56·4	+66	00	5·6–8·0	Semi-reg.	173	N
R	16	32·7	+66	45	6·7–13·0	Mira	245·5	M
T	17	56·4	+58	13	7·2–13·5	Mira	421·2	N
UW	17	57·5	+54	40	7·0–8·0	Irreg.	–	K
UX	19	21·6	+76	34	5·9–7·1	Semi-reg.	168	N

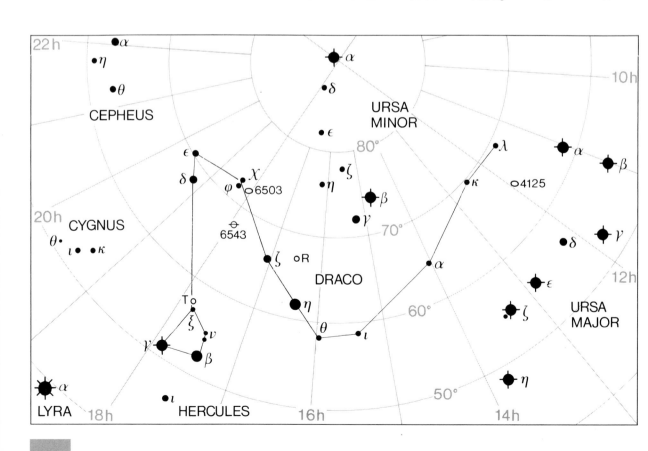

DRACO (continued)

DOUBLE STARS

	R.A. h	m	Dec. deg	min	P.A. deg	Sep. sec	Mags.	
η	16	24·0	+61	31	142	5.2	2·7, 8·7	
μ	17	05·3	+54	28	020	1·9	5·7, 5·7	Binary, 482 y
ν	17	32·2	+55	11	312	61·9	4·9, 4·9	
ψ	17	41·9	+72	09	015	30·3	4·9, 6·1	
ε	19	48·2	+70	16	015	3·1	3·8, 7·4	

PLANETARY NEBULA

M	NGC	R.A. h	m	Dec. deg	min	Dimensions sec	Mag.	Mag. of central star
	6543	17	58·6	+66	38	18 × 350	8·8	11·4

GALAXIES

M	NGC	R.A. h	m	Dec. deg	min	Mag.	Dimensions min	Type
	3147	10	16·9	+73	24	10·6	4·0 × 3·5	Sb
	4125	12	08·1	+65	11	9·8	5·1 × 3·2	E5p
	4236	12	16·7	+69	28	9·7	18·6 × 6·9	Sb
	5866	15	06·5	+55	46	10·0	5·2 × 2·3	E6p
	5879	15	09·8	+57	00	11·5	4·4 × 1·7	Sb
	5907	15	15·9	+56	19	10·4	12·3 × 1·8	Sb
	5985	15	39·6	+59	20	11·0	5·5 × 3·2	Sb
	6015	15	51·4	+62	19	11·2	5·4 × 2·3	Sc
	5907	15	15·9	+56	19	10·4	12·3 × 1·8	Sb
	5985	15	39·6	+59	20	11·0	5·5 × 3·2	Sb
	6503	17	49·4	+70	09	10·2	6·2 × 2·3	Sb

EQUULEUS

(Abbreviation: Eql).

A very obscure and small constellation, but one of the 'originals'. It represents a foal given by Mercury to Castor, one of the Heavenly Twins. There is only one star above the fourth magnitude: α. The name often applied to α has a decidedly modern flavour!

	Mag.	Luminosity Sun = 1	Dist lt-yrs
α	3·92	50	150

See chart for Aquila.

BRIGHTEST STARS

Star	R.A. h	m	s	Dec. deg	min	sec	Mag.	Abs. mag.	Spectrum	Dist pc	
5 γ	21	10	20·3	+10	07	53	4·69	1·0	F0p	64	
7 δ	21	14	28·7	+10	00	25	4·49	4·0	F8	15	
8 α	21	15	49·3	+05	14	52	3·92	0·6	G0	46	Kitalpha

VARIABLE STAR

	R.A. h	m	Dec. deg	min	Range	Type	Period, d.	Spectrum
S	20	57·2	+05	05	8·0–10·1	Algol	3·44	B + F

DOUBLE STARS

	R.A. h	m	Dec. deg	min	P.A. deg	Sep. sec	Mags.	
ε	20	59·1	+04	18	AB 285	1·0	6·0, 6·3	Binary, 101·4 y
					AB + C 070	10·7	7·1	
					AD 280	74·8	12·4	
γ	21	10·4	+10	08	AB 268	1·9	4·7, 11·5	
					AC 005	47·7	12·5	
δ	21	14·5	+10	00	029	0·3	5·2, 5·3	Binary, 5·7 y
β	21	22·9	+06	49	257	34·4	5·2, 13·7	

ERIDANUS

(Abbreviation: Eri).

An immensely long constellation, extending from Achernar in the far south as far as Kursa, near Orion. Achernar is the only brilliant star, though θ (Acamar) was ranked as the first magnitude in ancient times, and there are suggestions that it has faded. Mythologically, Eridanus is the river Po – and this was the river into which the youth Phaëthon was plunged when he had obtained permission to drive the Sun-chariot for a day, and had lost control of it, so that Jupiter was forced to strike him down with a thunderbolt.

BRIGHTEST STARS

Star	R.A. h	m	s	Dec. deg	min	sec	Mag.	Abs. mag.	Spectrum	Dist pc	
α	01	37	42·9	−57	14	12	0·46	−1·6	B5	26	Achernar
χ	01	55	57·5	−51	36	32	3·70	3·2	G5	15	
κ	02	26	59·1	−47	42	14	4·25	−2·2	B5	190	
φ	02	16	30·6	−51	30	44	3·56	−0·2	B8	37	
ι	02	40	40·0	−39	51	19	4·11	0·2	K0	61	
1 τ¹	02	45	06·1	−18	34	21	4·47	3·7	F6	15	
2 τ²	02	51	02·2	−21	00	15	4·75	0·2	K0	81	
3 η	02	56	25·6	−08	53	54	3·89	0·9	K1	23	Azha
θ	02	58	15·6	−40	18	17	2·92	0·6+1·7	A3 + A2	17	Acamar
11 τ³	03	02	23·4	−23	37	28	4·09	2·1	A5	23	
13 ζ	03	15	49·9	−08	49	11	4·80	1·8	A3	16	Zibal
16 τ⁴	03	19	30·9	−21	45	28	3·69	−0·5	gM3	69	Angetenar
e	03	19	55·7	−43	04	10	4·27	5·3	G5	6·2	
17 ν	03	30	37·0	−05	04	30	4·73	−0·2	B8	97	
18 ε	03	32	55·8	−09	27	30	3·73	6·1	K2	3·3	
19 τ⁵	03	33	47·2	−21	37	58	4·27	−0·2	B8	78	
γ	03	37	05·6	−40	16	29	4·58	0·2	K0	89	
h	03	42	50·0	−37	18	49	4·59	−0·3	gK5	95	
δ	03	43	14·8	−09	45	48	3·54	3·8	K0	9	Rana
26 π	03	46	08·4	−12	06	06	4·42	−1·1	gMa	50	
27 τ⁶	03	46	50·8	−23	14	59	4·23	3·1	F3	17	
g	03	48	35·3	−37	37	20	4·27	0·6	A0	50	

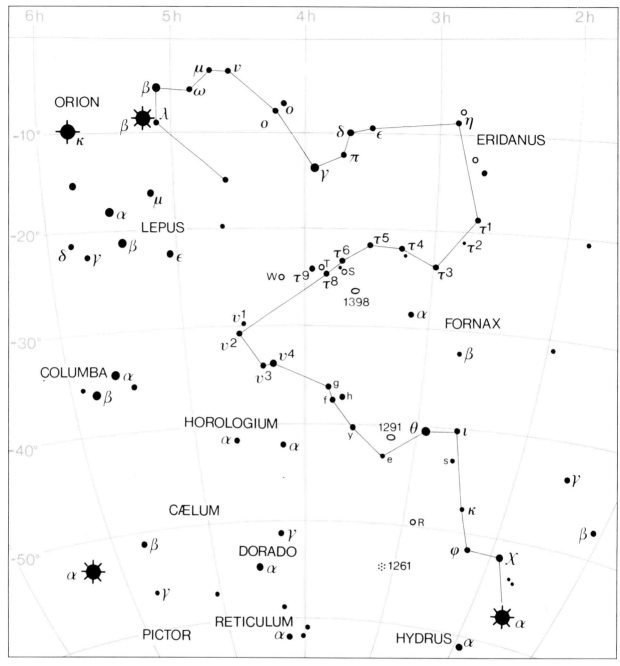

ERIDANUS (continued)

Star	R.A.			Dec.			Mag.	Abs. mag.	Spectrum	Dist pc	
	h	m	s	deg	min	sec					
33 τ⁸	03	53	42·6	−24	36	45	4·65	−1·1	B5	140	
32 w	03	54	17·4	−02	57	17	4·46	0·3	G8	68	
34 γ	03	58	01·7	−13	30	31	2·95	−0·4	M0	44	Zaurak
36 τ⁹	03	59	55·4	−24	00	59	4·66	−0·6	A0	110	
38 o¹	04	11	51·8	−06	50	15	4·04	−0·7	F2	85	Beid
40 o²	04	15	16·2	−07	39	10	4·43	6·0	K1	4·9	Keid
41 υ⁴	04	17	53·6	−33	47	54	3·56	0·0	B8·5	40	
43 υ³	04	24	02·1	−34	01	01	3·96	−0·5	M1	78	
50 υ¹	04	33	30·6	−29	46	00	4·51	0·3	gG6	63	
52 υ²	04	35	33·0	−30	33	45	3·82	0·2	K0	53	Theemini
48 v	04	36	19·1	−03	21	09	3·93	−3·6	B2	350	
53 l	04	38	10·7	−14	18	15	3·87	−0·1	K2	44	Sceptrum
54	04	40	26·4	−19	40	18	4·32	−0·5	gM4	92	
57 μ	04	45	30·1	−03	15	17	4·02	−1·6	B5	130	
61 ω	04	52	53·6	−05	27	10	4·39	1·6	A9	36	

ERIDANUS (continued)

There are 15 stars above magnitude 4:

	Mag.	Luminosity Sun = 1	Dist lt-yrs
α	0·46	400	85
β	2·79	82	100
θ	2·92	50+17	55
γ	2·95	110	144
δ	3·54	3	29
υ⁴	3·56	82	130
φ	3·56	105	120
τ⁴	3·69	120	225
χ	3·70	4·5	49
ε	3·73	0·3	10·7
υ²	3·82	60	173
53	3·87	96	144
η	3·89	37	75
ν	3·93	2200	1140
υ³	3·96	120	255

Star	R.A.			Dec.			Mag.	Abs. mag.	Spectrum	Dist pc	
	h	m	s	deg	min	sec					
67 β	05	07	50·9	−05	05	11	2·79	0·0	A3	28	Kursa
69 λ	05	09	08·7	−08	45	15	4·27	−3·0	B2	280	

Also above mag. 5:

	Mag.	Abs. mag.	Spectrum	Dist
45	4·91	−1·3	K3	170
39	4·87	−0·2	K3	100
15	4·88	0·3	G6	82
64 S	4·79	1·7	F0	42
65 ψ	4·81	−2·3	B2	290

VARIABLE STARS

	R.A.		Dec.		Range	Type	Period, d.	Spectrum
	h	m	deg	min				
Z	02	47·9	−12	28	7·0–8·6	Semi-reg.	80	M
RR	02	52·2	−08	16	7·4–8·6	Semi-reg.	97	M
T	03	55·2	−24	02	7·4–13·2	Mira	252·2	M
W	04	11·5	−25	08	7·5–14·5	Mira	376·7	M
RZ	04	43·8	−10	41	7·8–8·7	Algol	39·28	A + G

DOUBLE STARS

	R.A.		Dec.		P.A.	Sep.	Mags.	
	h	m	deg	min	deg	sec		
χ	01	56·0	−51	37	202	5·0	3·7, 10·7	
p	01	39·8	−56	12	194	11·2	5·5, 5·8	Binary, 484 y
θ	02	58·3	−40	18	088	8·2	3·4, 4·5	
τ⁴	03	19·5	−21	45	AB 288	5·7	3·7, 9·2	
					AC 112	39·2	10·7	
υ⁴	04	17·9	−33	48	013	49·2	3·6, 11·8	A is a close dble
ρ²	03	02·7	−07	41	075	1·8	5·3, 9·5	
o²	04	15·2	−07	39	107	83·4	4·4, 9·5	

PLANETARY NEBULA

M	NGC	R.A.		Dec.		Dimensions	Mag.	Mag. of central star
		h	m	deg	min	sec		
	1535	04	14·2	−12	44	18 × 44	9·6	12·2

GALAXIES

M	NGC	R.A.		Dec.		Mag.	Dimensions min	Type
		h	m	deg	min			
	1084	02	46·0	−07	35	10·6	2·9 × 1·5	Sc
	1179	03	02·6	−18	54	11·8	4·6 × 3·9	Sp
	1187	03	02·6	−22	52	10·9	5·0 × 4·1	SBc
	1291	03	17·3	−41	08	8·5	10·5 × 9·1	SBa
	1300	03	19·7	−19	25	10·4	6·5 × 4·3	SBb
	1332	03	26·3	−21	20	10·3	4·6 × 1·7	E7
	1337	03	28·1	−08	23	11·7	6·8 × 2·0	S
	1395	03	38·5	−23	02	11·3	3·2 × 2·5	E3
	1407	03	40·2	−18	35	9·8	2·5 × 2·5	E0
	1532	04	12·1	−32	52	11·1	5·6 × 1·8	Sb
	1637	04	41·5	−02	51	10·9	3·3 × 2·9	Sc

FORNAX

(Abbreviation: For).

A southern group, originally Fornax Chemica (the Chemical Furnace). It has no bright stars, but is notable for containing a large number of faint galaxies.

The only star above the fourth magnitude is α.

	Mag.	Luminosity Sun = 1	Dist lt-yrs
α	3·87	4	46

See chart for Eridanus.

BRIGHTEST STARS

Star	R.A.			Dec.			Mag.	Abs. mag.	Spectrum	Dist pc
	h	m	s	deg	min	sec				
ν	02	04	29·4	−29	17	49	4·69	−0·6	A0	110
β	02	49	05·4	−32	24	22	4·46	0·3	G6	61
α	03	12	04·2	−28	59	13	3·87	3·3	F8	14

Also above mag. 5:

	Mag.	Abs. mag.	Spectrum	Dist
δ	5·00		B5	
ω	4·90	0·2	B9	

VARIABLE STARS

	R.A.		Dec.		Range	Type	Period, d.	Spectrum
	h	m	deg	min				
R	02	29·3	−26	06	7·5–13·0	Mira	387·9	N
ST	02	44·4	−29	12	7·7–9·0	Semi-reg.	277	M
S	03	46·2	−24	24	?5·6–8·5?	Suspected	?	M

DOUBLE STARS

	R.A.		Dec.		P.A.	Sep.	Mags.
	h	m	deg	min	deg	sec	
ω	02	33·8	−28	14	244	10·8	5·0, 7·7
γ¹	02	49·8	−24	34	AB 145	12·0	6·1, 12·5

FORNAX (continued)

	R.A.		Dec.		P.A.	Sep.	Mags.	
	h	m	deg	min	deg	sec		
					AC 143	40·9	10·5	
η^2	02	50·2	−35	51	014	5·0	5·9, 10·1	
α	03	12·1	−28	59	298	4·0	4·0, 7·0	Binary, 314 y
χ^3	03	28·2	−35	51	248	6·3	6·5, 10·5	

PLANETARY NEBULA

M	NGC	R.A.		Dec.		Dimensions	Mag.	Mag. of central star
		h	m	deg	min	sec		
	1360	03	33·3	−25	51	390	−	11·3

GALAXIES

M	NGC	R.A.		Dec.		Mag.	Dimensions	Type
		h	m	deg	min		min	
	986	02	33·6	−39	02	11·0	3·7 × 2·8	SBb
	1097	02	46·3	−30	17	9·2	9·3 × 6·6	SBb
	1201	03	04·1	−26	04	10·6	4·4 × 2·8	Sa
	1255	03	13·5	−25	44	11·1	4·1 × 2·8	Sa
	1302	03	19·9	−26	04	11·5	4·4 × 4·2	SBa
	1316	03	22·7	−37	12	8·8	7·1 × 5·5	SB0p
	1326	03	23·9	−36	28	10·5	4·0 × 3·0	SB0
	1344	03	28·3	−31	04	10·3	3·9 × 2·3	E3
	1350	03	31·1	−33	38	10·5	4·3 × 2·4	SBb
	1365	03	33·6	−36	08	9·5	9·8 × 5·5	SBb
	1371	03	35·0	−24	56	11·5	5·4 × 4·0	SBa
	1380	03	36·5	−34	59	11·1	4·9 × 1·9	S0
	1385	03	37·5	−24	30	11·2	3·0 × 2·0	Sc
	1399	03	38·5	−35	27	9·9	3·2 × 3·1	E1p
	1398	03	38·9	−26	20	9·7	6·6 × 5·2	SBb
	1404	03	38·9	−35	35	10·2	2·5 × 2·3	E1
	1425	03	42·2	−29	54	11·7	5·4 × 2·7	Sb

GEMINI

(Abbreviation: Gem).

A brilliant Zodiacal constellation. In mythology, Castor and Pollux were the twin sons of the King and Queen of Sparta. Pollux was immortal, but Castor was not. When Castor was killed, Pollux pleaded with the gods to be allowed to share his immortality; so Castor was brought back to life, and both youths placed in the sky.

Today Pollux is the brighter of the two stars, but in ancient times Castor was recorded as being the more brilliant. If any change has occurred (and this is by no means certain) it is more likely to have been in the late-type Pollux than in Castor, which is a multiple system. There are 13 stars in Gemini above the fourth magnitude:

	Mag.	Luminosity Sun = 1	Dist lt-yrs
β	1·14	60	36
α	1·58	36	46
γ	1·93	82	85
μ	2·88v	120	230
ε	2·98	5200	680
η	3·1 (max)	120v	186
ξ	3·36	46	75
δ	3·53	14	59
κ	3·57	58	147
λ	3·58	17	81
θ	3·60	82	166
ζ	3·7 (max)	5200	1400
ι	3·79	60	163

BRIGHTEST STARS

Star	R.A.			Dec.			Mag.	Abs. mag.	Spectrum	Dist pc	
	h	m	s	deg	min	sec					
ι	06	04	07·2	+23	15	48	4·16	0·3	gG5	59	
7 η	06	14	52·6	+22	30	24	var.	−0·5v	M3	57	Propus
13 μ	06	22	57·6	+22	30	49	2·88v	−0·5	M3	71	Tejat
18 ν	06	28	57·7	+20	12	43	4·15	−1·0	B7	110	
24 γ	06	37	42·7	+16	23	57	1·93	0·0	A0	26	Alhena
27 ε	06	43	55·9	+25	07	52	2·98	−4·5	G8	210	Mebsuta
30	06	43	59·2	+13	13	40	4·49	0·0	K1	70	

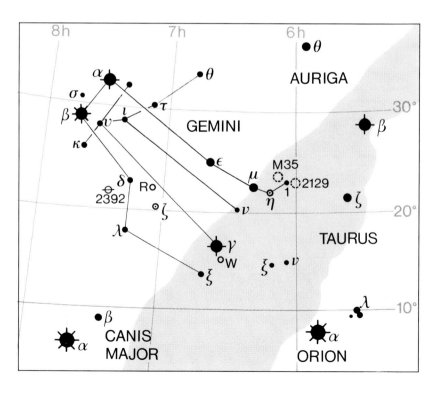

GEMINI (continued)

Star	R.A. h	m	s	Dec. deg	min	sec	Mag.	Abs. mag.	Spectrum	Dist pc	
31 ξ	06	45	17·3	+12	53	44	3·36	0·7	F5	23	Alzirr
34 θ	06	52	47·3	+33	57	40	3·60	0·0	A3	51	
38 e	06	54	38·6	+13	10	40	4·65	2·6	F0	26	
43 ζ	07	04	06·5	+20	34	13	var.	−4·5	G0	430	Mekbuda
46 τ	07	11	08·3	+30	14	43	4·41	−0·1	K2	67	
54 λ	07	18	05·5	+16	32	25	3·58	1·7	A3	25	
55 δ	07	20	07·3	+21	58	56	3·53	1·9	F2	18	Wasat
60 ι	07	25	43·5	+27	47	53	3·79	0·2	K0	50	
62 ρ	07	29	06·6	+31	47	03	4·18	2·6	F0	19	
66 α	07	34	35·9	+31	53	18	1·58	1·2	A0	14	Castor
69 υ	07	35	55·3	+26	53	45	4·06	−0·4	M0	78	
75 σ	07	43	18·7	+28	53	01	4·28	0·0	K1	40	
77 κ	07	44	26·8	+24	23	52	3·57	0·3	G8	45	
78 β	07	45	18·9	+28	01	34	1·14	0·2	K0	11	Pollux

Also above mag. 5:

	Mag.	Abs. mag.	Spectrum	Dist
51	5·00	−0·5	M4	120
71 o	4·90	0·6	F3	70
81	4·88	−0·3	K5	110
83 φ	4·97	1·7	A3	45

VARIABLE STARS

	R.A. h	m	Dec. deg	min	Range	Type	Period, d.	Spectrum
BU	06	12·3	+22	54	5·7–7·5	Irreg.	−	M
η	06	14·9	+22	30	3·2–3·9	Semi-reg.	233	M
W	06	35·0	+15	20	6·5–7·4	Cepheid	7·91	F–G
X	06	47·1	+30	17	7·5–13·6	Mira	263·7	M
ζ	07	04·1	+20	34	3·7–4·1	Cepheid	10·15	F–G
R	07	07·4	+22	42	6·0–14·0	Mira	369·8	S
V	07	23·2	+13	06	7·8–14·9	Mira	275·1	M
T	07	49·3	+23	44	8·0–15·0	Mira	287·8	S
U	07	55·1	+22	00	8·2–14·9	SS Cygni	±103	M + WD

DOUBLE STARS

	R.A. h	m	Dec. deg	min	P.A. deg	Sep. sec	Mags.	
η	06	14·9	+22	30	266	1·4	var., 8·8	Binary, 474 y
μ	06	22·9	+22	31	077	72·7	3·0, 9·8	
ν	06	29·0	+20	13	329	112·5	4·2, 8·7	
ε	06	43·9	+25	08	094	110·3	3·0, 9·0	
38	06	54·6	+13	11	147	7·0	4·7, 7·7	Binary, 3190 y
ζ	07	04·1	+20	34	AB 084	87·0	var., 10·5	
					AC 350	96·5	8·0	
λ	07	18·1	+16	32	033	9·6	3·6, 10·7	
δ	07	20·1	+21	59	223	6·0	3·5, 8·2	Binary, 1200 y
α	07	34·6	+31	53	AB 088	2·5	1·9, 2·9	Binary, 420 y
					AC 164	72·5	1·6, 8·8	
κ	07	44·4	+24	24	240	7·1	3·6, 8·1	

OPEN CLUSTERS

M	NGC	R.A. h	m	Dec. deg	min	Diameter min	Mag.	No. of stars
	2129	06	01·0	+23	18	7	6·7	40
	IC 2157	06	05·0	+24	00	7	8·4	20
	2169	06	08·4	+13	57	7	5·9	30
35	2168	06	08·9	+24	20	28	5·0	200
	2266	06	43·2	+26	58	7	9·5	30
	2355	07	16·9	+13	47	9	9·7	40
	2395	07	27·1	+13	35	12	8·0	30 Asterism?

PLANETARY NEBULA

M	NGC	R.A. h	m	Dec. deg	min	Dimensions sec	Mag.	Mag. of central star	
	2392	07	29·2	+20	55	13 × 44	10	10·5	Eskimo Nebula

GRUS

(Abbreviation: Gru).

The most distinctive of the 'Southern Birds'. The contrast in colour between α and β is very marked.

BRIGHTEST STARS

Star	R.A. h	m	s	Dec. deg	min	sec	Mag.	Abs. mag.	Spectrum	Dist pc	
γ	21	53	55·6	−37	21	54	3·01	−1·2	B8	70	
λ	22	06	06·7	−39	32	36	4·46	−0·4	M0	94	
α	22	08	13·8	−46	57	40	1·74	−1·1	B5	21	Alnair
δ¹	22	29	15·9	−43	29	45	3·97	0·3	gG5	43	
δ²	22	29	45·3	−43	44	58	4·11		M4	27	
β	22	42	39·9	−46	53	05	2·11v	−2·4	M3	53	Al Dhanab

GRUS (continued)

There are six stars above the fourth magnitude:

	Mag.	Luminosity Sun = 1	Dist lt-yrs
α	1·74	230	68
β	2·11v	800	173
γ	3·01	250	228
ε	3·49	26	81
ι	3·90	60	173
δ¹	3·97	58	140

β has a very small range (2·11 to 2·23), so that its fluctuations are not detectable with the naked eye.

Star	R.A. h	m	s	Dec. deg	min	sec	Mag.	Abs. mag.	Spectrum	Dist pc
ε	22	48	33·1	−51	19	01	3·49	1·4	A2	25
ζ	23	00	52·6	−52	45	15	4·12	0·3	G5	50
θ	23	06	52·6	−43	31	14	4·28	2·2	F6	26
ι	23	10	21·4	−45	14	48	3·90	0·2	K0	53

Also above mag. 5:

	Mag.	Abs. mag.	Spectrum	Dist
μ	4·79	0·3	G4	79
ρ	4·85	0·2	K0	83
η	4·85	0·2	K0	62

VARIABLE STARS

	R.A. h	m	Dec. deg	min	Range	Type	Period, d.	Spectrum
RS	21	43·1	−48	11	7·9–8·5	Delta Scuti	0·15	A–F
R	21	48·5	−46	55	7·4–14·9	Mira	331·9	M
π¹	22	22·7	−45	57	5·4–6·7	Semi-reg.	150	S
T	22	25·7	−37	34	7·8–12·3	Mira	136·5	M
S	22	26·1	−48	26	6·0–15·0	Mira	401·4	M

DOUBLE STARS

	R.A. h	m	Dec. deg	min	P.A. deg	Sep. sec	Mags.
θ	23	06·9	−43	31	075	1·1	4·5, 7·0
υ	23	06·9	−38	54	211	1·1	5·7, 8·0

GALAXIES

M	NGC	R.A. h	m	Dec. deg	min	Mag.	Dimensions min	Type
	7144	22	52·7	−48	15	10·7	3·5 × 3·5	E0
	7213	22	09·3	−47	10	10·4	1·9 × 1·8	Sa
	7410	22	55·0	−39	40	10·4	5·5 × 2·0	SBa
	7412	22	55·8	−42	39	11·4	4·0 × 3·1	SBb
	7418	22	56·6	−37	02	11·4	3·3 × 2·8	SBc
	IC 1459	22	57·2	−36	28	10·0	–	E3
	IC 5267	22	57·2	−43	24	10·5	5·0 × 4·1	S0
	7424	22	57·3	−41	04	11·0	7·6 × 6·8	SBc
	IC 5273	22	59·5	−37	42	11·4	2·9 × 2·1	SBc
	7456	23	02·1	−39	35	11·9	5·9 × 1·8	Sc
	7496	23	09·8	−43	26	11·1	3·5 × 2·8	SBb
	7531	23	14·8	−43	36	11·3	3·5 × 1·5	Sb
	7552	23	16·2	−42	35	10·7	3·5 × 2·5	SBb
	7582	23	18·4	−42	22	10·6	4·6 × 2·2	SBG
	7599	23	19·3	−42	15	11·4	4·4 × 1·5	Sc

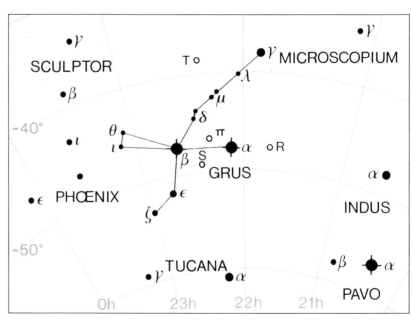

HERCULES

(Abbreviation: Her).

A large but by no means brilliant constellation, commemorating the great hero of mythology. The most interesting objects are the red supergiant α and the globular clusters M.13 and M.92; M.13 is the brightest globular in the northern hemisphere of the sky, and is surpassed only by the southern ω Centauri and 47 Tucanæ.

Hercules contains fifteen stars above the fourth magnitude:

	Mag.	Luminosity	Dist
		Sun = 1	lt-yrs
β	2·77	58	100
ζ	2·81	5·2	31
α	3·0 (max)	700v	220
δ	3·14	37	91
π	3·16	700	390
μ	3·42	2·5	26
η	3·53	16	68
ξ	3·70	60	163
γ	3·75	50	137
ι	3·80	450	420
o	3·83	60	170
109	3·84	96	108
θ	3·86	650	420
τ	3·89	400	425
ε	3·92	50	85

BRIGHTEST STARS

Star	R.A.			Dec.			Mag.	Abs.	Spectrum	Dist	
	h	m	s	deg	min	sec		mag.		pc	
1 χ	15	52	40·4	+42	27	05	4·62	4·2	F9	15	
11 φ	16	08	46·0	+44	56	05	4·26	0·0	B9p	23	
22 τ	16	19	44·3	+46	18	48	3·89	−1·6	B5	130	
20 γ	16	21	55·1	+19	09	11	3·75	0·6	A9	42	
24 ω	16	25	24·8	+14	02	00	4·57	1·8	A0p	33	Cujam
27 β	16	30	13·1	+21	29	22	2·77	0·3	G8	31	Kornephoros
35 σ	16	34	06·0	+42	26	13	4·20	0·2	B9	61	
40 ζ	16	41	17·1	+31	36	10	2·81	3·0	G0	9·6	Rutilicus
44 η	16	42	53·7	+38	55	20	3·53	1·8	G8	21	
58 ε	17	00	17·2	+30	55	35	3·92	0·6	A0	26	
64 α	17	14	38·8	+14	23	25	var.	−2·3v	M5	67	Rasalgethi
65 δ	17	15	01·8	+24	50	21	3·14	0·9	A3	28	Sarin
67 π	17	15	02·6	+36	48	33	3·16	−2·3	K3	120	
68 u	17	17	19·4	+33	06	00	var.	−2·9v	B3	350	
69 e	17	17	40·1	+37	17	29	4·65	1·4	A2	45	
75 ρ	17	23	40·8	+37	08	45	4·17	0·6	A0	52	
76 λ	17	30	44·1	+26	06	39	4·41	−0·3	K4	85	Masym
85 ι	17	39	27·7	+46	00	23	3·80	−1·7	B3	130	
86 μ	17	46	27·3	+27	43	15	3·42	3·9	G5	8·1	
91 θ	17	56	15·1	+37	15	02	3·86	−2·2	K1	130	
92 ξ	17	57	45·7	+29	14	52	3·70	0·2	K0	50	
94 ν	17	58	30·0	+30	11	22	4·41	−2·0	F2	190	
93	18	00	03·2	+16	45	03	4·67	−0·9	K0	100	
95	18	01	30·3	+21	35	44	4·27	0·5	A7	48	
103 o	18	07	32·4	+28	45	45	3·83	0·2	B9	52	
102	18	08	45·4	+20	48	52	4·36	−2·5	B2	240	
109	18	23	41·7	+21	46	11	3·84	−0·1	K2	33	
110	18	45	39·6	+20	32	47	4·19	3·7	F6	15	
111	18	47	01·1	+18	10	53	4·36	1·7	A3	34	
113	18	54	44·7	+22	38	43	4·56	1·7	A3	36	

Also above mag. 5:

		Mag.	Abs. mag.	Spectrum	Dist
6 υ		4·76		B9	
7 κ		5·00	−0·1	K2	87
29		4·84	−0·3	K5	70
g		5·0v	var.	M6	23
52		4·82	0·9	A2p	
60		4·91	0·9	A3	62
104		4·97	−0·5	M3	120
106		4·95	−0·4	M0	110

VARIABLE STARS

	R.A.		Dec.		Range	Type	Period, d.	Spectrum
	h	m	deg	min				
X	16	02·7	+47	14	7·5–8·6	Semi-reg.	95	M
R	16	06·2	+18	22	7·8–15·0	Mira	318·4	M
RU	16	10·2	+25	04	6·8–14·3	Mira	485·5	M
U	16	25·8	+18	54	6·5–13·4	Mira	406·0	M
g (30)	16	28·6	+41	53	5·7–7·2	Semi-reg.	70	M
W	16	35·2	+37	21	7·6–14·4	Mira	280·4	M
S	16	51·9	+14	56	6·4–13·8	Mira	307·4	M
α	17	14·6	+14	23	3–4	Semi-reg.	±100?	M
u (68)	17	17·3	+35	06	4·6–5·3	Beta Lyræ	2·05	B + B
RS	17	21·7	+22	55	7·0–13·0	Mira	219·6	M
Z	17	58·1	+15	08	7·3–8·1	Algol	3·99	F + K
T	18	09·1	+31	01	6·8–13·9	Mira	165·0	M
AC	18	30·3	+21	52	7·4–9·7	RV Tauri	75·5	F + K
RX	18	30·7	+12	37	7·2–7·8	Algol	1·78	A + A

DOUBLE STARS

	R.A.		Dec.		P.A.	Sep.	Mags.	
	h	m	deg	min	deg	sec		
κ	16	08·1	+17	03	012	28·4	5·3, 6·5	
γ	16	21·9	+19	09	233	41·6	3·8, 9·8	
ω	16	25·4	+14	02	AB 223	1·0	4·6, 11·6	
					AC 096	28·4	11·1	
37	16	40·6	+04	13	230	69·8	5·8, 7·0	
ζ	16	41·3	+31	36	089	1·6	2·9, 5·5	Binary, 34·5 y
54	16	55·4	+18	26	183	2·5	5·4, 12·7	
α	17	14·6	+14	23	107	4·7	var., 5·4	Binary, 3600 y
δ	17	15·0	+24	50	236	8·9	3·7, 8·2	Optical pair
u (68)	17	17·3	+35	06	060	4·4	var., 10·2	
ρ	17	23·7	+37	09	316	4·1	4·6, 5·6	
μ	17	46·5	+27	43	247	33·8	3·4, 10·1	

HERCULES (continued)

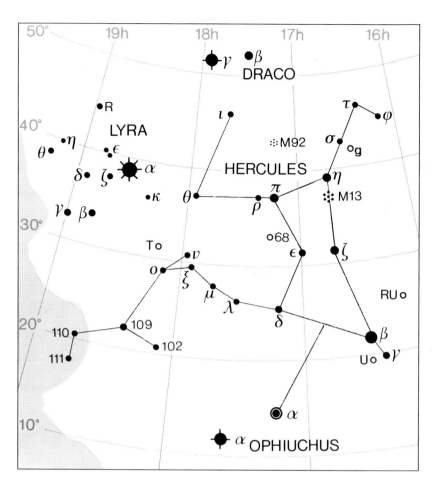

GLOBULAR CLUSTERS

M	NGC	R.A.		Dec.		Diameter	Mag.
		h	m	deg	min	min	
92	6341	17	17·1	+43	08	11·2	6·5
13	6205	16	41·7	+36	28	16·6	5·9

PLANETARY NEBULÆ

M	NGC	R.A.		Dec.		Dimensions	Mag.	Mag. of central star
		h	m	deg	min	sec		
	6058	16	04·4	+40	41	23	13·3	13·8
	IC 4593	16	12·2	+12	04	12 × 120	10·9	11·3
	6210	16	44·5	+23	49	14	9·3	12·9

HOROLOGIUM

(Abbreviation: Hor).

A very obscure southern constellation. The only star above the fourth magnitude is α. The only other star above the fifth magnitude is δ; magnitude 4·93.

	Mag.	Luminosity	Dist
		Sun = 1	lt-yrs
α	3·86	82	192

See chart for Eridanus.

BRIGHTEST STAR

Star	R.A.			Dec.			Mag.	Abs.	Spectrum	Dist
	h	m	s	deg	min	sec		mag.		pc
α	04	14	00·0	−42	17	40	3·86	0·0	K1	59

Also above mag. 5:

	Mag.	Abs. mag.	Spectrum	Dist
δ	4·93	2·6	F0	28

VARIABLE STARS

	R.A.		Dec.		Range	Type	Period, d.	Spectrum
	h	m	deg	min				
R	02	53·9	−49	53	4·7–14·3	Mira	404·0	M
T	03	00·9	−50	39	7·2–13·7	Mira	217·7	M
V	03	03·5	−58	56	7·8–8·9	Semi-reg.	?	M
U	03	52·8	−45	50	7·8–15·1	Mira	348·4	M

GLOBULAR CLUSTER

M	NGC	R.A.		Dec.		Diameter	Mag.
		h	m	deg	min	min	
	1261	03	12·3	−55	13	6·9	8·4

HOROLOGIUM
(continued)

GALAXIES

M	NGC	R.A.		Dec.		Mag.	Dimensions	Type
		h	m	deg	min		min	
	1249	03	10·0	−53	21	11·7	5·2 × 2·7	SBc
	1411	03	38·8	−44	05	11·9	2·8 × 2·3	S0
	1433	03	42·0	−47	13	10·0	6·8 × 6·0	SBa
	1448	03	44·5	−44	39	11·3	8·1 × 1·8	Sc
	1493	03	57·5	−46	12	11·8	2·6 × 2·3	SBc
	1512	04	03·9	−43	21	10·6	4·0 × 3·2	SBa

HYDRA

(Abbreviation: Hya).

The largest constellation in the sky, representing the hundred-headed monster which lived in the Lernæan marshes until it was killed by Hercules. It extends for over 6 hours of R.A. α (Alphard) is often known as 'the Solitary One', because there are no other bright stars anywhere in the neighbourhood.

Hydra contains 10 stars above the fourth magnitude:

	Mag.	Luminosity	Dist
		Sun = 1	lt-yrs
α	1·98	105	85
γ	3·00	58	104
ζ	3·11	60	124
ν	3·11	96	127
π	3·27	96	153
ε	3·38	50	110
ξ	3·54	58	144
λ	3·61	60	150
μ	3·81	110	192
θ	3·88	50	147

BRIGHTEST STARS

Star	R.A.			Dec.			Mag.	Abs. mag.	Spectrum	Dist pc	
	h	m	s	deg	min	sec					
4 δ	08	37	39·3	+05	42	13	4·16	0·6	A0	43	
5 σ	08	38	45·4	+03	20	29	4·44	−0·1	K2	75	Al Minliar al Shuja
7 η	08	43	13·4	+03	23	55	4·30	−1·7	B3	160	
12 D	08	46	22·4	−13	32	52	4·32	0·3	G8	64	
11 ε	08	46	46·5	+06	25	07	3·38	0·6	G0	34	
13 ρ	08	48	25·9	+05	50	16	4·36	0·6	A0	57	
16 ζ	08	55	23·6	+05	56	44	3·11	0·2	K0	38	
22 θ	09	14	21·8	+02	18	51	3·88	0·6	A0	45	
30 α	09	27	35·2	−08	39	31	1·98	−0·2	K3	26	Alphard
31 τ¹	09	29	08·8	−02	46	08	4·60	3·7	F6	15	
32 τ²	09	31	58·9	−01	11	06	4·57	0·0	A3	82	
35 ι	09	39	51·3	−01	08	34	3·91	−0·2	K3	63	
39 υ¹	09	51	28·6	−14	50	48	4·12	0·3	G8	58	
40 υ²	10	05	07·4	−13	03	53	4·60	−1·2	B8	140	
41 λ	10	10	05·3	−12	21	15	3·61	0·2	K0	46	
42 μ	10	26	05·3	−16	50	11	3·81	−0·3	K4	59	
ν	10	49	37·4	−16	11	37	3·11	−0·1	K2	39	
ξ	11	33	00·1	−31	51	27	3·54	0·3	G7	44	
o	11	40	12·9	−34	44	40	4·70	0·3	B9	?	
β	11	52	54·5	−33	54	28	4·28	−0·3	B9	82	
46 γ	13	18	55·2	−23	10	17	3·00	0·3	G5	32	
R	13	29	42·7	−23	16	52	var.	var.	Md	100	(Very red)
49 π	14	06	22·2	−26	40	56	3·27	−0·1	K2	47	
58 E	14	50	17·2	−27	57	37	4·41	−0·3	gK4	88	

Also above mag. 5:

	Mag.	Abs. mag.	Spectrum	Dist.
6	4·98	−0·3	K4	110
9	4·88	0·0	K1	95

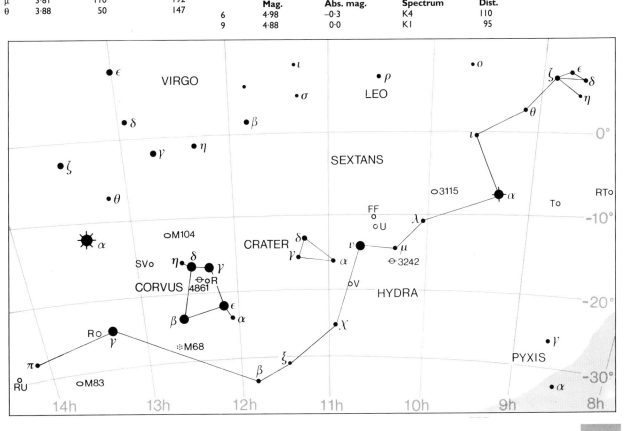

HYDRA (continued)

	Mag.	Abs. mag.	Spectrum	Dist.
18 ω	4·97	−1·1	K2	170
26	4·79	0·3	G8	79
27 P	4·80	1·8	G8	39
F	4·70	−4·1	G4	(=31 Mon)
HD 83953	4·77	−1·1	B5	150
φ	4·91	0·2	K0	88
χ¹	4·94	3·3	F4	21
45 ψ	4·95	0·2	K0	81
51 κ	4·77	−0·3	K5	48
52	4·97	−0·6	B8	130
54	4·94	0·6	F0	51

VARIABLE STARS

	R.A. h	m	Dec. deg	min	Range	Type	Period, d.	Spectrum
RT	08	29·7	−06	19	7·0–11·0	Semi-reg.	253	M
S	08	53·6	+03	04	7·4–13·3	Mira	256·4	M
T	08	55·7	−09	08	6·7–13·2	Mira	289·2	M
X	09	35·5	−14	42	8·0–13·6	Mira	301·4	M
U	10	37·6	−13	23	4·8–5·8	Semi-reg.	450	N
R	13	29·7	−23	17	4·0–10·0	Mira	389·6	M
TT	11	13·2	−26	28	7·5–9·5	Algol	6·95	A + G
HZ	11	26·3	−25	45	7·6–8·2	Semi-reg.	95	M
W	13	49·0	−28	22	7·7–11·6	Semi-reg.	397	M
RU	14	11·6	−28	53	7·2–14·3	Mira	333·2	M

DOUBLE STARS

	R.A. h	m	Dec. deg	min	P.A. deg	Sep. sec	Mags.	
ε	08	46·8	+06	25	AB 295	0·2	3·8, 4·7	Binary, 890 y
					AB + C 281	2·8	6·8	
θ	09	14·4	+02	19	197	29·4	3·9, 9·9	
α	09	17·6	−08	40	153	283·1	2·0, 9·5	
β	11	52·9	−33	54	008	0·9	4·7, 5·5	
R	13	29·7	−23	17	324	21·2	var., 12·0	
52	14	28·2	−29	30	AB 130	0·1	5·8, 5·8	
					AB + C 279	4·2	10·0	
					AB + D 282	140·8	12·0	
59	14	58·7	−27	39	335	0·8	6·3, 6·6	

OPEN CLUSTER

M	NGC	R.A. h	m	Dec. deg	min	Diameter min	Mag.	No. of stars
48	2548	08	13·8	−05	48	54	5·8	80

GLOBULAR CLUSTER

M	NGC	R.A. h	m	Dec. deg	min	Diameter min	Mag.
68	4590	12	39·5	−26	45	12·0	8·2

GALAXIES

M	NGC	R.A. h	m	Dec. deg	min	Mag.	Dimensions min	Type
	2784	09	12·3	−24	10	10·1	5·1 × 2·3	S0
	2835	09	17·9	−22	21	11·1	6·3 × 4·4	Sp
	3109	10	03·1	−26	09	10·4	14·5 × 3·5	Irr.
	3585	11	13·3	−26	45	10·0	2·9 × 1·6	E5
	3621	11	18·3	−32	49	9·9	10·0 × 6·5	Sc
	3923	11	51·0	−28	48	10·1	2·9 × 1·9	E3
	5078	13	19·8	−27	24	12·0	3·2 × 1·7	Sa
	5085	13	20·3	−24	26	11·9	3·4 × 3·0	Sb
	5061	13	18·1	−26	50	11·7	2·6 × 2·3	E2
	5101	13	21·8	−27	26	11·7	5·5 × 4·9	Sa
83	5236	13	37·0	−29	52	8·2	11·2 × 10·2	Sc

PLANETARY NEBULA

	R.A. h	m	Dec. deg	min	Dimensions sec	Mag.	
NGC 3242	10	24·8	−18	38	16 × 1250	12·0	Ghost of Jupiter

HYDRUS

Abbreviation: Hyi).

, constellation in the far south – remark-
bly lacking in interesting objects. There are
ree stars above the fourth magnitude:

Mag.	Luminosity Sun = 1	Dist lt-yrs
2·80	3	20·5
2·86	7	36
3·24	115	160

is the closest star to the south celestial pole
/hich is above the third magnitude.

BRIGHTEST STARS

Star	R.A. h	m	s	Dec. deg	min	sec	Mag.	Abs. mag.	Spectrum	Dist pc
β	00	25	46·0	−77	15	15	2·80	3·8	G1	6·3
η²	01	54	56·1	−67	38	50	4·69	0·3	G5	69
α	01	58	46·2	−61	34	12	2·86	2·6	F0	11
δ	02	21	45·0	−68	39	34	4·09	1·4	A2	35
ε	02	29	35·5	−68	16	00	4·11	−0·8	B9	91
ν	02	50	28·7	−75	04	00	4·75	−0·3	gK6	100
γ	03	47	14·5	−74	14	20	3·24	−0·4	M0	49

Also above mag. 5:

	Mag.	Abs. mag.	Spectrum	Dist
ζ	4·84	−1·6	A2	

VARIABLE STAR

	R.A. h	m	Dec. deg	min	Range	Type	Period, d.	Spectrum
VW	04	09·1	−71	18	8·4–14·4	SS Cygni	100	M

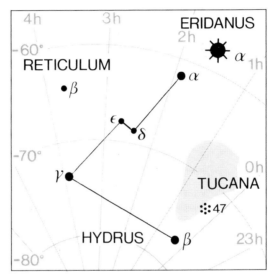

INDUS

Abbreviation: Ind).

An undistinguished little constellation, but
here are two stars above the fourth magni-
ude:

Mag.	Luminosity Sun = 1	Dist lt-yrs
3·11	60	124
3·65	60	121

ee chart for Grus.

BRIGHTEST STARS

Star	R.A. h	m	s	Dec. deg	min	sec	Mag.	Abs. mag.	Spectrum	Dist pc	
α	20	37	33·9	−47	17	29	3·11	0·2	K0	38	Persian
η	20	44	02·2	−51	55	16	4·51	2·4	dA7	26	
β	20	54	48·5	−58	27	15	3·65	0·2	K0	37	
θ	21	19	51·1	−53	26	57	4·39	2·1	A5	28	
δ	21	57	55·0	−54	59	34	4·40	1·7	F0	35	
ε	22	03	21·5	−56	47	10	4·69	7·0	K5	3·4	

Also above mag. 5:

	Mag.	Abs. mag.	Spectrum	Dist
ζ	4·89	−0·5	M1	120

VARIABLE STARS

	R.A. h	m	Dec. deg	min	Range sec	Type	Period, d.	Spectrum
S	20	56·4	−54	19	7·4–14·5	Mira	399·9	M
T	21	20·2	−45	01	7·7–9·4	Semi-reg.	320	N

DOUBLE STARS

	R.A. h	m	Dec. deg	min	P.A. deg	Sep. sec	Mags.	
θ	21	19·9	−53	27	275	6·0	4·5, 7·0	
δ	21	57·9	−55	00	323	0·1	5·3, 5·3	Binary, 12 y

GALAXIES

M	NGC	R.A. h	m	Dec. deg	min	Mag.	Dimensons min	Type
	7049	21	19·0	−48	34	10·7	2·8 × 2·2	S0
	7083	21	35·7	−63	54	11·8	4·5 × 2·9	Sb
	7090	21	36·5	−54	33	11·1	7·1 × 1·4	SBc
	7168	22	02·1	−51	45	12·6	2·0 × 1·6	E3
	7205	22	08·5	−57	25	11·4	4·3 × 2·2	Sb

LACERTA

(Abbreviation: Lac)

An obscure constellation with only one star above the fourth magnitude (α).

	Mag.	Luminosity	Dist
		Sun = 1	lt-yrs
α	3·77	26	98

BRIGHTEST STARS

Star		R.A.			Dec.			Mag.	Abs.	Spectrum	Dist
		h	m	s	deg	min	sec		mag.		pc
HD 211073	1H	22	13	52·5	+39	42	54	4·49	−0·2	K3	77
1		22	15	58·1	+37	44	56	4·13	−1·3	K3	100
2		22	21	01·4	+46	32	12	4·57	−1·3	B6	150
3 β		22	23	33·4	+52	13	44	4·43	0·2	G9	66
4		22	24	30·8	+49	28	35	4·57	−6·5	B9	1500
5		22	29	31·7	+47	42	25	4·36	−2·4	M0	230
6		22	30	29·1	+43	07	25	4·51	−3·0	B2	280
7 α		22	31	17·3	+50	16	57	3·77	1·4	A2	30
9		22	37	22·3	+51	32	43	4·63	1·5	A7	41
11		22	40	30·7	+44	16	35	4·46	−0·2	K3	84

Also above mag. 5:

	Mag.	Abs. mag.	Spectrum	Dist
15	4·94	−0·4	M0	120
10	4·88	−4·8	O9	780

VARIABLE STARS

	R.A.		Dec.		Range	Type	Period, d.	Spectrum
	h	m	deg	min				
S	22	29·0	+40	19	7·6–13·9	Mira	241·8	M
Z	22	40·9	+56	50	7·9–8·8	Cepheid	10·89	F–G

OPEN CLUSTERS

M	NGC	R.A.		Dec.		Diameter	Mag.	No. of stars
		h	m	deg	min	min		
	7243	22	15·3	+49	53	21	6·4	40
	7296	22	28·2	+52	17	4	9·7	20

LEO

(Abbreviation: Leo)

An important Zodiacal constellation – mythologically, the Nemæan lion which was killed by Hercules. There are 12 stars above the fourth magnitude:

	Mag.	Luminosity	Dist
		Sun = 1	lt-yrs
α	1·38	130	85
γ	1·99	60	90
β	2·14	17	39
δ	2·56	14	52
ε	2·98	520	310
θ	3·34	26	78
ζ	3·44	50	117
η	3·52	9500	1800
o	3·52	11	55
ρ	3·85	16 000	2500
μ	3·88	96	200
ι	3·94	14	78

BRIGHTEST STARS

Star		R.A.			Dec.			Mag.	Abs.	Spectrum	Dist	
		h	m	s	deg	min	sec		mag.		pc	
1 κ		09	24	39·2	+26	10	56	4·46	−0·1	K2	73	Al Minliar al Asad
4 λ		09	31	43·1	+22	58	04	4·31	−0·3	K5	79	Alterf
14 o		09	41	09·0	+09	53	32	3·52	2·1	A5	17	Subra
17 ε		09	45	51·0	+23	46	27	2·98	−2·0	G0	95	Asad Australis
24 μ		09	52	45·8	+26	00	25	3·88	−0·1	K2	55	Rassalas
29 π		10	00	12·7	+08	02	39	4·70	−0·5	M2	110	
30 η		10	07	19·9	+16	45	45	3·52	−5·2	A0	560	
31 A		10	07	54·2	+09	59	51	4·37	−0·3	K4	81	
32 α		10	08	22·2	+11	58	02	1·35	−0·6	B7	26	Regulus
36 ζ		10	16	41·4	+23	25	02	3·44	0·6	F0	36	Adhafera
41 γ		10	19	58·3	+19	50	30	1·99	0·2	K0 + G7	28	Algieba
47 ρ		10	32	48·6	+09	18	24	3·85	−5·7	B1	770	
54		10	55	37·2	+24	44	55	4·32	1·2	A1	42	
61		11	01	49·6	−02	29	04	4·74	−0·3	K5	86	
60 b		11	02	19·7	+20	10	47	4·42	0·4	A0	22	
63 χ		11	05	01·0	+07	20	10	4·63	1·3	F2	47	
68 δ		11	14	06·4	+20	31	25	2·56	1·9	A4	16	Zosma
70 θ		11	14	14·3	+15	25	46	3·34	1·4	A2	24	Chort
72		11	15	12·2	+23	05	43	4·63	−0·5	gM2	100	
74 φ		11	16	39·6	−03	39	06	4·47	1·5	A7	39	

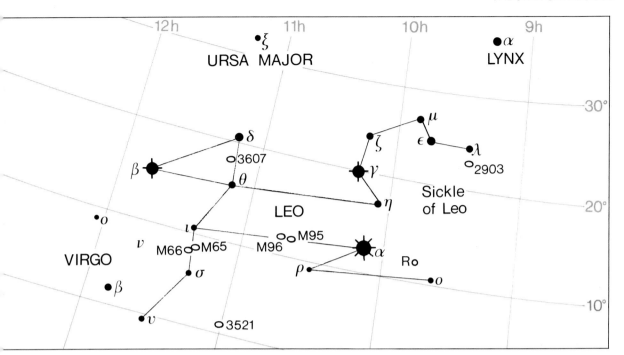

LEO (continued)

β (Denebola) was ranked as of the first magnitude by Ptolemy and others, and there is therefore a suspicion that it has faded, though the evidence is not conclusive.

Star	R.A.			Dec.			Mag.	Abs.	Spectrum	Dist	
	h	m	s	deg	min	sec		mag.		pc	
77 σ	11	21	08·1	+06	01	45	4·05	0·2	B9	59	
78 ι	11	23	55·4	+10	31	45	3·94	1·9	F2	24	
91 υ	11	36	56·9	−00	49	26	4·30	0·2	G9	64	
93	11	47	59·0	+20	13	08	4·53	2·2	A0	22	
94 β	11	49	03·5	+14	34	19	2·14	1·7	A3	12	Denebola

Also above mag. 5:

	Mag.	Abs. mag.	Spectrum	Dist
5 ξ	4·97	0·2	K0	84
10	5·00	0·0	K1	100
40	4·79	2·2	F6	29
58	4·84	0·0	K1	83
84 τ	4·95	−0·9	G8	150

VARIABLE STAR

	R.A.		Dec.	Range	Period, d.	Spectrum		
	h	m	deg	min				
R	09	47·6	+11	25	4·4–11·3	Mira	312·4	M

DOUBLE STARS

	R.A.		Dec.		P.A.	Sep.	Mags.	
	h	m	deg	min	deg	sec		
ω	09	28·5	+09	03	053	0·5	5·9, 6·5	Binary, 118 y
α	10	08·4	+11	58	307	176·9	1·4, 7·7	
γ	10	20·0	+19	51	AB 124	4·3	2·2, 3·5	Binary, 619 y
					AC 291	259·9	9·2	
					AD 302	333·0	9·6	
TX	10	35·0	+08	39	157	2·4	5·8, 8·5	
ι	11	23·9	+10	32	131	1·5	4·0, 6·7	Binary, 192 y
τ	11	27·9	+02	51	176	91·1	4·9, 8·0	

GALAXIES

M	NGC	R.A.		Dec.		Mag.	Dimensions	Type
		h	m	deg	min		min	
	3190	10	18·1	+21	50	11·0	4·6 × 1·8	SG
95	3351	10	44·0	+11	42	9·7	7·4 × 5·1	SBb
96	3368	10	46·8	+11	49	9·2	7·1 × 5·1	Sb
	3377	10	47·7	+13	59	10·2	4·4 × 2·7	E5
105	3379	10	47·8	+12	35	9·3	4·5 × 4·0	E1
	3384	10	48·3	+12	38	10·0	5·9 × 2·6	E7
	3412	10	50·9	+13	25	10·6	3·6 × 2·0	E5
	3489	11	00·3	+13	54	10·3	3·7 × 2·1	E6
	3521	11	05·8	−00	02	8·9	9·5 × 5·0	Sb
	3593	11	14·6	+12	49	11·0	5·8 × 2·5	Sb

LEO (continued)

M	NGC	R.A.		Dec.		Mag.	Dimensions	Type	
		h	m	deg	min		min		
	3596	11	15·1	+14	47	11·6	4·2 × 4·1	Sc	
	3607	11	16·9	+18	03	10·0	3·7 × 3·2	E1	
65	3623	11	18·9	+13	05	9·3	10·0 × 3·3	Sb	
	3626	11	20·1	+18	21	10·9	3·1 × 2·2	Sb	
66	3627	11	20·2	+12	59	9·0	8·7 × 4·4	Sb	
	3628	11	20·3	+13	36	9·5	14·8 × 3·6	Sb	Arp 317
	3630	11	20·3	+02	58	12·8	2·3 × 0·9	E7	
	3640	11	21·1	+03	14	10·3	4·1 × 3·4	E1	
	3646	11	21·7	+20	10	11·2	3·9 × 2·6	Sc	
	3686	11	27·7	+17	13	11·4	3·3 × 2·6	Sc	
	3810	11	41·0	+11	28	10·8	4·3 × 3·1	Sc	

LEO MINOR

(Abbreviation: LMi).

A small and obscure constellation. The only star above the fourth magnitude is 46.

	Mag.	Luminosity	Dist
		Sun = 1	lt-yrs
46	3·83	17	75

See chart for Ursa Major.

BRIGHTEST STARS

Star	R.A.			Dec.			Mag.	Abs.	Spectrum	Dist	
	h	m	s	deg	min	sec		mag.		pc	
10	09	34	13·3	+36	23	51	4·55	0·3	G8	71	
21	10	07	25·7	+35	14	41	4·48	2·4	A7	26	
30	10	25	54·8	+33	47	45	4·74	2·6	F0	27	
31 β	10	27	52·9	+36	42	26	4·21	1·8	G8	31	
37	10	38	43·1	+31	58	34	4·71	−2·1	G2	230	
46	10	53	18·6	+34	12	53	3·83	1·7	K0	23	Præcipua

VARIABLE STARS

	R.A.		Dec		Range	Type	Period, d.	Spectrum
	h	m	deg	min				
R	09	45·6	+34	31	6·3–13·2	Mira	371·9	M
S	09	53·7	+34	55	7·9–14·3	Mira	233·8	M
RW	10	16·1	+30	34	6·9–10·1	Mira	?	N

DOUBLE STAR

	R.A.		Dec.		P.A.	Sep.	Mags.	
	h	m	deg	min	deg	sec		
β	10	27·9	+36	42	250	0·2	4·4, 6·1	Binary, 37·2 y

GALAXIES

M	NGC	R.A.		Dec.		Mag.	Dimensions	Type
		h	m	deg	min		min	
	3003	09	48·6	+33	25	11·7	5·9 × 1·7	SBc
	3245	10	27·3	+28	30	10·8	3·2 × 1·9	E5
	3254	10	29·3	+29	30	11·5	5·1 × 1·9	Sb
	3294	10	36·3	+37	20	11·7	3·3 × 1·8	Sc
	3344	10	43·5	+24	55	9·9	6·9 × 6·5	Sc
	3414	10	51·3	+27	59	10·7	3·6 × 2·7	SBa
	3430	10	52·2	+32	57	11·5	3·9 × 2·3	Sc
	3432	10	52·5	+36	37	11·2	6·2 × 1·5	SB
	3486	11	00·4	+28	58	10·3	6·9 × 5·4	Sc

LEPUS

(Abbreviation: Lep).

An original constellation. In mythology, it was said that Orion was particularly fond of hunting hares – and so a hare was placed beside him in the sky.

Lepus is fairly distinctive. There are 8 stars above the fourth magnitude:

	Mag.	Luminosity	Dist
		Sun = 1	lt-yrs
α	2·58	7500	945
β	2·84	600	316
ε	3·19	110	160
μ	3·31	180	215
ζ	3·55	17	78
γ	3·60	2	26
η	3·71	17	65
δ	3·81	58	156

BRIGHTEST STARS

Star	R.A.			Dec.			Mag.	Abs.	Spectrum	Dist	
	h	m	s	deg	min	sec		mag.		pc	
2 ε	05	05	27·6	−22	22	16	3·19	−0·3	K5	50	
3 ι	05	12	17·8	−11	52	09	4·45	−0·2	B8	85	
5 μ	05	12	55·8	−16	12	20	3·31	−0·8	B9	66	
4 κ	05	13	13·8	−12	56	30	4·36	−1·2	B8	130	
6 λ	05	19	34·4	−13	10	36	4·29	−4·5	B0·5	520	(Measure uncertain)
HD 34968	05	20	26·8	−21	14	23	4·71	0·6	A0	66	
9 β	05	28	14·7	−20	45	35	2·84	−2·1	G2	97	Nihal
11 α	05	32	43·7	−17	49	20	2·58	−4·7	F0	290	Arneb
13 γ	05	44	27·7	−22	26	55	3·60	4·1	F6	8·1	
14 ζ	05	46	57·2	−14	49	20	3·55	1·7	A3	24	
15 δ	05	51	19·2	−20	52	45	3·81	0·3	G8	48	
16 η	05	56	24·2	−14	10	04	3·71	1·7	F0	20	
18 θ	06	06	09·3	−14	56	07	4·67	0·8	A0	18	

Also above mag. 5:

	Mag.	Abs. mag.	Spectrum	Dist
17	4·93		A2	

VARIABLE STARS

	R.A.		Dec.		Range	Type	Period, d.	Spectrum
	h	m	deg	min				
R	04	59·6	−14	48	5·5–11·7	Mira	432·1	N
T	05	04·8	−21	54	7·4–13·5	Mira	368·1	M
RX	05	11·4	−11	51	5·0–7·0	Irreg.	–	M
S	06	05·8	−24	12	7·1–8·9	Semi-reg.	90	M

LEPUS (continued)

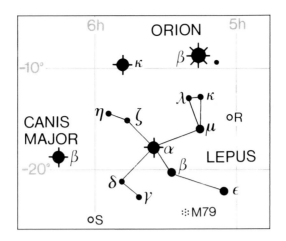

DOUBLE STARS

	R.A.		Dec.		P.A.	Sep.	Mags.
	h	m	deg	min	deg	sec	
ι	05	12·3	−11	52	337	12·7	4·5, 10·8
κ	05	13·2	−12	56	358	2·6	4·5, 7·4
β	05	28·2	−20	46	AB 330	2·5	2·8, 7·3
					AC 145	64·3	11·8
					AD 075	206·4	10·3
					AE 058	241·5	10·3
γ	05	44·5	−22	27	350	96·3	3·7, 6·3

GLOBULAR CLUSTER

M	NGC	R.A.		Dec.		Diameter	Mag.
		h	m	deg	min	min	
79	1904	05	24·5	−24	33	8·7	8·0

PLANETARY NEBULA

M	NGC	R.A.		Dec.		Dameter	Mag.	Mag. of central star
		h	m	deg	min	sec		
	IC 418	05	27·5	−12	42	12	10·7	10·7

GALAXIES

M	NGC	R.A.		Dec.		Mag.	Dimensions	Type
		h	m	deg	min		min	
	1744	05	00·0	−26	01	11·2	6·8 × 4·1	SBc
	1964	05	33·4	−21	57	10·8	6·2 × 2·5	Sb

LIBRA

(Abbreviation: Lib).

One of the Zodiacal constellations. It is, however, decidedly obscure. It was originally known as Chelæ Scorpionis (the Scorpion's Claws). Some Greek legends associate it, though rather vaguely, with Mochis, the inventor of weights and measures.

There are six stars above the fourth magnitude:

	Mag.	Luminosity	Dist
		Sun = 1	lt-yrs
β	2·61	105	121
α²	2·75	28	72
σ	3·29	120	166
υ	3·58	110	127
τ	3·66	310	326
γ	3·91	16	75

BRIGHTEST STARS

Star	R.A.			Dec.			Mag.	Abs. mag.	Spectrum	Dist pc	
	h	m	s	deg	min	sec					
9 α²	14	50	52·6	−16	02	30	2·75	1·2	A3	22	Zubenel-genubi
16	14	57	10·9	−04	20	47	4·49	1·7	F0	31	
20 σ	15	04	04·1	−25	16	55	3·29	−0·5	M4	51	Zubenal-gubi
24 ι	15	12	13·2	−19	47	30	4·54	−0·3	B9	93	
27 β	15	17	00·3	−09	22	58	2·61	−0·2	B8	37	Zubenel-chemale
38 γ	15	35	31·5	−14	47	23	3·91	1·8	G8	23	Zubenel-hakrabi
39 υ	15	37	01·4	−28	08	06	3·58	−0·3	K5	39	
40 τ	15	38	39·3	−29	46	40	3·66	−1·4	B4	100	
43 κ	15	41	56·7	−19	40	44	4·74	−0·3	K5	91	
46 θ	15	53	49·4	−16	43	46	4·15	1·8	G8	25	

σ Libræ was formerly known as γ Scorpii. β Libræ is said to be the only single star to show a greenish tint, though most people will certainly call it white!

Also above mag. 5:

	Mag.	Abs. mag.	Spectrum	Dist
δ	4·8 (max)	0·6	A2	73
ν	4·83	−0·3	A0	120
31 ε	4·94	3·4	F5	23
42	4·96	−0·3	K4	110
48	4·88	−0·8	B9	120

LIBRA (continued)

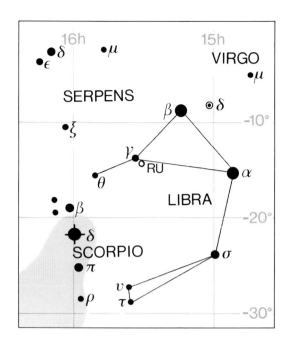

VARIABLE STARS

	R.A.		Dec.		Range	Type	Period, d.	Spectrum
	h	m	deg	min				
δ	15	01·1	−08	31	4·9–5·9	Algol	2·33	B
Y	15	11·7	−06	01	7·6–14·7	Mira	275·0	M
S	15	21·4	−20	23	7·5–13·0	Mira	192·4	M
RS	15	24·3	−22	55	7·0–13·0	Mira	217·7	M
RU	15	33·3	−15	20	7·2–14·2	Mira	316·6	M
RR	15	56·4	−18	18	7·8–15·0	Mira	277·0	M

DOUBLE STARS

	R.A.		Dec.		P.A.	Sep.	Mags.
	h	m	deg	min	deg	sec	
μ	14	49·3	−14	09	355	1·8	5·8, 6·7
α	14	50·9	−16	02	314	231·0	2·8, 5·2
ι	15	12·2	−19	47	111	57·8	5·1, 9·4
κ	15	41·9	−19	41	279	172·0	4·7, 9·7

GLOBULAR CLUSTER

M	NGC	R.A.		Dec.		Diameter	Mag.
		h	m	deg	min	min	
	5897	15	17·4	−21	01	12·6	8·6 H.IV. 19

LUPUS

(Abbreviation: Lup).

An original constellation, though no definite legends seem to be attached to it. There
are 11 stars above the fourth magnitude:

	Mag.	Luminosity	Dist
		Sun = 1	lt-yrs
α	2·30	5000	6800
β	2·68	830	360
γ	2·78	450	258
δ	3·22	1320	587
ε	3·37	700	456
ζ	3·41	58	137
η	3·41	830	490
φ¹	3·56	110	183
κ	3·72	60	127
π	3·89	400	424
χ	3·95	110	220

BRIGHTEST STARS

Star	R.A.			Dec.			Mag.	Abs. mag.	Spectrum	Dist pc	
	h	m	s	deg	min	sec					
ι	14	19	24·1	−46	03	28	3·55	−1·7	B3	110	
τ¹	14	26	08·1	−45	13	17	4·56	−3·0	B2	310	
τ²	14	26	10·7	−45	22	45	4·35	4·0	dF8	28	
σ	14	32	36·8	−50	27	25	4·42	−2·5	B2	130	
ρ	14	37	53·1	−49	25	32	4·05	−1·1	B5	110	
α	14	41	55·7	−47	23	17	2·30	−4·4	B1	210	Men
o	14	51	38·3	−43	34	31	4·32	−1·9	B6	180	
β	14	58	31·8	−43	08	02	2·68	−2·5	B2	110	KeKouan
π	15	05	07·1	−47	03	04	3·89	−1·6	B5	130	
λ	15	08	50·5	−45	16	47	4·05	−2·3	B3	190	
κ	15	11	56·0	−48	44	16	3·72	0·2	B9	39	
ζ	15	12	17·0	−52	05	57	3·41	0·3	G8	42	
2 f	15	17	49·7	−30	08	55	4·34	0·1	gK0	55	
μ	15	18	31·9	−47	51	30	4·27	−0·2	B8	77	
δ	15	21	22·2	−40	38	51	3·22	−3·0	B2	180	
φ¹	15	21	48·3	−36	15	41	3·56	−0·3	K5	56	
ε	15	22	40·7	−44	41	21	3·37	−2·3	B3	140	
φ²	15	23	09·2	−36	51	30	4·54	−2·3	B3	220	
k	15	25	20·1	−38	44	01	4·60	0·0	A0	80	
γ	15	35	08·4	−41	10	00	2·78	−1·7	B3	79	

LUPUS (continued)

Star	R.A.			Dec.			Mag.	Abs.	Spectrum	Dist
	h	m	s	deg	min	sec		mag.		pc
ω	15	38	03·1	−42	34	02	4·33	−0·4	M0	69
3 ψ¹	15	39	45·9	−34	24	42	4·67	0·3	gG5	63
g	15	41	11·2	−44	39	40	4·64	2·8	F5	22
4 ψ²	15	42	40·9	−34	42	37	4·75	−0·9	B6	140
5 χ	15	50	57·4	−33	37	38	3·95	−0·3	B9	68
η	16	00	07·1	−38	23	48	3·41	−2·5	B2	150
θ	16	06	35·4	−36	48	08	4·23	−2·3	B3	200

Also above mag. 5:

	Mag.	Abs. mag.	Spectrum	Dist
l	4·91	−6·6	F0	1800
ν¹	5·00	3·4	F3	20
ξ	4·6 (5·1 + 5·6)		A+A	

VARIABLE STARS

	R.A.		Dec.		Range	Type	Period, d.	Spectrum
	h	m	deg	min				
S	14	53·4	−46	37	7·8–13·5	Mira	342·7	S
GG	15	18·9	−40	47	5·4–6·0	Beta Lyræ	2·16	B + A

DOUBLE STARS

	R.A.		Dec.		P.A.	Sep.	Mags.
	h	m	deg	min	deg	sec	
τ¹	14	26·1	−45	13	204	148·2	4·6, 9·3
π	15	05·1	−47	03	073	1·4	4·6, 4·7
κ	15	11·9	−48	44	144	26·8	3·9, 5·8
μ	15	18·5	−47	53	AB 142	1·2	5·1, 5·2
					AC 130	23·7	7·2
ε	15	22·7	−44	41	247	0·6	3·7, 7·2
υ	15	24·7	−39	43	038	1·4	5·4, 10·9
ξ	15	56·9	−33	58	049	10·4	5·3, 5·8
η	16	00·1	−38	24	020	15·0	3·6, 7·8

OPEN CLUSTERS

M	NGC	R.A.		Dec.		Diameter	Mag.	No. of stars
		h	m	deg	min	min		
	5749	14	48·9	−54	31	8	8·8	30
	5822	15	05·2	−54	21	40	6·5	150

GLOBULAR CLUSTERS

M	NGC	R.A.		Dec.		Diameter	Mag.
		h	m	deg	min	min	
	5824	15	04·0	−33	04	6·2	9·0
	5927	15	28·0	−50	40	12·0	8·3
	5986	15	46·1	−37	47	9·8	7·1

PLANETARY NEBULÆ

M	NGC	R.A.		Dec.		Diameter	Mag.	Mag. of central star
		h	m	deg	min	sec		
	IC 4406	14	22·4	−44	09	28	10·6	14·7
	5882	15	16·8	−45	39	7	10·5	12·0

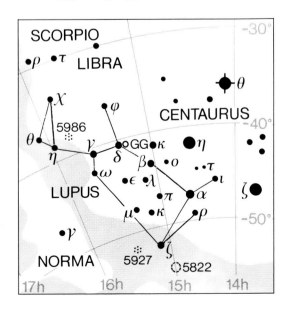

LUPUS (continued)

GALAXY

M	NGC	R.A.		Dec.		Mag.	Dimensions	Type
		h	m	deg	min		min	
	5643	14	32·7	−44	10	10·7	4·6 × 4·1	SB0

LYNX

(Abbreviation: Lyn).

A very ill-defined and obscure northern constellation. It was added to the sky by Hevelius, and has no mythological associations.

There are two stars above the fourth magnitude:

	Mag.	Luminosity	Dist
		Sun = 1	lt-yrs
α	3·13	115	166
38	3·92	17	88

See chart for Ursa Major.

BRIGHTEST STARS

Star	R.A.			Dec.			Mag.	Abs.	Spectrum	Dist	
	h	m	s	deg	min	sec		mag.		pc	
2	06	19	37·3	+59	00	39	4·48	1·4	A2	35	
15	06	57	16·5	+58	25	21	4·35	1·8	G5	32	
21	07	26	42·8	+49	12	42	4·64	0·3	A1	74	
31	08	22	50·1	+43	11	17	4·25	−0·3	K5	76	Alsciaukat
HD 77912	09	06	31·7	+38	27	08	4·56	−3·3	G8	370	
38	09	18	50·6	+36	48	09	3·92	1·7	A3	27	
40 α	09	21	03·2	+34	23	33	3·13	−0·4	M0	51	

Also above mag. 5:

	Mag.	Abs. mag.	Spectrum	Dist
12	4·87	1·2	A2	
16	4·90	1·4	A2	50
24	4·99	0·0	A3	100
27	4·84	1·4	A2	60

VARIABLE STARS

	R.A.		Dec.		Range	Type	Period, d.	Spectrum
	h	m	deg	min				
RR	06	26·4	+56	17	5·6–6·0	Algol	9·95	A
R	07	01·3	+55	20	7·2–14·5	Mira	378·7	S
Y	07	28·2	+45	59	7·8–10·3	Semi-reg.	110	M

DOUBLE STARS

	R.A.		Dec.		P.A.	Sep.	Mags.
	h	m	deg	min	deg	sec	
4	06	22·1	+59	22	124	0·8	6·2, 7·7
12	06	45·2	+59	27	AB 070	1·7	5·4, 6·0
					AC 308	8·7	7·3
					AD 256	170·0	10·6
19	07	22·9	+55	17	AB 315	14·8	5·6, 6·5
					AD 003	214·9	8·9
					BC 287	74·2	10·9
38	09	18·8	+36	48	AB 229	2·7	3·9, 6·6
					BC 212	87·7	10·8
					BD 256	177·9	10·7

GLOBULAR CLUSTER

M	NGC	R.A.		Dec.		Diameter	Mag.	
		h	m	deg	min	min		
	2419	7	38·1	+38	53	4·1	10·4	'Intergalactic Tramp'

GALAXIES

M	NGC	R.A.		Dec.		Mag.	Dimensions	Type	
		h	m	deg	min		min		
	2537	08	13·2	+46	00	12·3	1·7 × 1·5	S	Bear Paw Galaxy
	2541	08	14·7	+49	04	11·7	6·6 × 3·5	S	
	2683	08	52·7	+33	25	9·7	9·3 × 2·5	Sb	
	2776	09	12·2	+44	57	11·6	2·9 × 2·7	Sc	

LYRA

(Abbreviation: Lyr).

A small constellation, but a very interesting one; it is graced by the presence of the brilliant blue Vega, as well as the prototype eclipsing binary β Lyrae, and the quadruple ε Lyra, and the 'Ring Nebula' M.57. Mythologically it represents the harp which Apollo gave to the great musician Orpheus.

There are several stars above the fourth magnitude. Of these, one (β) is the famous variable; the combined magnitude of the quadruple ε is about 3·9, though keen-

BRIGHTEST STARS

Star	R.A.			Dec.			Mag.	Abs.	Spectrum	Dist	
	h	m	s	deg	min	sec		mag.		pc	
1 κ	18	19	51·5	+36	03	52	4·33	−0·1	K2	77	
3 α	18	36	56·2	+38	47	01	0·03	0·5	A0	8·1	Vega
4 ε¹	18	44	20·1	+39	40	15	4·67	1·7	A3	38	
5 ε²	18	44	22·7	+39	36	46	5·1	2·1	A5	38	
6 ζ¹	18	44	46·2	+37	36	18	4·36	1·2	A3	64	}
7 ζ²	18	44	48·0	+37	35	46	5·73	2·8	F0	64	}
10 β	18	50	04·6	+33	21	46	var.	−0·6	B7	92	Sheliak
12 δ²	18	54	30·0	+36	53	56	4·30v	−2·4	M4	220	
13 R	18	55	19·9	+43	56	46	var.	var.	M5	40	
14 γ	18	58	56·4	+32	41	22	3·24	−0·8	B9	59	Sulaphat
20 η	19	13	45·3	+39	08	46	4·39	−3·0	B2	270	Aladfar
21 θ	19	16	21·9	+38	08	01	4·36	−2·1	K0	170	

Also above mag. 5:

	Mag.	Abs. mag.	Spectrum	Dist
15 λ	4·93	−2·3	K3	260

LYRA (continued)

sighted people can see the two main components as separated. The stars are:

	Mag.	Luminosity Sun = 1	Dist lt-yrs
α	0·03	52	26
γ	3·24	180	192
β	3·3 (max)	130	300
ε	3·9 (combined)	17+11	124

VARIABLE STARS

	R.A. h	m	Dec. deg	min	Range	Type	Period, d.	Spectrum
W	18	14·9	+36	40	7·3–13·0	Mira	196·5	M
T	18	32·3	+37	00	7·8–9·6	Irreg.	–	R
β	18	50·1	+33	22	3·3–4·3	Beta Lyræ	12·94	B + A
R	18	55·3	+43	57	3·9–5·0	Semi-reg.	46	M
RR	19	25·5	+42	47	7·1–8·1	RR Lyræ	0·57	A–F

DOUBLE STARS

	R.A. h	m	Dec. deg	min	P.A. deg	Sep. sec	Mags.
ε	18	44·3	+39	40	AB + CD 173	207·7	4·7, 5·1
					ε¹ = AB 357	2·6	5·0, 5·1
					ε² = CD 094	2·3	5·2, 5·5
ζ	18	44·8	+37	36	150	43·7	4·3, 5·9
β	18	50·1	+33	22	149	45·7	var., 8·6
δ¹	18	53·7	+36	58	020	174·6	5·6, 9·3
δ²	18	54·5	+36	54	349	86·2	4·5, 11·2
η	19	13·8	+39	09	082	28·1	4·4, 9·1

OPEN CLUSTER

M	NGC	R.A. h	m	Dec. deg	min	Diameter min	Mag.	No. of stars
	6791	19	20·7	+37	51	16	9·5	300

GLOBULAR CLUSTER

M	NGC	R.A. h	m	Dec. deg	min	Diameter min	Mag.
56	6779	19	16·6	+30	11	7·1	8·2

PLANETARY NEBULA

M	NGC	R.A. h	m	Dec. deg	min	Dimensions sec	Mag.	Mag. of central star	
57	6720	18	53·6	+33	02	70 × 150	9·7	14·8	Ring Nebula

MENSA

(Abbreviation: Men).

A very dim constellation, introduced by Lacaille in 1752 under the name of Mons Mensæ (the Table Mountain). A small part of the Large Magellanic Cloud extends into it. There are no stars brighter than the fifth magnitude, and no objects to be listed. For the record, the brightest star is α (5·09). Next comes γ: R.A. 5 h 31 m 53 s·1, dec. –76 deg 20 min 28 sec, mag. 5·19, absolute mag. –0·3. Spectrum K4. Distance 130 pc. It has an optical companion of mag. 11, at P.A. 107 deg, distance 38 sec ·2.

See chart for Musca.

VARIABLE STARS

	R.A.		Dec.		Range	Type	Period, d.	Spectrum
	h	m	deg	min				
U	04	09·6	–81	51	8·0–10·9	Mira	407	M
TY	05	26·9	–81	35	7·7–8·2	W UMa	0·46	A
TZ	05	30·2	–84	47	6·2–6·9	Algol	8·57	B

MICROSCOPIUM

(Abbreviation: Mic).

A small southern constellation. γ (4·67) is the brightest star.

See chart for Grus.

BRIGHTEST STARS

Star	R.A.			Dec.			Mag.	Abs. mag.	Spectrum	Dist pc
	h	m	s	deg	min	sec				
γ	21	01	17·3	–32	15	28	4·67	0·3	G4	70
ε	21	17	56·1	–32	10	21	4·71	2·1	A2p	

γ was formerly known as I PsA and ε as 4 PsA.

Also above mag. 5:

	Mag.	Abs. mag.	Spectrum	Dist
α	4·90	0·3	G6	73
θ¹	4·82	–0·6	A2p	

VARIABLE STARS

	R.A.		Dec.		Range	Type	Period, d.	Spectrum
	h	m	deg	min				
T	20	27·9	–28	16	7·7–9·6	Semi-reg.	344	M
U	20	29·2	–40	25	7·0–14·4	Mira	334·2	M
S	21	26·7	–29	51	7·8–14·3	Mira	208·9	M

DOUBLE STARS

	R.A.		Dec.		P.A.	Sep.	Mags.
	h	m	deg	min	deg	sec	
α	20	50·0	–33	47	166	20·5	5·0, 10·0
θ²	21	24·4	–41	00	AB 267	0·5	6·4, 7·0
					AC 066	78·4	10·5

GALAXIES

M	NGC	R.A.		Dec.		Mag.	Dimensions	Type
		h	m	deg	min		min	
	6923	20	31·7	–30	50	12·1	2·5 × 1·4	Sb
	6925	20	34·3	–31	59	11·3	4·1 × 1·6	Sb

MONOCEROS

(Abbreviation: Mon).

Not an ancient constellation, and though it represents the fabled unicorn there are no definite legends attached to it. It is crossed by the Milky Way, and the general area is decidedly rich. There are 4 stars above the fourth magnitude; of these, β is a double, and the magnitude is combined. The stars are:

	Mag.	Luminosity Sun = I	Dist lt-yrs
β	3·7 (combined)	900+450	720
30	3·90	50	150
α	3·93	60	176
γ	3·98	105	215

See Chart for Orion.

BRIGHTEST STARS

Star	R.A.			Dec.			Mag.	Abs. mag.	Spectrum	Dist pc	
	h	m	s	deg	min	sec					
5 γ	06	14	51·3	–06	16	29	3·98	–0·2	K3	66	
8 ε	06	23	46·0	+04	35	34	4·33	0·3	A5	54	
11 β	06	28	48·9	–07	01	58	3·7	–2·6 / –1·7	B2 / B3	220	
13	06	32	54·2	+07	19	58	4·50	–5·2	A0	860	
15 S	06	40	58·6	+09	53	45	var.	–5·5	O7	920	
18	06	47	51·6	+02	24	44	4·47	0·2	K0	59	
22 δ	07	11	51·8	–00	29	34	4·15	0·0	A0	64	
26 α	07	41	14·8	–09	33	04	3·93	0·2	K0	54	
28	08	01	13·2	–01	23	33	4·68	–0·3	K4	88	
29 ζ	08	08	35·6	–02	59	02	4·34	–4·5	G2	560	
30	08	25	39·5	–03	54	23	3·90	0·6	A0	46	
31	08	43	40·4	–07	14	01	4·62	–4·5	G2	670	= F Hydræ

Also above mag. 5:

	Mag.	Abs. mag.	Spectrum	Dist
17	4·77	–0·3	K4	100
19	4·99	–3·5	B1	500
27	4·93	–0·1	K2	97

MONOCEROS (continued)

VARIABLE STARS

	R.A. h	m	Dec. deg	min	Range	Type	Period, d.	Spectrum
V	06	22·7	−02	12	6·0–13·7	Mira	333·8	M
T	06	25·2	+07	05	6·0–6·6	Cepheid	27·02	F–K
S	06	41·0	+09	54	4–5?	Irreg.	–	07
X	06	57·2	−09	04	6·9–10·0	Semi-reg.	156	M
RY	07	06·9	−07	33	7·7–9·2	Semi-reg.	466	N
U	07	30·8	−09	47	6·1–8·1	RV Tauri	92·3	F–K

DOUBLE STARS

	R.A. h	m	Dec. deg	min	P.A. deg	Sep. sec	Mags.
ε	06	23·8	+04	36	027	13·4	4·5, 6·5
β	06	28·8	−07	02	AB 132	7·3	4·7, 5·2
					AC 124	10·0	6·1
					AD 056	25·9	12·2
S (15)	06	41·0	+09	54	AB 213	2·8	4·7v, 7·5
					AC 013	16·6	9·8
					AD 308	41·3	9·6
					AE 139	73·9	9·9
					AF 222	156·0	7·7
					AK 056	105·6	8·1

OPEN CLUSTERS

M	NGC	R.A. h	m	Dec. deg	min	Diameter min	Mag.	No. of stars	
	2215	06	21·0	−07	17	11	8·4	40	
	2244	06	32·4	+04	52	24	4·8	100	In Rosette Neb.
	2251	06	34·7	+08	22	10	7·3	30	
	2286	06	47·6	−03	10	15	7·5	50	
	2301	06	51·8	+00	28	12	6·0	80	
50	2323	07	03·2	−08	20	16	5·9	80	
	2335	07	06·6	−10	05	12	7·2	35	
	2343	07	08·3	−10	39	7	6·7	20	
	2353	07	14·6	−10	18	20	7·1	30	
	2506	08	00·2	−10	47	7	7·6	150	

NEBULÆ

M	NGC	R.A. h	m	Dec. deg	min	Dimensions min	Mag. of illuminating star	
	2149	06	03·5	−09	44	3 × 2	9	
	2237–9	06	32·3	+05	03	80 × 60		Rosette Nebula
	2261	06	39·2	+08	44	2 × 1	10v	R Monocerotis
	2264	06	40·9	+09	54	60 × 30	4v	S Monocerotis. Cone Nebula

MUSCA AUSTRALIS

(Abbreviation: Mus).

This small but fairly distinctive constellation, not far from Crux, is generally known simply as 'Musca'. There are five stars above the fourth magnitude:

	Mag.	Luminosity Sun = 1	Dist lt-yrs
α	2·69v	700	326
β	3·05	450	290
δ	3·62	96	176
λ	3·64	11	52
γ	3·87	230	192

α is actually variable over a very small range (2·66–2·73).

BRIGHTEST STARS

Star	R.A. h	m	s	Dec. deg	min	sec	Mag.	Abs. mag.	Spectrum	Dist pc
λ	11	45	36·4	−66	43	43	3·64	2·1	A5	16
μ	11	48	14·4	−66	48	53	4·72	−0·5	gM2	110
ε	12	17	34·2	−67	57	38	4·11	2·2	gM6	12
γ	12	32	28·1	−72	07	58	3·87	−1·1	B5	59
α	12	37	11·0	−69	08	07	2·69v	−2·3	B3	100
β	12	46	16·9	−68	06	29	3·05	−1·7	B3	89
δ	13	02	16·3	−71	32	56	3·62	−0·1	K2	54

Also above mag. 5:

	Mag.	Abs. mag.	Spectrum	Dist
η	4·80	−0·2	B8	100

VARIABLE STARS

	R.A. h	m	Dec. deg	min	Range	Type	Period, d.	Spectrum
S	12	12·8	−70	09	5·9–6·4	Cepheid	9·66	F
B0	12	34·9	−67	45	6·0–6·7	Irreg.	–	M
R	12	42·1	−69	24	5·9–6·7	Cepheid	7·48	F
T	13	21·2	−74	27	7·0–9·0	Semi-reg.	93	N

DOUBLE STARS

	R.A. h	m	Dec. deg	min	P.A. deg	Sep. sec	Mags.
ζ²	12	22·1	−67	31	130	32·4	5·2, 10·6
α	12	37·2	−69	08	316	29·6	2·7, 12·8
β	12	46·3	−68	06	014	1·4	3·7, 4·0
θ	13	08·1	−65	18	187	5·3	5·7, 7·3

MUSCA AUSTRALIS
(continued)

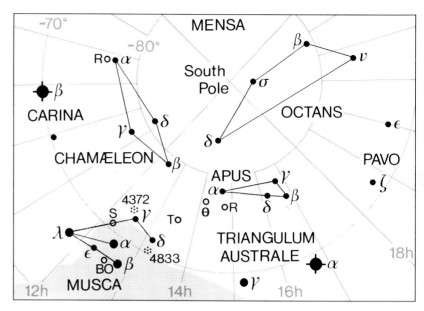

OPEN CLUSTER

M	NGC	R.A.		Dec.		Diameter	Mag.	No. of stars
		h	m	deg	min	min		
	4463	12	30·0	−64	48	5	7·2	30

GLOBULAR CLUSTERS

M	NGC	R.A.		Dec.		Diameter	Mag.
		h	m	deg	min	min	
	4372	12	25·8	−72	40	18·6	7·8
	4833	12	59·6	−70	53	13·5	7·3

PLANETARY NEBULÆ

M	NGC	R.A.		Dec.		Diameter	Mag.	Mag. of central star
		h	m	deg	min	sec		
	IC 4191	13	08·8	−67	39	5	12·0	−
	5189	13	33·5	−65	59	153	10	14 Gum 47

NORMA

(Abbreviation: Nor).

An obscure constellation, once known as Quadra Euclidis (Euclid's Quadrant); it was added to the sky by Lacaille. It contains no star above the fourth magnitude.

See charts for Ara and Lupus.

BRIGHTEST STARS

Star	R.A.			Dec.			Mag.	Abs.	Spectrum	Dist
	h	m	s	deg	min	sec		mag.		pc
η	16	03	12·6	−49	13	47	4·65	0·3	gG4	64
ι¹	16	03	31·9	−57	46	31	4·63	2·1	A5	30
δ	16	06	29·3	−45	10	24	4·72	0·6	A0	18
γ²	16	19	50·3	−50	09	20	4·02	0·3	G8	40
ε	16	27	10·9	−47	33	18	4·47	−1·7	B3	150

Also above mag. 5:

	Mag.	Abs. mag.	Spectrum	Dist
κ	4·94	0·3	G4	61
γ¹	4·99	−6·3	F8	1700

VARIABLE STARS

	R.A.		Dec.		Range	Type	Period, d.	Spectrum
	h	m	deg	min				
R	15	36·0	−49	30	6·5–13·9	Mira	492·7	M
T	15	44·1	−54	59	6·2–13·6	Mira	242·6	M
S	16	18·9	−57	54	6·1–6·8	Cepheid	9·75	F–G

DOUBLE STARS

	R.A.		Dec.		P.A.	Sep.	Mags.	
	h	m	deg	min	deg	sec		
ι¹	16	03·5	−57	47	100	0·2	5·3, 5·5	Binary, 26·9 v
ε	16	27·2	−47	33	335	22·8	4·8, 7·5	

OPEN CLUSTERS

M	NGC	R.A.		Dec.		Diameter	Mag.	No. of stars
		h	m	deg	min	min		
	5925	15	27·7	−54	31	15	8·4	120
	5999	15	52·2	−56	28	5	9·0	40
	6031	16	07·6	−54	04	2	8·5	20

NORMA (continued)

M	NGC	R.A. h m	Dec. deg min	Diameter min	Mag.	No. of stars	
	6067	16 13·2	−54 13	13	5·6	100	
	6087	16 18·9	−57 54	12	5·4	40	S Normæ cluster
	H.10	16 19·9	−54 59	30	–	30	
	6134	16 27·7	−49 09	7	7·2	–	
	6152	16 32·7	−52 37	30	8·1	70	
	6167	16 34·4	−49 36	8	6·7	–	

PLANETARY NEBULA

M	NGC	R.A. h m	Dec. deg min	Diameter sec	Mag.	Mag. of central star
	Sp-1	15 51·7	−51 31	76	13·6	13·8

OCTANS

Abbrevations: Oct).

The south polar constellation. It is very obscure, and contains only one star above the fourth magnitude: ν.

	Mag.	Luminosity Sun = 1	Dist lt-yrs
ν	3·76	60	104

See chart for Musca.

BRIGHTEST STARS

Star	R.A. h m s	Dec. deg min sec	Mag.	Abs. mag.	Spectrum	Dist pc
δ	14 26 55·0	−83 40 04	4·32	−0·1	gK2	60
ν	21 41 28·6	−77 23 24	3·76	0·2	K0	32
β	22 46 03·1	−81 23 54	4·15	2·6	dF0	20

The south pole star is σ Octantis; R.A. 21 h 08 m 44 s ·9, dec. −88 deg 57 min 24 sec (epoch 2000), magnitude 5·47, absolute magnitude 2·7, spectrum A7, distance 37 parsecs (121 light-years). The polar distance was 45 sec in 1900, but has now increased to over one degree.

Also above mag. 5:

	Mag.	Abs. mag.	Spectrum	Dist
θ	4·78	−0·1	K2	78

VARIABLE STARS

	R.A. h m	Dec. deg min	Range	Type	Period, d.	Spectrum
R	05 26·1	−86 23	6·4–13·2	Mira	405·6	M
U	13 24·5	−84 13	7·1–14·1	Mira	302·6	M
S	18 08·7	−86 48	7·3–14·0	Mira	258·9	M
ε	22 20·0	−80 26	4·9–5·4	Semi-reg.	55	M

DOUBLE STARS

	R.A. h m	Dec. deg min	P.A. deg	Sep. sec	Mags.
ι	12 55·0	−85 07	230	0·6	6·0, 6·5
μ²	20 41·7	−75 21	017	17·4	7·1, 7·6
λ	21 50·9	−82 43	070	3·1	5·4, 7·7

OPHIUCHUS

Abbreviations: Oph).

This constellation is also sometimes known as Serpentarius. It commemorates Æsculapius, son of Apollo and Coronis, who became so skilled in medicine that he was even able to restore the dead to life. To avoid depopulation of the Underworld, Jupiter reluctantly disposed of Æsculapius with a thunderbolt, but relented sufficiently to place him in the sky!

Ophiuchus is a very large constellation, with 13 stars above the fourth magnitude:

	Mag.	Luminosity Sun = 1	Dist lt-yrs
α	2·08	58	62
η	2·43	26	59
ζ	2·56	5000	550
δ	2·74	120	140
β	2·77	96	121
κ	3·20	96	117
ε	3·24	58	104
θ	3·27	1320	590
ν	3·34	60	137
72	3·73	14	91
γ	3·75	50	114
λ	3·82	28	108
67	3·97	16 000	2400

BRIGHTEST STARS

Star	R.A. h m s	Dec. deg min sec	Mag.	Abs. mag.	Spectrum	Dist pc	
1 δ	16 14 20·6	−03 41 39	2·74	−0·5	M1	43	Yed Prior
2 ε	16 18 19·1	−04 41 33	3·24	0·3	G8	32	Yed Post
4 ψ	16 24 06·0	−20 02 15	4·50	0·2	K0	72	
5 ρ	16 25 34·9	−23 26 46	4·59	−2·5	B2	230	
7 χ	16 27 01·3	−18 27 23	4v	−2·5v	B2p	150	
3 υ	16 27 48·1	−08 22 18	4·63	1·4	A2	21	
10 λ	16 30 54·7	+01 59 02	3·82	1·2	A1	33	Marfik
8 φ	16 31 08·2	−16 36 46	4·28	0·3	G8	63	
9 ω	16 32 08·0	−21 27 59	4·45	1·7	A7	19	
13 ζ	16 37 09·4	−10 34 02	2·56	−4·4	09·5	170	Han
20	16 49 49·9	−10 46 59	4·65	0·7	F6	56	
25 ι	16 54 00·4	+10 09 55	4·38	−0·6	B8	99	
27 κ	16 57 40·0	+09 22 30	3·20	−0·1	K2	36	
35 η	17 10 22·5	−15 43 30	2·43	1·4	A2	18	Sabik
36	17 15 20·7	−26 36 04	4·31	6·4	K0	5·4	
41	17 16 36·5	−00 26 43	4·73	−0·1	K2	93	
40 ξ	17 21 00·0	−21 06 46	4·39	3·0	F2	19	
42 θ	17 22 00·4	−24 59 58	3·27	−3·0	B2	180	
44 b	17 26 22·1	−24 10 31	4·17	2·5	A9	25	
49 σ	17 26 30·7	+04 08 25	4·34	−2·3	K3	190	
47	17 26 37·7	−05 05 12	4·54	3·1	F3	21	
45 d	17 27 21·1	−29 52 01	4·29	2·1	F5	27	
55 υ	17 34 55·9	+12 33 36	2·08	0·3	A5	19	Rasalhague
57 μ	17 37 50·5	−08 07 08	4·62	−0·2	B8	73	
60 β	17 43 28·2	+04 34 02	2·77	−0·1	K2	37	Cheleb
62 γ	17 47 53·4	+02 42 26	3·75	0·6	A0	35	
64 ν	17 59 01·4	−09 46 25	3·34	0·2	K0	42	
66	18 00 15·5	+04 22 07	4·64	−2·5	B2	250	
67	18 00 38·5	+02 55 53	3·97	−5·7	B5	740	
68	18 01 45·0	+01 18 19	4·45	1·2	A1	45	
70 p	18 05 27·2	+02 29 58	4·03	5·7	K0	5·1	

OPHIUCHUS (continued)

Barnard's Star (Munich 15040) lies in Ophiuchus near 67 and 70; R.A. 17 h 55 m ·4, dec +04 deg 33 min. The apparent magnitude is 9·54, spectral type M5. The annual proper motion is 10 sec ·29, and the star is moving almost due north; it covers one degree in 351 years.

The region of the celebrated binary 70 Ophiuchus was once separated out into the constellation Taurus Poniatowski, which has, however, been deleted from modern maps.

Star	R.A. h	m	s	Dec. deg	min	sec	Mag.	Abs. mag.	Spectrum	Dist pc
71	18	07	18·2	+08	44	02	4·64	1·8	G8	34
72	18	07	20·8	+09	33	50	3·73	1·9	A4	28

Also above mag. 5:

	Mag.	Abs. mag.	Spectrum	Dist
30	4·82	−0·3	K4	93
58	4·87	3·4	F5	20
69 τ	4·79	2·6	F0	22
74	4·86	0·3	G8	82

VARIABLE STARS

	R.A. h	m	Dec. deg	min	Range	Type	Period, d.	Spectrum	
V	16	26·7	−12	26	7·3–11·6	Mira	298·0	N	
χ	16	27·0	−18	27	4·2–5·0	Irreg.	–	B	
SS	16	57·9	−02	46	7·8–14·5	Mira	180·0	M	
R	17	07·8	−16	06	7·0–13·8	Mira	302·6		
U	17	16·5	+01	13	5·9–6·6	Algol	1·68	B + B	
Z	17	19·5	+01	31	7·6–14·0	Mira	348·7	K–M	
RS	17	50·2	−06	43	5·3–12·3	Recurrent nova	–	O + M	(Outbursts 1933, 1958, 1987)
Y	17	52·6	−06	09	5·9–6·4	Cepheid	17·12	F–G	
RY	18	16·6	+03	42	7·5–13·8	Mira	150·5	M	
X	18	38·3	+08	50	5·9–9·2	Mira	334·4	M + K	

DOUBLE STARS

	R.A. h	m	Dec. deg	min	P.A. deg	Sep. sec	Mags.	
ρ	16	25·6	−23	27	344	3·1	5·3, 6·0	
υ	16	27·8	−08	22	095	1·0	4·6, 7·8	
λ	16	30·9	+01	59	AB 022	1·5	4·2, 5·2	Binary, 129·9 y
					AB + C 170	119·2	11·1	
					AD 246	313·8	9·9	
φ	16	31·1	−16	37	037	34·4	4·3, 12·8	
19	16	47·2	+02	04	089	23·4	6·1, 9·4	
η	17	10·4	−15	43	247	0·5	3·0, 3·5	Binary, 84·3 y
36	17	15·3	−26	36	150	4·7	5·1, 5·1	Binary, 549 y
41	17	16·6	−00	27	346	1·0	4·8, 7·8	
53	17	34·6	+09	35	191	41·2	5·8, 8·5	
τ	18	03·1	−08	11	AB 280	1·8	5·2, 5·9	Binary, 280 y
					AC 127	100·3	9·3	
70	18	05·5	+02	30	224	1·5	4·2, 6·0	Binary, 88·1 y
73	18	09·6	+04	00	300	0·4	6·1, 7·0	Binary, 270 y
X	18	38·3	+08	50	150	0·4	var., 8·6	Binary, 485 y

OPEN CLUSTERS

M	NGC	R.A. h	m	Dec. deg	min	Diameter min	Mag.	No. of stars
	IC 4665	17	46·3	+05	43	41	4·2	30
	6633	18	27·7	+06	34	27	4·6	30

GLOBULAR CLUSTERS

M	NGC	R.A. h	m	Dec. deg	min	Diameter min	Mag.
107	6171	16	32·5	−13	03	10·0	8·1
12	6218	16	47·2	−01	57	14·5	6·6
10	6254	16	57·1	−04	06	15·1	6·6
62	6266	17	01·2	−30	07	14·1	6·6
19	6273	17	02·6	−26	16	13·5	7·1
	6304	17	14·5	−29	28	6·8	8·4
	6316	17	16·6	−28	08	4·9	9·0
9	6333	17	19·2	−18	31	9·3	7·9
	6356	17	23·6	−17	49	7·2	8·4
	6355	17	24·0	−26	21	5·0	9·6
14	6402	17	37·6	−03	15	11·7	7·6
	6401	17	38·6	−23	55	5·6	9·5

PLANETARY NEBULÆ

M	NGC	R.A. h	m	Dec. deg	min	Dimensions sec	Mag.	Mag. of central star
	6309	17	14·1	−12	55	14 × 66	10·8	14·4
	6572	18	12·1	+06	51	8 × 8	9·0	13·6

GALAXY

M	NGC	R.A. h	m	Dec. deg	min	Mag.	Dimensions min	Type
	6384	17	32·4	+07	04	10·6	6·0 × 4·3	Sb

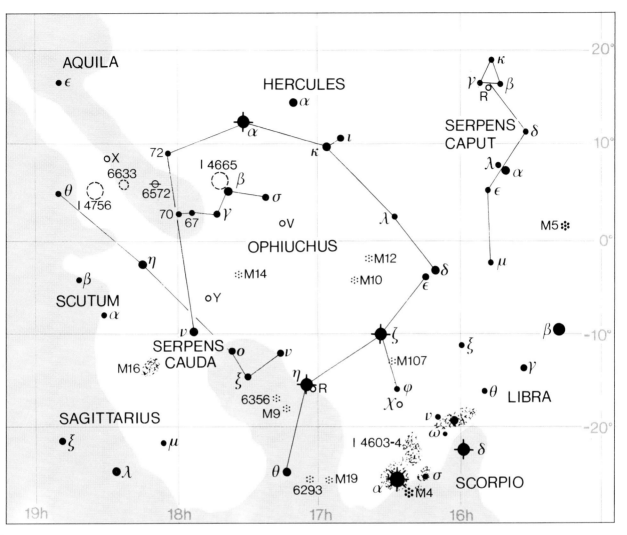

ORION

(Abbreviation: Ori).

One of the most magnificent constellations in the sky; it represents the mythological hunter who boasted that he could kill any creature on earth, but who was fatally stung by a scorpion. The two leading stars are Rigel, which is actually variable over a very small range (0·08 to 0·20) and the red variable Betelgeux – a name which may also be spelled Betelgeuse or Betelgeuze. The distance of Betelgeux is 310 light-years according to the Cambridge catalogue, but other catalogues increase this to over 500 light-years. The gaseous nebula M.42, in the Sword, is the most famous example of its type, and is easily visible with the naked eye. Altogether there are fifteen stars above the fourth magnitude:

BRIGHTEST STARS

Star	R.A.			Dec.			Mag.	Abs. mag.	Spectrum	Dist pc	
	h	m	s	deg	min	sec					
1 π³	04	49	50·3	+06	57	41	3·19	3·8	F6	7·7	
2 π²	04	50	36·6	+08	54	01	4·36	0·6	A0	55	
3 π⁴	04	51	12·3	+05	36	18	3·69	−3·6	B2	280	
4 o¹	04	52	31·9	+14	15	02	4·74		M3	32	
8 π⁵	04	54	15·0	+02	26	26	3·72v	−3·6	B2	290	
7 π¹	04	54	53·7	+10	09	03	4·65	0·9	A0p	83	
9 o²	04	56	22·2	+13	30	52	4·07	−0·1	K2	68	
10 π⁶	04	58	32·8	+01	42	51	4·47	−2·2	K2	190	
11	05	04	34·1	+15	24	14	4·68	0·0	A0p	26	
17 ρ	05	13	17·4	+02	51	40	4·46	−0·2	K3	86	
19 β	05	14	32·2	−08	12	06	0·12v	−7·1	B8	280	Rigel
20 τ	05	17	36·3	−06	50	40	3·60	−2·2	B5	130	
22 o	05	21	45·7	−00	22	57	4·73	−3·0	B2	340	
29 e	05	23	56·7	−07	48	29	4·14	0·3	G8	58	
28 η	05	24	28·6	−02	23	50	3·36	−3·5	B1	230	Algjebbah
24 γ	05	25	07·8	+06	20	59	1·64	−3·6	B2	110	Bellatrix
30 ψ	05	26	50·2	+03	05	44	4·59	−3·0	B2	330	
31 Cl	05	29	43·9	−01	05	32	4·71	−0·3	K5	93	
32 A	05	30	47·0	+05	56	53	4·20	−1·6	B5	150	
36 υ	05	31	55·8	−07	18	05	4·62	−4·1	B0	560	Thabit
34 δ	05	32	00·3	−00	17	57	2·23v	−6·1	O9·5	720	Mintaka
37 φ¹	05	34	49·2	+09	29	22	4·41	−4·6	B0	570	
39 λ	05	35	08·2	+09	56	02	3·39	−5·1	O8	550	Heka
42 c	05	35	23·1	−04	50	18	4·59	−3·6	B2	135	(near θ)
44 ι	05	35	25·9	−05	54	36	2·76	−6·0	O9	570	Hatysa
46 ε	05	36	12·7	−01	12	07	1·70	−6·2	B0	370	Alnilam
40 φ²	05	36	54·3	+09	17	27	4·09	0·2	K0	60	

ORION (continued)

	Mag.	Luminosity Sun = 1	Dist lt-yrs
β	0·12v	60 000	910
α	0·5v	15 000v	310
γ	1·64	2200	360
ε	1·70	23 000	1200
ζ	1·77	19 000	1100
κ	2·06	49 000	2100
δ	2·23v	22 000	2350
ι	2·76	20 900	1860
π³	3·19	37	270
η	3·36v	2100	750
λ	3·39	9000	1800
τ	3·60	650	420
π⁴	3·69	2200	910
π⁵	3·72v	2200	940
σ	3·73	5000	1800

Star	R.A. h	m	s	Dec. deg	min	sec	Mag.	Abs. mag.	Spectrum	Dist pc	
48 σ	05	38	44·7	−02	36	00	3·73	−4·4	O9·5	550	
47 ω	05	39	11·0	+04	07	17	4·57	−2·9	B3	310	
50 ζ	05	40	45·5	−01	56	34	1·77	−5·9	O9·5	340	Alnitak
53 κ	05	47	45·3	−09	40	11	2·06	−6·9	B0·5	650	Saiph
54 χ¹	05	54	22·9	+20	16	34	4·41	4·4	G0	9·9	
58 α	05	55	10·2	+07	24	26	var.	−5·6v	M2	95	Betelgeux
HD 40657	06	00	03·3	−03	04	27	4·53	−0·1	K2	80	
61 μ	06	02	22·9	+09	38	51	4·12	1·2	A0	37	
62 χ²	06	03	55·1	+20	08	18	4·63	−6·8	B2	1000	
67 ν	06	07	34·2	+14	46	06	4·42	−1·7	B3	170	
70 ξ	06	11	56·3	+14	12	31	4·48	−1·7	B3	170	

δ has long been classed as a variable, but is in fact an eclipsing star with a range of only 2·20 to 2·35 (period 5·73 days). η is also an eclipsing star with a range of only 0 m·2. π⁵ is an ellipsoidal variable with a range of 0·1 mag.

Also above mag. 5:

	Mag.	Abs. mag.	Spectrum	Dist
15	4·82	1·9	F2	41
23 m	5·00	−3·5	B1	470
25	4·95	−3·5	B1	490
49 d	4·80	1·0	A4	57
56	4·78	−2·2	K2	240
51	4·91	0·0	K1	87
69	4·95	−1·1	B5	160

VARIABLE STARS

	R.A. h	m	Dec. deg	min	Range	Type	Period, d.	Spectrum
W	05	05·4	+01	11	5·9–7·7	Semi-reg.	212	N
S	05	29·0	−04	42	7·5–13·5	Mira	419·2	M
CK	05	30·3	+04	12	5·9–7·1	Semi-reg.	120	K
α	05	55·2	+07	24	0·1–0·9	Semi-reg.	2110	M
U	05	55·8	+20	10	4·8–12·6	Mira	372·4	M

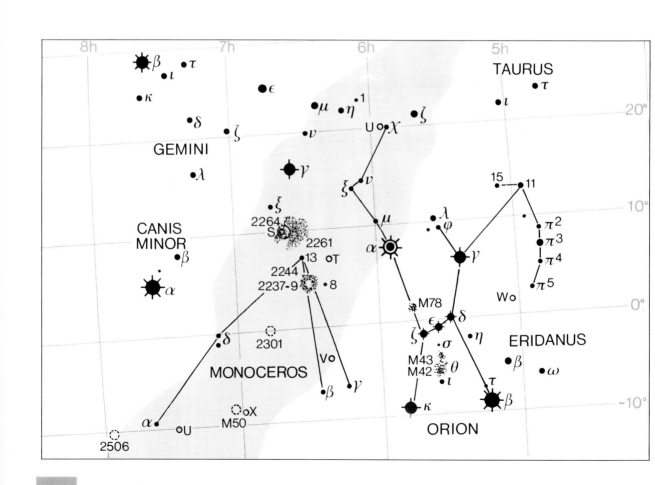

NEWTONIAN

EQUATORIAL MOUNT

FOCAL LENGTH   900mm

FOCAL RATIO   f/8

SR 4   H12

ORION (continued)

DOUBLE STARS

	R.A. h	m	Dec. deg	min	P.A. deg	Sep. sec	Mags.	
π^3	04	49·8	+06	58	138	94·6	3·2, 8·7	
β	05	14·5	−08	12	202	9·5	0·1, 6·8	
ρ	05	15·3	+02	54	064	7·0	4·5, 8·3	
η	05	24·5	−02	24	AB 080	1·5	3·8, 4·8	
					AC 051	115·1	9·4	
δ	05	32·0	−00	18	359	52·6	2·2v, 6·3	
λ	05	35·1	+09	56	043	4·4	3·6, 5·5	
σ	05	38·7	−02	36	AB 137	0·2	4·0, 6·0	Binary, 170 y
					AB + C 238	11·4	10·3	
					AB + D 084	12·9	7·5	
					AB + E 061	42·6	6·5	
θ	05	35·3	−05	23	AB 031	8·8	6·7, 7·9	
					AC 132	12·8	5·1	
					AD 096	21·5	6·7	
ι	05	35·4	−05	55	141	11·3	2·8, 6·9	
ζ	05	40·8	−01	57	AB 162	2·4	1·9, 4·0	Binary, 1509 y
					AC 010	57·6	9·9	
μ	06	02·4	+09	39	023	0·4	4·4, 6·0	

OPEN CLUSTERS

M	NGC	R.A. h	m	Dec. deg	min	Diameter min	Mag.	No. of stars
	1981	05	35·2	−04	26	25	4·6	20
	2112	05	53·9	+00	24	11	9·1	5
	2175	06	09·8	+20	19	18	6·8	60
	2186	06	12·2	+05	27	4	8·7	30

NEBULÆ

M	NGC	R.A. h	m	Dec. deg	min	Dimensions min	Mag. of illuminating star	
42	1976	05	35·4	−05	27	66 × 60	5	Great Nebula
43	1982	05	35·6	−05	16	20 × 15	7	Extension of M.42
78	2068	05	46·7	+00	03	8 × 6	10	Nebula is mag. 8
	IC 434	05	41·0	−02	24	60 × 10	2	(ζ). Behind Horse's Head dark nebula. Barnard 33

PAVO

(Abbreviation: Pav).

One of the 'Southern Birds'. The brightest star, α, is somewhat isolated from the main pattern; the most celebrated object is κ, often called a Cepheid even though it is, strictly speaking, a W Virginis type variable.

Pavo includes 6 stars above the fourth magnitude:

	Mag.	Luminosity Sun = 1	Dist lt-yrs
α	1·94	700	230
β	3·42	28	100
δ	3·56	1·0	18·6
η	3·62	82	162
κ	3·9 (max)	3·8v	75
ε	3·96	50	134

BRIGHTEST STARS

Star	R.A. h	m	s	Dec. deg	min	sec	Mag.	Abs. mag.	Spectrum	Dist pc
η	17	45	43·8	−64	43	25	3·62	0·0	K1	45
π	18	08	34·6	−63	40	06	4·35	1·6	A	17
ξ	18	23	13·3	−61	29	38	4·36	−0·5	M1	94
ν	18	31	22·2	−62	16	42	4·64	−1·2	B8	150
ζ	18	43	02·0	−71	25	42	4·01	−0·1	K2	66
λ	18	52	12·8	−62	11	16	4·22v	−3·5	B1	350
κ	18	56	56·9	−67	14	01	var.	3·4v	F5v	23
ε	20	00	35·4	−72	54	38	3·96	0·6	A0	41
δ	20	08	43·3	−66	10	56	3·56	4·8	G5	5·7
α	20	25	38·7	−56	44	06	1·94	−2·3	B3	71
β	20	44	57·4	−66	12	12	3·42	1·2	A5	28
γ	21	26	26·6	−65	21	59	4·22	4·5	F6	86

Also above mag. 5:

	Mag.	Abs. mag.	Spectrum	Dist
φ¹	4·76		F0	16
ρ	4·88	3·4	F5	24

VARIABLE STARS

	R.A. h	m	Dec. deg	min	Range	Type	Period, d.	Spectrum
R	18	12·9	−63	37	7·5–13·8	Mira	229·8	M
λ	18	52·2	−62	11	3·4–4·3	Irreg.	–	B
κ	18	56·9	−67	14	3·9–4·7	W Virginis	9·09	F
T	19	50·7	−71	46	7·0–14·0	Mira	244·0	M
S	19	55·2	−59	12	6·6–10·4	Semi-reg.	386	M
Y	21	24·3	−69	44	5·7–8·5	Semi-reg.	233	N
SX	21	28·7	−69	30	5·4–6·0	Semi-reg.	50	M

DOUBLE STAR

	R.A. h	m	Dec. deg	min	P.A. deg	Sep. sec	Mags.
ξ	18	23·2	−61	30	154	3·3	4·4, 8·6

PAVO (continued)

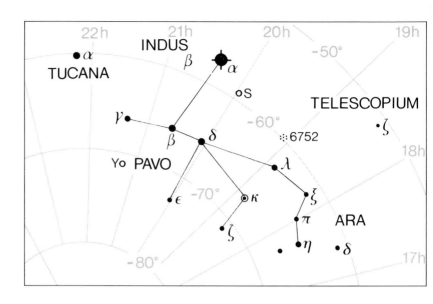

GLOBULAR CLUSTERS

M	NGC	R.A.		Dec.		Diameter	Mag.
		h	m	deg	min	min	
	6752	19	10·9	−59	59	20·4	5·4

GALAXIES

M	NGC	R.A.		Dec.		Mag.	Dimensions	Type
		h	m	deg	min		min	
	IC 4662	17	47·1	−64	38	11·4	2·2 × 1·4	Irreg.
	6684	18	49·0	−65	11	10·4	3·7 × 2·7	SB0
	6744	19	09·8	−63	51	9·0	15·5 × 10·2	SBb
	6753	19	11·4	−57	03	11·9	2·5 × 2·2	Sb

PEGASUS

(Abbreviation: Peg).

One of the most distinctive of the northern constellations. It commemorates the flying horse which the hero Bellerophon rode during an expedition to slay the fire-breathing Chimæra. The main stars of Pegasus make up a square; three of these are α, β and γ. The fourth is Alpheratz, which used to be included in Pegasus as δ Pegasi, but has been officially – and, frankly, illogically – transferred to Andromeda, as α Andromedæ. In Pegasus, excluding Alpheratz, there are nine stars above the fourth magnitude:

	Mag.	Luminosity	Dist
		Sun = 1	lt-yrs
ε	2·38	5000	520
β	2·4 (max)	310v	176
α	2·49	60	100
γ	2·83v	1320	490
η	2·94	200	173
ζ	3·40	82	156
μ	3·48	60	147
θ	3·53	26	82
ι	3·76	3·8	39

The celebrated group of galaxies known as Stephan's Quintet lies in Pegasus; NGC 7317, 7318 A and B, 7319 and 7320. All are faint, but are of special interest.

NGC	R.A.		Dec.		Dimensions	Mag.	Type
	h	m	deg	min	min		
7317	22	35·9	+33	57	1·0 × 0·8	14·6	E2
7318A	22	35·9	+33	58	1·0 × 1·0	14·3	E2p
7318B	22	36·0	+33	58	1·9 × 1·3	14·0	SBb + p
7319	22	36·1	+33	49	1·7 × 1·3	14·0	SBb + p
7320	22	36·1	+33	57	2·2 × 1·2	13·3	Sd

Though the galaxies seem to be connected, their red shifts are not the same; NGC 7320 has a red shift much less than the remaining members. Neither are the radial velocities the same

BRIGHTEST STARS

Star	R.A.			Dec.			Mag.	Abs.	Spectrum	Dist	
	h	m	s	deg	min	sec		mag.		pc	
1	21	22	05·0	+19	48	16	4·08	0·0	K1	63	
2	21	29	56·8	+23	38	20	4·57	−0·5	M1	97	
8 ε	21	44	11·0	+09	52	30	2·38	−4·4	K2	160	Enif
9	21	44	30·5	+17	21	00	4·34	−4·5	G5	490	
10 κ	21	44	38·5	+25	38	42	4·13	2·1	F5	27	
24 ι	22	07	00·5	+25	20	42	3·76	3·4	F5	13	
29 π	22	09	59·1	+33	10	42	4·29	−0·6	F5	96	
26 θ	22	10	11·8	+06	11	52	3·53	1·4	A2	25	Biham
42 ζ	22	41	27·6	+10	49	53	3·40	0·0	B8·5	48	Homan
44 η	22	43	00·0	+30	13	17	2·94	−0·9	G2	53	Matar
47 λ	22	46	31·7	+23	33	56	3·95	−0·9	G8	33	
46 ξ	22	46	41·4	+12	10	22	4·19	3·8	F7	14	Al Suud al Nujam
48 μ	22	50	00·0	+24	36	06	3·48	0·2	K0	45	Sadalbari
53 β	23	03	46·3	+28	04	58	var.	−1·4v	M2	54	Scheat
54 α	23	04	45·5	+15	12	19	2·49	0·2	B9	31	Markab
55	23	07	00·1	+09	24	34	4·52	−0·5	M2	100	
62 τ	23	20	38·1	+23	44	25	4·60	1·2	A5	47	Kerb
68 υ	23	25	22·7	+23	24	15	4·40	2·4	F8	22	
70 q	23	29	09·1	+12	45	38	4·55	0·3	G8	70	
84 ψ	23	57	45·4	+25	08	29	4·66	−0·5	M3	110	
88 γ	00	13	14·1	+15	11	01	2·83v	−3·0	B2	150	Algenib

PEGASUS (continued)

Also above mag. 5:

	Mag.	Abs. mag.	Spectrum	Dist
22 ν	4·84	−0·3	K4	99
32	4·81	−0·2	B8	90
35	4·79	0·2	K0	77
43 o	4·79	1·2	A1	52
50 ρ	4·90	1·2	A1	55
56	4·76	−2·1	K0	240
72	4·98	−0·3	K4	110
78	4·93	0·2	K0	88
89 χ	4·80	−0·5	M2	120

VARIABLE STARS

	R.A.		Dec.		Range	Type	Period, d.	Spectrum
	h	m	deg	min				
AG	21	51·0	+12	38	6·0–9·4	Z Andromedæ	830	WN + M
AW	21	52·3	+24	01	7·8–9·2	Algol	10·62	A + F
V	22	01·0	+06	07	7·0–15·0	Mira	302·3	M
TW	22	04·0	+28	21	7·0–9·2	Semi-reg.	956	M
RZ	22	05·9	+33	30	7·6–13·6	Mira	439·4	M
β	23	03·8	+28	05	2·3–2·8	Semi-reg.	38	M
R	23	06·6	+10	33	6·9–13·8	Mira	378·0	M
W	23	19·8	+26	17	7·9–13·0	Mira	344·9	M
S	23	20·6	+08	55	7·1–13·8	Mira	319·2	M
Z	00	00·1	+25	53	7·7–13·6	Mira	325·5	M

DOUBLE STARS

	R.A.		Dec.		P.A.	Sep.	Mags.	
	h	m	deg	min	deg	sec		
ε	21	44·2	+09	52	AB 325	81·8	2·4, 11·2	
					AC 320	142·5	9·4	
κ	21	44·6	+25	39	095	0·3	4·7, 5·0	Binary, 11·6 y
35	22	27·9	+04	42	AB 210	98·3	4·8, 9·8	
					AC 241	181·5	9·7	
37	22	30·0	+04	26	118	0·9	5·8, 7·1	Binary, 140 y
η	22	43·0	+30	13	339	90·4	2·9, 9·9	B is a close dble
β	23	03·8	+28	05	AB 211	108·5	2v, 11·6	
					AC 098	253·1	9·4	

GLOBULAR CLUSTER

M	NGC	R.A.		Dec.		Diameter	Mag.
		h	m	deg	min	min	
15	7078	21	30·0	+12	10	12·3	6·3

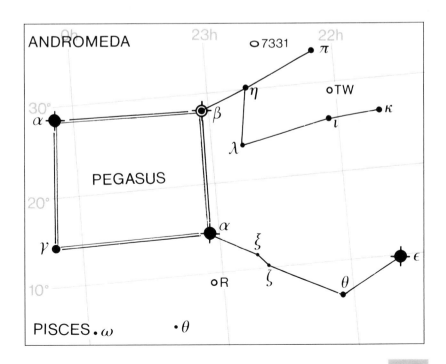

PEGASUS (continued)

GALAXIES

M	NGC	R.A.		Dec.		Mag.	Dimensions	Type
		h	m	deg	min		min	
	7331	22	37·1	+34	25	9·5	10·7 × 4·0	Sb
	7332	22	37·4	+23	48	11·8	4·2 × 1·3	E7
	7479	23	04·9	+12	19	11·0	4·1 × 3·2	SBb
	7814	00	03·3	+16	09	10·5	6·3 × 2·6	Sb

PERSEUS

(Abbreviation: Per).

A prominent constellation, containing the prototype eclipsing star Algol as well as the superb Sword-Handle cluster (H.VI.33–4). Mythologically, Perseus was the hero of one of the most famous of all legends; he killed the Gorgon, Medusa, and married Andromeda, daughter of Cepheus and Cassiopeia. The Gorgon's Head is marked by the winking 'Demon Star', Algol.

Perseus includes 12 stars above the fourth magnitude:

	Mag.	Luminosity	Dist
		Sun = 1	lt-yrs
α	1·80	6000	620
β	2·12 (max)	105	95
ζ	2·85	16000	1100
ε	2·89	2600	680
γ	2·93	58	110
δ	3·01	650	326
ρ	3·2 (max)	120v	196
η	3·76	5000	820
ν	3·77	520	460
κ	3·80	60	173
o	3·83	5000	1010
τ	3·95	58	176

BRIGHTEST STARS

Star		R.A.			Dec.			Mag.	Abs.	Spectrum	Dist	
		h	m	s	deg	min	sec		mag.		pc	
	φ	01	43	39·6	+50	41	20	4·07	−3·9	B1	350	
13	θ	02	44	11·9	+49	13	43	4·12	3·8	F7	13	
16		02	50	34·9	+38	19	07	4·23	0·6	F2	50	
15	η	02	50	41·8	+55	53	44	3·76	−4·4	K3	250	Miram
17		02	51	30·8	+35	03	35	4·53	−0·3	K5	150	
18	τ	02	54	15·4	+52	45	45	3·95	0·3	G4	54	Kerb
22	π	02	58	45·6	+39	39	46	4·70	1·4	A2	44	Gorgonea Secunda
23	γ	03	04	47·7	+53	30	23	2·93	0·3	G8	34	
25	ρ	03	05	10·5	+38	50	25	var.	−0·5v	M4	60	Gorgonea Terti
26	β	03	08	10·1	+40	57	21	2·12v	−0·2	B8	29	Algol
	ι	03	09	03·9	+49	36	49	4·05	3·7	G0	12	
27	κ	03	09	29·7	+44	51	27	3·80	0·2	K0	53	Misam
28	ω	03	11	17·3	+39	36	42	4·63	0·0	K0	180	Gorgonea Quarta
33	α	03	24	19·3	+49	51	40	1·80	−4·6	F5	190	Mirphak
34		03	29	22·0	+49	30	32	4·67	−2·3	B3	210	
35	σ	03	30	34·4	+47	59	43	4·35	−0·2	K3	71	
37	ψ	03	36	29·3	+48	11	34	4·23	−1·2	B5e	126	
39	δ	03	42	55·4	+47	47	15	3·01	−2·2	B5	100	
38	o	03	44	19·1	+32	17	18	3·83	−4·4	B1	310	Ati
41	ν	03	45	11·6	+42	34	43	3·77	−2·0	F5	140	
44	ζ	03	54	07·8	+31	53	01	2·85	−5·7	B1	340	Atik
45	ε	03	57	51·1	+40	00	37	2·89	−3·7	B0·5	208	
46	ξ	03	58	57·8	+35	47	28	4·04	−5·4	O7	210	Menkib
47	λ	04	06	35·0	+50	21	05	4·29	0·2	B9	62	
48	υ	04	08	39·6	+47	42	45	4·04	−1·7	B3	140	Nembus
52	f	04	14	53·3	+40	29	01	4·71	−4·5	G5 + A5	560	
51	μ	04	14	53·8	+48	24	33	4·14	−4·5	G0	460	
	b	04	18	14·6	+50	17	44	4·62v	var.	A2	55	

Also above mag. 5:

	Mag.	Abs. mag.	Spectrum	Dist
12	4·91	4·2	F9	15
24	4·93	−0·1	K2	92
32	4·95	1·4	A2	51
40	4·97	0·5	B1	

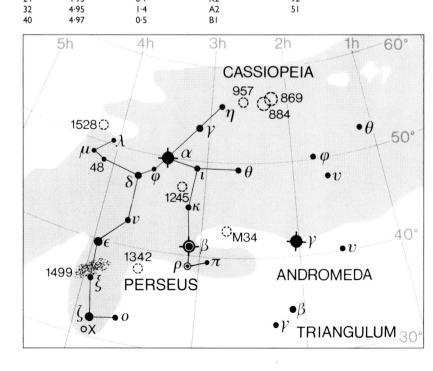

PERSEUS (continued)

	Mag.	Abs. mag.	Spectrum	Dist
54	4·93	0·3	G8	84
53	4·85	−1·7	B3	170

VARIABLE STARS

	R.A. h	m	Dec. deg	min	Range	Type	Period, d.	Spectrum
U	01	59·6	+54	49	7·4–12·3	Mira	321·0	M
S	02	22·9	+58	35	7·9–11·5	Semi-reg.	Long	M
ρ	03	05·2	+38	50	3–4	Semi-reg.	33 to 55	M
β	03	08·2	+40	57	2·1–3·4	Algol	2·87	B + G
R	03	30·1	+35	40	8·1–14·8	Mira	210·0	M
X	03	55·4	+31	03	6·0–7·0	Irreg.	–	09·5 X-ray source
AW	04	47·8	+36	43	7·1–7·8	Cepheid	6·46	F–G

DOUBLE STARS

	R.A. h	m	Dec. deg	min	P.A. deg	Sep. sec	Mags	
η	02	50·7	+55	54	300	28·3	3·3, 8·5	
θ	02	44·2	+49	14	215	19·8	4·1, 9·9	Binary, 2720 y
γ	03	04·8	+53	30	326	57·0	2·9, 10·6	
ζ	03	54·1	+31	53	AB 208	12·9	2·9, 9·5	
					AC 286	32·8	11·3	
					AD 195	94·2	9·5	
					AE 185	120·3	10·2	
ε	03	57·9	+40	01	010	8·8	2·9, 8·1	

OPEN CLUSTERS

M	NGC	R.A. h	m	Dec. deg	min	Diameter min	Mag.	No. of stars
	744	01	58·4	+55	29	11	7·9	20
	869	02	19·0	+57	09	30	4·3	200 } Sword
	884	02	22·4	+57	07	30	4·4	150 } Handle
	957	02	33·6	+57	32	11	7·6	30
34	1039	02	42·0	+42	47	35	5·2	60
	1245	03	14·7	+47	15	10	8·4	200
	1444	03	49·4	+52	40	4	6·6	–
	1513	04	10·0	+49	31	9	8·4	50
	1528	04	15·4	+51	14	24	6·4	40
	1545	04	20·9	+50	15	8	6·2	20

PLANETARY NEBULA

M	NGC	R.A. h	m	Dec. deg	min	Dimensions sec	Mag.	Mag. of central star
76	650–1	01	42·4	+51	34	65 × 290	12·2	17 Little Dumbbell

NEBULÆ

M	NGC	R.A. h	m	Dec. deg	min	Dimensions min	Mag. of illuminating star	
	1333	03	29·3	+31	25	9 × 7	9·5	(Near dark nebula B.205)
	1499	04	00·7	+36	37	145 × 40	4	California Nebula

GALAXIES

M	NGC	R.A. h	m	Dec. deg	min	Mag.	Dimensions min	Type	
	1003	02	39·3	+40	52	11·5	5·4 × 2·1	Sc	
	1023	02	40·4	+39	04	9·5	8·7 × 3·3	E7p	
	1275	03	19·8	+41	31	11·6	2·6 × 1·9	Pec.	Perseus A

PHŒNIX

(Abbreviation: Phe).

One of the 'Southern birds'. It is not very distinctive, and Ankaa is the only bright star; there are however seven stars above the fourth magnitude:

	Mag.	Luminosity Sun = 1	Dist lt-yrs
α	2·39	60	78
β	3·31	58	130
γ	3·41	5000	910
ζ	3·6 (max)	105	220

BRIGHTEST STARS

Star	R.A. h	m	s	Dec. deg	min	sec	Mag.	Abs. mag.	Spectrum	Dist. pc	
ι	23	35	04·4	−42	36	54	4·71	?	A	25?	
ε	00	09	24·6	−45	44	51	3·88	0·2	K0	23	
κ	00	26	12·1	−43	40	48	3·94	1·7	A3	19	
α	00	26	17·0	−42	18	22	2·39	0·2	K0	24	Ankaa
μ	00	41	19·5	−46	05	06	4·59	0·3	G8	70	
η	00	43	21·2	−57	27	48	4·36	0·0	A0	71	
β	01	06	05·0	−46	43	07	3·31	0·3	G8	40	
ζ	01	08	23·0	−55	14	45	var.	−0·2	B8	67	
γ	01	28	21·9	−43	19	06	3·41	−4·4	K5	280	
δ	01	31	15·0	−49	04	22	3·95	1·7	K0	28	
ψ	01	53	38·7	−46	18	09	4·41	−0·5	M4	96	

PHŒNIX (continued)

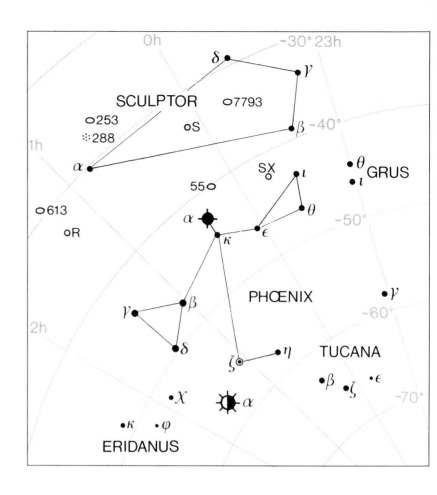

	Mag.	Luminosity Sun = 1	Dist lt-yrs
ε	3·88	60	75
κ	3·94	17	62
δ	3·95	17	91

Also above mag. 5:

	Mag.	Abs. mag.	Spectrum	Dist.
λ¹	4·77	0·6	A0	130
ν	4·96	4·0	F8	15
HD 12055	4·83	2·6	G5	90

VARIABLE STARS

	R.A. h m	Dec. deg min	Range	Type	Period, d.	Spectrum
SX	23 46·5	−41 35	6·8–7·5	Delta Scuti	0·055	A
R	23 56·5	−49 47	7·5–14·4	Mira	267·9	M
S	23 53·1	−56 35	7·4–8·2	Semi-reg.	141	M
ζ	01 08·4	−55 15	3·6–4·1	Algol	1·67	B + B

DOUBLE STARS

	R.A. h m	Dec. deg min	P.A. deg	Sep. sec	Mags.
ξ	00 41·8	−56 30	253	13·2	5·8, 10·2
η	00 43·4	−57 28	217	19·8	4·4, 11·4
β	01 06·1	−46 43	346	1·4	4·0, 4·2

PICTOR

(Abbreviation: Pic).

An unremarkable constellation near Canopus, known originally under the cumbersome name of Equuleus Pictoris (the Painter's Easel). There are two stars above the fourth magnitude:

BRIGHTEST STARS

Star	R.A. h m s	Dec. deg min sec	Mag.	Abs. mag.	Spectrum	Dist pc
β	05 47 17·1	−51 03 59	3·85	0·3	A5	24
γ	05 49 49·6	−56 10 00	4·51	0·0	K1	80
δ	06 10 17·9	−54 58 07	4·7v		B1	
α	06 48 11·4	−61 56 29	3·27	2·1	A5	16

VARIABLE STARS

	R.A. h m	Dec. deg min	Range	Type	Period, d.	Spectrum
R	04 46·2	−49 15	6·7–10·0	Semi-reg.	164	M
S	05 11·0	−48 30	6·5–14·0	Mira	426·6	M

PICTOR (continued)

	Mag.	Luminosity Sun = 1	Dist lt-yrs
α	3·27	11	52
β	3·85	58	78

See chart for Carina.

	R.A. h	m	Dec. deg	min	Range	Type	Period, d.	Spectrum
T	05	15·1	−46	55	7·9–14·4	Mira	200·6	M
δ	06	10·3	−54	58	4·7–4·9	Beta Lyræ	1·67	B

DOUBLE STARS

	R.A. h	m	Dec. deg	min	P.A. deg	Sep. sec	Mags.
ι	04	50·9	−53	28	058	12·3	5·6, 6·4
θ	05	24·8	−52	19	AB 152	0·2	6·9, 7·2
					AB + C 287	38·2	6·8
μ	06	32·0	−58	45	231	2·4	5·8, 9·0

PISCES

(Abbreviation: Psc).

A large but faint Zodiacal constellation; it now contains the Vernal Equinox. Its mythological associations are rather vague, but it may represent the fishes into which Venus and Cupid once changed themselves in order to escape from the monster Typhon. There are three stars above the fourth magnitude:

	Mag.	Luminosity Sun = 1	Dist lt-yrs
η	3·62	58	144
γ	3·69	58	156
α	3·79	26	98

BRIGHTEST STARS

Star	R.A. h	m	s	Dec. deg	min	sec	Mag.	Abs. mag.	Spectrum	Dist pc	
4 β	23	03	52·5	+03	49	12	4·53	−0·4	B5p	101	
6 γ	23	17	09·7	+03	16	56	3·69	0·3	G8	48	
10 θ	23	27	57·9	+06	22	44	4·28	0·0	K1	72	
17 ι	23	39	56·9	+05	37	35	4·13	3·8	F7	13	
18 λ	23	42	02·6	+01	46	48	4·50	2·4	A7	26	
19 TX	23	46	23·3	+03	29	13	4·3–5·1	−2·0v	N	330	
28 ω	23	59	18·5	+06	51	48	4·01	0·8	F4	26	
30	00	01	57·5	−06	00	51	4·41	0·4	M3	30	
33	00	05	20·1	−05	42	27	4·61	0·0	K1	84	
63 δ	00	48	40·9	+07	35	06	4·43	−0·3	K5	88	
71 ε	01	02	56·5	+07	53	24	4·28	0·2	K0	66	
74 ψ	01	05	40·9	+21	28	24	4·7	−0·1, 0·2	B9·5, B9	120	
84 χ	01	11	27·1	+21	02	05	4·66	0·2	K0	75	
83 τ	01	11	39·5	+30	05	23	4·51	1·7	K0	29	
85 φ	01	13	44·8	+24	35	01	4·65	0·2	K0	74	
99 η	01	31	28·9	+15	20	45	3·62	0·3	G8	44	Alpherg
106 ν	01	41	25·8	+05	29	15	4·44	−0·2	K3	42	
110 ο	01	45	23·5	+09	09	28	4·26	0·2	K0	65	Torcular
111 ξ	01	53	33·3	+03	11	15	4·62	0·2	K0	77	
113 α	02	02	02·7	+02	45	49	3·79	1·4	A2	30	Al Rischa

Also above mag. 5:

	Mag.	Abs. mag.	Spectrum	Dist
κ	4·94	2·2	A2	
27	4·86	0·2	G9	86
86 ζ	4·86	2·1	F6	33
90 υ	4·76	1·4	A2	47
98 μ	4·84	−0·3	K4	67
47 TV	4·8v	−0·5	M3	140

VARIABLE STARS

	R.A. h	m	Dec. deg	min	Range	Type	Period, d.	Spectrum
TX	23	46·4	+03	29	6·9–7·7	Irreg.	−	N
TV	00	28·0	+17	24	4·6–5·4	Semi-reg.	70	M
Z	01	16·1	+25	46	7·0–7·9	Semi-reg.	144	N
R	01	30·6	+02	53	7·1–14·8	Mira	344·0	M

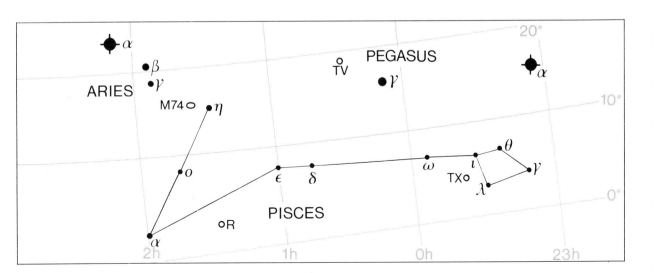

PISCES (continued)

DOUBLE STARS

	R.A.		Dec.		P.A.	Sep.	Mags.	
	h	m	deg	min	deg	sec		
ζ	01	13·7	+07	35	063	23·0	5·6, 6·5	
ψl	01	05·6	+21	28	AB 159	30·0	5·6, 5·8	
					AC 123	92·6	11·2	
α	02	02·0	+02	46	279	1·9	4·2, 5·1	Binary, 933 y

GALAXIES

M	NGC	R.A.		Dec.		Mag.	Dimensions	Type
		h	m	deg	min		min	
	470	01	19·7	+03	25	11·9	3·0 × 2·0	Sc
	474	01	20·1	+03	25	11·1	7·9 × 7·2	S0
	488	01	21·8	+05	15	10·3	5·2 × 4·1	Sb
	524	01	24·8	+09	32	10·6	3·2 × 3·2	E1
74	628	01	36·7	+15	47	9·2	10·2 × 9·5	Sc

PISCIS AUSTRALIS

(Abbreviation: PsA).

Also known as Piscis Austrinus. No specific mythological legends have been associated with it. It contains Fomalhaut, which is, incidentally, the most southerly of the first-magnitude stars to be visible from England, but there are no other stars above the fourth magnitude.

	Mag.	Luminosity Sun = 1	Dist lt-yrs
α	1·16	13	22

BRIGHTEST STARS

Star	R.A.			Dec.			Mag.	Abs. mag.	Spectrum	Dist pc	
	h	m	s	deg	min	sec					
9 ι	21	44	56·7	−33	01	33	4·34	0·6	A0	42	
14 μ	22	08	22·8	−32	59	19	4·50	1·4	A2	42	
17 β	22	31	30·1	−32	20	46	4·29	0·6	A0	53	Fum el Samakah
18 ε	22	40	39·2	−27	02	37	4·17	−0·2	B8	75	
22 γ	22	52	31·4	−32	52	32	4·46	0·6	A0	59	
23 δ	22	55	56·8	−32	32	23	4·21	0·3	gG4	49	
24 α	22	57	38·9	−29	37	20	1·16	2·0	A3	6·7	Fomalhaut

Also above mag. 5:

	Mag.	Abs. mag.	Spectrum	Dist
υ	4·99	−0·3	K5	110
15 τ	4·92	3·4	F5	11

VARIABLE STARS

	R.A.		Dec.		Range	Type	Period, d.	Spectrum
	h	m	deg	min				
S	22	03·8	−28	03	8·0–14·5	Mira	271·7	M
V	22	55·3	−29	37	8·0–9·0	Semi-reg.	148	M

DOUBLE STARS

	R.A.		Dec.		P.A.	Sep.	Mags	
	h	m	deg	min	deg	sec		
η	22	00·8	−28	27	115	1·7	5·8, 6·8	
β	22	31·5	−32	21	172	30·3	4·4, 7·9	(optical)
γ	22	52·5	−32	53	262	4·2	4·5, 8·0	
δ	22	55·9	−32	32	244	5·0	4·2, 9·2	

GALAXIES

M	NGC	R.A.		Dec.		Mag.	Dimensions	Type	
		h	m	deg	min		min		
	7172	22	02·0	−31	52	11·9	2·2 × 1·3	S	
	7174	22	02·1	−31	59	12·6	1·3 × 0·7	S	
	7314	22	35·8	−26	03	10·9	4·6 × 2·3	Sc	Arp 14

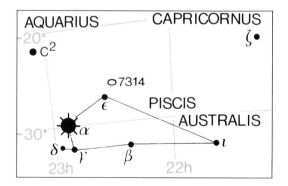

PUPPIS

(Abbreviation: Pup).

The poop of the dismembered ship, Argo Navis. There are 12 stars above the fourth magnitude.

	Mag.	Luminosity Sun = 1	Dist lt-yrs
ζ	2·25	60 000	2400
π	2·70	110	130
ρ	2·81v	525	300
τ	2·93	60	82
ν	3·17	250	245
σ	3·25	110	166
ξ	3·34	5200	750
L²	3·4 (max)	1400v	75
c	3·59	8200	1730
a	3·73	58	125
k	3·82	150+130	360
3	3·96	70 000	5540

ρ is a variable with very small range (2·72–2·87).

See chart for Carina.

BRIGHTEST STARS

Star	R.A. h	m	s	Dec. deg	min	sec	Mag.	Abs. mag.	Spectrum	Dist pc	
ν	06	37	45·6	−43	11	45	3·17	−1·2	B8	75	
τ	06	49	56·1	−50	36	53	2·93	0·2	K0	25	
I	07	12	33·6	−46	45	34	4·49	2·6	F0	24	
L²	07	13	13·3	−45	10	59	var.	−3·1v	M5	23	
π	07	17	08·5	−37	05	51	2·70	−0·3	K5	40	
υ¹	07	18	18·4	−36	44	03	4·66	−1·7	B3	180	
σ	07	29	13·8	−43	18	05	3·25	−0·3	K5	51	
HD 60532	07	34	03·1	−22	17	46	4·45	2·3	F7	26	
p	07	35	22·8	−28	22	10	4·64	0·1	B8	77	
f	07	37	22·0	−34	58	07	4·53	−0·2	B8	88	
m	07	38	17·9	−25	21	53	4·67	−0·1	B8	92	
k	07	38	49·7	−26	48	13	3·82	−0·7, −0·6	B8	110	
I	07	43	32·3	−28	24	40	4·59	−0·3	gK5	79	
3	07	43	48·4	−28	57	18	3·96	−7·5	A2	1700	
c	07	45	15·2	−37	58	07	3·59	−5·0	cK	530	
o	07	48	05·1	−25	56	14	4·50	−3·5	B1	390	
Q	07	48	20·2	−47	04	39	4·71	0·2	K0	72	
P	07	49	14·3	−46	22	24	4·11	−5·6	B0	820	
7 ξ	07	49	17·6	−24	51	35	3·34	−4·5	G3	230	Asmidiske
a	07	52	13·0	−40	34	33	3·73	0·3	G5	38	
J	07	53	03·7	−49	36	47	4·63	−3·6	B2	440	
11	07	56	51·5	−22	52	48	4·20	−2·0	F8	150	
V	07	58	14·3	−49	14	42	var.	−4·0v	B1 + B3	470	
232G	07	59	52·0	−18	23	58	4·61	1·7	A3	38	
ζ	08	03	35·0	−40	00	12	2·25	−7·1	O5·8	740	Suhail Hadar
ρ	08	07	32·6	−24	18	15	2·81v	−2·0	F6	92	Turais
16	08	09	01·5	−19	14	42	4·40	−1·1	B5	130	
19	08	11	16·2	−12	55	37	4·72	0·2	K0	45	
h¹	08	11	21·5	−39	37	07	4·45	−5·9	cK	850	
h²	08	14	02·8	−40	20	52	4·44	−0·1	gK2	81	
q	08	18	33·2	−36	39	34	4·45	0·5	A7	62	

Also above mag. 5:

	Mag.	Abs. mag.	Spectrum
Y	5·00	−0·2	G7
A	4·85		M3
d¹	4·91	−0·9	B3
r	4·77	−1·3	B3
W	4·94	1·8	M0

VARIABLE STARS

	R.A. h	m	Dec. deg	min	Range	Type	Period, d.	Spectrum
L²	07	13·5	−44	39	2·6–6·2	Semi-reg.	140	M
Z	07	32·6	−20	40	7·2–14·6	Mira	499·7	M
VX	07	32·6	−21	56	7·7–8·5	Cepheid	3·01	F
X	07	32·8	−20	55	7·8–9·2	Cepheid	25·96	F–G
W	07	46·0	−41	12	7·3–13·6	Mira	120·1	M
AP	07	57·8	−40	07	7·1–7·8	Cepheid	5·08	F
V	07	58·2	−49	15	4·7–5·2	Beta Lyræ	1·45	B + B
AT	08	12·4	−36	57	7·5–8·4	Cepheid	6·66	F–G
RS	08	13·1	−34	35	6·5–7·6	Cepheid	41·39	F–G

DOUBLE STAR

	R.A. h	m	Dec. deg	min	P.A. deg	Sep. sec	Mags.
σ	07	29·2	−43	18	074	22·3	3·3, 9·4

OPEN CLUSTERS

M	NGC	R.A. h	m	Dec. deg	min	Diameter min	Mag.	No. of stars	
	2383	07	24·8	−20	56	6	8·4	40	
	2421	07	36·3	−20	37	10	8·3	70	
47	2422	07	36·6	−14	30	30	4·4	30	
	Mel 71	07	37·5	−12	04	9	7·1	80	
	Mel 72	07	38·4	−10	41	9	10·1	40	
	2432	07	40·9	−19	05	8	10·2	50	
	2439	07	40·8	−31	39	10	6·9	80	R Puppis. Asterism
46	2437	07	41·8	−14	49	27	6·1	100	
93	2447	07	44·6	−23	52	22	6·2	80	
	2451	07	45·4	−37	58	45	2·8	40	
	2477	07	52·3	−38	33	27	5·8	160	
	2479	07	55·1	−17	43	7	9·6	45	
	2489	07	56·2	−30	04	8	7·9	45	

PUPPIS (continued)

M	NGC	R.A. h	m	Dec. deg	min	Diameter min	Mag.	No. of stars
	2509	08	00·7	−19	04	8	9·3	70
	2527	08	05·3	−28	10	22	6·5	40
	2533	08	07·0	−29	54	3·5	7·6	60
	2539	08	10·7	−12	50	22	6·5	50
	2546	08	12·4	−37	38	41	6·3	40
	2567	08	18·6	−30	38	10	7·4	40
	2571	08	18·9	−29	44	13	7·0	30
	2580	08	21·6	−30	19	8	9·7	50
	2587	08	23·5	−29	30	9	9·2	40

PLANETARY NEBULÆ

M	NGC	R.A. h	m	Dec. deg	min	Dimensions sec	Mag.	Mag. of central star	
	2438	07	41·8	−14	44	66	10·1	17·7	In cluster NGC 2437
	2440	07	41·9	−18	13	14 × 32	10·8	14·3	Protoplanetary?

NEBULA

M	NGC	R.A. h	m	Dec. deg	min	Dimensions min	Mag. of illuminating star	
	2467	07	52·5	−26	24	8 × 7	9·2	Gum 9

PYXIS

(Abbreviation: Pyx).

Also originally part of Argo. The only stars above the fourth magnitude are α and β.

	Mag.	Luminosity Sun = 1	Dist lt-yrs
α	3·68	5000	1340
β	3·97	58	150

See chart for Carina.

BRIGHTEST STARS

Star	R.A. h	m	s	Dec. deg	min	sec	Mag.	Abs. mag.	Spectrum	Dist pc
β	08	40	06·1	−35	18	30	3·97	0·3	G4	46
α	08	43	35·5	−33	11	11	3·68	−4·4	B2	410
γ	08	50	31·9	−27	42	36	4·01	−0·3	K4	73
κ	09	08	02·8	−25	51	30	4·58	−0·4	gM0	97
θ	09	21	29·5	−25	57	55	4·72	−0·5	M1	100
λ	09	23	12·1	−28	50	02	4·69	0·3	gG7	76

Also above mag. 5:

	Mag.	Abs. mag.	Spectrum	Dist
ζ	4·89	0·3	G4	75
δ	4·89	1·7	A3	42

VARIABLE STARS

	R.A. h	m	Dec. deg	min	Range	Type	Period, d.	Spectrum
TY	08	59·7	−27	49	6·9–7·5	Eclipsing	3·20	G + G
T	09	04·7	−32	23	6·3–14·0	Recurrent nova	–	Q Outbursts 1890, 1902 1920, 1944 1966
S	09	05·1	−23	05	8·0–14·2	Mira	206·4	M

DOUBLE STARS

	R.A. h	m	Dec. deg	min	P.A. deg	Sep. sec	Mags.
ζ	08	39·7	−29	34	061	52·4	4·9, 9·1
δ	08	55·5	−27	41	AB 268	23·8	4·9, 14·0
					CD 017	2·5	11·0, 11·0
ε	09	09·9	−30	22	A + BC 147	17·8	5·6, 10·5
					BC 088	0·3	10·5, 10·8
					AD 340	35·4	5·6, 13·5
κ	09	08·0	−25	52	263	2·1	4·6, 9·8

OPEN CLUSTERS

M	NGC	R.A. h	m	Dec. deg	min	Diameter min	Mag.	No. of stars
	2627	08	37·3	−29	57	11	8·4	60
	2658	08	43·4	−32	39	12	9·2	80

PLANETARY NEBULA

M	NGC	R.A. h	m	Dec. deg	min	Diameter sec	Mag.	Mag. of central star
	2818	09	16·0	−36	28	38	13·0	13·0

GALAXY

M	NGC	R.A. h	m	Dec. deg	min	Mag.	Dimensions min	Type
	2613	08	33·4	−22	58	10·4	7·2 × 2·1	Sb

RETICULUM

(Abbreviation: Ret).

Originally Reticulum Rhomboidalis (the Rhomboidal Net). A small but quite distinctive constellation of the far south. There are two stars above the fourth magnitude:

	Mag.	Luminosity Sun = 1	Dist lt-yrs
α	3·35	600	390
β	3·85	4·5	55

BRIGHTEST STARS

Star	R.A. h	m	s	Dec. deg	min	sec	Mag.	Abs. mag.	Spectrum	Dist pc
κ	03	29	22·7	−62	56	15	4·72	3·4	F5	20
β	03	44	12·0	−64	48	26	3·85	3·2	K0	17
δ	03	58	44·7	−61	24	01	4·56	−0·5	M2	97
γ	04	00	53·8	−62	09	34	4·51	?	Mb	?
α	04	14	25·5	−62	28	26	3·35	−2·1	G6	120
ε	04	16	28·9	−59	18	07	4·44	−0·3	gK5	21

Also above mag. 5:

	Mag.	Abs. mag.	Spectrum	Dist
ι	4·97	−0·4	M0	120

VARIABLE STAR

	R.A. h	m	Dec. deg	min	Range	Type	Period, d.	Spectrum
R	04	33·5	−63	02	6·5–14·0	Mira	278·3	M

DOUBLE STARS

	R.A. h	m	Dec. deg	min	P.A. deg	Sep. sec	Mags.	
β	03	44·2	−64	48	following	1440	3·9, 8·1	
ζ	03	18·2	−62	30	218	310·0	4·7, 5·2	Optical pair

GALAXIES

M	NGC	R.A. h	m	Dec. deg	min	Mag.	Dimensions min	Type
	1313	03	18·3	−66	30	9·4	8·5 × 6·6	SBd
	1559	04	17·6	−62	47	10·4	3·3 × 2·1	SBc

SAGITTA

(Abbreviation: Sge).

An original constellation; small though it is, it is quite distinctive. It has been identified with Cupid's bow, and also with the arrow used by Apollo against the one-eyed Cyclops. There are two stars above the fourth magnitude:

	Mag.	Luminosity Sun = 1	Dist lt-yrs
γ	3·47	110	166
δ	3·82	800	550

See chart for Aquila.

BRIGHTEST STARS

Star	R.A. h	m	s	Dec. deg	min	sec	Mag.	Abs. mag.	Spectrum	Dist pc	
5 α	19	40	05·6	+18	00	50	4·37	−2·0	G0	190	Alsahm
6 β	19	41	02·8	+17	28	33	4·37	−2·1	G8	200	
7 δ	19	47	23·0	+18	32	03	3·82	−2·4	M2	170	
12 γ	19	58	45·3	+19	29	32	3·47	−0·3	K5	51	

Also above mag. 5:

	Mag.	Abs. mag.	Spectrum	Dist
8 ζ	5·00	1·7	A3	46

VARIABLE STARS

	R.A. h	m	Dec. deg	min	Range	Type	Period, d.	Spectrum	
U	19	18·8	+19	37	6·6–9·2	Algol	3·38	B–K	
S	19	56·0	+16	38	5·3–6·0	Cepheid	8·38	F–G	
X	20	05·1	+20	39	7·9–8·4	Semi-reg.	196	N	
WZ	20	07·6	+17	42	7·0–15·5	Recurrent nova	–	Q	Outbursts 1913, 1946, 1978

SAGITTA (continued)

DOUBLE STAR

	R.A.		Dec.		P.A.	Sep.	Mags.	
	h	m	deg	min	deg	sec		
ζ	19	49·0	+19	09	AB + C 311	8·6	5·5, 8·7	
					AB 163	0·3	5·5, 6·2	Binary, 22·8 y

OPEN CLUSTER

M	NGC	R.A.		Dec.		Diameter	Mag.	No. of stars
		h	m	deg	min	min		
	H.20	19	53·1	+18	20	7	7·7	15

GLOBULAR CLUSTER

M	NGC	R.A.		Dec.		Diameter	Mag.
		h	m	deg	min	min	
71	6838	19	53·8	+18	47	7·2	8·3

PLANETARY NEBULÆ

M	NGC	R.A.		Dec.		Diameter	Mag.	Mag. of central star
		h	m	deg	min	sec		
	6879	20	10·5	+16	55	5	13·0	15
	IC 4997	20	20·2	+16	45	2	11·6	13 (v?)

SAGITTARIUS

(Abbreviation: Sgr)

The southernmost of the Zodiacal constellations, and not wholly visible from England. Mythologically it has been associated with Chiron, the wise centaur who was tutor to Jason and many others; but it would certainly be more logical to associate Chiron with Centaurus, and another version states that Chiron merely invented the constellation Sagittarius to help in guiding the Argonauts in their quest of the Golden Fleece. The centre of the Milky Way lies behind the star-clouds here, and the whole area is exceptionally rich; it abounds in Messier objects. It is worth commenting that the stars lettered α and β are relatively faint. There are 16 stars above the fourth magnitude:

	Mag.	Luminosity	Dist
		Sun = 1	lt-yrs
ε	1·85	110	85
σ	2·02	525	209
ζ	2·59	50	78
δ	2·70	96	81
λ	2·81	96	98
π	2·89	525	310
γ	2·99	60	117
η	3·11	800	420
φ	3·17	250	244
τ	3·32	82	130
ξ²	3·51	82	144
ο	3·77	58	140
μ	3·86v	60 000	3900
ρ¹	3·93	17	91
β¹	3·93	105	218
α	3·97	105	117

μ is an Algol binary with a very small range (3·8 to 3·9).

BRIGHTEST STARS

Star	R.A.			Dec.			Mag.	Abs.	Spectrum	Dist	
	h	m	s	deg	min	sec		mag.		pc	
3 X	17	47	33·4	−27	49	51	var.	−2·0	F7	200	
W	18	05	01·1	−29	34	48	var.	var.	F8	400	
10 γ	18	05	48·3	−30	25	26	2·99	0·2	K0	36	Alnasr
HD 165634	18	08	04·8	−28	27	25	4·57	1·0	Gp	52	
13 μ	18	13	45·6	−21	03	32	3·86v	−7·1	B8	1200	Polis
η	18	17	37·5	−36	45	42	3·11	−2·4	M3	130	
HD 167818	18	18	03·0	−27	02	33	4·65	−0·3	gK5	84	
19 δ	18	20	59·5	−29	49	42	2·70	−0·1	K2	25	Kaus Meridionalis
20 ε	18	24	10·2	−34	23	05	1·85	−0·3	B9	26	Kaus Australis
22 λ	18	27	58·1	−25	25	18	2·81	−0·1	K2	30	Kaus Borealis
27 φ	18	45	39·2	−26	59	27	3·17	−1·2	B8	75	
34 σ	18	55	15·7	−26	17	48	2·02	−2·0	B3	64	Nunki
37 ξ²	18	57	43·6	−21	06	24	3·51	0·0	K1	44	
38 ζ	19	02	36·5	−29	52	49	2·59	0·6	A2	24	Ascella
39 ο	19	04	40·8	−21	44	30	3·77	0·3	gG8	43	
40 τ	19	06	56·2	−27	40	13	3·32	0·0	K1	40	
41 π	19	09	45·6	−21	01	25	2·89	−2·0	F2	95	Albaldah
44 ρ¹	19	21	40·2	−17	50	50	3·93	1·7	F0	28	
46 υ	19	21	43·5	−15	57	18	4·61v	3·0	F2	21	

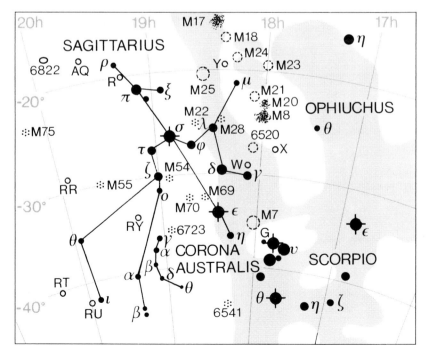

SAGITTARIUS
(continued)

Star	R.A. h	m	s	Dec. deg	min	sec	Mag.	Abs. mag.	Spectrum	Dist pc	
β¹	19	22	38·1	−44	27	32	3·93	−0·2	B8	67	Arkab
β²	19	23	12·9	−44	47	59	4·29	0·6	F0	52	
α	19	23	53·0	−40	36	58	3·97	−0·2	B8	36	Rukbat
52 h¹	19	36	42·3	−24	53	01	4·60	0·9	B9	55	
ι	19	55	15·5	−41	52	06	4·13	0·2	K0	55	
58 ω	19	55	50·2	−26	17	58	4·70	5·2	dG5	11	Terebellum
59	19	56	56·6	−27	10	12	4·52	−0·2	gK3	69	
θ¹	19	59	44·0	−35	16	35	4·37	−2·3	B3	210	
62 c	20	02	39·4	−27	42	35	4·58	−0·5	M4	100	

Also above mag. 5:

	Mag.	Abs. mag.	Spectrum	Dist
4	4·76	0·6	A0	21
1	4·98	0·2	K0	83
21	4·81	0·2	K0	57
32 ν¹	4·83	−6·0	K2	1400
36 ξ¹	5·00	0·6	A0	76
42 ψ	4·85	3·4	F5	21
43	4·90	−2·1	G8	210
60	4·83	0·3	G5	80

VARIABLE STARS

	R.A. h	m	Dec. deg	min	Range	Type	Period, d.	Spectrum
X	17	47·6	−27	50	4·2–4·8	Cepheid	7·01	F
W	18	05·0	−29	35	4·3–5·1	Cepheid	7·59	F–G
VX	18	08·1	−22	13	6·5–12·5	Semi-reg.	732	M
RS	18	17·6	−34	06	6·0–6·9	Algol	2·41	B + A
Y	18	21·4	−18	52	5·4–6·1	Cepheid	5·77	F
RV	18	27·9	−33	19	7·2–14·8	Mira	317·5	M
U	18	31·9	−19	07	6·3–7·1	Cepheid	6·74	F–G
YZ	18	49·5	−16	43	7·0–7·7	Cepheid	9·55	F–G
UX	18	54·9	−16	31	7·6–8·4	Semi-reg.	100	M
ST	19	01·5	−12	46	7·6–16·0	Mira	395·1	S
T	19	16·3	−16	59	7·6–12·9	Mira	392·3	S
RY	19	16·5	−33	31	6·0–15	R Coronæ	–	Gp
R	19	16·7	−19	18	6·7–12·8	Mira	268·8	M
AQ	19	34·3	−16	22	6·6–7·7	Semi-reg.	200	N
RR	19	55·9	−29	11	5·6–14·0	Mira	334·6	M
RU	19	58·7	−41	51	6·0–13·8	Mira	240·3	M
RT	20	17·7	−39	07	6·0–14·1	Mira	305·3	M

DOUBLE STARS

	R.A. h	m	Dec. deg	min	P.A. deg	Sep. sec	Mags.	
21	18	25·3	−20	32	289	1·8	4·9, 7·4	
ζ	19	02·6	−29	53	320	0·3	3·3, 3·4	Binary, 21·2 y
η	18	17·6	−36	46	105	3·6	3·2, 7·8	
π	19	09·8	−21	01	AB 150	0·1	3·7, 3·7	
					AB + C 122	0·4	5·9	
β¹	19	22·6	−44	28	077	28·3	3·9, 8·0	Wide naked-eye pair with β²
κ²	20	23·9	−42	25	234	0·8	6·9, 6·9	

OPEN CLUSTERS

M	NGC	R.A. h	m	Dec. deg	min	Diameter min	Mag.	No. of stars	
	6469	17	52·9	−22	21	12	8·2	50	
23	6494	17	56·8	−19	01	27	5·5	150	
	6520	18	03·4	−27	54	6	7·6	60	In M.20
21	6531	18	04·6	−22	30	13	5·9	70	
	6530	18	04·8	−24	20	15	4·6	–	In M.20
	6546	18	07·2	−23	20	13	5·9	70	
	6568	18	12·8	−21	36	13	8·6	50	
24	–	18	16·9	−18	29	90	4·5	–	Star-cloud; not a true cluster
18	6613	18	19·9	−17	08	9	6·9	20	
25	IC 4725	18	31·6	−19	15	32	4·6	30	υ Sagittarii cluster
	6645	18	32·6	−16	54	10	8·5	40	
	6716	18	54·6	−19	53	7	6·9	20	

GLOBULAR CLUSTERS

M	NGC	R.A. h	m	Dec. deg	min	Diameter min	Mag.
	6522	18	03·6	−30	02	5·6	8·6
	6544	18	07·3	−25	00	8·9	8·2
	6553	18	09·3	−25	54	8·1	8·2
	6558	18	10·3	−31	46	3·7	–

SAGITTARIUS
(continued)

M	NGC	R.A. h m	Dec. deg min	Diameter min	Mag.	
	6569	18 13·6	−31 50	5·8	8·7	
	6624	18 23·7	−30 22	5·9	8·3	H.I.50
28	6626	18 24·5	−24 52	11·2	6·9	
69	6637	18 31·4	−32 21	7·1	7·7	
	6638	18 30·9	−25 30	5·0	9·2	H.I.51
	6652	18 35·8	−32 59	3·5	8·9	
22	6656	18 36·4	−23 54	24·0	5·1	
54	6715	18 55·1	−30 29	9·1	7·7	
70	6681	18 43·2	−32 18	7·8	8·1	
55	6809	19 40·0	−30 58	19·0	6·9	
75	6864	20 06·1	−21 55	6·0	8·6	

PLANETARY NEBULÆ

M	NGC	R.A. h m	Dec. deg min	Diameter sec	Mag.	Mag. of central star
	6567	18 13·7	−19 05	8	11·1	15·0
	6629	18 25·7	−23 12	15	11·6	12·8
	6644	18 32·6	−25 08	3	12·2	15·9
	6818	19 44·0	−14 09	17	9·9	13·0

NEBULÆ

M	NGC	R.A. h m	Dec. deg min	Dimensions min	Mag. of illuminating star	
20	6514	18 02·6	−23 02	29 × 27	7·6	Trifid Nebula
8	6523	18 03·8	−24 23	90 × 40	−	Lagoon Nebula
17	6618	18 20·8	−16 11	46 × 37	−	Omega Nebula

GALAXY

M	NGC	R.A. h m	Dec. deg min	Mag.	Dimensions min	Type
	6822	19 44·9	−14 48	9·3	10·2 × 9·5	Irreg. Barnard's Galaxy

SCORPIUS

(Abbreviation: Sco).

Alternatively, and less correctly, known as Scorpio. Mythologically it is usually associated with the scorpion which Juno caused to attack and kill the great hunter Orion. Note that Orion and Scorpius are now on opposite sides of the sky – placed there, it is said, so that the creature could do Orion no further damage!

Scorpius is one of the most magnificent of all constellations, and one of the few which gives at least a vague impression of the creature it is meant to represent. It is dominated by Antares, but the whole area is exceptionally rich. The 'sting', which includes Shaula – only just below the first magnitude – is to all intents and purposes invisible from England.

There are 20 stars above the fourth magnitude. Antares itself is very slightly variable (range 0·86–1·02). The leaders are (see p. 241):

	Mag.	Luminosity Sun = 1	Dist lt-yrs
α	0·96v	7500	330
λ	1·63	1300	275
θ	1·87	14000	900
ε	2·29	96	65
δ	2·32	3800	550
κ	2·41	1300	390
β	2·64	2600	815
υ	2·69	16000	1560
τ	2·82	3800	780

BRIGHTEST STARS

Star	R.A. h m s	Dec. deg min sec	Mag.	Abs. mag.	Spectrum	Dist pc	
1 b	15 50 58·6	−25 45 05	4·64	−1·7	B3	170	
2 A	15 53 36·6	−25 19 38	4·59	−2·1	B2·5	220	
5 ρ	15 56 53·0	−29 12 50	3·88	−2·5	B2	190	
6 π	15 58 51·0	−26 06 50	2·89	−3·5	B1	190	
7 δ	16 00 19·9	−22 37 18	2·32	−4·1	B0	170	Dschubba

SCORPIUS (continued)

	Mag.	Luminosity Sun = 1	Dist lt-yrs
σ	2·89v	5000	590
π	2·89	2100	620
ι¹	3·03	200 000	5500
μ¹	3·04	1300	520
G	3·21	96	150
η	3·33	50	68
μ²	3·57	1300	680
ζ²	3·62	110	160
ρ	3·88	830	620
ω¹	3·96	2100	810
ν	4·00	1000+100	550

σ is very slightly variable (2·82–2·90).

Star		R.A. h	m	s	Dec. deg	min	sec	Mag.	Abs. mag.	Spectrum	Dist pc	
	ξ	16	04	22·0	−11	22	24	4·16	2·2	F6	26	
8	β	16	05	26·1	−19	48	19	2·64	−3·7	B0·5 + B2	250	Graffias
9	ω¹	16	06	48·3	−20	40	09	3·96	−3·5	B1	250	Jabhat al Akrab
10	ω²	16	07	24·2	−20	52	07	4·32	0·4	gG2	53	
14	ν	16	11	59·6	−19	27	38	4·00	0·0, −2·8	A0 + B2	170	Jabbah
13	c²	16	12	18·1	−27	55	35	4·58	−2·1	B2·5	220	
19	o	16	20	38·0	−24	10	10	4·55	−2·1	A5	92	
20	σ	16	21	11·2	−25	35	34	2·89v	−4·4	B1	180	Alniyat
21	α	16	29	24·3	−26	25	55	0·96v	−4·7	M1	100	Antares
	N	16	31	22·8	−34	42	15	4·23	−2·5	B2	220	
23	τ	16	35	52·8	−28	12	58	2·82	−4·1	B0	240	
	H	16	36	22·4	−35	15	21	4·16	0·6	M0	55	
26	ε	16	50	09·7	−34	17	36	2·29	−0·1	K2	20	Wei
	μ¹	16	51	52·1	−38	02	51	3·04	−3·0	B1·5	160	
	μ²	16	52	20·0	−38	01	03	3·57	−3·0	B2	210	
	ζ¹	16	53	59·6	−42	21	44	4·73	−6·7	B1·5	780	
	ζ²	16	54	34·9	−42	21	41	3·62	−0·3	K5	50	
	η	17	12	09·0	−43	14	21	3·33	0·6	F2	21	
34	υ	17	30	45·6	−37	17	45	2·69	−5·7	B3	480	Lesath
35	λ	17	33	36·4	−37	06	14	1·63	−3·0	B2	84	Shaula
	Q	17	36	32·6	−38	38	07	4·29	0·2	gK0	58	
	θ	17	37	19·0	−42	59	52	1·87	−5·6	F0	280	Sargas
	κ	17	42	29·0	−39	01	48	2·41	−3·0	B2	120	Girtab
	ι¹	17	47	34·9	−40	07	37	3·03	−8·4	F2	1700	
	G	17	49	51·3	−37	02	36	3·21	−0·1	K2	46	

Also above mag. 5:

	Mag.	Abs. mag.	Spectrum	Dist
22	4·79	−2·5	B2	260
ι²	4·81	−0·6	A2	
d	4·78	1·4	A2	44
k	4·87	−4·8	B1	
15 ψ	4·94	1·1	A0	40

VARIABLE STARS

	R.A. h	m	Dec. deg	min	Range	Type	Period, d.	Spectrum
RT	17	03·5	−36	55	7·0–16·0	Mira	449·0	M
FV	17	13·7	−32	51	7·9–8·6	Algol	5·72	B
RY	17	50·9	−33	42	7·5–8·4	Cepheid	20·31	F–G
RR	16	55·6	−30	35	5·0–12·4	Mira	279·4	M
RS	16	56·6	−45	06	6·2–13·0	Mira	320·0	M
RV	16	58·3	−33	37	6·6–7·5	Cepheid	6·06	F–G
BM	17	41·0	−32	13	6·8–8·7	Semi-reg.	850	K
RU	17	42·4	−43	45	7·8–13·7	Mira	369·2	M

DOUBLE STARS

	R.A. h	m	Dec. deg	min	P.A. deg	Sep. sec	Mags.	
2	15	53·6	−25	20	274	2·5	4·7, 7·4	
π	15	58·9	−26	07	132	50·4	2·9, 12·1	
ξ	16	04·4	−11	22	AB 040	0·8	4·8, 5·1	
					AC 051	7·6	7·3	
β	16	05·4	−19	48	AC 021	13·6	2·6, 4·9	A is a close dble
11	16	07·6	−12	45	257	3·3	5·6, 9·9	
ν	16	12·0	−19	28	AB 003	0·9	4·3, 6·8	Binary, 45·7 y
					AC 337	41·1	6·4	
12	16	12·3	−28	25	073	4·0	5·9, 7·9	
σ	16	21·2	−25	36	273	20·0	2·9, 8·5	
α	16	29·4	−26	26	273	2·7	1·2, 5·4	Binary, 878 y

OPEN CLUSTERS

M	NGC	R.A. h	m	Dec. deg	min	Diameter min	Mag.	No. of stars	
	6124	16	25·6	−40	40	29	5·8	100	
	6178	16	35·7	−45	38	4	7·2	12	
	6192	16	40·3	−43	22	8	8·5	60	
	6231	16	54·0	−41	48	15	2·6	–	
	6242	16	55·6	−39	30	9	6·4	–	
	6259	17	00·7	−44	40	10	8·0	120	
	6268	17	02·4	−39	44	6	9·5	–	
	6281	17	04·8	−37	54	8	5·4	–	
	6383	17	34·8	−32	34	5	5·5	40	(Nebulosity)
	6400	17	40·8	−36	57	8	8·8	60	
6	6405	17	40·1	−32	13	15	4·2	50	Butterfly Cluster
	6416	17	44·4	−32	21	18	5·7	40	
	6451	17	50·7	−30	13	8	8·2	80	
7	6475	17	53·9	−34	49	80	3·3	80	

SCORPIUS (continued)

GLOBULAR CLUSTERS

M	NGC	R.A. h	m	Dec. deg	min	Diameter min	Mag.
80	6093	16	17·0	−22	59	8·9	7·2
4	6121	16	23·6	−26	32	26·3	5·9
	6388	17	36·3	−44	44	8·7	6·8

PLANETARY NEBULÆ

M	NGC	R.A. h	m	Dec. deg	min	Diameter sec	Mag.	Mag. of central star
	6153	16	31·5	−40	15	25	11·5	−
	6302	17	13·7	−37	06	50	12·8	− Bug Nebula
	6337	17	22·3	−38	29	48	−	14·7

SCULPTOR

(Abbreviation: Scl).

Originally Apparatus Sculptoris. There is no star above the fourth magnitude.

See chart for Phœnix.

BRIGHTEST STARS

Star	R.A. h	m	s	Dec. deg	min	sec	Mag.	Abs. mag.	Spectrum	Dist pc
γ	23	28	49·3	−32	31	55	4·41	0·3	G8	47
β	23	32	58·0	−37	49	07	4·37	0·0	B9	77
δ	23	48	55·4	−28	07	49	4·51	0·6	A0	9·1
α	00	58	36·3	−29	21	27	4·31	−1·2	B8	130

Also above mag. 5:

	Mag.	Abs. mag.	Spectrum	Dist
η	4·81		M5	

VARIABLE STARS

	R.A. h	m	Dec. deg	min	Range	Type	Period, d.	Spectrum
Y	23	09·1	−30	08	7·5−9·0	Semi-reg.	300	M
S	00	15·4	−32	03	5·5−13·6	Mira	365·3	M
R	01	27·0	−32	33	5·8−7·7	Semi-reg.	370	N

DOUBLE STARS

	R.A. h	m	Dec. deg	min	P.A. deg	Sep. sec	Mags.
δ	23	48·9	−28	08	AB 243	3·9	4·5, 11·5
					AC 297	74·3	9·3
ζ	00	02·3	−29	43	320	3·0	5·0, 13·0
κ¹	00	09·3	−27	59	265	1·4	6·1, 6·2
λ¹	00	42·7	−38	28	003	0·7	6·7, 7·0
ε	01	45·6	−25	03	028	4·7	5·4, 8·6 Binary, 1192 y

GLOBULAR CLUSTER

M	NGC	R.A. h	m	Dec. deg	min	Diameter min	Mag.
	288	00	52·8	−26	35	13·8	8·1

GALAXIES

M	NGC	R.A. h	m	Dec. deg	min	Mag.	Dimensions min	Type
	IC 5332	23	34·5	−36	06	10·6	6·6 × 5·1	Sd
	7713	23	36·5	−37	56	11·6	4·3 × 2·0	SBd
	7755	23	47·9	−30	31	11·8	3·7 × 3·0	SBd
	7793	23	57·8	−32	35	9·1	9·1 × 6·6	Sd
	24	00	09·9	−24	58	11·5	5·5 × 1·6	Sb
	55	00	14·9	−39	11	8·2	32·4 × 6·5	SB
	134	00	30·4	−33	15	10·1	8·1 × 2·6	SBb
	253	00	47·6	−25	17	7·1	25·1 × 7·4	Scp
	300	00	54·9	−37	41	8·7	20·0 × 14·8	Sd
	613	01	34·3	−29	25	10·0	5·8 × 4·6	SBb

SCUTUM

(Abbreviation: Sct).

Originally Scutum Sobieskii or Clypeus Sobieskii (Sobieski's Shield). It has only one star brighter than the fourth magnitude, but is a rich area bordering Aquila, and contains the glorious 'Wild Duck' cluster M.11.

	Mag.	Luminosity Sun = 1	Dist lt-yrs
α	3·85	105	180

See chart for Aquila.

BRIGHTEST STARS

Star	R.A. h	m	s	Dec. deg	min	sec	Mag.	Abs. mag.	Spectrum	Dist pc
ζ	18	23	39·3	−08	56	03	4·68	0·2	K0	79
γ	18	29	11·7	−14	33	57	4·70	1·4	A2	45
α	18	35	12·1	−08	14	39	3·85	−0·2	K3	55
δ	18	42	16·2	−09	03	09	4·71v	1·3	F3	49
β	18	47	10·3	−04	44	52	4·22	−2·1	G5	150

Also above mag. 5:

	Mag.	Abs. mag.	Spectrum	Dist
ε	4·90	−2·1	G8	250

SCUTUM (continued)

VARIABLE STARS

	R.A.		Dec.		Range	Type	Period, d.	Spectrum
	h	m	deg	min				
RZ	18	26·6	−09	12	7·3–8·8	Algol	15·19	B
R	18	47·5	−05	42	4·4–8·2	RV Tauri	140	G–K
S	18	50·3	−07	54	7·0–8·0	Semi-reg.	148	N

OPEN CLUSTERS

M	NGC	R.A.		Dec.		Diameter	Mag.	No. of stars	
		h	m	deg	min	min			
	6664	18	36·7	−08	13	16	7·8	50	EV Scuti cluster
26	6694	18	45·2	−09	24	15	8·0	30	
	6704	18	50·9	−05	12	6	9·2	30	
11	6705	18	51·1	−06	16	14	5·8	500	Wild Duck cluster

GLOBULAR CLUSTER

M	NGC	R.A.		Dec.		Diameter	Mag.
		h	m	deg	min	min	
	6712	18	53·1	−08	42	7·2	8·2

NEBULA

M	NGC	R.A.		Dec.		Dimensions	Mag. of illuminating star
		h	m	deg	min	min	
	IC 1287	18	31·3	−10	50	4·4 × 3·4	5·5

SERPENS

(Abbreviation: Ser).

A curious constellation inasmuch as it is divided into two parts: Caput (the Head) and Cauda (the Body). It evidently represents the serpent with which Ophiuchus is struggling, since it has been pulled in half! Caput contains six stars above the fourth magnitude, and Cauda three.

	Mag.	Luminosity Sun = 1	Dist lt-yrs
Caput			
α	2·65	96	85
μ	3·54	50	144
β	3·67	50	121
ε	3·71	19	107
δ	3·8 (combined)	17	88
γ	3·85	3·2	39
Cauda			
η	3·26	17	52
θ	3·4 (combined)	12+12	102
ξ	3·54	17	75

See chart for Ophiuchus.

BRIGHTEST STARS

Star	R.A.			Dec.			Mag.	Abs. mag.	Spectrum	Dist pc	
	h	m	s	deg	min	sec					
CAPUT											
13 δ	15	34	48·0	+10	32	21	3·80	1·7	F0	27	Tsin
21 ι	15	41	33·0	+19	40	13	4·52	1·2	A1	36	
24 α	15	44	16·0	+06	25	32	2·65	−0·1	K2	26	Unukalhai
28 β	15	46	11·2	+15	25	18	3·67	0·6	A2	37	
27 λ	15	46	26·5	+07	21	12	4·43	4·4	G0	11	
35 κ	15	48	44·3	+18	08	29	4·09	−0·5	M1	78	
32 μ	15	49	37·1	−03	25	49	3·54	0·6	A0	44	
37 ε	15	50	48·9	+04	28	40	3·71	1·6	A2	33	
41 γ	15	56	27·1	+15	39	42	3·85	3·7	F6	12	
CAUDA											
53 ν	17	20	49·4	−12	50	48	4·33	1·2	A1	42	
55 ξ	17	37	35·0	−15	23	55	3·54	1·7	F0	23	
56 ο	17	41	24·7	−12	52	31	4·26	1·4	A2	36	
57 ζ	18	00	28·7	−03	41	25	4·62	3·1	F3	22	
58 η	18	21	18·4	−02	53	56	3·26	1·7	K0	16	Alava
63 θ	18	56	13·0	+04	12	13	4·5	2·1	A5	31	Combined mag. 3·4. Alya
	18	56	14·5	+04	12	07	4·5	2·1	A5		

Also above mag. 5:

	Mag.	Abs. mag.	Spectrum	Dist.
38 ρ	4·76	−0·3	K5	98
44 π	4·83	1·7	A3	42

VARIABLE STARS

	R.A.		Dec.		Range	Type	Period, d.	Spectrum
	h	m	deg	min				
S	15	21·7	+14	19	7·0–14·1	Mira	368·6	M
τ⁴	15	36·5	+15	06	7·5–8·9	Irreg.	–	M
R	15	50·7	+15	08	5·1–14·4	Mira	356·4	M
U	16	07·3	+09	56	7·8–14·7	Mira	237·9	M
d	18	27·2	+00	12	4·9–5·9	?	?	G + A

DOUBLE STARS

	R.A.		Dec.		P.A.	Sep.	Mags.	
	h	m	deg	min	deg	sec		
δ	15	34·8	+10	32	177	4·4	4·1, 5·2	Binary, 3168 y
β	15	46·2	+15	25	265	30·6	3·7, 9·9	
ν	17	20·8	−12	51	028	46·3	4·3, 8·3	
d	18	27·2	+00	12	318	3·8	5·3v, 7·6	
θ	18	56·2	+04	12	104	22·3	4·5, 4·5	

OPEN CLUSTERS

M	NGC	R.A.		Dec.		Diameter	Mag.	No. of stars
		h	m	deg	min	min		
	6611	18	18·8	−13	47	7	6·0	In M.16
	6604	18	18·1	−12	14	2	7·0	30

SERPENS (continued)

GLOBULAR CLUSTER

M	NGC	R.A.		Dec.		Diameter	Mag.
		h	m	deg	min	min	
5	5904	15	18·6	+02	05	17·4	5·8

NEBULA

M	NGC	R.A.		Dec.		Dimensions		Mag.
		h	m	deg	min	min		
16	6611	18	18·8	−13	47	35 × 28	Eagle Nebula	6·4

GALAXY

M	NGC	R.A.		Dec.		Mag.	Dimensions	Tab
		h	m	deg	min		min	
	6118	16	21·8	−02	17	12·3	4·7 × 2·3	Sb

SEXTANS

(Abbreviations: Sxt).

A very obscure constellation, with no star as bright as the fourth magnitude.

See chart for Hydra.

BRIGHTEST STAR

Star	R.A.			Dec.			Mag.	Abs.	Spectrum	Dist
	h	m	s	deg	min	sec		mag.		pc
15 α	10	07	56·2	−00	22	18	4·49	−1·1	B5	100

VARIABLE STAR

	R.A.		Dec.		Range	Type	Period, d.	Spectrum
	h	m	deg	min				
S	10	34·9	−00	20	8·2–13·5	Mira	261·0	M

DOUBLE STAR

	R.A.		Dec.		P.A.	Sep.	Mags.	
	h	m	deg	min	deg	sec		
γ	09	52·5	−08	06	AB 067	0·6	5·5, 6·1	Binary, 75·6 y
					AC 325	35·8	12·0	

GALAXIES

M	NGC	R.A.		Dec.		Mag.	Dimensions	Type	
		h	m	deg	min		min		
	2967	09	42·1	+00	20	11·6	3·0 × 2·9	Sc	
	3115	10	05·2	−07	43	9·1	8·3 × 3·2	E6	Spindle Galaxy
	3166	10	13·8	+03	26	10·6	5·2 × 2·7	SBa	
	3169	10	14·2	+03	28	10·4	4·8 × 3·2	Sb	

TAURUS

(Abbreviation: Tau).

One of the brightest of the Zodiacal constellations. Mythologically it has been said to represent the bull into which Jupiter transformed himself when he wished to carry off Europa, daughter of the King of Crete.

Taurus includes the reddish first-magnitude star Aldebaran, and also the two most famous open clusters in the sky: the Pleiades and the Hyades. Altogether there are 16 stars above the fourth magnitude:

	Mag.	Luminosity	Dist
		Sun = 1	lt-yrs
α	0·85v	100	68
β	1·65	400	130
η	2·87	400	238
ζ	3·00	1300	490
λ	3·3 (max)	450	330
θ²	3·42	52	125
ε	3·54	60	145
o	3·60	58	150
27	3·63	250	290
γ	3·63	60	166
17	3·70	500	390
ξ	3·74	96	98
δ¹	3·76	60	166
θ¹	3·85	60	176
20	3·88	400	390
ν	3·91	28	111

BRIGHTEST STARS

Star	R.A.			Dec.			Mag.	Abs.	Spectrum	Dist	
	h	m	s	deg	min	sec		mag.		pc	
1 o	03	24	48·7	+09	01	44	3·60	0·3	G8	46	
2 ξ	03	27	10·1	+09	43	58	3·74	−0·1	B8p	30	
5	03	30	52·3	+12	56	12	4·11	−0·9	K0	97	
10	03	36	52·3	+00	24	06	4·28	4·0	F8	13	
17	03	44	52·5	+24	06	48	3·70	−1·9	B6	120	Electra (Pleiades)
19	03	45	12·4	+24	28	02	4·30	−0·9	B6	110	Taygete (Pleiades)
20	03	45	49·5	+24	22	04	3·88	−1·6	B7	120	Maia (Pleiades)
23	03	46	19·5	+23	56	54	4·18	−1·3	B6	120	Merope (Pleiades)
25 η	03	47	29·0	+24	06	18	2·87	−1·6	B7	73	Alcyone (Pleiades)
27	03	49	09·7	+24	03	12	3·63	−1·2	B8	90	Atlas (Pleiades)
28 BU	03	49	11·1	+24	08	12	var.	var.	B8p	29	Pleione (Pleiades)
35 λ	04	00	40·7	+12	29	25	var.	−1·7	B3	100	
38 ν	04	03	09·3	+05	59	21	3·91	1·2	A1	34	
37 A¹	04	04	41·7	+22	04	55	4·36	0·2	K0	60	
49 μ	04	15	32·0	+08	53	32	4·29	−1·7	B3	140	
54 γ	04	19	47·5	+15	37	39	3·63	0·2	K0	51	Hyadum Primus (Hyades)
61 δ¹	04	22	56·0	+17	32	33	3·76	0·2	K0	51	(Hyades)
65 κ	04	25	22·1	+22	17	38	4·22	2·4	A7	23	
68 δ³	04	25	29·3	+17	55	41	4·30	1·4	A2	38	(Hyades)
69 υ	04	26	18·4	+22	48	49	4·29	1·1	F0	42	(Hyades)
71	04	26	20·7	+15	37	06	4·49	2·6	F0	24	
73 π	04	26	36·4	+14	42	49	4·69	0·3	G8	72	
77 θ¹	04	28	34·4	+15	57	44	3·85	0·2	K0	54	(Hyades)
74 ε	04	28	36·9	+19	10	49	3·54	0·2	K0	45	Ain. (Hyades)

TAURUS (continued)

Aldebaran has a very small range (0·78–0·93). β was formerly included in Auriga, as γ Aurigæ.

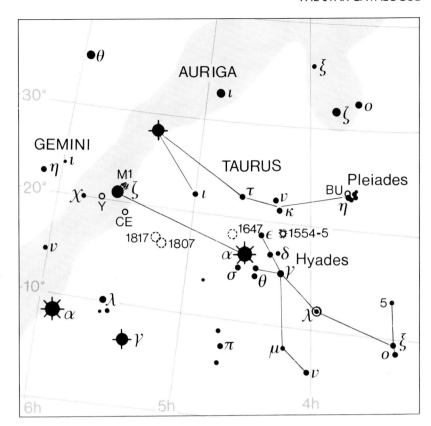

Star		R.A.			Dec.			Mag.	Abs.	Spectrum	Dist	
		h	m	s	deg	min	sec		mag.		pc	
78	θ²	04	28	39·6	+15	52	15	3·42	0·5	A7	38	(Hyades)
86	ρ	04	33	50·8	+14	50	40	4·65	2·6	F0	26	
88	d	04	35	39·2	+10	09	39	4·25	2·4	A3	25	
87	α	04	35	55·2	+16	30	33	0·85	−0·3	K5	21	Aldebaran
90	c¹	04	38	09·4	+12	30	39	4·27	2·1	A5	27	
92	σ²	04	39	16·4	+15	55	05	4·68	2·1	A5	33	
94	τ	04	42	14·6	+22	57	25	4·28	−1·7	B3	150	
102	ι	05	03	05·7	+21	35	24	4·64	2·4	A7	28	
112	β	05	26	17·5	+28	36	27	1·65	−1·6	B7	40	Al Nath.
												=γ Aurigæ
119	CE	05	32	12·7	+18	35	39	4·3v	−4·8v	M2	290	
123	ζ	05	37	38·6	+21	08	33	3·00	−3·0	B2	150	Alheka
136		05	53	19·6	+27	36	44	4·58	0·4	B9·5	64	

Also above mag. 5:

	Mag.	Abs. mag.	Spectrum	Dist
47	4·84	0·3	G5	81
50 ω	4·94	2·3	A5	
52 φ	4·95	0·0	K1	90
64 δ²	4·80	2·4	A7	30
75	4·97	−0·1	K2	100
104	4·92	5·0	G4	13
109	4·94	0·3	G8	85
111	4·98	4·0	F8	16
114	4·88	−1·7	B3	210
126	4·86	−2·3	B3	250
132	4·86	0·3	G8	73
134	4·91	0·3	B9	110
139	4·82	−5·7	B1	1100

VARIABLE STARS

	R.A.		Dec.		Range	Type	Period, d.	Spectrum
	h	m	deg	min				
BU	03	49·2	+24	08	4·8–5·5	Irreg.	–	Bp Pleione
λ	04	00·7	+12	29	3·3–3·8	Algol	3·95	B + A
T	04	22·0	+19	32	8·4–13·5	T Tauri	Irreg.	G–K
R	04	28·3	+10	10	7·6–14·7	Mira	323·7	M
HU	04	38·3	+20	41	5·9–6·7	Algol	2·06	A
ST	05	45·1	+13	35	7·8–8·6	W Virginis	4·03	F–G

TAURUS (continued)

	R.A. h m	Dec. deg min	Range	Type	Period , d.	Spectrum
TU	05 45·2	+24 25	5·9–8·6	Semi-reg.	190	N
SU	05 49·1	+19 04	9·1–16·0	R Coronæ	–	G0p

DOUBLE STARS

	R.A. h m	Dec. deg min	P.A. deg	Sep. sec	Mags.	
φ	04 20·4	+27 21	250	52·1	5·0, 8·4	
χ	04 22·6	+25 38	024	19·4	5·5, 7·6	
66	04 23·9	+09 28	265	0·1	5·8, 5·9	Binary, 51·6 y
κ + 67	04 25·4	+22 18	173	339	4·2, 5·3	
θ	04 28·7	+15 32	346	337·4	3·4, 3·8	
σ	04 39·3	+15 55	193	431·2	4·7, 5·1	
126	05 41·3	+16 32	238	0·3	5·3, 5·9	

OPEN CLUSTERS

M	NGC	R.A. h m	Dec. deg min	Diameter min	Mag.	No. of stars
45	1432/5	03 47·0	+24 07	110	1·2	300 + Pleiades
	–	04 27	+16	330	1	200 + Hyades
	1647	04 46·0	+19 04	45	6·4	200
	1746	05 03·6	+23 49	42	6·1	20
	1807	05 10·7	+16 32	17	7·0	20 Asterism?
	1817	05 12·1	+16 42	16	7·7	60

PLANETARY NEBULA

M	NGC	R.A. h m	Dec. deg min	Diameter sec	Mag.	Mag. of central star
	1514	04 09·2	+30 47	114	10	9·4

NEBULÆ

M	NGC	R.A. h m	Dec. deg min	Dimensions min	Mag. of illuminating star	
	1554–5	04 21·8	+19 32	var.	9v	Hind's Variable Nebula (T Tauri)
1	1952	05 34·5	+22 01	6 × 4	16	Crab Nebula: SNR

TELESCOPIUM

(Abbreviation: Tel).

A small constellation with only one star above the fourth magnitude: α.

	Mag.	Luminosity Sun = 1	Dist lt-yrs
α	3·51	1200	587

See chart for Ara.

BRIGHTEST STARS

Star	R.A. h m s	Dec. deg min sec	Mag.	Abs. mag.	Spectrum	Dist pc
ε	18 11 13·6	−45 57 15	4·53	0·3	G5	57
α	18 26 58·2	−45 58 06	3·51	−2·9	B3	180
ζ	18 28 49·7	−49 04 15	4·13	0·3	gG8	51

Also above mag. 5:

	Mag.	Abs. mag.	Spectrum	Dist
λ	5·00	−0·8	B9	140
L	4·90	0·2	G9	74
ξ	4·94	−0·5	M2	120

VARIABLE STARS

	R.A. h m	Dec. deg min	Range	Type	Period, d.	Spectrum
BL	19 06·6	−51 25	7·7–9·8	Eclipsing	778·1	F + M
RR	20 04·2	−55 43	6·5–16·5	Z Andromedæ	–	F5p
R	20 14·7	−46 58	7·6–14·8	Mira	461·9	M

PLANETARY NEBULA

M	NGC	R.A. h m	Dec. deg min	Diameter sec	Mag.
	IC 4699	18 18·5	−45 59	10	11·9

TRIANGULUM

(Abbreviation: Tri).

A small but original constellation – and its main stars really do form a triangle! There are two stars above the fourth magnitude:

	Mag.	Luminosity Sun = 1	Dist lt-yrs
β	3·00	58	114
α	3·41	10	59

See chart for Andromeda.

BRIGHTEST STARS

Star	R.A. h m s	Dec. deg min sec	Mag.	Abs. mag.	Spectrum	Dist pc	
2 α	01 53 04·8	+29 34 44	3·41	2·2	F6	18	Rasalmothallah
4 β	02 09 32·5	+34 59 14	3·00	0·3	A5	35	
9 γ	02 17 18·8	+33 50 50	4·01	0·6	A0	46	

Also above mag. 5:

	Mag.	Abs. mag.	Spectrum	Dist
6	4·94	0·3	G5	85
8 δ	4·87	4·4	G0	10

TRIANGULUM
(continued)

VARIABLE STAR

	R.A.		Dec.		Range	Type	Period, d.	Spectrum
	h	m	deg	min				
R	02	37·0	+34	16	5·4–12·6	Mira	266·5	M

DOUBLE STARS

	R.A.		Dec.		P.A.	Sep.	Mags.
	h	m	deg	min	deg	sec	
6	02	12·4	+30	18	071	3·9	5·3, 6·9
ι	02	15·9	+33	21	240	2·3	5·4, 7·0

GALAXIES

M	NGC	R.A.		Dec.		Mag.	Dimensions	Type
		h	m	deg	min		min	
33	598	01	33·9	+30	39	5·7	62 × 39	Sc
	925	02	27·3	+33	35	10·0	9·8 × 6·0	SBc

TRIANGULUM AUSTRALE

(Abbreviation: TrA).

This 'triangle' also merits its name. There are 4 stars above the fourth magnitude:

	Mag.	Luminosity	Dist
		Sun = 1	lt-yrs
α	1·92	96	55
β	2·85	5·2	33
γ	2·89	50	91
δ	3·85	600	360

BRIGHTEST STARS

Star	R.A.			Dec.			Mag.	Abs. mag.	Spectrum	Dist pc	
	h	m	s	deg	min	sec					
γ	15	18	54·5	−68	40	46	2·89	0·6	A0	28	
ε	15	36	43·1	−66	19	02	4·11	0·2	K0	44	
β	15	55	08·4	−63	25	50	2·85	3·0	F5	10	
δ	16	15	26·2	−63	41	08	3·85	−2·1	G2	110	
α	16	48	39·8	−69	01	39	1·92	−0·1	K2	17	Atria

Also above mag. 5:

	Mag.	Abs. mag.	Spectrum	Dist
ζ	4·91	4·4	G0	12

VARIABLE STARS

	R.A.		Dec.		Range	Type	Period, d.	Spectrum
	h	m	deg	min				
X	15	14·3	−70	05	8·1–9·1	Irreg.	–	N
R	15	19·8	−66	30	6·4–6·9	Cepheid	3·39	F–G
S	16	01·2	−63	47	6·1–6·8	Cepheid	6·32	F
U	16	07·3	−62	55	7·5–8·3	Cepheid	2·57	F

DOUBLE STARS

	R.A.		Dec.		P.A.	Sep.	Mags.
	h	m	deg	min	deg	sec	
ε	15	36·7	−66	19	218	83·2	4·1, 9·5
ι	16	28·0	−64	03	016	29·6	5·3, 10·3

OPEN CLUSTER

M	NGC	R.A.		Dec.		Diameter	Mag.	No. of stars
		h	m	deg	min	min		
	6025	16	03·7	−60	30	12	5·1	60

TUCANA

(Abbreviation: Tuc).

The dimmest of the 'Southern Birds', but graced by the presence of the glorious globular cluster 47 Tucanæ – inferior only to ω Centauri. The only star above the fourth magnitude is α (2·86), but the combined magnitude of β¹ and β² is 3·7.

	Mag.	Luminosity Sun = 1	Dist lt-yrs
α	2·86	105	114
β¹ + β²	3·7	105	108

BRIGHTEST STARS

Star	R.A. h	m	s	Dec. deg	min	sec	Mag.	Abs. mag.	Spectrum	Dist pc	
α	22	18	30·1	−60	15	35	2·86	−0·2	K3	35	
δ	22	27	19·9	−64	58	00	4·48	−0·2	B8	76	
γ	23	17	25·6	−58	14	08	3·99	0·6	F0	45	
ε	23	59	54·9	−65	34	38	4·50	0·0	B8·5	67	
ζ	00	20	04·4	−64	52	30	4·23	5·0	G0	7·1	
β	{ 00	32	32·7	−62	57	30	4·37	−0·2	B8 }	33	Combined
	{ 00	32	33·6	−62	57	57	4·53		A2 }		mag. 3·7

Also above mag. 5:

	Mag.	Abs. mag.	Spectrum	Dist
η	5·00		A2	

VARIABLE STARS

	R.A. h	m	Dec. deg	min	Range	Type	Period, d.	Spectrum
T	22	40·6	−61	33	7·7–13·8	Mira	250·8	M
S	00	23·1	−61	40	8·2–15·0	Mira	240·7	M
U	00	57·2	−75	00	8·0–14·8	Mira	259·5	M

DOUBLE STARS

	R.A. h	m	Dec. deg	min	P.A. deg	Sep. sec	Mags	
δ	22	27·3	−64	58	282	6·9	4·5, 9·8	
β¹	00	31·5	−62	58	169	27·1	4·4, 4·8	
β²	00	31·6	−62	58	295	0·6	4·8, 6·0	Binary, 44·4 y
κ	01	15·8	−68	53	336	5·4	5·1, 7·3	

GLOBULAR CLUSTERS

M	NGC	R.A. h	m	Dec. deg	min	Diameter min	Mag.	
	104		00	24·1	−72	05	30·9	4·0 47 Tucanæ
	362		01	03·2	−70	51	12·9	6·6

GALAXY

	R.A. h	m	Dec. deg	min	Mag.	Dimensions min
Small Cloud of Magellan		00	53	−72	50	2·3	280 × 160

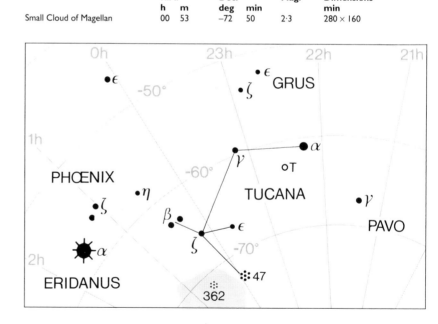

URSA MAJOR

(Abbreviation: UMa).

The most famous of all northern constellations; circumpolar in England and the northern United States. Mythologically it represents Callisto, daughter of King Lycaon of Arcadia. Her beauty surpassed that of Juno, which so infuriated the goddess that she ill-naturedly changed Callisto into a bear. Years later Arcas, Callisto's son, found the bear while out hunting, and was about to shoot it when Jupiter intervened, swinging both Callisto and Arcas – also transformed into a bear – up to the sky: Callisto as Ursa Major, Arcas as Ursa Minor. The sudden jolt explains why both bears have tails stretched out to decidedly unursine length!

The seven main stars of Ursa Major are often called the Plough; sometimes King Charles' Wain, and, in America, the Big Dipper. One of the Plough stars is Mizar, the most celebrated naked-eye double in the sky since it makes a pair with Alcor; Mizar is itself a compound system, and the two main components are easily separable with a small telescope.

Ursa Major contains nineteen stars above the fourth magnitude:

	Mag.	Luminosity Sun = 1	Dist lt-yrs
ε	1·77	60	62
α	1·79	60	75
η	1·86	450	108
ζ	2·09	56+11	59
β	2·37	28	62
γ	2·44	50	75
ψ	3·01	82	121
μ	3·05	115	156
ι	3·14	8	49
θ	3·17	10	46
δ	3·31	17	65
ο	3·36	200	230
λ	3·45	50	121
ν	3·48	105	150
κ	3·60	130	91
h	3·67	17	81
χ	3·71	60	121
ξ	3·79	0·9	25
υ	3·80v	14	85

Several of the Plough stars have alternative proper names; thus η may also be called Benetnasch, while γ may be Phekda or Phecda. However, Mizar is the only star whose proper name is generally used.

BRIGHTEST STARS

Star	R.A. h	m	s	Dec. deg	min	sec	Mag.	Abs. mag.	Spectrum	Dist pc	
1 ο	08	30	15·8	+60	43	05	3·36	−0·9	G4	71	Muscida
4 π²	08	40	12·9	+64	19	40	4·60	−0·1	K2	87	Ta Tsun
9 ι	08	59	12·4	+48	02	29	3·14	2·4	A7	15	Talita
10	09	00	38·3	+41	46	57	3·97	3·4	·F5	14	
12 κ	09	03	37·5	+47	09	23	3·60	−0·6	A0	28	Al Kaprah
15 f	09	08	52·2	+51	36	16	4·48	+2·1	A0	29	
14 τ	09	10	55·0	+63	30	49	4·67	0·5	F6 + A5	70	
23 h	09	31	31·7	+63	03	42	3·67	1·7	F0	25	
25 θ	09	32	51·3	+51	40	38	3·17	2·2	F6	14	
24 d	09	34	28·8	+69	49	49	4·56	3·0	G2	21	
26	09	34	49·4	+52	03	05	4·50	1·4	A2	42	
29 υ	09	50	59·3	+59	02	19	3·80v	1·9	F2	26	
30 φ	09	52	06·3	+54	03	41	4·59	0·9	A3	13	
33 λ	10	17	05·7	+42	54	52	3·45	0·6	A2	37	Tania Borealis
34 μ	10	22	19·7	+41	29	58	3·05	−0·4	M0	48	Tania Australis
48 β	11	01	50·4	+56	22	56	2·37	1·2	A1	19	Merak
50 α	11	03	43·6	+61	45	03	1·79	0·2	K0	23	Dubhe
52 ψ	11	09	39·7	+44	29	54	3·01	0·0	K1	37	
53 ξ	11	18	10·9	+31	31	45	3·79	4·9	G0	7·7	Alula Australis
54 ν	11	18	28·7	+33	05	39	3·48	−0·2	K3	46	Alula Borealis
63 χ	11	46	03·0	+47	46	45	3·71	0·2	K0	37	Alkafzah
64 γ	11	53	49·7	+53	41	41	2·44	0·6	A0	23	Phad
69 δ	12	15	25·5	+57	01	57	3·31	1·7	A3	20	Megrez
77 ε	12	54	01·7	+55	57	35	1·77v	0·2	A0	19	Alioth
79 ζ	13	23	55·5	+54	55	31	2·09	0·4, 2·1	A2 + A6	18	Mizar
80 g	13	25	13·4	+54	59	17	4·01	2·1	A5	25	Alcor
83	13	40	44·1	+54	40	54	4·66	−0·5	M2	100	
85 η	13	47	32·3	+49	18	48	1·86	−1·7	B3	33	Alkaid

Also above mag. 5:

	Mag.	Abs. mag.	Spectrum	Dist
8 ρ	4·76		M0	
13 σ²	4·80	3·1	F7	22
18	4·83	2·1	A5	34
55	4·78	1·4	A2	42
56	4·99	−2·1	G8	260
78	4·93	3·0	F2	24

VARIABLE STARS

	R.A. h	m	Dec. deg	min	Range	Type	Period, d.	Spectrum
X	08	40·8	+50	08	8·0–14·8	Mira	248·8	M
W	09	43·8	+55	57	7·9–8·6	W UMa	0·33	F + F
R	10	44·6	+68	47	6·7–13·4	Mira	301·7	M
TX	10	45·3	+45	34	7·1–8·8	Algol	3·06	B + F
VY	10	45·7	+67	25	5·9–6·5	Irreg.	–	N
VW	10	59·0	+69	59	6·8–7·7	Semi-reg.	125	M
ST	11	27·8	+45	11	7·7–9·5	Semi-reg.	81	M
CF	11	53·0	+37	43	8·5–12	Flare	–	B
Z	11	56·5	+57	52	6·8–9·1	Semi-reg.	196	M
RY	12	20·5	+61	19	6·7–8·5	Semi-reg.	311	M
T	12	36·4	+59	29	6·6–13·4	Mira	256·5	M
S	12	43·9	+61	06	7·0–12·4	Mira	225·0	S

DOUBLE STARS

	R.A. h	m	Dec. deg	min	P.A. deg	Sep. sec	Mags.	
ι	08	59·2	+48	02	100	1·8	3·1, 10·2	Binary, 818 y
κ	09	03·6	+47	09	258	0·1	4·2, 4·4	Binary, 70 y
σ²	09	10·4	+67	08	000	3·4	4·8, 8·2	Binary, 1067 y
φ	09	52·1	+54	04	188	0·2	5·3, 5·4	Binary, 105·5 y
α	11	03·7	+61	45	283	0·7	1·9, 4·8	Binary, 44·7 y
ξ	11	18·2	+31	32	060	1·3	4·3, 4·8	Binary, 59·8 y
ν	11	18·5	+33	06	147	7·2	3·5, 9·9	
78	13	00·7	+56	22	057	1·5	5·0, 7·4	Binary, 116 y
ζ	13	23·9	+54	56	AB 152	14·4	2·3, 4·0	
					AC 071	708·7	2·1, 4·0	

GALAXIES

M	NGC	R.A. h	m	Dec. deg	min	Mag.	Dimensions min	Type
	2681	08	53·5	+51	19	10·3	3·8 × 3·5	Sa
	2685	08	55·6	+58	44	11·0	5·2 × 3·0	Sbp

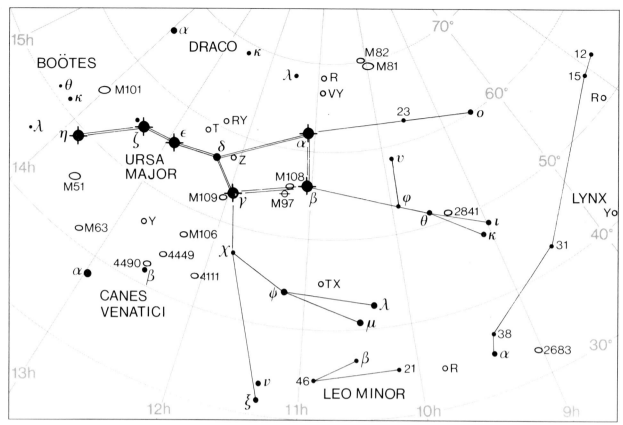

URSA MAJOR (continued)

M	NGC	R.A.		Dec.		Mag.	Dimensions	Type	
		h	m	deg	min		min		
	2768	09	11·6	+60	02	10·0	6·3 × 2·8	E5	
	2787	09	19·3	+69	12	10·8	3·4 × 2·3	Sap	
	2841	09	22·0	+50	58	9·3	8·1 × 3·8	Sb	
	2976	09	47·3	+67	55	10·1	4·9 × 2·5	Scp	
	2985	09	50·4	+72	17	10·5	4·5 × 3·4	Sb	
81	3031	09	55·6	+69	04	6·9	25·7 × 14·1	Sb	Bode's Nebula
82	3034	09	55·8	+69	41	8·4	11·2 × 4·6	Pec.	
	3077	10	03·3	+68	44	9·8	4·6 × 3·6	E2p	
	3079	10	02·0	+55	41	10·6	7·6 × 1·7	Sb	
	3184	10	18·3	+41	25	9·7	6·9 × 6·8	Sc	
	3198	10	19·9	+45	33	10·4	8·3 × 3·7	Sc	
	3310	10	38·7	+53	30	10·9	3·6 × 3·0	SBc	
108	3556	11	08·7	+55	57	10·0	8 × 1	Sc	
	3359	10	46·6	+63	13	10·4	6·8 × 4·3	SBc	
	3610	11	18·4	+58	47	10·7	3·2 × 2·5	E2p	
	3631	11	21·0	+53	10	10·4	4·6 × 4·1	Sc	
	3675	11	26·1	+43	35	10·9	5·9 × 3·2	Sb	
	3687	11	28·0	+29	31	12·6	2·0 × 2·0	Sb	
	3718	11	32·6	+53	04	10·5	8·7 × 4·5	SBap	
	3726	11	33·3	+47	02	10·4	6·0 × 4·5	Sc	
	3877	11	46·1	+47	30	11·6	5·4 × 1·5	Sb	
	3898	11	49·2	+56	05	10·8	4·4 × 2·6	Sb	
	3945	11	53·2	+60	41	10·6	5·5 × 3·6	SBa	
	3949	11	53·7	+47	52	11·0	3·0 × 1·8	Sb	
	3953	11	53·8	+52	20	10·1	6·6 × 3·6	Sb	
	3998	11	57·9	+55	27	10·6	3·1 × 2·5	E2p	
	4026	11	59·4	+50	58	11·7	5·1 × 1·4	S0	
	4036	12	01·4	+61	54	10·6	4·5 × 2·0	E6	
	4041	12	02·2	+62	08	11·1	2·8 × 2·7	Sc	
	4051	12	03·2	+44	32	10·3	5·0 × 4·0	Sc	
	4062	12	04·1	+31	54	11·2	4·3 × 2·0	Sb	
	4088	12	05·6	+50	33	10·5	5·8 × 2·5	Sc	
	4096	12	06·0	+47	29	10·6	6·5 × 2·0	Sc	
	4100	12	06·2	+49	35	11·5	5·2 × 1·9	Sb	
	4605	12	40·0	+61	37	11·0	5·5 × 2·3	SBcp	
	5308	13	47·0	+60	58	11·3	3·5 × 0·8	S0	
	5322	13	49·3	+60	12	10·0	5·5 × 3·9	E2	
101	5457	14	03·2	+54	21	7·7	26·9 × 26·3	Sc	Pinwheel
	5475	14	05·2	+55	45	12·4	2·2 × 0·6	Sa	

URSA MAJOR (continued)

PLANETARY NEBULA

M	NGC	R.A.		Dec.		Diameter	Mag.	Mag. of central star
		h	m	deg	min	sec		
97	3587	11	14·8	+55	01	194	12·0	15·9 Owl Nebula

URSA MINOR

(Abbreviation: UMi).

The north polar constellation. Mythologically it represents Arcas, son of Callisto (see Ursa Major). There are three stars above the fourth magnitude:

	Mag.	Luminosity Sun = 1	Dist lt-yrs
α	1·99v	6000	680
β	2·08	110	95
γ	3·05	230	225

The closest approach of Polaris to the celestial pole will be on 24 March 2100. Its declination will then be +89 deg 32 min 51 sec.

BRIGHTEST STARS

Star	R.A.			Dec.			Mag.	Abs. mag.	Spectrum	Dist pc	
	h	m	s	deg	min	sec					
2	01	01	31	+85	59	24	4·52	0·0	K0	77	
1 α	02	31	50·4	+89	15	51	1·99v	−4·6	F8	208	Polaris
4 π²	08	40	12·9	+64	19	40	4·60	−0·1	K2	87	
5	14	27	31·4	+75	41	45	4·25	−0·3	K4	79	
7 β	14	50	42·2	+74	09	19	2·08	−0·3	K4	29	Kocab
13 γ	15	20	43·6	+71	50	02	3·05	−1·1	A3	69	Pherkad Major
16 ζ	15	44	03·3	+77	47	40	4·32	1·7	A3	33	Alifa
22 ε	16	45	57·8	+82	02	14	4·23	0·3	G5	61	
23 δ	17	32	12·7	+86	35	11	4·36	1·2	A1	44	Yildun

Also above mag. 5:

	Mag.	Abs. mag.	Spectrum	Dist	
4	4·82	−0·2	K3	90	
21 η	4·95	2·6	F0	28	Alasco

VARIABLE STARS

	R.A.		Dec.		Range	Type	Period, d.	Spectrum
	h	m	deg	min				
T	13	34·7	+73	26	8·1–15·0	Mira	313·9	M
V	13	38·7	+74	19	7·4–8·8	Semi-reg.	72	M
U	14	17·3	+66	48	7·4–12·7	Mira	326·5	M
RR	14	57·6	+65	56	6·0–6·5	Semi-reg.?	40?	M
S	15	29·6	+78	38	7·7–12·9	Mira	326·2	N
R	16	30·0	+72	17	8·8–11·0	Semi-reg.	324	M

DOUBLE STAR

	R.A.		Dec.		P.A.	Sep.	Mags.
	h	m	deg	min	deg	sec	
α	02	31·8	+89	16	218	18·4	2·0, 9·0

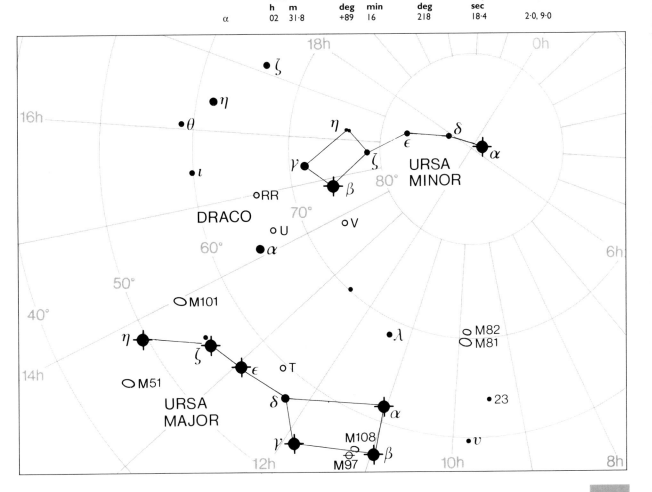

VELA

(Abbreviation: Vel).

The Sails of the dismembered ship Argo. There are 14 stars above the fourth magnitude:

	Mag.	Luminosity Sun = 1	Dist lt-yrs
γ	1·78	3800	520
δ	1·96	50	68
λ	2·21v	5000	490
κ	2·50	1320	390
μ	2·69	58	98
N	3·13	110	147
φ	3·54	20 900	2510
ψ	3·60	14	62
o	3·62v	450	390
c	3·75	96	186
p	3·84	2	75
b	3·84	200 000	6200
q	3·85	26	102
a	3·91	130	245

δ and κ make up the 'False Cross' with ε and ι Carinæ.

Vela contains the Gum Nebula; R.A. 8h 30m, dec. –45 deg. It measures 1200 × 720 minutes of arc. This is the remnant of the Vela supernova and contains the pulsar PSR 0833–45.

See chart for Carina.

BRIGHTEST STARS

Star	R.A. h	m	s	Dec. deg	min	sec	Mag.	Abs. mag.	Spectrum	Dist pc	
γ	08	09	31·9	–47	20	12	1·78	–4·1	WC7	160	Regor
e	08	37	38·6	–42	59	21	4·14	–5·8	A9	950	
o	08	40	17·6	–52	55	19	3·62v	–1·7	B3	120	
δ	08	44	42·2	–54	42	30	1·96	0·6	A0	21	Koo She
b	08	40	37·6	–46	38	55	3·84	–8·4	F2	1900	
a	08	46	01·7	–46	02	30	3·91	–0·6	A0	75	
W	09	00	05·4	–41	15	14	4·45	2·4	F8	20	
c	09	04	09·2	–47	05	02	3·75	–0·1	K2	57	
λ	09	07	59·7	–43	25	57	2·21v	–4·4	K5	150	Al Suhail al Wazn
k²	09	15	45·1	–37	24	47	4·62	2·5	F3	22	
κ	09	22	06·8	–55	00	38	2·50	–3·0	B2	120	Markeb
ψ	09	30	41·9	–40	28	00	3·60	1·9	F2	19	
N	09	31	13·3	–57	02	04	3·13	–0·3	K5	45	
M	09	36	49·7	–49	21	18	4·35	2·1	dA5	28	
m	09	51	40·7	–46	32	52	4·58	0·3	gG6	46	
φ	09	56	51·7	–54	34	03	3·54	–6·0	B5	770	
q	10	14	44·1	–42	07	19	3·85	1·4	A2	31	
HD 89682	10	19	36·8	–55	01	46	4·57	–5·9	cK	910	
HD 89890	10	20	55·4	–56	02	36	4·50	–2·3	B3	210	
p	10	37	18·0	–48	13	32	3·84	2·0	F4	23	
x	10	39	18·3	–55	36	12	4·28	–2·1	G2	150	
μ	10	46	46·1	–49	25	12	2·69	0·3	G5	30	
i	11	00	09·2	–42	13	33	4·39	0·6	A2	52	

Also above mag. 5:

	Mag.	Abs. mag.	Spectrum	Dist
HD 74272	4·77	–2·3	A3	260
HD 92036	4·89	–0·3	K5	93

VARIABLE STARS

	R.A. h	m	Dec. deg	min	Range	Type	Period, d.	Spectrum
AH	08	12·0	–46	39	5·5–5·9	Cepheid	4·23	F
AI	08	14·1	–44	34	6·4–7·1	Delta Scuti	0·11	A–F
RZ	08	37·0	–44	07	6·4–7·6	Cepheid	20·40	G
T	08	37·7	–47	22	7·7–8·3	Cepheid	4·64	F
SW	08	43·6	–47	24	7·4–9·0	Cepheid	23·47	K
SX	08	44·9	–46	21	8·0–8·6	Cepheid	9·55	G
CV	09	00·6	–51	33	6·5–7·3	Algol	6·89	B + B
SY	09	12·4	–43	47	7·6–8·1	Semi-reg.	63	M
RW	09	20·3	–49	31	7·8–12·0	Mira	451·7	M
V	09	22·3	–55	58	7·2–7·9	Cepheid	4·37	F
S	09	33·2	–45	13	7·7–9·5	Algol	5·93	A + K
U	09	33·2	–45	31	7·9–8·2	Semi-reg.	37	M
Z	09	52·9	–54	11	7·8–14·8	Mira	421·6	M
SV	10	44·9	–56	17	7·9–9·1	Cepheid	14·10	F–G

DOUBLE STARS

	R.A. h	m	Dec. deg	min	P.A. deg	Sep. sec	Mags.	
γ	08	09·5	–47	20	AB 220	41·2	1·9, 4·2	
					AC 151	62·3	8·2	
					AD 141	93·5	9·1	
					DE 146	1·8	12·5	
b	08	40·6	–46	39	058	37·5	3·8, 10·2	
δ	08	44·7	–54	43	AB 153	2·6	2·1, 5·1	
					AC 061	69·2	11·0	
					CD 102	6·2	13·5	
μ	10	46·8	–49	25	055	2·3	2·7, 6·4	Binary, 116 y

OPEN CLUSTERS

M	NGC	R.A. h	m	Dec. deg	min	Diameter min	Mag.	No. of stars	
	2547	08	10·7	–49	16	20	4·7	80	
	IC 2391	08	40·2	–53	04	50	2·5	30	o Velorum cluster
	IC 2395	08	41·1	–48	12	8	4·6	40	
	2669	08	44·9	–52	58	12	6·1	40	
	2670	08	45·5	–48	47	9	7·8	30	
	IC 2488	09	27·6	–56	59	15	7·4	70	
	2910	09	30·4	–52	54	5	7·2	30	
	2925	09	33·7	–53	26	12	8·3	40	
	2972	09	40·3	–50	20	4	9·9	25	
	3033	09	48·8	–56	25	5	8·8	50	
	3228	10	21·8	–51	43	18	6·0	15	

VELA (continued)

GLOBULAR CLUSTER

M	NGC	R.A.		Dec.		Diameter	Mag.
		h	m	deg	min	min	
	3201	10	17·6	−46	25	18·2	6·7

PLANETARY NEBULA

M	NGC	R.A.		Dec.		Diameter	Mag.	Mag. of central star
		h	m	deg	min	sec		
	3132	10	07·7	−40	26	47	8·2	10·1

VIRGO

(Abbreviation: Vir).

A very large constellation, representing Astræa – the goddess of justice, daughter of Jupiter and Themis. The 'bowl' of Virgo abounds in faint galaxies. Of the 9 stars down to the fourth magnitude, Spica is an eclipsing variable over a small range (0·91–1·01), and γ (Arich) used to be one of the most spectacular binary pairs in the sky, though it is closing up and will have become very difficult to separate by the end of the 20th century. The leading stars are:

	Mag.	Luminosity Sun = 1	Dist lt-yrs
α	0·98v	2100	260
γ	2·75	7	36
ε	2·83	60	104
ζ	3·37	17	75
δ	3·38	120	147
β	3·61	3·4	33
109	3·72	50	124
μ	3·88	14	85
η	3·89	26	104

BRIGHTEST STARS

Star	R.A.			Dec.			Mag.	Abs. mag.	Spectrum	Dist pc	
	h	m	s	deg	min	sec					
3 ν	11	45	51·5	+06	31	45	4·03	−0·5	M1	51	
5 β	11	50	41·6	+01	45	53	3·61	3·6	F8	10	Zavijava
8 π	12	00	52·3	+06	36	51	4·66	1·7	A3	37	
9 o	12	05	12·5	+08	43	59	4·12	0·3	G8	34	
15 η	12	19	54·3	−00	40	00	3·89	1·4	A2	32	Zaniah
26 χ	12	39	14·7	−07	59	45	4·66	−0·1	K2	78	
29 γ	12	41	39·5	−01	26	57	2·75	2·6	F0 + F0	11	Arich
43 δ	12	55	36·1	+03	23	51	3·38	−0·5	M3	45	Minelauva
47 ε	13	02	10·5	+10	57	33	2·83	0·2	G9	32	Vindemiatrix
51 θ	13	09	56·9	−05	32	20	4·38	1·2	A1	43	Apami-Atsa
61	13	18	24·2	−18	18	21	4·74	5·1	G6	8·4	
67 α	13	25	11·5	−11	09	41	0·98v	−3·5	B1	79	Spica
74	13	31	57·8	−06	15	21	4·69	−0·5	gM3	110	
79 ζ	13	34	41·5	−00	35	46	3·37	1·7	A3	340	Heze
93 τ	14	01	38·7	+01	32	40	4·26	1·7	A3	32	
98 κ	14	12	53·6	−10	16	25	4·19	−0·2	K3	71	
99 ι	14	16	00·8	−06	00	02	4·08	0·7	F6	22	Syrma
100 λ	14	19	06·5	−13	22	16	4·52	2·3	A0	20	Khambalia
107 μ	14	43	03·5	−05	39	30	3·88	1·9	F3	26	Rijl al Awwa
109	14	46	14·9	+01	53	34	3·72	0·6	A0	38	
110	15	02	53·9	+02	05	28	4·40	0·2	K0	65	

Also above mag. 5:

	Mag.	Abs. mag.	Spectrum	Dist
2 ξ	4·85		A3	
16	4·96	0·2	K0	67
30 ρ	4·88	0·6	A0	63
40 ψ	4·80	−0·5	M3	120
60 σ	4·80	−0·5	M2	100
69	4·76	0·0	K1	32
78	4·94	0·9	A2	26

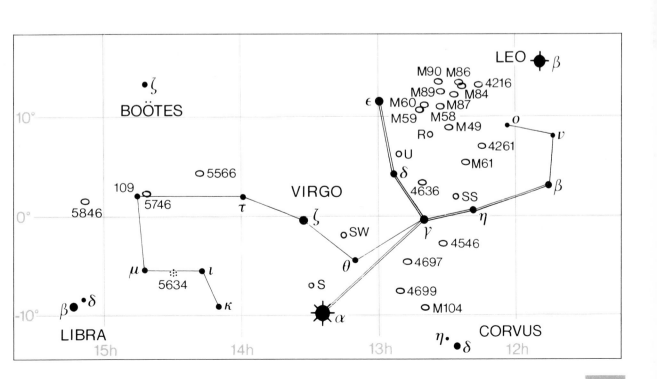

VIRGO (continued)

VARIABLE STARS

	R.A.		Dec.		Range	Type	Period, d.	Spectrum
	h	m	deg	min				
X	12	01·9	+09	04	7·3–11·2	?	–	F
SS	12	25·3	+00	48	6·0–9·6	Mira	354·7	N
R	12	38·5	+06	59	6·0–12·1	Mira	145·6	M
U	12	51·1	+05	33	7·5–13·5	Mira	206·8	M
S	13	33·0	–07	12	6·3–13·2	Mira	377·4	M
RS	14	27·3	+04	41	7·0–14·4	Mira	352·8	M

DOUBLE STARS

	R.A.		Dec.		P.A.	Sep.	Mags.	
	h	m	deg	min	deg	sec		
17	12	22·5	+05	18	337	20·0	6·6, 9·4	
γ	12	41·7	–01	27	287	3·0	3·5, 3·5	Binary, 171·4 y
θ	13	09·9	–05	32	343	7·1	4·4, 9·4	
73	13	32·0	–18	44	183	0·1	6·7, 6·9	
84	13	43·1	+03	32	229	2·9	5·5, 7·9	
τ	14	01·6	+01	33	290	80·0	4·3, 9·6	
φ	14	28·2	–02	14	110	4·8	4·8, 9·3	

GALAXIES

M	NGC	R.A.		Dec.		Mag.	Dimensions	Type	
		h	m	deg	min		min		
	4216	12	15·9	+13	09	10·0	8·3 × 2·2	Sb	
	4261	12	19·4	+05	49	10·3	3·9 × 3·2	E2	
61	4303	12	21·9	+04	28	9·7	6·0 × 5·5	Sc	
84	4374	12	25·1	+12	53	9·3	5·0 × 4·4	E1	
	4429	12	27·4	+11	07	10·2	5·5 × 2·6	S0	
	4438	12	27·8	+13	01	10·1	9·3 × 3·9	Sap	
	4442	12	28·1	+09	48	10·5	4·6 × 1·9	E5p	
86	4406	12	26·2	+12	57	9·2	7·4 × 5·5	E3	
49	4472	12	29·8	+08	00	8·4	8·9 × 7·4	E4	
87	4486	12	30·8	+12	24	8·6	7·2 × 6·8	E1	Virgo A
	4699	12	49·0	–08	40	9·6	3·5 × 2·7	Sa	
	4535	12	34·3	+08	12	9·8	6·8 × 5·0	SBc	
	4546	12	35·5	–03	48	10·3	3·5 × 1·7	E6	
	4527	12	34·1	+02	39	11·3	6·3 × 2·3	Sb	
	4536	12	34·5	+02	11	11·0	7·4 × 3·5	Sc	
	4546	12	35·5	–03	48	11·3	3·5 × 1·7	E6	
89	4552	12	35·7	+12	33	9·8	4·2 × 4·2	E0	
90	4569	12	36·8	+13	10	9·5	9·5 × 4·7	Sb	
58	4579	12	37·7	+11	49	9·8	5·4 × 4·4	Sb	
104	4594	12	40·0	–11	37	8·3	8·9 × 4·1	Sb	Sombrero Hat
	4596	12	39·9	+10	11	10·5	3·9 × 2·8	SBa	
59	4621	12	42·0	+11	39	9·8	5·1 × 3·4	E3	
60	4649	12	43·7	+11	33	8·8	7·2 × 6·2	E1	
	4654	12	44·0	+13	08	10·5	4·7 × 3·0	Sc	
	4636	12	42·8	+02	41	9·6	6·2 × 5·0	E1	
	4660	12	44·5	+11	11	11·9	2·8 × 1·9	E5	
	4697	12	48·6	–05	48	9·3	6·0 × 3·8	E4	
	4699	12	49·0	–08	40	9·6	3·5 × 2·7	Sa	
	4753	12	52·4	–01	12	9·9	5·4 × 2·9	Pec.	
	4762	12	52·9	+11	14	10·2	8·7 × 1·6	SB0	
	4856	12	59·3	–15	02	10·4	4·6 × 1·6	SBa	
	5247	13	38·1	–17	53	10·5	5·4 × 4·7	Sb	
	5363	13	56·1	+05	15	10·2	4·2 × 2·7	Ep	
	5364	13	56·2	+05	01	10·4	7·1 × 5·0	SB + p	
	5068	13	18·9	–21	02	10·8	6·9 × 6·3	SBc	
	5850	15	07·1	+01	33	11·7	4·3 × 3·9	SBb	

VOLANS

(Abbreviation: Vol).

Originally Piscis Volans. A small constellation, intruding into Carina. There are four stars above the fourth magnitude:

	Mag.	Luminosity	Dist
		Sun = 1	lt-yrs
γ	3·6 (combined)	37+6	75
β	3·77	96	192
ζ	3·95	60	176
δ	3·98	12 000	2380

See chart for Carina.

BRIGHTEST STARS

Star	R.A.			Dec.			Mag.	Abs.	Spectrum	Dist	
	h	m	s	deg	min	sec		mag.		pc	
γ	{ 07	08	42·3	–70	29	50	5·7	2·8	dF4 }	23	Combined
	{ 07	08	45·0	–70	29	57	3·8	0·9	G8 }		mag. 3·6
δ	07	16	49·8	–67	57	27	3·98	–5·4	F8	730	
ζ	07	41	49·3	–72	36	22	3·95	0·2	K0	54	
ε	08	07	55·9	–68	37	02	4·35	–1·1	B5	120	
β	08	25	44·3	–66	08	13	3·77	–0·1	K2	59	
α	09	02	26·9	–66	23	46	4·00	2·1	A5	24	

VARIABLE STAR

	R.A.		Dec.		Range	Type	Period, d.	Spectrum
	h	m	deg	min				
S	07	29·8	–73	23	7·7–13·9	Mira	395·8	M

VOLANS (continued)

DOUBLE STARS

	R.A.		Dec.		P.A.	Sep.	Mags.
	h	m	deg	min	deg	sec	
γ^2	07	08·8	−70	30	300	13·6	4·0, 5·9
ζ	07	41·8	−72	36	116	16·7	4·0, 9·8
ε	08	07·9	−68	37	024	6·1	4·4, 8·0
κ	08	19·8	−71	31	AB 057	65·0	5·4, 5·7
					BC 030	37·7	8·5
θ	08	39·1	−70	23	108	45·0	5·3, 10·3

GALAXY

M	NGC	R.A.		Dec.		Mag.	Dimensions	Type
		h	m	deg	min		min	
	2442	07	36·4	−69	32	11·2	6·0 × 5·5	SBb

VULPECULA

(Abbreviation: Vul).

Originally Vulpecula et Anser, the Fox and Goose – nowadays the goose has disappeared (possibly the fox has eaten it). Vulpecula contains no star above magnitude 4·4, and is notable only because of the presence of the Dumbbell Nebula and the fact that several novæ have appeared within the boundaries of the constellation.

See chart for Cygnus.

BRIGHTEST STARS

Star	R.A.			Dec.			Mag.	Abs.	Spectrum	Dist
	h	m	s	deg	min	sec		mag.		pc
6 α	19	28	42·2	+24	39	54	4·44	0·0	M0	26
13	19	53	27·5	+24	04	47	4·58	−0·6	A0	110
15	20	01	05·9	+27	45	13	4·64	1·7	A0	18
23	20	15	49·1	+27	48	51	4·52	−0·2	K3	88

Also above mag. 5:

	Mag.	Abs. mag.	Spectrum	Dist
12	4·95	−1·7	B3	210

VARIABLE STARS

	R.A.		Dec.		Range	Type	Period, d.	Spectrum
	h	m	deg	min				
RS	19	17·7	+22	26	6·9–7·6	Algol	4·46	B + A
Z	19	21·7	+25	34	7·4–9·2	Algol	2·45	B + A
U	19	36·6	+20	20	6·8–7·5	Cepheid	7·99	F–G
T	20	51·5	+28	15	5·4–6·1	Cepheid	4·44	F–G
SV	19	51·5	+27	28	6·7–7·7	Cepheid	45·03	F–K
R	21	04·4	+23	49	7·0–14·3	Mira	136·4	M

DOUBLE STARS

	R.A.		Dec.		P.A.	Sep	Mags.
	h	m	deg	min	deg	sec	
2 (ES)	19	17·7	+23	02	127	1·8	5·4, 9·2
α–8	19	28·7	+24	40	028	413·7	4·4, 5·8

OPEN CLUSTERS

M	NGC	R.A.		Dec.		Diameter	Mag.	No. of stars	
		h	m	deg	min	min			
	Cr 399	19	25·4	+20	11	60	3·6	40	Coat-hanger (Brocchi's Cluster)
	6823	19	43·1	+23	18	12	7·1	30	
	6830	19	51·0	+23	04	12	7·9	20	
	6885	20	12·0	+26	29	7	5·7	30	
	6940	20	34·6	+28	18	31	6·3	60	

PLANETARY NEBULA

M	NGC	R.A.		Dec.		Dimensions	Mag.	Mag. of central star	
		h	m	deg	min	sec			
27	6853	19	59·6	+22	43	350 × 910	7·6	13·9	Dumbbell Nebula

TELESCOPES AND OBSERVATORIES

OBSERVATORIES

*T*he oldest observatory building now standing is the Chomsong-dae, in Kyongju, South Korea, AD 632.

✳ *The oldest observatory to be associated with a national body* is at Leiden University, Holland, dating from 1632.

✳ *The oldest national observatory* was that at Copenhagen, Denmark. Unfortunately, the original buildings were destroyed by fire.

✳ *The first national British observatory* is the Royal Greenwich Observatory, founded in 1675 by order of King Charles II. The original observatory in Greenwich Park, some buildings of which were designed by Wren, is now a museum (Flamsteed House) and the research equipment was shifted to Herstmonceux, Sussex as well as La Palma in the Canary Islands. Unfortunately, Herstmonceux has now been closed, and the RGO has been transferred to an office block in Cambridge. In 1992 it was announced that the two British Royal Observatories (Greenwich and Edinburgh) are in future to have a single controller, with combined administration.

✳ *The world's highest observatory* is at Boulder, Colorado (University of Denver High Altitude Observatory) at 4297 m. It was opened in 1973. The main telescope is the 60·48 cm Ealing Beck reflector. Only

Deployment of the Hubble Space Telescope, 25 April 1990, from the payload bay of the Shuttle *Discovery*; photograph by the STS-31 crew.

slightly lower is the Mauna Kea Observatory in Hawaii, close to the summit of a volcano which is (we hope!) extinct; altitude 4205 m. The main telescopes are the Keck (1000 cm), the UKIRT (United Kingdom Infra-Red Telescope) (375 cm), the CFH (Canada–France–Hawaii Telescope) (366 cm), the University of Hawaii reflector (224 cm) and the NASA infra-red telescope (305 cm). Also on Mauna Kea are the JCMT (James Clerk Maxwell Telescope) designed for microwave observations, and the nearby NASA microwave telescope. A twin of the first Keck telescope is now being constructed (1995), and there are also to be very large Japanese and Italian telescopes.

✳ *The world's lowest observatory* is at Homestake Mine, South Dakota, at a depth of 1·5 km below ground level. The 'telescope' is a 100 000 gallon tank of 'cleaning fluid' to trap solar neutrinos. The experiment was started in 1965, and is still continuing.

✳ *The only flying observatory* is the KAO (Kuiper Airborne Observatory) which is really a modified Lockheed C–141 aircraft carrying a 91 cm reflector. At peak altitude it flies above 85 per cent of the atmosphere. The telescope is installed in an open cavity recessed into the port side of the aircraft, immediately ahead of the wing.

Telescopes

*T*he first large space telescope, the Hubble Space Telescope (HST) was launched from the Space Shuttle on 25 April 1990. It has a 240-cm mirror, and orbits at an altitude of 600 km in a period of 95 minutes, with orbital inclination 28·5 degrees. The total weight is 11 360 kg, and the length 13·3 m; its diameter is 12 m with the solar arrays extended. Initially, the telescope had a faulty mirror, but a 'servicing mission' by astronauts introduced corrective optics, and the telescope is now just as good as had been originally hoped. With perfect seeing conditions all the time, it can out-perform any Earth-based telescope.

✳ *The world's largest single-mirror optical telescope* is the 600 cm reflector at Mount Semirodriki, near Zelenchukskaya in the Caucasus; altitude 2080 m. The 70-tonne mirror was completed in 1974, and regular observations began on 7 February 1976. It must be admitted that the telescope has

never been a real success. The only other single-mirror telescope over 500 cm aperture is the Hale Reflector at Palomar, USA, completed in 1948.

✳ *The world's largest optical telescope* is the 1000 cm Keck reflector on Mauna Kea, Hawaii. The mirror is made of 36 segments, fitted together to make the correct optical curve. Twelve of the segments were in place by the end of 1991, so that the light-collecting area was already greater than that of the Hale reflector at Palomar. A twin Keck telescope is being set up 85 m from the first. When completed, Keck I and Keck II will be able to work together as an interferometer. Theoretically, they would be able to distinguish a car's headlights separately from a range of over 25 000 km.

✳ *The first telescope* was probably built by Leonard Digges, in England, some time between 1545 and 1559. Research by the English historian C. A. Ronan seems to have shown that the Digges telescope used a combination of mirrors and lenses. Whether or not it was ever turned to the sky is not known, but it is a possibility.

✳ *The first refracting telescope* whose existence can be definitely proved was made by H. Lippershey (Holland) in 1608.

✳ *The world's largest refractor* is the 101·6 cm telescope at Yerkes Observatory, USA. (The largest object-glass ever made was the 124·5 cm lens shown at the Paris Exhibition of 1900, but it was never used seriously.)

✳ *The first reflecting telescope* whose existence can be definitely proved was made by Newton, and presented to the Royal Society in 1671, so that the actual construction was a year or two before this, possibly in 1668 or 1669.

✳ *The first multiple-mirror telescope* (MMT) was that at the Whipple Observatory on Mount Hopkins, Arizona. It used six 183 cm mirrors in conjunction, so that the total light-grasp was equal to a single 442 cm mirror. It performed well, but the six mirrors have now been replaced by a single 6·5 m mirror constructed by Roger Angel, using his new 'spin-casting' technique.

✳ *The largest speculum metal mirror ever made for an optical telescope* was 182·9 cm in diameter. It was made by the third Earl of Rosse for his observatory at Birr Castle,

SOME OF THE WORLD'S GREAT OBSERVATORIES

Location	Name	Latitude			Longitude			Altitude
		deg	min	sec	deg	min	sec	metres
Aarhus, Denmark	Ole Rømer Obs.	56N	07	40·0	10E	11	48	50
Alma Ata, Kazakhstan	Mountain Obs. of							
	Academy of Sciences	43N	11	16·9	76E	57	24	1450
Ann Arbor, Mich., USA	Univ. of Michigan	42N	16	48·7	83W	43	48	282
Arcetri (Florence), Italy	Astrophysical Obs.	43N	45	44·7	11E	15	18	184
Armagh, N Ireland	Armagh Obs.	54N	21	1·1	06W	38	54	64
Athens, Greece	National Obs.	37N	58	19·7	23E	43	0	110
Auckland, N Zealand	Auckland Public Obs.	36S	54	28·0	174E	46	36	80
Barcelona, Spain	Fabra Obs.	41N	24	59·3	02E	07	36	415
Beijing, China	Beijing University Obs.	39N	57	24	116E	21	36	76
Beirut, Lebanon	American Univ. Obs.	33N	54	22·0	35E	28	12	38
Belgrade, Yugoslavia	Obs. of Academy of							
	Science	44N	48	13·2	20E	30	48	253
Belo Horizonte, Brazil	Piedade Obs.	19S	49	18	43W	30	42	1746
Berlin, Germany	Wilhelm Förster Obs.	52N	28	30·0	13E	25	30	40
Berlin-Treptow, Germany	Archenhold Obs.	52N	29	07·0	13E	28	36	38
Berne, Switzerland	Astr. Inst. of Univ.	46N	57	12·7	07E	25	42	563
Big Bear Lake, Calif., USA	Big Bear Solar Obs.	34N	15	12	116W	54	54	2067
Black Birch, New Zealand	Carter Obs. Station	41S	44	42	173W	48	12	1366
Bloemfontein, S Africa	Boyden Obs.	29S	02	18·0	26E	24	18	1387
Bochum, Germany	Astronomical Station	51N	27	54·8	07E	13	24	132
Bogotá, Colombia	National Astr. Obs.	04N	35	54	74W	04	54	2640
Bologna, Italy	San Vittore Obs.	44N	28	06	11E	20	30	280
Bombay, India	Government Obs.	18N	53	36·2	72E	48	54	14
Bonn, Germany	University Obs.	50N	43	45·0	07E	05	48	62
Bordeaux, France	Obs. of Univ. of							
	Bordeaux	44N	50	07	00E	31	36	73
Bosque Alegro, Argentina	Cordoba Obs. Station	31S	35	54	64W	32	48	1250
Boulder, Colo., USA	Sommers-Bausch Obs.	40N	00	13·0	105W	15	42	1648
Brno, Czech Republic	Nicolaus Copernicus Obs.	49N	12	18	16E	35	18	310
Bro, Sweden	Kvistoberg Obs.	59N	30	06	17E	36	24	0
Brussels, Belgium	Astronomy and							
	Astrophys. Inst.	50N	48	48	04E	23	00	147
Bucharest, Romania	National Obs.	44N	24	49·4	26E	05	48	83
Budapest, Hungary	Konkoly Obs.	47N	29	58·6	18E	57	54	474
Buenos Aires, Arg.	Naval Obs.	34S	37	18·3	58W	21	18	6
Calar Alto, Spain	German-Spanish Ast.							
	Centre	37N	13	48	02E	32	12	2168
Cambridge, England	University Obs.	52N	12	51·6	00E	05	42	28
Cambridge, England	Royal Greenwich Obs.	52N	12	51·6	00E	05	42	28
Cambridge, Mass., USA	Harvard College Obs.	42N	22	47·6	71W	07	48	24
Cape, South Africa	S African Astr. Obs.	33S	56	02·5	18E	28	36	10
Caracas, Venezuela	Cagigal Obs.	10N	30	24·3	66W	55	42	1042
Castleknock, Ireland	Dunsink Obs.	53N	23	18	06W	20	12	85
Catania, Sicily	Astrophysical Obs.	37N	31	.42·0	15E	04	42	193
Cerro Las Campanas,								
Chile	Las Campanas Obs.	29S	00	30	70W	42	00	2282
Cerro La Silla, Chile	European Southern Obs.	29S	15	24	70W	43	48	2347
Cerro Tololo, Chile	Cerro Tololo Inter-							
	American Obs.	30S	09	54	70W	48	54	2215
Chung-li, Taiwan	National Central							
	Univ. Obs.	24N	58	12	121E	11	12	152
Cincinnati, Ohio, USA	Cincinnati Obs.	39N	08	19·8	84W	25	18	247
Cocoa, Florida, USA	Brevard Community							
	Astr. Obs.	28N	33	06	80W	45	42	17
Copenhagen, Denmark	Urania Obs.	55N	41	19·2	12E	32	18	10
Cordoba, Argentina	National Obs.	31S	25	16·4	64W	11	48	434
Crimea, Ukraine	Crimean Astro. Obs.	44N	43	42·0	34E	01	00	550
Edinburgh, Scotland	Royal Obs.	55N	55	30·0	03W	11	00	146
Flagstaff, Arizona, USA	US Naval Obs.	35N	11	00·0	111W	44	24	2310
Flagstaff, Ariz.ona USA	Lowell Obs.	35N	12	06·0	111W	39	48	2210
Göttingen, Germany	University Obs.	51N	31	48·2	09E	56	36	161
Greenbelt, Md., USA	Goddard Research Obs.	39N	01	11·5	76W	49	36	49
Groningen, Holland	Kapteyn Astron. Lab.	53N	13	13·8	06E	33	48	4
Haleakala, Hawaii, USA	Univ. of Hawaii	20N	42	22·0	156W	15	24	3054
Hamburg, Germany	Bergedorf Obs.	53N	28	46·9	10E	14	24	41
Hartebeespoort, S. Africa	Republic Obs. Annexe	25S	46	22·4	27E	52	36	1220
Helsinki, Finland	University Obs.	60N	09	42·3	24E	57	18	33
Helwan, Egypt	Helwan Obs.	29N	51	31·1	31E	20	30	115
Hyderabad, India	Nizamiah Obs.	17N	25	53	78E	27	12	293
Jena, Germany	Karl Schwarzschild Obs.	50N	58	51·0	11E	42	48	331

Ireland in 1845. He later made a second mirror for the same material. The telescope has not been used since 1909, but the tube is still at Birr Castle, and it is hoped to have the telescope working again within the next few years, though its importance is of course purely historical.

✳ *The last large speculum metal mirror made* was the 122 cm for the Great Melbourne Reflector, in 1868. For many years the telescope was a relative failure, but it has now been completely rebuilt and has been transferred to the Mount Stromlo Observatory.

By now electronic devices have more or less superseded photography; indeed, *the last photographic plate taken with the Hale reflector at Palomar* was exposed on 29 September 1989. Plates have been succeeded by CCDs (Charge-Coupled Devices), which are vastly more sensitive. A CCD is a silicon-integrated circuit consisting of an oxide-covered silicon substate upon which is formed an array of closely spaced electrodes.

Another new technique involves fibre optics; *the first practical experiment in fibre optics for astronomical research* was carried out by John Hill, a postgraduate student of Roger Angel, in 1979. An optical fibre is a solid piece of very pure silica glass, only about ⅛ mm across. The fibre is made of two types of glass. There is the core glass, at the centre of the fibre and around $\frac{1}{100}$ to $\frac{1}{200}$ mm in diameter. Round this is the cladding glass, which coats the outside of the core glass; the whole fibre is then encased in a plastic cover. Light can be transmitted along the optical fibre thread, and is confined to the centre of the fibre, passing only through the glass core. The purity of the glass means that very little light is lost – and of course the light can be 'bent', and the tubes are flexible. Hill's initial device, Medusa, was used on the 2·3 m telescope at the Steward Observatory in Tucson, Arizona. Hill used an aluminium plate with holes in it, each hole matching the position of a galaxy or group of galaxies; this made it possible to obtain many spectra simultaneously, and since then the principle has been extended and elaborated.

Yet another new technique is being used by J. Davis at Narrabri, Australia. He has built SUSI (the Sydney University Stellar Interferometer) to measure star diameters. SUSI consists of twelve 200 mm telescopes, separated by between 5 and 640 m. By combining the light from the mirrors, two at a time, it is possible to make measurements which would otherwise require a 640 m mirror. SUSI was completed in 1991.

✳ *The world's largest infra-red telescope* is the UKIRT, on Mauna Kea. It was made in

Sheffield, and has proved to be so good that it is used for optical research as well as for infra-red.

✳ *The world's largest telescope designed specially for microwave observations* is the James Clerk Maxwell Telescope, on Mauna Kea. It has a 'dish' 15 metres in diameter, made up of 276 separate panels which have been fitted together to form the correct shape.

✳ *The largest telescope now being planned* is the VLT (Very Large Telescope) of the European Southern Observatory. It will be made up of four 8m telescopes working together, providing a light-grasp equal to a single 16m mirror. It is to be set up at Paranal, in the Atacama Desert, north of La Silla. The first units should be working in 1995, and the complete telescope by 2000.

✳ *The largest Schmidt telescope* is at Tautenburg, Germany; aperture 134 cm with a 200 cm mirror; focal length 4 m.

✳ *The largest telescope ever set up in Britain* was the INT or Isaac Newton Telescope, with a mirror 244 cm in diameter. It was completed in 1967, but has now been moved to the new Canary Island observatory – El Observatorio des Roques de los Muchachos (the Observatory of the Rocks of the Boys) on La Palma (2423 m altitude). It has a new 256 cm mirror. The 4·2-metre William Herschel Telescope on La Palma was brought into operation in 1987.

✳ *The largest balloon-borne telescope in the world* is the 101·6 cm reflector designed and used by Dr. R. Joseph, Imperial College, University of London.

✳ *The largest solar telescope* is at Kitt Peak, USA. It uses a 203 cm heliostat, and has an inclined tunnel 146 m long. It was completed in 1962. Though designed for solar work, it is also capable of being used for other types of observation as well – an initially unexpected bonus.

PLANETARIA

*T*he first orrery was made in the 18th century by George Graham, for Prince Eugene of Savoy. *The first American orrery* was made by David Rittenhouse in 1772.

✳ *The ancestor of the modern planetarium* is the Gottorp Globe, made by H. Busch in Denmark about 1654–64. It was 4 m in diameter. The audience sat inside it; the stars were painted on the inside of the globe. *The first modern-type planetarium* (due to Dr. W. Bauersfelt of the Zeiss optical works) was opened at the Deutsches Museum in Bonn in 1923. *The world's largest planetarium* is at the Miyazumi Science Centre

Location	Name	Latitude			Longitude			Altitude
		deg	min	sec	deg	min	sec	metres
Johannesburg, S Africa	Republic Obs.	26S	10	55·3	28E	04	30	1806
Juvisy, France	Flammarion Obs.	48N	41	37·0	02E	22	18	92
Kharkov, Ukraine	Kharkov Univ. Obs.	50N	00	02	36E	13	54	138
Kiev, Ukraine	Kiev Univ. Obs.	50N	27	12	30E	29	54	184
Kiso, Japan	Kiso Obs.	35N	47	36	137E	37	42	1130
Kitt Peak, Ariz., USA	Kitt Peak National Obs.	31N	57	48	111W	36	00	2120
Kitt Peak, Ariz., USA	Steward Obs. Station	31N	57	48	111W	36	00	2071
Kodaikanal, India	Astrophysical Obs.	10N	13	50·0	77E	28	06	2343
Kracow, Poland	University Obs.	50N	03	52·0	19E	57	36	221
Kyoto, Japan	Kwasan Obs.	34N	59	40·8	135E	47·	36	234
La Palma, Canary Is.	Roque de los Muchachos Obs.	28N	45	36	17W	52	54	2326
La Plata, Argentina	La Plata Obs.	34S	54	30	57W	55	54	17
Las Cruces, N Mex., USA	Corralitos Obs.	32N	22	48	107W	02	36	1453
Leyden, Holland	University Obs.	52N	09	19·8	04E	29	00	6
Lisbon, Portugal	Lisbon Astr. Obs.	38N	42	42	09N	11	12	111
Los Angeles, Calif., USA	Griffith Obs.	34N	06	46·8	118W	18	06	357
Lund, Sweden	Royal University Obs.	55N	41	51·6	13E	11	12	34
Madison, Wis., USA	Washburn Obs.	43N	04	36	89W	24	30	292
Madrid, Spain	Astronomical Obs.	40N	24	30·0	03W	41	18	655
Mauna Kea, Hawaii, USA	Mauna Kea Obs.	19N	46	36	155W	28	18	4215
Meudon, France	Obs. of Physical Astr.	48N	48	18·0	02E	13	54	162
Milan, Italy	Brera Obs.	45N	27	59·2	09E	11	30	120
Mill Hill, London, Eng.	Univ. of London Obs.	51N	36	46·3	00W	14	24	82
Minneapolis, Minn., USA	Univ. of Minnesota Obs.	44N	58	40·0	93W	14	18	260
Mitaka, Japan	National Astr. Obs.	35N	40	18	139E	32	30	58
Montevideo, Uruguay	National Obs.	34S	54	33·0	56W	12	42	24
Montreal, Quebec, Can.	McGill Univ. Obs.	45N	30	20·0	73W	34	42	57
Moscow, Russia	Sternberg Inst. Obs.	55N	45	19·8	37E	34	12	166
Mt Aragatz, Armenia	Byurakan Astrophys. Obs.	40N	20	06	44E	17	30	1500
Mt Bigelow, Ariz., USA	Steard Obs., Catalina Stn	32N	25	00	110W	43	54	2510
Mt Hamilton, Calif., USA	Lick Obs.	37N	20	25·3	121W	38	42	1283
Mt Hopkins, Ariz., USA	Fred L. Whipple Obs.	31N	41	18	110W	53	06	2608
Mt John, New Zealand	Mt John Univ. Obs.	43S	59	12	170E	27	54	1027
Mt Lemmon, Ariz., USA	Steward Obs., Catalina Station	32N	26	30	110W	47	30	2776
Mt Locke, Texas, USA	McDonald Obs.	30N	40	18	104W	01	06	2075
Mt Semirodriki, Caucasus, Russia	Zelenchukskaya	43N	49	32·0	41E	35	24	973
Mt Stromlo, Canberra, Australia	Mount Stromlo Obs.	35S	19	12	149E	00	30	767
Mt Wilson, Calif., USA	Hale Obs.	34N	12	59·5	118W	03	36	1742
Nanking, China	Purple Mountain Obs.	32N	03	59·9	118E	49	18	367
Naples, Italy	Capodimonte Obs.	40N	51	45·7	14E	15	24	164
Nice, France	Nice Obs.	43N	43	17·0	07E	18	00	376
Ondřejov, Czech Rep.	Astrophysical Obs.	49N	54	38·1	14E	47	00	533
Ottawa, Ontario, Canada	Dominion Obs.	45N	23	38·1	75W	43	00	87
Palermo, Sicily	University Astron. Obs.	38N	06	43·6	13E	21	30	72
Palomar, Calif., USA	Hale Obs.	33N	21	22·4	116W	51	48	1706
Pic du Midi, France	Obs. of University of Toulouse	42N	56	12·0	00E	08	30	2862
Piikkiö, Finland	Turku Univ. Obs.	60N	25	00	22E	26	48	40
Pittsburgh, Penn., USA	Allegheny Obs.	40N	20	58·1	80W	01	18	370
Potsdam, Germany	Astrophysical Obs.	52N	22	56·0	13E	04	00	107
Pulkovo, Russia	Astron. Obs. of Academy of Sciences	59N	46	18·5	30E	19	36	75
Purple Mountain, China	Purple Mountain Obs.	32N	00		118E	39	18	367
Quezon City, Philippines	Manila Obs.	14N	38	12	121E	04	36	58
Quito, Ecuador	National Obs.	00S	14	00·0	78W	29	36	2908
Richmond Hill, Ontario, Canada	David Dunlap Obs.	43N	51	46·0	79W	25	18	244
Riga, Latvia	Latvian State Univ. Astr. Obs.	56N	57	06	24E	07	00	39
Rio de Janeiro, Brazil	National Obs.	22S	53	42·2	43W	13	24	33
Rome, Italy	Vatican Obs.	41N	44	48	12E	39	06	450
St Andrews, Scotland	University Obs.	56N	20	12·0	02W	48	54	30
St Michel, France	Obs. of Haute-Provence	43N	55	54	05E	42	48	665
St Petersburg, Russia	Pulkovo Obs.	59N	46	05·5	30E	19	24	70
São Paulo, Brazil	Astronomical and Geophysical Inst.	23S	39	06·9	46W	37	24	800
Siding Spring, NSW, Australia	Siding Spring Obs.	31S	16	37·3	149E	04	00	1165
Stockholm, Sweden	Saltsjöbaden Obs.	59N	16	18·0	18E	18	30	55
Sunspot, N Mex., USA	Sacr. Peak Natl. Solar Obs.	32N	47	12	105W	49	12	2811
Sutherland, CP, S.A.	S.A. Astr. Obs	32S	22	46·5	20E	48	36	1830
Sydney, Australia	Government Obs.	33S	51	41·1	151E	12	18	44
Tartu, Estonia	Wilhelm Struve Astr. Obs.	58N	16	00	26E	28	00	0
Tautenberg, Germany	Karl Schwarzschild Obs.	50N	58	54	11E	42	48	331

Location	Name	Latitude deg	min	sec	Longitude deg	min	sec	Altitude metres
Tokyo, Japan	Tokyo Obs. at Mitaka	35N	40	21·4	139E	32	30	59
Tonantzintla, Mexico	National Astro. Obs.	19N	01	57·9	98W	18	48	2150
Toruń, Poland	Copernicus Univ. Obs.	53N	05	47·7	18E	33	18	90
Trieste, Italy	Trieste Astr. Obs.	45W	38	30	13E	52	30	400
Tucson, Arizona	Steward Obs.	32N	14	00	110W	56	54	757
Uccle, Belgium	Royal Obs.	50N	47	55·0	04E	21	30	105
Uppsala, Sweden	Univ. Astron. Obs.	59N	51	29·4	17E	37	30	21
Utrecht, Holland	Sonnenborgh Obs.	52N	05	09·6	05E	07	48	14
Victoria, BC, Canada	Dominion Astro. Obs.	48N	31	15·7	123W	25	00	229
Vienna, Austria	University Obs.	48N	13	55·1	16E	20	18	240
Vilnius, Lithuania	Vilnius Astr. Obs.	54N	41	00	25E	17	12	122
Washington, DC, USA	US Naval Obs.	38N	55	14·0	77W	03	54	86
Wellington, N Zealand	Carter Obs.	41S	17	03·9	174E	45	54	129
Williams Bay, Wisconsin, USA	Yerkes Obs.	42N	34	13·4	88W	33	24	334
Zürich, Switzerland	Swiss Federal Obs.	47N	22	36	08E	33	06	469

RADIO OBSERVATORIES

Location	Name	Latitude deg	min	sec	Longitude deg	min	sec	Altitude metres
Arcetri (Florence), Italy	Astrophysical Obs.	43N	45	14·4	11E	15	18	184
Arecibo, Puerto Rico	Arecibo Obs., Cornell University	18N	20	36·6	66W	45	12	496
Big Pine, Calif., USA	Owens Valley Radio Obs.	37N	13	54	118W	16	54	1236
Bochum, Germany	Radio Telescope Stn.	51N	25	43·0	07E	11	30	160
Boulder, Colorado	High Altitude Obs.	40N	04	42·0	105W	16	30	1692
Cambridge, England	Mullard Radio Astr. Obs	52N	09	45·0	00E	02	24	26
Cassel, Calif., USA	Hat Creek Radio Astr. Obs.	40N	49	06	121W	28	24	1043
Cebrerps, Spain	Deep Space Stn.	40N	27	18	04E	22	00	789
Columbia, S.C., USA	Univ. of S Carolina Radio Obs.	33N	59	48	81W	01	54	127
Crimea, Ukraine	Crimean Astrophys. Obs.	44N	43	42·0	34E	01	00	550
Culgoora, NSW, Aus.	Aus. Tel. Nat. Facility	30S	18	54	149E	33	42	217
Delaware, Ohio	Ohio State Obs.	40N	15	04·7	83W	02	54	282
Dwingeloo, Holland	Dwingeloo Radio Obs.	52N	48	48	06E	23	48	25
Effelsberg, Germany	Max Planck Inst. for Radio Astr.	50N	31	36	06E	53	06	369
Eschweiler, Germany	Stockert (Bonn Univ.)	50N	34	14·0	06E	43	24	435
Fort Irwin, Calif., USA	Goldstone Complex	35N	23	24	116W	50	54	1036
Gauribidanur, India	Gauribidanur Radio Obs.	13N	36	12	77E	26	06	686
Green Bank, W Virginia	Nat. Radio Astr. Obs.	38N	26	17·0	79W	50	12	823
Harestua, Norway	Univ. of Oslo Obs.	60N	12	30·0	10E	45	30	585
Hartebeeshoek, S Africa	Hartebeeshoek Radio Astr. Obs.	25S	53	24	27E	41	06	1391
Hoskinstown, NSW, Aus.	Molongo Radio Obs.	35S	22	18	149E	25	24	732
Jodrell Bank, England	Nuffield Rad. Astr. Lab.	53N	14	11·0	02W	18	24	70
Kitt Peak, Arizona	Nat. Radio Astr. Obs.	31N	57	11·0	111W	36	48	1920
Kunming, China	Yunnan Obs.	25N	01	30	102E	47	18	1940
Mauna Kea, Hawaii, USA	Caltech Submillimetre Obs.	19S	49	36	155W	28	18	4060
Mitaka, Japan	National Astr. Obs.	35N	40	18	139E	32	30	58
Nagoya, Japan	Nagoya Univ. Radio Astr. Lab.	35N	08	54	136E	58	24	75
Nançay, France	Paris Ons. Radio Astr. Stn	47N	22	48	02E	11	48	150
Nançay, France	Rad. Obs. of Nançay	47N	22	48·0	02E	11	48	150
Nederhorst den Berg, Holland	Rad. Astr. Obs.	52N	14	03·0	05E	04	36	0
Parkes, NSW, Australia	Australian Nat. Rad. Obs.	33S	00	00·4	148E	15	42	392
Pulkovo, Russia	Astron. Obs. Acad. Sciences	59N	46	05·5	30E	19	24	70
Richmond Hill, Ontario, Canada	David Dunlap Obs.	43N	51	44·0	79W	25	12	244
St Michel, France	Nat. Centre of Scientific Research	43N	55	00·0	05E	42	30	614
Socorro, N Mex., USA	National Radio Astr. Obs.	34N	04	42	107W	37	06	81
Sugar Grove, W Va., USA	Naval Research Lab. Radio Station	38N	31	12	79W	16	24	705
Tidbinbilla, Australia	Deep Space Station	35S	24	06	148E	58	48	656
Tokyo, Japan	Tokyo Obs. at Mitaka	35N	40	18·2	139E	32	24	70
Tremsdorf, Germany	Tremsdorf Radio Astr. Obs.	52N	17	06	13E	08	12	35
Uchinoura, Japan	Kagoshima Space Centre	31N	13	42	131E	04	00	228
Washington, DC, USA	Radio Astr. Obs. National Lab.	38N	49	16·6	77W	01	36	30
Westerbork, Holland	Westerbork Radio Astr. Obs.	52N	55	00	06E	36	18	16
Westford, Mass., USA	Haystack Obs.	42N	37	24	71W	29	18	146

in Japan; diameter 27 m; opened in 1987. The Osaka Planetarium has a 26·5 m dome, and the Moscow Planetarium a 25·15 m dome. *The first British planetarium* was at Madame Tussaud's, Baker Street, London; the dome is 21·33 m in diameter. The planetarium was opened in 1958. *The first Irish planetarium* was at Armagh, opened in 1968. *The most northerly planetarium* in the world is the Northern Lights planetarium at Tromsø in Norway, lat 69 deg 40 min.

'INVISIBLE' ASTRONOMY

Until comparatively modern times, astronomers were limited to studying radiations in the visible range of the electromagnetic spectrum. Then, in 1931, came the detection of radio waves from the Milky Way, and by now it is possible to examine almost all wavelengths, from long radio waves down to the very short gamma-rays.

Wavelengths are measured in nanometres or Ångströms – one nanometre being one thousand-millionth of a metre, and 1 Ångström being one ten thousand millionth of a metre, so that 1 mm = 10 Å. The accepted regions are as follows:

Below 0·01 nm	Gamma-rays
0·01–10 nm	X-rays
	Hard: 0·01–0·1
	Soft: 0·1–10
10–400 nm	Ultra-violet
	EUV (Extreme ultra-violet)
	10–120 nm
400–700 nm	Visible light
	(=4000–7000 Å)
700 nm–1 mm	Infra-red
0·3 nm–1 mm	Sub-millimetre
1 mm–0·3 m	Microwaves
Over 0·3 m	Radio waves

The World's Largest Telescopes

The following list contains telescopes which were in use at the beginning of 1995. The final list contains telescopes which are in an advanced stage of planning, though of course completion by the dates given here cannot be guaranteed!

✳ *The first really giant telescope* was built by William Herschel in 1789; it had a 49 in (124·5 cm) mirror. This was surpassed in 1845 by Lord Rosse's 72 in (189·9 cm), which was actually the largest in the world until 1917, when the Hooker reflector at Mount Wilson was completed.

TELESCOPES AND OBSERVATORIES

REFLECTORS

Telescope	Observatory	Aperture m	in	Latitude deg	min	Longitude deg	min	Elevation m	Completion
Keck I	W. M. Keck Observatory, Mauna Kea, Hawaii, USA	9·82	387	19	49N	155	28W	4150	1991
Mono-Mirror Tel. (MMT)	Mount Hopkins Observatory, Arizona, USA	6·50	256	31	41N	110	53W	2608	1995
Bolshoi Teleskop Azimutalnyi	Special Astrophysical Obs., Mt Pastukhov, Russia	6·00	236	43	39N	41	26E	2100	1975
Hale Telescope	Palomar Observatory, Palomar Mtn, California, USA	5·08	200	33	21N	116	52W	1706	1948
William Herschel Telescope	Obs. del Roque de los Muchachos, La Palma	4·20	165	28	46N	17	53W	2332	1987
4 m Telescope	Cerro Tololo Interamerican Observatory, Chile	4·00	158	30	10S	70	49W	2215	1976
Anglo-Australian Tel. (AAT)	Anglo-Australian Telescope, Siding Spring, Australia	3·89	153	31	17S	149	04E	1149	1975
Nicholas U. Mayall Reflector	Kitt Peak National Observatory, Arizona, USA	3·81	150	31	58N	111	36W	2120	1973
United Kingdom Infra-Red Tel. (UKIRT)	Joint Astronomy Centre, Mauna Kea, Hawaii, USA	3·80	150	19	50N	155	28W	4194	1978
Canada–France–Hawaii Tel. (CFH)	Dominion Astrophysical Obs., Mauna Kea, Hawaii, USA	3·58	141	19	49N	155	28W	4200	1979
3·6 m Telescope	European Southern Observatory, La Silla, Chile	3·57	141	29	16S	70	44W	2387	1977
3·5 m Telescope	Calar Alto Observatory, Calar Alto, Spain	3·50	138	37	13N	02	32W	2168	1984
New Technology Tel. (NTT)	European Southern Observatory, La Silla, Chile	3·50	138	29	16S	70	44W	2353	1989
Astrophys. Research Consortium (ARC)	Apache Point, New Mexico, USA	3·50	138	32	47N	105	49W	2800	1993
Wisconsin–Indiana–Yale–NOAO Telescope (WIYN)	WIYN Observatory, Kitt Peak, Arizona, USA	3·50	138	31	57N	111	36W	2089	1994
C. Donald Shane Telescope	Lick Observatory, Mount Hamilton, California, USA	3·05	120	37	21N	121	38W	1290	1959
NASA Infra-Red Telescope Facility (IRTF)	Mauna Kea Observatory, Hawaii, USA	3·00	118	19	50N	155	28W	4208	1979
107 inch Telescope	McDonald Observatory, Mount Locke, Texas, USA	2·72	107	30	40N	104	01W	2075	1969
UBC–Laval Telescope (LMT)	Univ. of Brit. Col. & Laval Univ., Vancouver, Canada	2·70	106	49	07N	122	35W	50	1992
Shajn 2·6 m Reflector	Crimean Astrophysical Observatory, Ukraine	2·64	104	44	44N	34	00E	550	1960
Byurakan 2·6 m Reflector	Byurakan Observatory, Mount Aragatz, Armenia	2·64	104	40	20N	44	18E	1500	1976
Nordic Optical Tel. (NOT)	Observatory del Roque de los Muchachos, La Palma	2·56	101	28	45N	17	53W	2382	1989
Isaac Newton Telescope (INT)	Observatory del Roque de los Muchachos, La Palma	2·56	101	28	46N	17	53W	2336	1984
Irénée du Pont Telescope	Las Campanas Observatory, Las Campanas, Chile	2·54	100	29	00S	70	42W	2282	1976
Hooker Telescope (100 in)	Mount Wilson Observatory, California, USA	2·50	100	34	13N	118	03W	1742	1917
Hubble Space Telescope (HST)	Space Telescope Science Inst., Baltimore, Md., USA	2·40	94	orbital				orbital	1990
Hiltner Telescope	Michigan–Dartmouth–MIT Obs., Kitt Peak, Ariz., USA	2·34	92	31	57N	111	37W	1938	1986
2·3 m Telescope	Vainu Bappu Observatory, Kavalur, Tamil Nadu, India	2·30	91	12	35N	78	50E	725	1985
Mount Stromlo 2·3 m Telescope	Mt Stromlo & Siding Spring Obs., Siding Spring, Aus.	2·30	91	31	16S	149	03E	1149	1984
Wyoming Infra-Red Telescope	Wyoming Infra-Red Obs., Jelm Mtn, Wyoming, USA	2·29	90	41	06N	105	59W	2943	1977

TELESCOPES IN PREPARATION

Telescope	Observatory	Aperture m	in	Latitude deg	min	Longitude deg	min	Elevation m	Completion
VLT (Very Large Telescope)	ESO, Cerro Paranal, Chile	16	630	24	51S	70	27W	1640	2000
LBT (Large Binocular Telescope)	Mount Graham, Arizona, USA	11·80	465	32	42N	109	51N	3170	1997
SST (Spectroscopic Survey Tel.)	McDonald Observatory, Texas, USA	11·00	433	30	40N	104	01W	2075	1996
Keck II	W. M. Keck Observatory, Mauna Kea, Hawaii, USA	9·82	387	19	49N	155	28W	4150	1996
Suburu Telescope	National Astr. Obs. (Japan), Mauna Kea, Hawaii, USA	8·30	327	19	49N	155	28W	4214	1999
Gemini Telescope North	Joint Astronomy Centre, Mauna Kea, Hawaii, USA	8·10	319	19	49N	155	28W	4100	1999
Gemini Telescope South	Cerro Tololo Interamerican Observatory, Chile	8·10	319	30	21S	70	49W	2725	2001
Magellan Project	Las Campanas Observatory, Las Campanas, Chile	6·50	256	29	00S	70	42W	2300	1998

REFRACTORS

Telescope	Observatory	Aperture m	in	Latitude deg	min	Longitude deg	min	Elevation m	Completion
Yerkes 40 in Telescope	Yerkes Observatory, Williams Bay, Wisconsin, USA	1·01	40	42	34N	88	33W	334	1897
36 in Refractor	Lick Observatory, Mount Hamilton, California, USA	0·89	36	37	20N	121	39W	1290	1888
33 in Meudon Refractor	Paris Observatory, Meudon, France	0·83	33	48	48N	02	14E	162	1889
Potsdam Refractor	Potsdam Observatory, Germany	0·80	31	52	23N	13	04E	107	1899
Thaw Refractor	Allegheny Observatory, Pittsburgh, Pennsylvania, USA	0·76	30	40	29N	80	01W	380	1985
Lunette Bischoffscheim	Nice Observatory, France	0·74	29	43	43N	07	18E	372	1886
Grosser Refraktor	Archenold Observatory, Treptow, Germany	0·68	27	52	29N	13	29E	41	1896
Grosser Refraktor	Vienna Observatory, Austria	0·67	26	48	14N	16	20E	241	1880
McCormick Refractor	Leander McCormick Obs., Charlottesville, Va., USA	0·67	26	38	02N	78	31W	264	1883
26 in Equatorial	US Naval Observatory, Washington, DC, USA	0·66	26	38	55N	77	04W	92	1873
Thompson Refractor	Royal Greenwich Observatory, England	0·66	26	51	29N	00	00	50	1897
Innes Telescope	Republic Observatory, Johannesburg, South Africa	0·66	26	26	10S	55	33E	1806	1926

SCHMIDT TELESCOPES

Telescope	Observatory	Aperture m	in	Latitude deg	min	Longitude deg	min	Elevation m	Completion
2 m Telescope	Karl Schwarzschild Obs., Tautenberg, Germany	1·34	53	50	59N	11	43E	331	1960
Oschin 48 in Telescope	Palomar Observatory, California, USA	1·24	49	33	21N	116	51W	1706	1948
United Kingdom Schmidt Telescope (UKS)	Royal Observatory Edinburgh, Siding Spring, Australia	1·24	49	31	16S	149	04E	1145	1973
Kiso Schmidt Telescope	Kiso Observatory, Kiso, Japan	1·05	41	35	48N	137	38E	1130	1975
3TA–10 Schmidt Telescope	Byurakan Astrophys. Obs., Mt Aragatz, Armenia	1·00	39	40	20N	44	30E	1450	1961
Kvistaberg Schmidt Telescope	Uppsala University Observatory, Kvistaberg, Sweden	1·00	39	59	30N	17	36E	33	1963
ESO 1 m Schmidt Telescope	European Southern Observatory, La Silla, Chile	1·00	39	29	15S	70	44W	2318	1972

Telescope	Observatory	Aperture		Latitude		Longitude		Elevation	Completion
		m	in	deg	min	deg	min	m	
Venezuela 1 m Schmidt Tel.	Centro F. J. Duarte, Merida, Venezuela	1·00	39	08	47N	70	52W	3610	1978
Télescope de Schmidt	Observatoire de Calern, Calern, France	0·90	35	43	45N	06	56W	1270	1981
Télescope Combiné de Schmidt	Royal Observatory, Brussels, Belgium	0·84	33	50	48N	04	21E	105	1958
Schmidt Telescope	Radiophysical Observatory, Riga, Latvia	0·80	31	56	47N	24	24E	75	1968
Calar-Alto-Schmidtspiegel	Calar Alto Observatory, Spain	0·80	31	37	13N	02	32W	2168	1980

COSMIC RAY ASTRONOMY

Cosmic rays were discovered in 1912 by the Austrian physicist V. Hess, who was anxious to find out why electrometers at ground level always recorded a certain amount of background. He therefore flew electrometers in a balloon, and this led to the discovery that cosmic rays are coming constantly from all directions in space.

They are not rays at all, but particles. Most cosmic-ray primaries are protons although there is, in fact, a tiny gamma-ray component, with many α-particles (helium nuclei) and some heavier atomic nuclei up to uranium and perhaps beyond.

About 3 per cent of the primaries are electrons. The primaries cannot pass directly through the Earth's atmosphere, but break up both themselves and the air-particles, so that only the secondary cosmic rays reach ground level.

The origin of cosmic rays is still very much of a problem. Some come from the Sun, most from the Galaxy, and some from beyond the Galaxy. Pulsars, supernova outbursts and other violent phenomena may well be involved. Unfortunately the particles are affected by the magnetic fields in the Galaxy, so that ordinarily it is not possible to determine the directions from which they originally come.

Cosmic rays of low energy are detected by special 'telescopes', mainly carried in satellites. When a cosmic ray particle passes through a gas, it may produce an electrical signal, or a flash of light from a particle of gas which has been struck. At the highest energies dense showers of particles are generated in the atmosphere.

GAMMA-RAY ASTRONOMY

The ultra-short, highly-penetrating gamma-rays are difficult to detect. Spark chambers carried in balloons and artificial satellites are used, though care must be taken to distinguish gamma-rays from the much more plentiful cosmic rays.

✴ *The first detection of high-energy gamma-rays from space* was achieved by the Explorer XI satellite in 1961.

✴ *The first gamma-ray bursts* were detected by the Vela satellite in 1967.

✴ *The first gamma-rays from the direction of the centre of the Galaxy* were recorded by OSO 3 (Orbiting Solar Observatory 3) in 1968.

✴ *The first discrete gamma-ray source identified* was in 1969 (in Sagittarius).

✴ *The first interplanetary gamma-ray detectors* were carried on Apollos 15 and 16; they were designed to study the background gamma-ray emission from the Milky Way.

✴ *The first gamma-ray satellites* were SAS 2 (1972) and COS B (1975). SAS 2 failed after six months but COS B operated until 1982.

✴ *The first gamma-ray map of the Milky Way* was made from SAS 2 (Small Astronomy Satellite 2) in 1972.

✴ *The first low-energy gamma-ray survey of the sky* was made in 1977 by HEAO-1 (High-Energy Astronomical Observatory 1).

✴ *Gamma-ray bursts* come from neither the Sun nor the Earth, and their origin is still obscure. Explanations range from flare stars to neutron stars. White Dwarfs, and even (less plausibly) comets hitting neutron stars! *The greatest gamma-ray burst so far recorded* was that of 5 March 1979, by the satellite ISEE-3 (International Sun-Earth Explorer 3). The burst seems to have come from a supernova remnant, L49 in the Large Magellanic Cloud. It was far too violent for a conventional thermonuclear outburst, and so its origin remains unclear.

✴ *The most advanced American gamma-ray observatory to date* is GRO, the Gamma-Ray Observatory (now named after Arthur Holly Compton), launched from the Shuttle in 1991. It weighs nearly 17 tonnes, and fills two-thirds of the Shuttle bay; the instruments weigh 6 tonnes. Its instruments are capable of detecting gamma-ray photons, measuring their energies, and determining the directions from which they come. Over 600 gamma-ray bursts coming in from all areas have been recorded, and it has been found that the Milky Way glows at gamma-ray wavelengths. *The first all-sky map of gamma-ray sources* was completed from the Compton satellite in 1993. A big surprise was the existence of many Active Galactic Nuclei (AGN) amongst the sources.

✴ *The most powerful gamma-ray source in the sky* is the Vela pulsar. The second most powerful source is the Crab pulsar, which was also the first source to be detected optically.

Very high-energy gamma-rays may be detected indirectly from observations made at ground level. In passing through the atmosphere they generate electrons which travel faster than the speed of light in air, resulting in blue Čerenkov radiation which may be detected with special equipment. *The largest gamma-ray telescope* of this kind is the 10-metre reflector at the Whipple Observatory, Mount Hopkins, Arizona. Intensive researches are being carried out by scientists at the University of Durham, where main equipment has been set up at Narrabri in Australia.

Gamma-rays come from objects of various kinds. Discrete sources include pulsars, such as those of Vela and the Crab, and the curious source known as Geminga, now known to be a neutron star.

Amongst the Active Galactic Nuclei (AGN) the quasar 3C-273 is prominent as a gamma-ray source. The 'isotropic diffuse flux' is probably mainly composed of AGN which are too far away to be identified individually. By far the main contributor to the gamma-ray flux at the Earth is our own Galaxy – the bulk of the gamma-rays coming from cosmic rays interacting with the gas in the interstellar medium.

Much progress has been made during the last few years, but it is fair to say that gamma-ray astronomy is still in a comparatively early stage of development.

X-RAY ASTRONOMY

The first observations of X-rays from the sky were made on 5 August 1948, when R. Burnright of the US Naval Research Laboratory detected solar X-rays from the darkening of a photographic emulsion carried on a V2 rocket. On 29 September 1948 H. Friedman and Hulbert, also using a V2, detected intense X-radiation from the Sun. In 1956 Friedman's team recorded results which could have been celestial X-rays, but they could not be sure, as they always had the Sun in their field of view.

In 1959 R. Giacconi and his colleagues published a paper in which they predicted X-rays from very hot stars and supernova remnants. They referred particularly to the Crab Nebula.

On 24 October 1961 a rocket was launched from White Sands to search for X-rays from the Moon. It was thought that these could be due to incident X-rays striking the lunar surface and causing X-ray fluorescence, together with X-rays due to the surface being struck by energetic electrons from the solar wind. The equipment failed, as the protective covers refused to open. A second rocket – an Aerobee – was launched on 18 June 1962. No lunar X-rays were found, but a discrete source was detected, later found to be Scorpius X-1. On 12 October 1962 a new launch recorded X-rays from the Crab Nebula, with an intensity 15 per cent of those from Scorpius X-1, and also obtained a good value for the position of Scorpius X-1. By 1966, 30 sources had been identified, including the first galaxy, M.87.

* *The first X-ray source to be optically identified* was the Crab Nebula (1962). Scorpius X-1 was optically identified in March 1966 by Japanese observers, using the Tokyo 178 cm reflector; the possibility had been suggested by Minoru Odo.

* *The first X-ray picture of the Sun* using a rocket-borne grazing incidence telescope was taken on 17 March 1965.

* *The first accurate positions of X-ray sources other than Scorpius X-1 and the Crab* were determined in October 1966. Among these was Cygnus X-3, made up of a supergiant star, HDE 226868, and a companion which is very probably a black hole.

* *The first X-ray satellites* were those of the OAO series, of which the most successful was OAO 3 (the Copernicus satellite).

* *The first purely X-ray satellite, Uhuru,* was launched from Kenya on 12 December 1970. It discovered 300 sources. In 1971 Uhuru results showed that Centaurus X-3 is a binary system; the intensity of the X-radiation varied rapidly in a period of 4·8 seconds, suggesting a neutron star, and a period of 2·087 days indicated a binary companion eclipsing the neutron star. The larger component was then optically identified. In 1971 a second binary, Hercules X-1, was identified.

* *The first X-ray nova,* Centaurus X-4, was discovered in May 1969 by an X-ray detector carried in a Vela satellite. An X-ray nova seen in December 1974, near the known source Centaurus A (though certainly unconnected with it) reached its maximum on Christmas Day, and was inevitably nicknamed CenXmas; another, in May 1975 close

to the Crab Nebula (again, unconnected with it) was equally inevitably nicknamed Fresh Crab. X-ray novæ seem to be binaries with low-mass optical companions.

* *The most successful British X-ray satellite to date* has been Ariel 5, launched in October 1974; it operated for five years.

* *The Einstein Observatory* (HEAO 2) was launched on 13 November 1978, and operated until 1980, observing 5 per cent of the entire sky. It carried the first large X-ray telescope, consisting of four nested paraboloids and four nested hyperboloids, with an outer diameter of 58 cm. X-rays were detected from sources of various kinds, including stars, galaxies, clusters of galaxies, and quasars (previously, only three X-ray quasars had been known).

* *The first X-ray satellite launched by the European Space Agency* was Exosat (26 May 1983). It has a $356 \times 19\,581$ km orbit, with an inclination of 72 degrees. This very eccentric polar orbit means that it remains in sunlight for most of the time, maintaining the sensors at stable temperatures.

* *The first real link between studies of the sky in X-radiation and in EUV (Extreme Ultra-Violet) radiation* was achieved by Rosat (the Röntgen satellite), named after Wilhelm Röntgen, the discoverer of X-rays. It was launched on 1 June 1990, carrying a German X-ray telescope and a British Wide-Field Camera. It was still operating well in 1995.

* *The first spatially-resolved X-ray image of the corona round a solar-type star* was obtained by Rosat in 1993; the star was the eclipsing binary Alphekka (Alpha Coronæ Borealis). As the A-type primary passed in front of the G-type secondary, Rosat detected small 'jumps' in the X-ray curve, showing that the X-ray emission was not evenly distributed over the surface of the G-star; almost half came from one active region. The coronal density appeared to be about twice that of the Sun's corona.

* *The most ambitious Japanese X-ray satellite, Yohkoh* ('Sunlight') was launched on 30 August 1991. It is 1 m square, 2 m high and weighs 400 kg. It is programmed mainly for studies of the Sun, and observations are combined with ground-based work in Europe, Japan, China and the United States. Special attention is paid to solar flares.

* *The first observation of a transit of Mercury at X-ray wavelengths* was made from Yohkoh on 6 November 1993, when the satellite recorded high-resolution images every 16 seconds during the transit. Since Mercury absorbs X-rays it appeared perfectly black; its size is well known, and so its passage across a features on the Sun's disk

enables these features to be accurately measured.

* *The first X-ray image of the Moon* was obtained by the 84 cm X-ray camera on Rosat in 1990. The Moon shows up to Rosat because lunar material fluoresces when irradiated by energetic emissions from the Sun.

ULTRA-VIOLET ASTRONOMY

*T*he existence of ultra-violet radiation from the Sun was demonstrated in 1801 by J. Ritter, by producing a spectrum with a prism and noting the darkening of paper soaked in sodium chloride held in the region beyond the violet. This was possible because ultra-violet radiation of between 300 and 400 nm can penetrate the Earth's atmosphere.

* *The first ultra-violet spectrograph* was launched on 28 June 1946, from White Sands (New Mexico), by a V2 rocket. It crashed, and the film was lost, but on 10 October 1946 a V2 soared to 88 km, and R. Tousey recorded the first recoverable solar ultra-violet spectra from above the atmosphere.

Ultra-violet observations were subsequently made from the American satellites of the OAO (Orbiting Astronomical Observatory) series. OAO 1 (8 April 1966) failed. OAO 2 (7 December 1968) operated until 13 February 1973; it had an orbit extending between 770 and 780 km above the ground. Its 11 ultra-violet telescopes viewed 1930 objects. Discoveries included a hydrogen cloud round a comet (Tago-Sato–Kosaka), magnetic fields of stars, and ultra-violet radiation from a supernova. OAO 3 (the Copernicus satellite) was launched on 21 August 1972 (orbit 748×740 km) and discovered many new sources.

The most durable ultra-violet satellite so far is the International Ultra-Violet Explorer (IUE), launched on 26 January 1978, carrying a 45-cm aperture telescope which can feed one of two spectrographs (one in the 190–320 nm range, the other 115–200 nm). Though its planned life expectancy was three years, it was still functioning in 1995. It has examined the ultra-violet spectra of all the principal bodies of the Solar System, including Halley's Comet and some asteroids; observations have been made of the interstellar medium, stars of all kinds and also external galaxies.

* *The first systematic surveys of the sky in EUV* (Extreme Ultra-Violet) were carried out by the Rosat satellite, launched on 1 June 1990. (Previously it had – wrongly –

been thought that no worthwhile observations in this part of the electromagnetic spectrum could be made, because of strong absorption by neutral hydrogen in the interstellar medium.) *The first object to be studied in EUV by Rosat* was the binary system WFC1, made up of a very hot white dwarf, HZ43, which has a temperature of 200 000 degrees and is one of the hottest stars known, together with a red dwarf. Rosat also obtained *the first EUV record of the Moon.*

✳ *The brightest star in the sky at EUV wavelengths* is ε Canis Majoris (Adhara). It is almost 500 light-years aways, and had the interstellar medium been as opaque to EUV as was originally thought, it would have meant that radiations from that range would have been blocked. However, it now seems that the Solar System lies in a relatively dense, cool part of the interstellar medium, and is surrounded by more rarefied material at a higher temperature – so that the atoms of hydrogen are ionized, and do not absorb EUV.

The Extreme Ultra-Violet Explorer satellite was launched on 7 June 1992, and within a year had completed a full map of the sky between 70 and 760 Ångströms. Over 600 sources were recorded, ranging from Solar System objects (notably Jupiter, Io and Mars), to cool stars with hot coronæ, white dwarfs sometimes with temperatures up to 100 000°C, and the photospheres and stellar winds of very hot stars. The nuclei of some active galaxies were also recorded, showing that in our own Galaxy there are 'tunnels' through which EUV can pass.

INFRA-RED ASTRONOMY

*T*he discovery of infra-red radiation from the Sun was made in 1801 by William Herschel, by placing a thermometer beyond the red end of the solar spectrum. It extends from a micron (0·001 mm) to several hundreds of microns, beyond which comes the microwave region.

✳ *The first major survey of the infra-red sky* was undertaken in the 1960s by G. Neugebauer and R. Leighton; they discovered 6000 discrete sources.

✳ *The world's largest infra-red telescope* is UKIRT (the United Kingdom Infra-Red Telescope) on Mauna Kea, above most of the atmospheric water vapour. The aperture is 380 cm. Superficially it looks like an ordinary reflector, and is in fact so good that it can be used at optical wavelengths as well as in infra-red.

✳ *The infra-red astronomical satellite IRAS* was planned in 1974, and launched on 25 January 1983; the reflector was of 60 cm aperture, enclosed in a cooling vessel containing 500 litres of liquid helium, holding the temperature not far above absolute zero (–273°C). IRAS operated for 300 days, and could have picked up the radiation from a speck of dust several kilometres away. There were eight staggered rows of detectors, and observations were made in four main bands (12, 25, 60 and 100 microns).

IRAS continued to function until 21 November 1983, when it ran out of coolant. By then it had surveyed 97 per cent of the sky, and had made many major discoveries, among them:

✳ *The first major infra-red catalogue of the sky*, between 8 and 120 microns.

✳ *The greatest number of discrete infra-red sources* (245 389), increasing the number of known sources by a factor of more than 100.

✳ *The first comet to be detected in infra-red* by IRAS was IRAS-Araki-Alcock, on 26 April 1983. It by-passed the Earth at only 5 000 000 km on 11 May, closer than any comet previously detected apart from Lexell's Comet of 1770.

✳ *The first cometary dust-tail detected in infra-red* was that of Tempel 2. The tail was 30 000 000 km long.

✳ *The first dust-ring in the Solar System* was found, taking the form of a torus. The distance is equal to that of the main asteroid belt, but is inclined by 10 degrees to the main plane of the Solar system.

✳ *The first asteroid to be discovered from IRAS* was 3200 Phæthon, on 11 October 1983, by Green.

✳ *The first studies of infra-red 'cirrus'* were made by IRAS. This produces faint emission from cool clouds of dust in the Galaxy, and when displayed on computerized maps shows up as wispy clouds resembling cirrus – hence the name. They are made up of tiny dust-grains, mainly graphite.

✳ *The first cloud of cool material round a star* (Vega) was found by Aumann and Gillett from IRAS results. Other similar infra-red excesses were subsequently discovered, notably round Fomalhaut and β Pictoris. These clouds may well indicate planet-forming material.

Infra-red studies can provide information which could not be obtained in any other way. For example, many star-forming regions are hidden at optical wavelengths. Active galaxies and quasars are also of special importance to infra-red astronomers. In the Solar System, too, infra-red work is of the utmost value. Fortunately, a considerable amount of infra-red observation can be carried out from the ground, though in some areas of research it is essential to use satellite-borne equipment.

MICROWAVE ASTRONOMY

*T*his part of the electromagnetic spectrum extends from 0·3 mm to 30 cm, longer than infra-red but shorter than radio radiation; sub-millimetre from 0·3 mm, millimetre from 1 mm.

The microwave background at 3 degrees above absolute zero was detected in 1964 by A. Penzias and R. Wilson at the Bell Telephone Laboratories in Holmdel, New Jersey, fortuitously when they were calibrating a 7·35 cm wavelength receiver built for satellite communications – at first the effect was thought to be due to pigeon droppings in the equipment!

It is assumed to be the last manifestation of the 'Big Bang' in which the universe was created; the overall temperature has now fallen to 2·7° above absolute zero, agreeing with theory. *The first slight irregularities in the microwave background* were detected in April 1992 with COBE, the Cosmic Background Explorer Satellite. There were 'ripples', or rarefied wisps of material, which are said to be the largest and most ancient structures known in the universe. This discovery was hailed as being among the most important of modern times, and confirmed that it was indeed possible for the initial material to collect into galaxies of the kind we know today.

✳ *The largest sub-millimetre telescopes* are the JCMT (James Clerk Maxwell Telescope) on Mauna Kea, the SEST (Swedish Sub-millimetre Telescope) at La Silla in Chile, and the CSO (California Institute of Technology Sub-millimetre Telescope) on Mauna Kea. The first two have 15 m segmented mirrors, and the CSO has a 10·4 m segmented mirror.

Conditions for sub-millimetre astronomy are particularly good in Antarctica, the plateau lies at almost 4300 m and the water vapour content of the atmosphere is low – advantageous for many branches of research. A sub-millimetre telescope at the South Pole will be in operation by the end of 1995.

RADIO ASTRONOMY

*T*he first detection of radio waves from the sky was due to the American radio engineer

K. Jansky in 1931. The discovery was fortuitous; he was investigating 'static' on behalf of the Bell Telephone Company with a homemade aerial nicknamed the 'Merry-go-Round' (part of the mounting was made from a dismantled Ford car). Jansky's first paper was published in 1932. *The first international radio telescope* was made by the American amateur Grote Reber, whose first paper appeared in 1940; Reber's telescope was a 9·5 metre 'dish'.

✳ *Radio waves from the Sun* were first detected in 1942 by a British team led by J. S. Hey; originally the effects were thought to be due to German jamming of British radar! In February 1946 it was found that the giant sunspot of that month was a strong radio source.

At Jodrell Bank, meteor work began in 1945, when radar was used to measure meteor trails. (*Radio echoes from the Moon* were also detected in 1945, independently by an American team led by J. H. de Witt and a Hungarian team led by Z. Bay.) The first Jodrell Bank radio telescope was a fixed 66 metre paraboloid, due to B. Lovell (now Sir Bernard Lovell).

✳ *The first discrete radio source detected beyond the Solar System* was Cygnus A, in 1946. In 1948 M. Ryle and F. G. Smith (now Sir Francis Graham-Smith) detected Cassiopeia A. *The first optical identifications of radio sources beyond the Solar System* were made in 1949; Taurus A (the Crab Nebula), Virgo A (the galaxy M.87) and Centaurus A (the galaxy NGC 5128).

The first planet found to be a radio source was Jupiter, in 1955, by two Americans, B. F. Burke and K. Franklin – again fortuitously. *The first quasar* was identified in 1963 because of its radio emissions, and the first pulsar by Jocelyn Bell-Burnell, at Cambridge, in 1967.

✳ *The oldest major radio telescope still operating* is the Dwingeloo telescope in Holland, completed in 1956. It is a 'dish' 25 m across. In 1994 it was used to discover a barred spiral galaxy, 10 000 000 light-years away, not far beyond the Local Group.

✳ *The Jodrell Bank 250 ft radio telescope* came into operation in 1957. In 1987, at a special ceremony, it was renamed The Lovell Telescope.

Originally it had been known as the Mark I, and from 1970, after structural modifications, as the Mark IA.

A second large telescope, the Mark II, was completed at Jodrell Bank in 1964. It is designed to operate at the shorter-centimetre wavelengths beyond the range of the Lovell Telescope itself. The bowl of the Mark II is elliptical, measuring 38·2 × 25·4 m; it can be used at wavelengths down to 3 cm.

✳ *The first long-baseline interferometry measurements* began at Jodrell Bank in 1960. The radio outputs of two (or more) widely separated telescopes can be combined to achieve great angular resolution. Today, a network of radio telescopes over England, ranging from Jodrell Bank to Cambridge, makes up the MERLIN (Multi-Element Radio Link Interferometer Network) network. The Lovell Telescope has also been used for VLBI (Very Long Baseline Interferometry) with other radio telescopes in Europe and in North America. Angles as small as 0·001 arc second can now be measured. This is the same as the angle subtended by a cricket ball at a distance of 16 000 km.

Today, *the largest steerable radio astronomy* 'dish' is at the Max Planck Institute, Bonn, West Germany; it was completed in 1971, and has a dish 100 metres in diameter. *The largest non-steerable dish* is at Arecibo, Puerto Rico; it is 304·8 metres in diameter, and was completed in 1963. *The largest array* is the VLA or Very Large Array, 80 km west of Socorro in New Mexico; it is Y-shaped, each arm being 20·9 km long, with 27 movable antennæ, each 25 metres across. It was completed in 1981.

✳ *The most extensive radio astronomy array* is the Australia Telescope, completed in 1988. It has six 22 m antennæ, used together with the Parkes 64 m dish and the dish at Coonabarabran. The six 22 m antennæ are at Narrabri in north-west New South Wales. The full telescope synthesizes a single 320 km instrument. At times it is further extended by using the 70 m dish at Tidbinbilla and the 24 m dish at Hobart in Tasmania.

Radio astronomy is a young science, but its importance in modern-day research cannot be over-estimated. It is strange now to recall that before the war, Grote Reber was the only intentional radio astronomer in the world!

Part of the VLA or Very Large Array of radio telescopes at Socorro, New Mexico, 1994.

THE HISTORY OF ASTRONOMY

To give every date of importance in the history of astronomy would be a mammoth undertaking. What I have therefore tried to do is to make a judicious selection, separating our purely space-research advances and discoveries.

It is impossible to say just when astronomy began, but even the earliest men capable of coherent thought must have paid attention to the various objects to be seen in the sky, so that it may be fair to say that astronomy is as old as *Homo sapiens*. Among the earliest peoples to make systematic studies of the stars were the Mesopotamians, the Egyptians and the Chinese, all of whom drew up constellation patterns. (There have also been suggestions that the constellations we use as a basis today were first worked out in Crete, but this is speculation only.) It seems that some constellation-systems date back to 3000 BC, probably earlier, but of course all dates in these very ancient times are uncertain.

The first essential among ancient civilizations was the compilation of a good calendar. Probably the first reasonably accurate value of the length of the year (365 days) was given by the Egyptians. (The first recorded monarch of all Egypt was Menes, who seems to have reigned around 3100 BC; he was eventually killed by a hippopotamus – possibly the only sovereign ever to have met with such a fate!) They paid great attention to the star Sirius (Sothis), because its 'heliacal rising', or date when it could first be seen in the dawn sky, gave a reliable clue to the time of the annual flooding of the Nile, upon which the Egyptian economy depended. The Pyramids are, of course, astronomically aligned, and arguments about the methods by which they were constructed still rage as fiercely as ever.

Obviously the Egyptians had no idea of the scale of the universe, and they believed the flat Earth to be all-important. So too did the Chinese, who also made observations. It has been maintained that a conjunction of the five naked-eye planets recorded during the reign of the Emperor Chuan Hsü refers to either 2449 or 2446 BC. There is also the legend of the Court Astronomers, Hsi and Ho, who were executed in 2136 BC (or, according to some authorities, 2159 BC) for their failure to predict a total solar eclipse; since the Chinese believed eclipses to be due to attacks on the Sun by a hungry dragon, this was clearly a matter of extreme importance! However, this legend is discounted by modern scholars.

The earliest data-collectors were the Assyrians; all students of ancient history know of the Library of Ashurbanipal (668–626 BC). This included the 'Venus Tablet', discovered by Sir Henry Layard and deciphered in 1911 by F. X. Kugler. It claims that when Venus appears, 'rains will be in the heavens'; when it returns after an absence of three months 'hostility will be in the land; the crops will prosper'. Early attempts at drawing up tables of the movements of the Moon and planets may well date from pre-Greek times, largely for astrological reasons; until relatively modern times astrology was regarded as a true science, and all the ancient astronomers (even Ptolemy) were also astrologers.

Babylonian astronomy continued well into Greek times, and some of the astronomers, such as Naburiannu (about 500 BC) and Kidinnu (about 380 BC) may have made great advances; but we know relatively little about them, and reliable dating begins only with the rise of Greek science.

Little progress was made in the following centuries, though there were some interesting Indian writings (Aryabhāta, 5th century AD), and in AD 570 Isidorus, Bishop of Seville, was the first to draw a definite distinction between astronomy and astrology. The revival of astronomy was due to the Arabs. In 813 Al-Ma'mūn founded the Baghdad school of astronomy, and various star catalogues were drawn up, the most notable being that of Al-Sūfī (born about 903). During this period two supernovæ were observed by Chinese astronomers; the star of 1006 (in Lupus) and 1054 (in Taurus, the remnant of which is today seen as the Crab Nebula).

The improved 'Alphonsine Tables' of planetary motions were published in 1270 by order of Alphonso X of Castile. In 1433 Ulūgh Beigh set up an elaborate observatory at Samarkand, but unfortunately he was a firm believer in astrology, and was told that his eldest son Abdallatif was destined to kill him. He therefore banished his son, who duly returned at the head of an army and had Ulūgh Beigh murdered. This marked the end of the Arab school of astronomy, and subsequent developments were mainly European. Some of the important dates in the history of astronomy are as follows:

1543 Publication of Copernicus' book *De Revolutionibus Orbium Cælestium*. This sparked off the 'Copernican revolution' which was not really complete until the publication of Newton's *Principia* in 1687.
1545–1559 First telescope built by Leonard Digges.
1572 Tycho Brahe observed the supernova in Cassiopeia.
1576–96 Tycho worked at Hven, drawing up the best star catalogue of pre-telescopic times.
1600 Giordano Bruno burned at the stake in Rome, partly because of his defence of the theory that the Earth revolves round the Sun.
1603 Publication of Johann Bayer's star catalogue, *Uranometria*.
1604 Appearance of the last supernova to be observed in our Galaxy (Kepler's Star, in Ophiuchus).
1608 Telescope built by H. Lippershey, in Holland.
1609 First telescopic lunar map, drawn by Thomas Harriot. **Serious** telescopic work begun by Galileo, who made a series of spectacular discoveries in 1609–10 (phases of Venus, satellites of Jupiter, stellar nature of the Milky Way). **Publication** of Kepler's first two Laws of Planetary Motion.
1618 Publication of Kepler's third Law of Planetary Motion.
1627 Publication by Kepler of improved planetary tables (the Rudolphine Tables).
1631 First transit of Mercury observed, by Gassendi (following Kepler's prediction of it).
1632 Publication of Galileo's *Dialogue*, which amounted to a defence of the Copernican system. In 1633 he was condemned by the Inquisition in Rome, and forced into a completely hollow recantation. **Founding** of the first official observatory (the tower observatory at Leiden in Holland).
1637 Founding of the first national observatory (Copenhagen, Denmark).
1638 Identification of the first variable star (Mira Ceti, by Phocylides Holwarda in Holland).
1639 First transit of Venus observed (by Horrocks and Crabtree, in England, following Horrocks' prediction of it).
1647 Publication of Hevelius' map of the Moon.
1651 Publication of Riccioli's map of the Moon, introducing the modern-type lunar nomenclature.
1655 Discovery of Saturn's main satellite, Titan, by C. Huygens, who announced the correct explanation of Saturn's ring system in the same year.
1656 Founding of the second Copenhagen Observatory.
1659 Markings on Mars seen for the first time (by Huygens).
1663 First description of the principle of the reflecting telescope, by the Scottish mathematician James Gregory.
1665 Newton's pioneer experiments on light and gravitation, carried out at Woolsthorpe in Lincolnshire while Cambridge University was temporarily closed because of the Plague.
1666 First observation of the Martian polar caps, by G. D. Cassini.
1667 Founding of the Paris Observatory, with Cassini as Director. (It was virtually in action by 1671.)
1668 First reflector made, by Newton. (This is the probable date. It was presented to the Royal Society in 1671, and still exists.)
1675 Founding of the Royal Greenwich Observatory. **Velocity** of light measured, by O. Rømer (Denmark).
1676 First serious attempt at cataloguing the southern stars, by Edmond Halley from St Helena.
1685 First astronomical observations made from South Africa (Father Guy Tachard, at the Cape).
1689 Publication of Newton's *Principia*, finally

proving the truth of the theory that the Sun is the centre of the Solar System.

1704 Publication of Newton's other major work, *Opticks*.

1705 Prediction of the return of a comet, by Halley (for 1758).

1723 Construction of the first really good reflecting telescope (a 6-in, by Hadley).

1725 Publication of the final version of the star catalogue by Flamsteed, drawn up at Greenwich. (Publication was posthumous.)

1728 Discovery of the aberration of light, by James Bradley.

1750 First extensive catalogue of the southern stars, by Lacaille at the Cape. (His observations extended from 1750 to 1752.) **Wright's** theory of the origin of the solar system.

1758 First observation of a comet at a predicted return (Halley's Comet, discovered on 25 Dec by Palitzsch. Perihelion occurred in 1759). **Principle** of the achromatic refractor discovered by Dollond. (Previously described, by Chester More Hall in 1729, but his basic theory was erroneous, and his discovery had been forgotten.)

1761 Discovery of the atmosphere of Venus, during the transit of that year, by M. V. Lomonosov in Russia.

1762 Completion of a new star catalogue by James Bradley; it contained the measured positions of 60 000 stars.

1767 Founding of the *Nautical Almanac*, by Nevil Maskelyne.

1769 Observations of the transit of Venus made from many stations all over the world, including Tahiti (the expedition commanded by James Cook).

1774 First recorded astronomical observation by William Herschel.

1779 Founding of Johann Schröter's private observatory at Lilienthal, near Bremen.

1781 Publication of Charles Messier's catalogue of clusters and nebulæ. **Discovery** of the planet Uranus, by William Herschel.

1783 First explanation of the variations of Algol, by Goodricke. (Algol's variability had been discovered by Montanari in 1669.)

1784 First Cepheid variable discovered; δ Cephei itself, by Goodricke.

1786 First reasonably correct description of the shape of the Galaxy given by William Herschel.

1789 Completion of Herschel's great reflector, with a mirror 49 in (124·5 cm) in diameter and a focal length of 40 ft (12·2 m).

1796 Publication of Laplace's 'Nebular Hypothesis' of the origin of the Solar System.

1799 Great Leonid meteor shower, observed by W. Humboldt.

1800 Infra-red radiation from the Sun detected by W. Herschel.

1801 First asteroid discovered (Ceres, by Piazzi at Palermo).

1802 Second asteroid discovered (Pallas, by Olbers). **Existence** of binary star systems established by W. Herschel. **Dark** lines in the solar spectrum observed by W. H. Wollaston.

1804 Third asteroid discovered (Juno, by Harding).

1807 Fourth asteroid discovered (Vesta, by Olbers).

1814–18 Founding of the Calton Hill Observatory, Edinburgh.

1815 Fraunhofer's first detailed map of the solar spectrum (324 lines).

1820 Foundation of the Royal Astronomical Society.

1821 Arrival of F. Fallows at the Cape, as Director of the first observatory in South Africa. **Founding** of the Paramatta Observatory by Sir Thomas Brisbane, Governor of New South Wales. (This was the first Australian observatory. It was dismantled in 1847.)

1822 First calculated return of a short-period comet (Encke's, recovered by Rümker at Paramatta).

1824 First telescope to be mounted equatorially, with clock drive (the Dorpat refractor, made by Fraunhofer).

1827 First calculation of the orbit of a binary star (ξ Ursæ Majoris, by Savary).

1829 Completion of the Royal Observatory at the Cape.

1834–8 First really exhaustive survey of the southern stars, carried out by John Herschel at Feldhausen (Cape).

1835 Second predicted return of Halley's Comet.

1837 Publication of the famous lunar map by Beer and Mädler. **Publication** of the first good catalogue of double stars (W. Struve's *Mensuræ Micrometricæ*).

1838 First announcement of the distance of a star (61 Cygni, by F. W. Bessel).

1839 Pulkovo Observatory completed.

1840 First attempt to photograph the Moon (by J. W. Draper).

1842 Important total solar eclipse, from which it was inferred that the corona and prominences are solar rather than lunar. First attempt to photograph totality (by Majocci), though he recorded only the partial phase.

1843 First daguerreotype of the solar spectrum obtained (by Draper).

1844 Founding of the Harvard College Observatory (first official observatory in the United States). The 15 in refractor was installed in 1847.

1845 Completion of Lord Rosse's 72 in reflector at Birr Castle, and the discovery with it of the spiral forms of galaxies ('spiral nebulæ'). **Daguerreotype** of the Sun taken by Fizeau and Foucault, in France. **Discovery** of the fifth asteroid (Astræa, by Hencke).

1846 Discovery of Neptune, by Galle and D'Arrest at Berlin, from the prediction by Le Verrier. The large satellite of Neptune (Triton) was discovered by W. Lassell in the same year.

1850 First photograph of a star (Vega, from Harvard College Observatory). Castor was also photographed, and the image was extended, though the two components were not shown separately. **Discovery** of Saturn's Crêpe Ring (Bond, at Harvard).

1851 First photograph of a total solar eclipse (by Berkowski). **Schwabe's** discovery of the solar cycle established by W. Humboldt.

1857 Clerk Maxwell proved that Saturn's rings must be composed of discrete particles. **Founding** of the Sydney Observatory. **First** good photograph of a double star (Mizar, with Alcor, by Bond, Whipple and Black).

1858 First photograph of a comet (Donati's, photographed by Usherwood).

1859 Explanation of the absorption lines in the solar spectrum given by Kirchhoff and Bunsen. Discovery of the Sun's differential rotation (by Carrington).

1860 Total solar eclipse. Final demonstration that the corona and prominences are solar rather than lunar.

1861–2 Publication of Kirchhoff's map of the solar spectrum.

1862 Construction of the first great refractors, including the Newall 25 in made by Cooke. (It was for many years at Cambridge, and is now in Athens.) **Discovery** of the Companion of Sirius (by Clark, at Washington). **Completion** of the *Bonner Durchmusterung*.

1863 Secchi's classification of stellar spectra published.

1864 Huggins' first results in his studies of stellar spectra. **First** spectroscopic examination of a comet (Tempel's, by Donati). First spectroscopic proof that 'nebulæ' are gaseous (by Huggins). **Found-** ing of the Melbourne Observatory. (The 'Great Melbourne Reflector' completed 1869.)

1866 Association between comets and meteors established (by G. V. Schiaparelli). **Great Leonid** meteor shower. **Announcement** by J. Schmidt of an alteration in the lunar crater Linné. (Though the reality of change is now discounted, regular lunar observation dates from this time.)

1867 Studies of 'Wolf-Rayet' stars by Wolf and Rayet, at Paris.

1868 First description of the method of observing the solar prominences at times of non-eclipse (independently by Janssen and Lockyer). **Publication** of a detailed map of the solar spectrum, by A. Ångström.

1870 First photograph of a solar prominence (by C. Young).

1872 First photograph of the spectrum of a star (Vega, by H. Draper).

1874 Transit of Venus; solar parallax redetermined. (Another transit occurred in 1882, but the overall results were disappointing.) **Founding** of observatories at Meudon (France) and Adelaide (Australia).

1876 First use of dry gelatine plates in stellar photography; spectrum of Vega photographed by Huggins.

1877 Discovery of the two satellites of Mars (by Hall, at Washington). Observations of the 'canals' of Mars (by Schiaparelli, at Milan).

1878 Publication of the elaborate lunar map by J. Schmidt (from Athens). **Completion** of the Potsdam Astrophysical Observatory.

1879 Founding of the Brisbane Observatory.

1880 First good photograph of a gaseous nebula (M.42, by Draper).

1882 Gill's classic photograph of the Great Comet of 1882, showing so many stars that the idea of stellar cataloguing by photography was born.

1885 Founding of the Tokyo Observatory. (An earlier naval observatory in Tokyo had been established in 1874.) **Supernova** in M.31, the Andromeda Galaxy (S Andromedæ). This was the only recorded extragalactic supernova to reach the fringe of naked-eye visibility until 1987.

1886 Photograph of M.31 (the Andromeda Galaxy) by Roberts, showing spiral structure. (A better photograph was obtained by him in 1888.)

1887 Completion of the Lick 36 in refractor.

1888 Publication of J. L. E. Dreyer's *New General Catalogue* of clusters and nebulæ. **Vogel's** first spectrographic measurements of the radial velocities of stars.

1889 Spectrum of M.31 photographed by J. Scheiner, from Potsdam. Discovery at Harvard of the first spectroscopic binaries (ζ Ursæ Majoris and β Aurigæ). First photographs of the Milky Way taken (by E. E. Barnard).

1890 Foundation of the British Astronomical Association. **Unsuccessful** attempts to detect radio waves from the Sun, by Edison. (Sir Oliver Lodge was equally unsuccessful in 1896.) **Publication** of the Draper Catalogue of stellar spectra.

1891 Completion of the Arequipa southern station of Harvard College Observatory. **Spectroheliograph** invented by G. E. Hale. **First** photographic discovery of an asteroid (by Max Wolf, from Heidelberg).

1892 First photographic discovery of a comet (by E. E. Barnard).

1893 Completion of the 28 in Greenwich refractor.

1894 Founding of the Lowell Observatory at Flagstaff, in Arizona.

1896 Publication of the first lunar photographic atlas (Lick). **Founding** of the Perth Observatory. **Completion** of the Meudon 33 in (83 cm) refractor. **Completion** of the new Royal Observatory at Blackford Hill, Edinburgh. **Discovery** of the predicted Companion to Procyon (by Schaeberle).

1897 Completion of the Yerkes Observatory.

1898 Discovery of the first asteroid to come well within the orbit of Mars (433 Eros, discovered by Witt at Berlin).

1899 Spectrum of the Andromeda Galaxy (M.31) photographed by Scheiner.

1900 Publication of Burnham's catalogue of 1290 double stars. **Horizontal** refractor, of 49 in aperture, focal length 197 ft (60 m), shown at the Paris Exhibition. (It was never used for astronomical research.)

1905 Founding of the Mount Wilson Observatory (California).

1908 Giant and dwarf stellar divisions described by E. Hertzsprung (Denmark). **Completion** of the Mount Wilson 60 in reflector. **Fall** of the Siberian meteorite.

1912 Studies of short-period variables in the Small Magellanic Cloud, by Miss H. Leavitt, leading on to the period-luminosity law of Cepheids.

1913 Founding of the Dominion Astrophysical Observatory, Victoria (British Columbia). **H. N. Russell's** theory of stellar evolution announced.

1915 W. S. Adams' studies of Sirius B, leading to the identification of White Dwarf stars.

1917 Completion of the 100 in Hooker reflector at Mount Wilson (the largest until 1948).

1918 Studies by H. Shapley leading him to the first accurate estimate of the size of the Galaxy.

1919 Publication of Barnard's catalogue of dark nebulæ.

1920 The Red Shifts in the spectra of galaxies announced by V. M. Slipher.

1923 Proof given (by E. Hubble) that the **galaxies** are true independent systems rather than parts of our Milky Way system. **Invention** of the spectrohelioscope, by Hale.

1925 Establishment of the Yale Observatory at Johannesburg. (It was finally dismantled in 1952, its work done.)

1927 Completion of the Boyden Observatory at Bloemfontein, South Africa.

1930 Discovery of Pluto, by Clyde Tombaugh at Flagstaff. **Invention** of the Schmidt camera, by Bernhard Schmidt (Estonia).

1931 First experiments by K. Jansky at Holmdel, New Jersey, with an improvised aerial, leading on to the founding of radio astronomy. Jansky published his first results in 1932, and in 1933 found that the radio emission definitely came from the Milky Way.

1932 Discovery of carbon dioxide in the atmosphere of Venus (by T. Dunham).

1933–5 Completion of the David Dunlap Observatory near Toronto (Canada).

1937 First intentional radio telescope built (by Grote Reber); it was a 'dish' 31 ft (9·4 m) in diameter.

1938 New (and correct) theory of stellar energy proposed by H. Bethe and, independently, by C. von Weizsäcker.

1942 Solar radio emission detected by M. H. Hey and his colleagues (27–28 Feb). The emission had previously been attributed to intentional jamming by the Germans!

1944 Suggestion, by H. C. van de Hulst, that interstellar hydrogen must emit radio waves at a wavelength of 21·2 cm.

1945 Thermal radiation from the Moon detected at radio wavelengths (by R. H. Dicke).

1945–6 First radar contact with the Moon, by Z. Bay (Hungary) and independently by the US Army Signal Corps Laboratory.

1946 Work at Jodrell Bank begun (radar reflections from the Giacobinid meteor trails, 10 Oct). **Beginning** of radio astronomy in Australia (solar work by a team led by E. G. Bowen). **Identification** of the radio source Cygnus A by Hey, Parsons and Phillips.

1947–8 Photoelectric observations of variable stars

in the infra-red carried out by Lenouvel, using a Lallemand electronic telescope.

1948 Completion of the 200 in Hale reflector at Palomar (USA). **Identification** of the radio source Cassiopeia A, by M. Ryle and F. G. Smith.

1949 Identification of further radio sources; Taurus A (the Crab Nebula), Virgo A (M.87), and Centaurus A (NGC 5128). These were the first radio sources beyond the Solar System to be identified with optical objects.

1950 M.31 (the Andromeda Galaxy) detected at radio wavelengths by M. Ryle, F. G. Smith and B. Elsmore. **Funds** for the building of the great Jodrell Bank radio telescope obtained by Sir Bernard Lovell.

1951 Discovery by H. Ewen and E. Purcell of the 21 cm emission from interstellar hydrogen thus confirming van de Hulst's prediction. **Optical** identification of Cygnus A and Cassiopeia A (by Baade and Minkowski, using the Palomar reflector, from the positions given by Smith).

1952 W. Baade's announcement of an error in the Cepheid luminosity scale, showing that the galaxies are about twice as remote as had been previously thought. **Electronic** images of Saturn and θ Orionis obtained by Lallemand and Duchesne (Paris). Tycho's supernova of 1572 identified at radio wavelengths by Hanbury Brown and Hazard.

1953 I. Shklovskii explains the radio emission from the Crab Nebula as being due to synchroton radiation.

1955 Completion of the 250 ft radio 'dish' at Jodrell Bank. First detection of radio emissions from Jupiter (by Burke and Franklin). **Construction** of a radio interferometer by M. Ryle, and also the completion of the 2nd Cambridge catalogue of radio sources. (The 3rd Cambridge catalogue was completed in 1959.)

1958 Observations of a red event in the lunar crater Alphonsus, by N. A. Kozyrev (Crimean Astrophysical Observatory, USSR). **Venus** detected at radio wavelengths (by Mayer).

1959 Radar contact with the Sun (Eshleman, at the Stanford Research Institute, USA).

1960 Aperture synthesis method developed by M. Ryle and A. Hewish.

1961 Completion of the Parkes radio telescope, 330 km west of Sydney.

1962 Thermal radio emission detected from Mercury, by Howard, Barrett and Haddock, using the 85 ft radio telescope at Michigan. First radar contact with Mercury (Kotelnikov, USSR). **First X-ray** source detected (in Scorpius). **Sugar Grove** fiasco; the US attempt to make a 600 ft fully steerable radio 'dish'. Work had been begun in 1959, and when discontinued had cost $96 000 000.

1963 Announcement, by P. van de Kamp, of a planet attending Barnard's Star. **Identification** of quasars (M. Schmidt, Palomar).

1965–6 Identification of the 3°K microwave radiation, as a result of theoretical work by Dicke and experiments by Penzias and Wilson.

1967 Completion of the 98 in Isaac Newton reflector at Herstmonceux. **Identification** of the first pulsar, CP 1919, by Miss Jocelyn Bell at Cambridge.

1968 Identification of the Vela pulsar (Large, Vaughan and Mills).

1969 First optical identification of a pulsar; the pulsar in the Crab Nebula by Cocke, Taylor and Disney at the Steward Observatory, USA.

1970 Completion of the 100 m radio 'dish' at Bonn (Germany). **Completion** of the large reflectors for Kitt Peak (Arizona) and Cerro Tololo (Chile); each 158 in (401 cm) aperture. First large reflector to be erected on Mauna Kea, Hawaii; an 88 in (224 cm).

1973 Opening of the Sutherland station of the South African Astronomical Observatories.

1974 Completion of the 153 in (389 cm) reflector at the Siding Spring Observatory, Australia.

1976 Completion of the 236 in (600 cm) reflector at Mount Semirodriki (USSR).

1977 Optical identification of the Vela pulsar (at Siding Spring). **Discovery** of Chiron (by C. Kowal, USA). **Discovery** of the rings of Uranus.

1978 Completion of the new Russian underground neutrino telescope. **Reported discovery** of a satellite of Pluto (J. Christy, USA). **Discovery** of the first satellite of an asteroid (Herculina). **Rings of Uranus** recorded from Earth (Matthews, Neugebauer, Nicholson). **Discovery** of X-rays from SS Cygni (HEAO 1).

1979 Official opening of the observatory at La Palma. **Pluto** and Charon recorded separately (Bonneau and Foy, Mauna Kea, thereby confirming Charon's independent existence). **First comet** observed to hit the Sun.

1980 Discovery of the first scintar (SS 433).

1981 Five asteroids contacted by radar from Arecibo, including two Apollos (Apollo itself, and Quetzalcoatl).

1982 Discovery of the remote quasar, PKS 2000–330 (Wright and Jauncey, Parkes). **Recovery** of Halley's Comet.

1983 Discovery of the fastest-vibrating pulsar, PKS 1937+215 in Vulpecula: period 1·557806449022 milliseconds – twenty times shorter than the Crab pulsar. It spins 642 times per second.

1984 Isaac Newton Telescope installed on La Palma.

1986 Return of Halley's Comet.

1987 Completion of the William Herschel telescope on La Palma. Completion of the James Clerk Maxwell telescope on Mauna Kea. Supernova seen in the Large Cloud of Magellan.

1988 **Completion of the Australia Telescope** (radio astronomy network). Collapse of the Green Bank radio telescope.

1989 NTT (**New Technology Telescope**) brought into action at the European Southern Observatory, La Silla. **Identification** of the 'Great Wall' of galaxies.

1990 **First brown dwarf** identified by M. Hawkins. **First surface details on a star** (Betelgeux) detected from La Palma Observatory. **First light** on the Keck Telescope (Mauna Kea). **White spot discovered on Saturn** (24 September). End of the Pluto-Charon mutual phenomena (24 September). **Sir Francis Graham-Smith** retires as Astronomer Royal.

1991 **Professor Arnold Wolfendale** appointed Astronomer Royal. **First really reliable measurement made** of the distance of the Large Magellanic Cloud. **Outburst of Halley's Comet** (12 February). Fall of the Glatton Meteorite (5 May). First space image obtained of an asteroid, Gaspra (13 November).

1992 **Completion** of the Keck I telescope on Mauna Kea, Hawaii. **Discovery** of the first Kuiper Belt object (Jewitt and Luu, 31 August).

1993 **Discovery** of the first asteroidal satellite, Dactyl (25 August). **Start** of SETI, the Search for Extra-terrestrial Intelligence (12 October). Galileo exonerated of heresy by the Pope (30 October), (Galileo had been condemned for heresy on 22 June 1633!)

1994 Hooker telescope on Mt Wilson reopened. SETI cancelled by US Congress (14 March). **Impact** of Comet Shoemaker-Levy 9 on Jupiter (17 to 22 July). **Discovery** of the nearest galaxy, 80 000 light-years away (Ibata, 4 August).

1995 January **Professor Sir Martin Rees** succeeds Sir Arnold Wolfendale as Astronomer Royal.

Astronomy and space research

It is no longer possible to separate what may be called 'pure' astronomy from space research. The Space Age began on 4 October 1957, with the launching of Russia's first artificial satellite, Sputnik 1; about a decade later Neil Armstrong and Edwin Aldrin stepped out on to the surface of the Moon, and there seems little doubt that elaborate space-stations will be in orbit during the 1990s. To talk of a lunar base before AD 2000 is no longer in the least far-fetched.

The following list has been restricted to the more 'astronomical' events. Full lists of all lunar and planetary probes are not given, because they will be found in the appropriate sections elsewhere in this book.

Pre-1957

c. 150 Lucian of Samosata's *True History* about a journey to the Moon – possibly the first of all science-fiction stories.

1232 Military rockets used in a battle between the Chinese and the Mongols.

1865 Publication of Jules Verne's novel *From the Earth to the Moon*.

1881 Early rocket design by N. I. Kibaltchitch. (Unwisely, he made the bomb used to kill the Czar of Russia, and was predictably executed.)

1891 Public lecture about space-flight by the eccentric German inventor Hermann Ganswindt.

1895 First scientific papers about space-flight by K. E. Tsiolkovskii. He published important papers in 1903 and subsequent years. The Russians refer to him as the 'father of space-flight'.

1919 Monograph, *A Method of Reaching Extreme Altitudes*, published by R. H. Goddard in America. This included a suggestion of sending a small vehicle to the Moon, and adverse Press comments made Goddard disinclined to expose himself to further ridicule.

1924 Publication of *The Rocket into Interplanetary Space*, by H. Oberth. This was the first truly scientific account of space-research techniques.

1926 First liquid-propelled rocket launched, by Goddard.

1927 Formation of the German rocket group, *Verein für Raumschiffahrt*.

1931 First European firing of a liquid-propelled rocket (Winkler, in Germany).

1937 First rocket tests at the Baltic research station at Peenemünde. One of the leaders of the team was Wernher von Braun, who died in 1977.

1942 First firing of the A4 rocket (better known as the V2) from Peenemünde. 1944–5 many V2s fell upon Southern England.

1945 White Sands proving ground established in New Mexico. **Idea of synchronous** artificial satellites for communications purposes proposed by Arthur C. Clarke.

1949 First step-rocket fired from White Sands; it reached an altitude of almost 400 km. **Rocket** testing ground established at Cape Canaveral, Florida.

1955 Announcement of the US 'Vanguard' project for launching artificial satellites.

The Space Age

1957 4 Oct; launching of the first artificial satellite, *Sputnik 1* (USSR).

1958 First successful US artificial satellite (*Explorer 1*). Instruments carried in it were responsible for the detection of the Van Allen radiation zones surrounding the Earth.

1959 First lunar probes; *Lunas 1, 2* and *3* (all USSR). *Luna 1* by-passed the Moon, *Luna 2* crash-landed there, and *Luna 3* went on a round trip, sending back pictures of the Moon's far side.

1960 First television weather satellite (*Tiros 1*, USA).

1961 First attempted Venus probe (USSR); contact with it lost. **First manned space-flight** (Yuri Gagarin, USSR). **First manned US** space-flight (A. Shepard; sub-orbital).

1962 First American to orbit the Earth (J. Glenn). **First** British-built satellite (*Ariel 1*, launched from Cape Canaveral). **First** transatlantic television pictures relayed by satellite (*Telstar*). **First** attempted Mars probe (USSR; contact lost). **First** successful planetary probe: *Mariner 2* to Venus.

1963 First occasion when two manned spacecraft were in orbit simultaneously (Nikolayev and Popovich, USSR). **First space-woman:** Valentina Tereshkova-Nikolayeva (USSR).

1964 First good close-range photographs of the Moon (*Ranger 7*, USA).

1965 First 'space-walk' (A. Leonov, USSR). **First** successful Mars probe (*Mariner 4*, USA).

1966 First soft landing on the Moon by an automatic probe (*Luna 9*, USSR). **First** landing of a probe on Venus (*Venera 3*, USSR), though contact with it was lost. **First** soft landing of an American probe on the Moon (*Surveyor 1*, USA). **First** circumlunar probe (*Luna 10*, USSR). **First** really good close-range lunar pictures (*Orbiter 1*, USA).

1967 First soft landing of an unmanned probe on Venus (*Venera 4*, USSR).

1968 First recovery of a circum-lunar probe (*Zond 5*, USSR). **First** manned Apollo orbital flight (*Apollo 7*; Schirra, Cunningham, Eisele, USA). **First** manned flight round the Moon; *Apollo 8* (Borman, Lovell, Anders, USA).

1969 First testing of the lunar module in orbit round the Moon (*Apollo 10*; Stafford, Cernan, Young, USA). **21 July,** first lunar landing (*Apollo 11*; N. Armstrong, E. Aldrin, USA).

1970 First Chinese and Japanese artificial satellites.

1971 First capsule landed on Mars, from the USSR probe *Mars 2*.

1971–2 First detailed pictures of Martian volcanoes, obtained from the probe *Mariner 9*, which entered orbit round the planet. (USA).

1972 End of the Apollo programme, with *Apollo 17* (Cernan, Schmitt, Evans, USA).

1973–4 Operational 'life' of the US space-station Skylab, manned by 3 successive three-man crews, and from which much pioneer astronomical work was carried out.

1973 First close-range information from Jupiter (including pictures) obtained from the fly-by probe *Pioneer 10*. (*Pioneer 11* repeated the experiments in 1974. Pioneer 10 was also the first probe to escape from the Solar System, while *Pioneer 11* was destined to be the first probe to bypass Saturn (in 1979). (USA).

1974 First pictures of the cloud-tops of Venus from close range, from the two-planet probe *Mariner 10*; the probe then encountered Mercury, and sent back the first pictures of the cratered surface.

1975 First pictures received from the surface of Venus, from the Russian probes *Venera 9* and *Venera 10*.

1976 First successful soft landings on Mars (*Vikings 1* and *2*) sending back direct pictures and information from the surface of the planet. (USA).

1977 Launching of *Voyagers 1* and *2* to the outer planets. (USA).

1978 Launch of two Pioneer probes to Venus – the first American attempts to put a vehicle into orbit round the planet and to land capsules there. **Launching** of the X-ray 'Einstein Observatory', which operated successfully for over 2 years.

1979 Fly-by of Jupiter by *Voyagers 1* and *2* (USA). **Decay** of *Skylab* in the Earth's atmosphere (11 July).

1980 *Voyager 1* fly-by of Saturn, obtaining data from Titan and other satellites during its pass.

1981 Successful *Voyager 2* pass of Saturn. (USA).

1982 Landings of *Veneras 13* and *14* on Venus, obtaining improved pictures and data (USSR). **Longest** space-mission undertaken by the Russians (2 months in orbit; A. Berezevoy and V. Lebedev in *Salyut 7*).

1983 Launch of IRAS (Infra-Red Astronomical Satellite).

1986 *Voyager 2* fly-by of Uranus. Shuttle disaster, with the destruction of the *Challenger*. Five probes to Halley's Comet; two Japanese, two Russian, and one European (Giotto).

1987 Longest space mission (326 days) completed by Yuri Romanenko, on the space-station Mir.

1988 **Longest** space mission (366 days on the Mir station) completed by V. Titov and M. Manorov. **Failure** of the *Phobos 1* probe (contact lost, 29 August).

1989 **Failure** of the *Phobos 2* probe (contact lost, 29 March). *Magellan* probe to Venus launched from the Shuttle *Atlantis* (5 May). *Hipparcos* astrometric satellite launched (8 August). *Voyager 2* passes Neptune (24 August). *Galileo* probe launched to Jupiter (18 October).

1990 **First Japanese launch to the Moon** (Muses-A, 24 January). **Hubble Space Telescope** launched from the Shuttle *Discovery* (25 April). **Launch** of Ulysses (6 October).

1991 **Gamma-Ray Observatory** (Arthur Holly Compton Observatory) launched, 5 April. **First** survey of the sky in Extreme Ultra-Violet (Rosat satellite). Spectacular images of *Venus* sent back by the Magellan probe. **Launch** of Yohkoh (30 August).

1992 Ulysses (solar polar probe) flew past Jupiter (8 February). Giotto encountered comet Grigg-Skjellerup (10 July). Discovery in April of slight variations in the microwave background radiation (COBE). First space-walk by 3 astronauts simultaneously (13 May to repair of Intelsat-6 satellite). Mars Observer launched (25 September). Pioneer Venus ceased transmitting (9 October).

1993 First gravitational wave test using satellites (21 March–12 April). Japanese probe Hiten impacts on the Moon (10 April), Hipparcos completes an all-sky survey at extreme ultra-violet wavelengths (18 November), repair to the Hubble Space Telescope by astronauts (2–13 December). COBE switched off (23 December).

1994 Launch of lunar probe Clementine (24 January); it entered lunar orbit in February, but failed to complete its programme by going on to a rendezvous with the asteroid Geographos. Mars Observer lost (2 August). Decay of Magellan probe in Venus' atmosphere (10 October).

1995 Record for the longest time spent in space broken by Dr Valeri Poliakov (Russia). He was launched from Baikonur on 8 January 1994, and by 9 January 1995 had been in space for a year – breaking the previous record by Musa Manarov and Vladimir Titov, also of Russia (365 days). In 1988–9 Dr Poliakov had spent 242 days in space.

ASTRONOMERS

Selecting a limited number of astronomers for short biographical notes may be somewhat invidious. However, the list given here includes most of the great pioneers and researchers. No astronomers still living at the time of writing are included. All dates are AD unless otherwise stated.

Abul Wafa, Mohammed. 959–88. Last of the famous Baghdad school of astronomers. He wrote a book called *Almagest*, a summary of Ptolemy's great work also called the *Almagest*, in Arabic.

Adams, John Couch, 1819–92. English astronomer, *b.* Lidcot, Cornwall. He graduated brilliantly from Cambridge in 1843, but had already formulated a plan to search for a new planet by studying the perturbations of Uranus. By 1845 his results were ready, but no quick search was made, and the actual discovery was due to calculations by U. Le Verrier. Later he became Director of the Cambridge Observatory, and worked upon lunar acceleration, the orbit of the Leonid meteor shower, and upon various other investigations.

Airy, George Biddell, 1801–92. English astronomer. Born in Northumberland, he graduated from Cambridge 1823; and was Professor of Astronomy there 1826–35. On becoming Astronomer Royal (1835–81) he totally reorganized Greenwich Observatory and raised it to its present eminence. He re-equipped the Observatory and ensured that the best use was made of its instruments; it is ironical that he is probably best remembered for his failure to instigate a prompt search for Neptune when receiving Adams' calculations.

Aitken, Robert Grant, 1864–1951. American astronomer, *b.* Jackson, Cal. In 1895 he joined the staff of Lick Observatory, and specialized in double star work; he discovered 31 000 new pairs, and wrote a standard book on the subject. From 1930 until his retirement in 1935 he was Director of the Lick Observatory.

Albategnius. *c.* 850–929. This is the Latinized form of the name of the Arab prince Al Battani. *b.* Batan, Mesopotamia; he drew up improved tables of the Sun and Moon, and found a more accurate value for the precession of the equinoxes. His *Movements of the Stars* enabled Hevelius, in the 17th century, to discover the secular variation in the Moon's motion. Albategnius was also a pioneer mathematician; in trigonometry, he introduced the use of sines.

Alfonso X. 1223–84. King of Castile. At Toledo he assembled many of the leading astronomers of the world, and drew up the famous Alphonsine Tables, which remained the standard for three centuries.

Alhazen (Abu Ali al Hassan). 987–1038. Arab mathematician, *b.* Basra. He went to Cairo, where he made his observations and also wrote the first important book on optics since the time of Ptolemy.

Allen, Clabon. 1905–88. Australian astronomer who specialized in studies of the Sun. For 20 years he lived in London, and was Director of the University of London Observatory. His book *Astronomical Quantities* (1955) is a classic.

Allen, David Anthony. 1946–94. British astronomer who graduated from Cambridge and spent much of his career at the Anglo-Australian Telescope in New South Wales. He was a specialist in infra-red research, and made many contributions; he was also a particularly good writer of popular astronomy.

Al-Ma'mūn, Abdalla. ?–833. Often refered to as Almanon. He was Caliph of Baghdad, son of Harun al Raschid; he collected and translated many Greek and Persian works, and built a major observatory in 829.

Al-Sūfī. 903–86. A Persian nobleman, who compiled an invaluable catalogue of 1018 stars, giving their approximate positions, magnitudes and colours.

Anaxagoras. 500–428 BC. *b.* Clazomenæ, Ionia. In Athens he became a friend of Pericles, and it was because of this friendship that he was merely banished, rather than being condemned to death, for teaching that the Moon contains plains, valleys and mountains, while the Sun is a blazing stone larger than the Peloponnesus (the peninsula upon which Athens stands).

Anaximander. *c.* 611–547 BC. Greek philosopher, *b.* Miletus. He believed the Earth to be a cylinder, suspended freely in the centre of a spherical universe. He attempted to draw up a map of the world, and introduced the gnomon into Greece.

Anaximenes. *c.* 585–525 BC. Greek philosopher, *b.* Miletus. He believed the Sun to be hot because of its quick motion round the Earth, that the stars were too remote to send us detectable heat, and the stars were fastened on to a crystal sphere.

Ångström, Anders. 1814–74. Swedish physicist, who graduated from Uppsala. He mapped the solar spectrum, and was the first to examine the spectra of aurorae. The Ångström unit (100-millionth part of a centimetre) is named in his honour.

Antoniadi, Eugenios. 1870–1944. Greek astronomer, who spent most of his life in France and became a naturalized Frenchman. He worked mainly at the Juvisy Observatory (with Camille Flammarion) and at Meudon, nr Paris, where he used the 83 cm refractor to make classic observations of the planets. Before the Space Age, his maps of Mars and Mercury were regarded as the standard works. He died in Occupied France during World War II.

Apian, Peter Bienewitz, 1495–1552. *b.* Leisnig, Saxony. Became professor of mathematics at Ingolstädt. He observed five comets, and was the first to note that their tails always point away from the Sun. The 1531 comet is known to be Halley's, and his observations of it enabled Edmond Halley to identify it with the comets of 1607 and 1682.

Apollonius. *c.* 250–200 BC. *b.* Perga, Asia Minor, but lived in Alexandria. An expert mathematician, he was one of the first to develop the theory of epicycles to represent the movement of the Sun, Moon and planets.

Arago, François Jean Dominique. 1786–1853. Director of the Paris Observatory from 1830. He made many important contributions, including a recognition of the importance of photography in astronomy. He made an exhaustive study of the great total solar eclipse of 1842, and maintained (correctly!) that the Sun is wholly gaseous.

Argelander, Friedrich Wilhelm August. 1799–1875. German astronomer, who became Director of the Bonn Observatory 1836. Here he drew up his atlas of the northern heavens (the *Bonn Durchmusterung*), containing the positions of 324 198 stars down to the ninth magnitude. This standard work was published in 1863.

Aristarchus. *c.* 310–250 BC. Greek astronomer, *b.* Samos. One of the first (quite possibly the very first) to maintain that the Earth moves round the Sun, and he also tried to measure the relative distances of the Sun and Moon by a method which was sound in theory, though inaccurate in practice.

Aristotle (384–322 BC) believed in a finite, spherical universe. He further developed the theory of concentric spheres, and gave the first practical proofs that the Earth cannot be flat.

Baade, Wilhelm Heinrich Walter. 1893–1959. German astronomer. In 1920, while assistant at Hamburg Observatory, he discovered the unique asteroid 944 Hidalgo. In 1931 he went to America, and joined the staff of Mount Wilson. In 1952 his work upon the two classes of 'Cepheid' short-period variables enabled him to show that the galaxies are approximately twice as remote as had previously been thought.

Bailey, Solon Irving. 1854–1931. American astronomer, *b.* New Hampshire. He joined the Harvard staff in 1879, and was for many years in charge of the Harvard southern station at Arequipa, Peru. His studies of globular clusters led to the discovery of 'cluster variables', now known as RR Lyræ stars.

Bappu, Manali Kallat Vainu. 1927–82. One of the most distinguished of Indian astronomers; he was Director of the Kodaikanal Observatory, and modernized it. He was a specialist in stellar spectroscopy.

Barnard, Edward Emerson. 1857–1923. American astronomer, *b.* Nashville, Tennessee. He was self-taught, but joined the staff at Lick Observatory in 1888, moving to Yerkes in 1897. He was a renowned comet-hunter; he discovered the fifth satellite of Jupiter (Amalthea) and the swift-moving star in Ophiuchus now called Barnard's Star. He also specialized in studies of dark nebulæ.

Barrow, Isaac. 1630–77. English mathematician. He made various important contributions to science, but is perhaps best known because in 1669 he resigned his post as Lucasian Professor at Cambridge so that his pupil Isaac Newton could succeed him.

Bayer, Johann. 1572–1625. German astronomer; a lawyer by profession and an amateur in science. He is remembered for his 1603 star catalogue, in which he introduced the system of allotting Greek letters to the stars in each constellation – the system still in use today.

Beer, Wilhelm. 1797–1850. A Berlin banker, who set up a private observatory and collaborated with Mädler in the great map of the Moon published in 1837–8. This map remained the standard for many years. Beer was the brother of Meyerbeer, the famous composer.

Belopolsky, Aristarch. 1854–1934. Russian astronomer, who went from Moscow to the Pulkova Observatory in 1888. He became Director of the Observatory in 1916, but resigned in 1918. He specialized in spectroscopic astronomy and in studies of variable stars.

Bessel, Friedrich Wilhelm, 1784–1846. German astronomer. He went to Lilienthal as assistant to Schröter, but in 1810 became Director of the Königsberg Observatory, retaining the post until his death. He determined the position of 75 000 stars by reducing Bradley's observations; was the first to obtain a parallax value for a star (61 Cygni, in 1838), and predicted the positions of the then-unknown companions of Sirius and Procyon.

Biela, Wilhelm von. 1782–1856. Austrian army officer, and amateur astronomer, who is remembered for his discovery (in 1826) of the now-defunct periodical comet which bears his name.

Bode, Johann Elert. 1747–1826. German astronomer. *b.* Hamburg. Appointed director of Berlin Observatory 1772. In the same year he drew attention to the 'law' of planetary distances which had been discovered by Titius of Wittenberg; rather unfairly, perhaps, this is known as Bode's Law. He published a star catalogue, did much to popularize astronomy, and for 50 years edited the *Berlin Astronomisches Jahrbuch*.

Bolton, John. 1922–93. Bolton was British, but spent most of his career in Australia. He was one of the great pioneers of radio astronomy; from 1961 to 1971 he was Director of the Australian National Radio Astronomy Laboratory, which controls the 64 mm dish at Parkes.

Bond, George Phillips. 1825–65. Son of W. C. Bond. *b.* Massachusetts. In 1859 succeeded his father as Director of the Harvard Observatory. He was a pioneer of planetary and cometary photography, and was the first to assert upon truly scientific principles that Saturn's rings could not be solid.

Bond, William Cranch. 1789–1859. American astronomer, *b.* Maine. He began his career as a watchmaker, but his fame as an amateur astronomer led to his appointment as Director of the newly-founded Harvard Observatory. In 1848 he discovered Saturn's satellite Hyperion, and in 1850 he discovered Saturn's Crêpe Ring. He was also a pioneer of astronomical photography.

Bouvard, Alexis. 1767–1843. A shepherd boy, born in a hut at Chamonix. He went to Paris, taught himself mathematics, and was appointed assistant to Laplace. He made contributions to lunar theory and drew up tables of the motions of the outer planets, as well as discovering several comets.

Bradley, James. 1692–1762. English astronomer (Astronomer Royal, 1742–62). He was educated in Gloucestershire, and entered the Ministry, becoming Vicar of Bridstow in 1719; in 1721 he went to Oxford as Professor of Astronomy, and remained there until his appointment to Greenwich, mainly on the recommendation of his close friend Halley. He discovered the aberration of light and the nutation of the Earth's axis, but his greatest work was his catalogue of the positions of 60 000 stars.

Brorsen, Theodor. 1819–95. Danish astronomer, who discovered several comets, and in 1854 made the first scientific observations of the Gegenschein.

Brown, Ernest William. 1866–1938. English astronomer, who graduated from Cambridge and then went to USA. His chief work was on lunar theory, and his tables of the Moon's motion are still recognized as the standard.

Burnham, Sherburne Wesley. 1838–1921. American astronomer, who began as an amateur and then went successively to Lick (1888) and Yerkes (1897). He specialized in double star work, and discovered over 1300 new pairs. His *General Catalogue of Double Stars* remains a standard reference work.

Campbell, William Wallace, 1862–1938. American astronomer, *b.* Ohio. He joined the staff at Lick Observatory in 1891, and was Director from 1900 until his retirement in 1930. His main work was in spectroscopy; he discovered 339 spectroscopic binaries (among them Capella), and determined the radial velocities of stars and of 125 nebulæ as well as carrying out spectroscopic observations of the planets.

Cannon, Annie Jump. 1863–1941. Outstanding American woman astronomer, *b.* Delaware. In 1896 she joined the staff at Harvard College Observatory, where she worked unceasingly on the classification of stellar spectra; the present system is due largely to her. She also discovered five novæ and over 300 variable stars. From 1938 she was William Cranch Bond Astronomer.

Carrington, Richard Christopher. 1826–75. English amateur astronomer, who had his Observatory at Redhill, Surrey. He concentrated upon the Sun, and made many contributions, including the first observation of a solar flare and the independent discovery of Spörer's Law concerning the distribution of sunspots throughout a cycle.

Cassini, Giovanni Domenico. 1625–1712. Italian astronomer; Professor of Astronomy, Bologna 1650–69, when he went to Paris as the first Director of the Observatory there. He discovered 4 of Saturn's satellites as well as the main division in the rings; he drew up new tables of Jupiter's satellites, made pioneer observations of Mars, and made the first reasonably good measurement of the distance of the Sun.

Cassini, Jacques J. 1677–1756. Son of G. D. Cassini; *b.* Paris. He succeeded his father as Director of the Paris Observatory. He confirmed Halley's discovery of the proper motions of certain stars, and played an important part in measuring an arc of meridian from Dunkirk to the Pyrenees in order to determine the figure of the Earth.

Challis, James. 1803–62. English astronomer; Professor of Astronomy at Cambridge from 1836. He accomplished much useful work, but, unfortunately, is remembered as the man who failed to discover Neptune before the success by Galle and D'Arrest at Berlin.

Charlier, Carl Vilhelm Ludwig. 1862–1934. Swedish cosmologist; Professor of Astronomy at Lund from 1897. He accomplished outstanding work with regard to the distribution of stars in our Galaxy.

Christie, William Henry Mahoney. 1845–1922. English astronomer (Astronomer Royal 1881–1910). He modernized Greenwich Observatory, and fully maintained its great reputation, achieved under Airy.

Clairaut, Alexis Claude. 1713–65. French mathematical genius, who published his first important paper at the age of 12. He studied the motion of the Moon, and worked out the perihelion passage of Halley's Comet in 1759 to within a month of the actual date.

Clavius, Christopher Klau. 1537–1612. German Jesuit mathematical teacher, who laid down the calendar reform of 1582 at the request of Pope Gregory.

Copernicus, Nicolaus. 1473–1543. The Latinized name of Mikołaj Kopernik, *b.* Toruń, Poland. He entered the Church, and became Canon of Frombork. He had a varied career, including medicine and also the defence of his country against the Teutonic Knights, but is remembered for his great book *De Revolutionibus Orbium Cœlestium*, finally published during the last days of his life (he had previously withheld it because he was well aware of Church opposition). It was this book which revived the heliocentric theory according to which the Earth moves round the Sun, and sparked off the 'Copernican revolution' which came to its end with the work of Newton more than a century later.

Curtis, Heber Doust. 1872–1942. American astronomer, who worked at Lick, Allegheny and Michigan Observatories. He was an outstanding spectroscopist, and in 1920 took part in the 'Great Debate' with Shapley about the size of the Galaxy and the status of the resolvable nebulæ; Curtis was wrong about the size of the Galaxy, but correct in maintaining that the spiral nebulæ were independent galaxies. He also played a major rôle in the establishment of the McMath-Hulbert Observatory, renowned for its solar research.

D'Arrest, Heinrich Ludwig. 1822–75. German astronomer, *b.* Berlin. While Assistant at the Berlin Observatory, he joined Galle in the successful search for Neptune. From 1857 he worked at Copenhagen Observatory. He specialized in comet and asteroid work, and also published improved positions for about 2000 nebulæ.

Darwin, George Howard. 1845–1912. Son of Charles Darwin. From 1883 he was Profesor of Astronomy at Cambridge, and drew up his famous though now rejected tidal theory of the origin of the planets. He was knighted in 1906.

Dawes, William Rutter. 1799–1868. English clergyman, and a keen-eyed amateur observer who specialized in observations of the Sun, planets and double stars. He discovered Saturn's Crêpe Ring independently of Bond.

Delambre, Jean-Baptiste Joseph. 1749–1822. French astronomer, best remembered for his work on the history of the science but also a skilled computer of planetary tables.

De la Rue, Warren. 1815–89. English astronomer, born in Guernsey. He was a pioneer of astronomical photography; in 1852 he obtained the first good photographs of the Moon, and in 1857 of the Sun. His photographs of the total solar eclipse of 1860 finally proved that the prominences are solar rather than lunar.

Delaunay, Charles. 1816–72. French astronomer, who specialized in studies of the Moon's motion. He became Director of the Paris Observatory in 1870, but was drowned in a boating accident two years later.

Democritus. *c.* 460–360 BC. Greek philosopher, *b.* Abdera, Thrace. He adopted Leucippus' atomic theory, and was the first to claim that the Milky Way is made up of stars.

Descartes, René. 1596–1650. French astronomer, author of the theory that matter originates as vortices in an all-pervading ether. He also made great improvements in optics. His books were published in Holland, but he died in Sweden.

De Sitter, Willem. 1872–1934. Dutch astronomer and cosmologst, *b.* Friesland; he went to the Cape, and from 1908 was Professor of Astronomy at Leiden. He studied the motions of Jupiter's satellites and also the rotation of the Sun, but is best remembered for his pioneer work in relativity theory. The 'De Sitter universe', finite but unbounded, was calculated to be 2000 million light-years in radius and to contain 80 000 million galaxies.

Deslandres, Henri Alexander. 1853–1948. French astronomer (originally an Army officer); from 1907 Director of the Meudon Observatory, and from 1927 Director of the Paris Observatory also. He was a pioneer spectroscopist, and developed the spectroheliograph independently of Hale.

Dollond, John. 1706–61. English optician, who reinvented the achromatic lens in 1758 and thus improved refractors beyond all recognition.

Donati, Giovanni Battista. 1826–73. Italian astronomer who discovered the great comet of 1858, and was the first to obtain the spectrum of a comet (Tempel's of 1864). From 1859 Director of the observatory at Florence, in 1872 he was largely responsible for the creation of the now-celebrated observatory at Arcetri.

Dreyer, John Louis Emil. 1852–1926. Danish astronomer, *b.* Copenhagen, who went to Ireland as astronomer to Lord Rosse at Birr Castle and became Director of the Armagh Observatory in 1882. He was a great astronomical historian, but is best remembered for his *New General Catalogue* of Clusters and Nebulæ (the N.G.C.) still regarded as a standard work. In 1916 he retired from Armagh and went to Oxford, where he lived for the rest of his life.

Dyson, Frank Watson. 1868–1939. English astronomer (Astronomer Royal, 1910–33). A great administrator as well as an energetic observer of eclipses; he also carried out important work in the field of astrophysics and stellar motions.

Eddington, Arthur Stanley. 1882–1945. English astronomer, *b.* Kendal. After working at Cambridge and Greenwich, he was appointed Professor of Astronomy at Cambridge, 1913. He was a pioneer of the theory of the evolution and constitution of the stars, and an outstanding relativist; in 1919 he confirmed Einstein's prediction of the displacement of star positions near the eclipsed Sun. He was knighted in 1930. In addition to his outstanding work, Eddington was one of the best of all writers of popular scientific books, and was a splendid broadcaster.

Einstein, Albert. 1879–1955. German Jew, whose name will be remembered as long as Newton's; in 1905 he laid down the Special Theory of Relativity, and from 1915–17 he developed the General Theory. In 1933 he left Germany, fearing persecution of the Jews, and settled in USA.

Elger, Thomas Gwyn. 1838–97. English amateur astronomer; he was first Director of the Lunar Section of the British Astronomical Association, and in 1895 published an excellent outline map of the Moon.

Empedocles of Agrigentum (c. 490–450 BC) believed the Sun to be a reflection of fire, but is credited with being the first to maintain that light has a finite velocity.

Encke, Johann Franz. 1791–1865. German astronomer; from 1825, Director of the Berlin Observatory. He was responsible for compiling the star maps which enabled Galle and D'Arrest to locate Neptune. In 1818 he computed the orbit of a faint comet, and successfully predicted its return; this was Encke's Comet, which has the shortest period of any known comet (3·3 years).

Eratosthenes. c. 276–196 BC. Greek philosopher, b. Cyrene; he became Librarian at Alexandria, and made a remarkably accurate measurement of the circumference of the Earth.

Eudoxus. c. 408–355 BC. Greek astronomer, b. Cnidus. He went to Athens, and attended lectures by Plato. Finally he settled in Sicily. He developed the theory of concentric spheres – the first truly scientific attempt to explain the movements of the celestial bodies.

Euler, Leonhard, 1707–83. Brilliant Swiss mathematician, b. Basle. He pioneered studies of the lunar theory, the movements of planets, comets, and the tides. He lost his sight in 1766, but this did not stop him from working; he undertook the complicated calculations mentally!

Fabricius, David. 1564–1617. Dutch minister and amateur astronomer, who observed Mira Ceti in 1596 (though without recognizing it as a variable) and made pioneer telescopic observations, notably of the Sun. In 1617 he announced from the pulpit that he knew the identity of a member of his congregation who had stolen one of his geese – and he was presumably correct, since he was assassinated before he could divulge the name of the culprit!

Fabricius, Johann. 1587–1616. Son of David Fabricius, and also a pioneer observer of the Sun by telescopic means; he discovered sunspots independently of Galileo and Scheiner.

Fallows, Fearon. 1789–1831. English astronomer, b. Cumberland. He went to South Africa in 1821 as the first Director of the Cape Observatory. He established the observatory, working under almost incredible difficulties, but the primitive living conditions undermined his health. The reduction of his Cape observations was undertaken by Airy.

Fauth, Philipp Johann Heinrich. 1867–1943. German astronomer, who compiled a large map of the Moon. Unfortunately he believed in the absurd theory that the Moon is ice-covered, and this influenced all his work.

Ferguson, James. 1710–76. Scottish popularizer of astronomy, who began life as a shepherd-boy but whose books gained great influence. He was also one of the first to suggest an evolutionary origin of the Solar System.

Flammarion, Camille. 1842–1925. French astronomer, renowned both for his observations of Mars and for his popular books. He set up his own observatory at Juvisy, and founded the Société Astronomique de France.

Flamsteed, John. 1646–1720. English astronomer (Astronomer Royal, 1675–1720, though at first the title was 'unofficial'). His main work was the compilation of a new star catalogue, the final version of which was published posthumously. Flamsteed was also Rector of Burstow, Surrey.

Fleming, Wilhelmina. 1857–1911. Scottish woman astronomer, who emigrated to America and worked at Harvard College Observatory, where she was in charge of the famous Draper star-catalogue. She discovered 10 novæ and 222 variable stars.

Fontana, Francisco. 1585–1656. Italian amateur (a lawyer by profession). He left sketches of Mars and Venus, though the 'markings' which he recorded were certainly illusory.

Fowler, Alfred. 1868–1940. English astronomer, whose spectroscopic work in connection with the Sun, stars and comets was of great importance.

Franklin-Adams, John. 1843–1912. English businessman who took up astronomy as a hobby at the age of 47, and compiled a photographic chart of the stars which is still regarded as a standard work.

Fraunhofer, Joseph von. 1787–1826. Outstanding German optical worker, orphaned in early childhood and rescued from poverty by the Elector of Bavaria. He joined the Physical and Optical Institute of Munich, and was Director from 1823. He invented the diffraction grating, constructed the best lenses in the world, and studied the dark lines in the solar spectrum (the 'Fraunhofer Lines'). He made the Dorpat refractor for Struve (the first telescope to be clock-driven) and also the Königsberg heliometer. His comparatively early death was a tragedy for science.

Galilei, Galileo. 1564–1642. The first great telescopic observer – and also the true founder of experimental mechanics. He worked successively at Pisa, Padua and Florence. The story of his remarkable telescopic discoveries (including the satellites of Jupiter, the phases of Venus and the gibbous aspect of Mars, the starry nature of the Milky Way and many more), and of how his defence of the Copernican theory brought him into conflict with the Church, is one of the most famous in scientific history. He was condemned by the Inquisition in 1633, and was kept a virtual prisoner in his villa at Arcetri; in his last years he also lost his sight.

Galle, Johann Gottfried. 1812–1910. German astronomer, best remembered as being the first (with D'Arrest) to locate Neptune in 1846. He discovered three comets, and in 1872, while director of the Breslau Observatory, was the first to use an asteroid for measuring solar parallax.

Gassendi, Pierre. 1592–1655. French mathematician and astronomer. In 1631 he made the first of all observations of a transit of Mercury.

Gauss, Karl Friedrich. 1777–1855. German mathematical genius. In 1801 he calculated the orbit of the first asteroid, Ceres, from a few observations, and enabled Olbers to recover it in the following year. He invented the 'method of least squares', known to every mathematician.

Gill, David. 1843–1914. Scottish astronomer. In 1877 he used observations of Mars to redetermine the solar parallax, and in 1879 went to South Africa as HM Astronomer at the Cape. It was his photograph of the comet of 1882 which showed him the importance of mapping the sky photographically – since his plate showed many stars as well as the comet. He was also deeply involved in cataloguing the southern stars. He was knighted in 1900.

Goldschmidt, Hermann. 1802–66. German astronomer, who settled in Paris. Using small telescopes poked through his attic window, he discovered 14 asteroids between 1852–61.

Goodacre, Walter. 1856–1938. English amateur astronomer, who published an excellent map of the Moon in 1910.

Goodricke, John. 1764–86. Born of English parents in Holland. He was a deaf-mute, but with a brilliant brain. It was he who found that Algol is an eclipsing binary rather than true variable, and he also discovered the fluctuations of the intrinsic variable δ Cephei.

Gould, Benjamin Apthorp. 1824–96. American astronomer, who founded the *Astrophysical Journal*. From Cordoba Observatory, Argentina, he compiled the *Uranimetria Argentina*, the first major catalogue of the southern stars.

Green, Charles. 1735–71. English astronomer who went with Captain Cook to study the 1769 transit of Venus. He died on the return voyage.

Gregory, James. 1638–75. Scottish mathematician. In 1663 he described the principle of the reflecting telescope, but never actually made one.

Grimaldi, Francesco Maria. 1618–63. Italian Jesuit, who made observations of the Moon used in the lunar map compiled by his friend Riccioli. Grimaldi discovered the refraction of light.

Gruithuisen, Franz von Paula. 1744–1852. German astronomer; from 1826 Professor of Astronomy, Munich. He was an assiduous observer of the Moon and planets, but his vivid imagination tended to discredit his work; at one stage he even reported the discovery of artificial structures on the Moon. He also proposed the impact theory of lunar crater formation.

Gum, Colin. 1924–60. Australian astronomer, who carried out work of vital importance in the surveying of southern radio sources. The famous 'Gum Nebula' in Vela/Puppis is named after him. He was killed in a skiing accident at Zermatt in Switzerland.

Hadley, John. 1682–1743. English astronomer; friend of Bradley. He made the first really good reflecting telescope (6 in aperture) in 1723, and 1731 constructed his 'reflecting quadrant', which replaced the astrolabe and the cross-staff in navigation.

Hale, George Ellery. 1868–1938. American astronomer. A pioneer solar observer, who invented the spectroheliograph and discovered the magnetic fields of sunspots. In 1897 he became Director of Yerkes Observatory, and transferred to Mount Wilson in 1905; he master-minded the building of the 60 in and 100 in reflectors, as well as the Yerkes refractor. He was mainly responsible for the building of the Palomar 200 in reflector, unfortunately not completed in his lifetime.

Hall, Asaph. 1829–1907. American astronomer, noted for his planetary work. At Washington, in 1877, he discovered the two satellites of Mars. From 1896 Professor of Astronomy at Harvard.

Halley, Edmond. 1656–1742. English astronomer (Astronomer Royal, 1720–42). Though best known for his prediction of the return of the great comet which now bears his name, Halley accomplished much other valuable work; he catalogued the southern stars from St Helena, studied star-clusters and nebulæ, and discovered the proper motions of some of the bright stars. More importantly, he was responsible for the writing of Newton's *Principia* and personally financed its publication.

Harding, Karl Ludwig. 1765–1834. German astronomer, who was at first assistant to Schröter and then was appointed Professor of Astronomy at Göttingen. In 1804 he discovered the third asteroid, Juno.

Haro, Guillermo. (1900–90). Possibly Mexico's most famous astronomer, celebrated for his studies of flare stars. He noted bright nebulæ with spectra showing emission lines, independently of G. Herbig; these are now known as Herbig-Haro Objects. He was Director of the Mexican Institute of Astronomy.

Harriot, Thomas. 1560–1621. English scholar, once tutor to Sir Walter Raleigh. He compiled the first telescopic map of the Moon, and completed it some months before Galileo began his work.

Harrison, John. 1693–1776. English clockmaker, who invented the marine chronometer which revolutionized navigation. Several of his original chronometers are now on display in London.

Hartmann, Johannes Franz. 1865–1936. German astronomer. Director Göttingen Observatory 1909–21, when he went to Argentina to superintend the National Observatory there. His important work was connected with stellar and nebular radial velocities, in the course of which he discovered interstellar absorption lines in the spectrum of δ Orionis.

Hay, William Thompson. 1888–1949. 'Will Hay' was probably the only skilled amateur astronomer who was by profession a stage and screen comedian! In 1933 he discovered the famous white spot on Saturn – the most prominent ever seen on that planet.

Heis, Eduard. 1806–77. German astronomer; Professor at Münster from 1852. He was a leading authority on the Zodiacal Light, meteors and variable stars, and published a valuable star catalogue. He was renowned for his keen eyesight, and is said to have counted 19 naked-eye stars in the Pleiades.

Hencke, Karl Ludwig. 1793–1866. German amateur astronomer-postmaster at Driessen. In 1845, after 15 years' search, he discovered the fifth asteroid, Astræa.

Henderson, Thomas. 1798–1844. Scottish astronomer. 1832–3 HM Astronomer at the Cape. While there, he made the measurements which enabled him to measure the parallax of α Centauri. In 1834 he became the first Astronomer Royal for Scotland.

Heraclides of Pontus (c. 388–315 BC) declared that the apparent daily rotation of the sky is due to the real rotation of the Earth. He also discovered that Mercury and Venus revolve round the Sun, not round the Earth.

Heraclitus of Ephesus (born c. 544 BC) took fire to be the principal element, and maintained that the diameter of the Sun was about one foot!.

Herschel, Friedrich Wilhelm (always known as William Herschel). 1738–1822. Probably the greatest observer of all time. He was born in Hanover, but spent most of his life in England. He was the best telescope maker of his day, and in 1781 became famous by his discovery of the planet Uranus. He made innumerable discoveries of double stars, clusters and nebulæ; he found that many doubles are physically-associated or binary systems, and he was the first to give a reasonable idea of the shape of the Galaxy. He was knighted in 1816, and received every honour that the scientific world could bestow. George III appointed him King's Astronomer (not Astronomer Royal).

Herschel, Caroline. 1750–1848. William Herschel's sister and constant assistant in his astronomical work. She discovered 8 comets.

Herschel, John Frederick William. 1792–1871. William Herschel's son. He graduated from Cambridge 1813 and from 1832–8 took a large telescope to the Cape to make the first really systematic observation of the southern heavens. He discovered 3347 double stars and 525 nebulæ, and may be said to have completed his father's pioneering work.

Hertzsprung, Ejnar. 1873–1967. Danish astronomer, who worked successively at Frederiksberg, Copenhagen, Göttingen, Mount Wilson and Leiden (Director Leiden Observatory from 1935). In 1905 he discovered the giant and dwarf subdivisions of late-type stars, and this led on to the compilation of H-R or Hertzsprung-Russell Diagrams, which are of fundamental importance in astronomy.

Hevelius. 1611–87. The Latinized name of

Johannes Hewelcke of Danzig (now Gdańsk). From his private observatory he drew up a catalogue of 1500 stars, and observed planets, the Moon and comets, using the unwieldy long-focus, small-aperture refractors of his day. His observatory was burned down in 1679, but he promptly constructed another. His original map of the Moon has been lost; tradition says that the copper engraving was melted down and made into a teapot after his death.

Hind, John Russell. 1823–95. English astronomer, who discovered 11 asteroids, the 1848 nova in Ophiuchus, and his 'variable nebula' round T Tauri. He also computed many cometary orbits, and from 1853 was superintendent of the *Nautical Almanac*.

Hipparchus. Fl. 140 BC. Great Greek astronomer, who lived in Rhodes. He drew up a star catalogue, later augmented by Ptolemy. Among his many discoveries was that of precession; he also constructed trigonometric tables. Unfortunately all his original works have been lost.

Hoffmeister, Cuno. 1892–1967. He was born in Sonneberg, Thuringia, and after working at two German observatories founded the Sonneberg Observatory, in 1925. He was a specialist in variable star work, and discovered almost 10 000 new variables. He was also an authority on meteoritic astronomy.

Hooke, Robert. 1653–1703. English scientific genius, contemporary with (though no friend of!) Newton. He built various astronomical instruments, and made some useful observations, including sketches of lunar craters.

Horrocks, Jeremiah, 1619–41. English astronomer who, with his friend Crabtree, was the first to observe a transit of Venus (1639). He also worked on lunar theory. His early death was a great tragedy for science.

Hubble, Edwin Powell. 1889–1953. American astronomer, who served in the Army during World War I and was also a boxing champion. In 1923, using the Mount Wilson 100 in reflector, he discovered short-period variables in the Andromeda Spiral, and proved the Spiral to be an independent galaxy. He also established the velocity/distance relationship known as Hubble's Law.

Huggins, William. 1824–1910. Pioneer English spectroscopist, who had his private observatory at Tulse Hill, nr London. Pioneer of stellar spectroscopy; he established that the irresolvable nebulæ are gaseous; he was the first to determine stellar radial motions by means of the Doppler shifts in their spectral lines, and he carried out important solar and planetary work. He was knighted in 1897.

Humason, Milton La Salle. 1891–1972. b. Minnesota. He was mainly self-taught, but joined the staff of Mount Wilson Observatory in 1920, and from then on worked closely with Hubble, studying the forms, spectra, radial motions and nature of the galaxies; he also photographed the spectra of supernovæ in external systems. In 1919 he carried out a photographic search for a trans-Neptunian planet at the request of W. H. Pickering, who had made independent calculations similar to Lowell's. Humason took several plates, but failed to locate the planet. When the plates were re-examined years later, after Pluto had been discovered at Flagstaff, it was found that Humason had recorded the planet twice – but once the image was masked by a star, and on the other occasion it fell on a flaw in the plate!

Huygens, Christiaan. 1629–95. Dutch astronomer; probably the best telescopic observer of his time. He discovered Saturn's brightest satellite (Titan) in 1655, and was the first to realize that the curious appearance of the planet was due to a system of rings. He was also the first to see markings on Mars. His activities extended into many fields of

science; in particular, he invented the pendulum clock.

Innes, Robert Thorburn Ayton. 1861–1933. Scottish astronomer, who emigrated first to Australia (becoming a wine merchant) and then went to South Africa, as director of the Observatory at Johannesburg. He specialized in double star work, discovering more than 1500 new pairs; he also discovered Proxima Centauri, the nearest star beyond the Sun.

Janssen, Pierre Jules César. 1824–1907. French astronomer, who specialized in solar work (in 1870 he escaped from the besieged city of Paris by balloon to study a total eclipse). Independently of Lockyer, he discovered the means of observing the Sun's chromosphere and prominences without waiting for an eclipse. From 1876 he was Director of the Meudon Observatory, and in 1904 published an elaborate solar atlas, containing more than 8000 photographs. The square at the entrance to the Meudon Observatory is still called the Place Janssen, and his statue is to be seen there.

Jansky, Karl Guthe. 1905–49. American radio engineer, of Czech descent. He joined the Bell Telephone Laboratories, and was using an improvised aerial to investigate problems of static when he detected radio waves which he subsequently showed to come from the Milky Way. This was, in fact, the beginning of radio astronomy; but for various reasons Jansky paid little attention to it after 1937, and virtually abandoned the problem.

Jeans, Sir James Hopwood. 1877–1946. English astronomer. He elaborated the plausible but now rejected theory of the tidal origin of the planets, but his major work was in connection with stellar constitution, in which he made notable advances. He was also an expert writer of popular scientific books, and was famous as a lecturer and broadcaster.

Jeffreys, Sir Harold. 1892–1989. Though primarily a geophysicist, Jeffreys also made many important advances in astronomy, and it was he who first showed that the giant planets are not miniature suns. His great book, *The Earth; its Origin, History and Physical Constitution* (1924) was immensely influential.

Jones, Sir Harold Spencer. 1890–1960. English astronomer (Astronomer Royal 1933–55). A Cambridge graduate, he was HM Astronomer at the Cape from 1923 until his appointment to Greenwich. From the Cape he carried out much im-portant work, mainly in connection with star catalogues and stellar radial velocities. While Astronomer Royal he redetermined the solar parallax by means of the world-wide observations of Eros, and published several excellent popular books as well as technical papers. He played a major rôle in the removal of the main equipment from Greenwich to the new site at Herstmonceux, in Sussex, and himself transferred to Herstmonceux in 1948, though it was not until 1958 that the move was completed. He was knighted in 1943.

Kant, Immanuel. 1724–1804. German philosopher, remembered astronomically for proposing a theory of the origin of the Solar System which had some points of resemblance to Laplace's later Nebular Hypothesis.

Kapteyn, Jacobus Cornelius. 1851–1922. Dutch astronomer and cosmologist. His most celebrated discovery was that of 'star-streaming'.

Kepler, Johannes. 1571–1630. German astronomer, b. Württemberg. He was the last assistant to Tycho Brahe, and after Tycho's death used the mass of observations to establish his three Laws of Planetary Motion. He observed the 1604 supernova, and also several comets, as well as making improvements to the refracting telescope, but his main achievements were theoretical. He ranks with Copernicus and Galileo as one of the main figures in the story of the 'Copernican revolution'.

Kirch, Gottfried. 1639–1710. German astronomer; Director of the Berlin Observatory from 1705. He was one of the earliest of systematic observers of comets, star-clusters and variable stars. In 1686 he discovered the variability of χ Cygni.

Kirchhoff, Gustav Robert. 1824–87. Professor of physics at Heidelberg. One of the greatest of German physicists, who explained the dark lines in the Sun's spectrum. His great map of the solar spectrum was published from Berlin in 1860.

Kirkwood, Daniel. 1814–95. American astronomer; an authority on asteroids and meteors. He drew attention to gaps in the asteroid belt, known today as the Kirkwood Gaps; they are due to the gravitational influence of Jupiter.

Kuiper, Gerard P. 1905–73. Dutch-American astronomer, who made notable advances in planetary and lunar work and was deeply involved with the programmes of sending probes beyond the Earth. The first crater to be identified on Mercury from *Mariner 10* was named in his honour.

Kulik, Leonid. 1883–1942. Russian scientist, trained as a forester, who achieved fame because of his work in meteorite research. In particular, he led several expeditions to study the Tunguska object of 1908. He died in a German prison camp in 1942.

Lacaille, Nicolas Louis de. 1713–62. French astronomer, who went to the Cape to draw up the first good southern-star catalogue.

Lagrange, Joseph Louis de. 1736–1813. French mathematical genius, and author of the classic *Mécanique Analytique.* He wrote numerous astronomical papers, dealing, among other topics, with the Moon's libration and the stability of the Solar System.

Laplace, Pierre Simon. 1749–1827. French mathematician who made great advances in dynamical astronomy. In 1796 he wrote *Système du Monde,* in which he outlined his Nebular Hypothesis of the origin of the planets. Discarded in its original form, but modern theories have many points of resemblance to it.

Lassell, William. 1799–1880. English astronomer. He discovered Triton, Neptune's larger satellite, and (independently of Bond) Hyperion, the 7th satellite of Saturn, as well as two satellites of Uranus (Ariel and Umbriel). He set up a 24 in reflector in Malta, and with it discovered 600 nebulæ.

Leavitt, Henrietta Swan. 1868–1921. American woman astronomer, best remembered for her observations of Cepheids in the Small Magellanic Cloud (1912), based on photographs taken in S America; these led on to the discovery of the vital period-luminosity law for Cepheids. She also discovered 4 novæ, several asteroids, and over 2400 variable stars.

Lemaître, Georges. 1894–1966. Belgian priest, who was a leading mathematician; from 1927, Professor at Louvain University. His most important paper, leading to what is now called the 'Big Bang' theory of the universe, appeared in 1927, but did not become well-known until publicized by Eddington three years later. During World War I Lemaître served in the Belgian Army, and won the Croix du Guerre.

Le Monnier, Pierre Charles. 1715–99. French astronomer who was concerned in star cataloguing. He observed the planet Uranus several times, but did not check his observations, and missed the chance of a classic discovery. It was said that he never failed to quarrel with anyone whom he met!

Le Verrier, Urbain Jean Joseph. 1811–77. French astronomer, whose calculations led in 1846 to the discovery of Neptune. He was an authority on meteors, and in 1867 computed the orbit of the Leonids. He also developed solar and planetary theory, and believed in the existence of a planet (Vulcan) closer to the Sun than Mercury – now known to be a myth. He was forced to resign the Directorship of the Paris Observatory in 1870 because of his irritability, but was reinstated on the death by drowning of his successor, Delaunay.

Levin, Boris Yuljevich. 1912–89. Born in Moscow. At first he was concerned with meteors, but then worked with O. Schmidt in developing his theories of the origin of the Solar System. He also made valuable contributions to cometary astronomy.

Lexell, Anders John. 1740–84. Finnish astronomer, *b.* Abö. He became Professor of Mathematics at St Petersburg. He discovered the periodical comet of 1770 (now lost), and was one of the first to prove that the object discovered by Herschel in 1781 was a planet rather than a comet.

Lindsay, Eric Mervyn. 1907–74. Irish astronomer, Director of the Armagh Observatory from 1936 until his death. His main work was in connection with the Magellanic Clouds and with quasars. He had close connections with the Boyden Observatory in South Africa (where he had previously been assistant astronomer) and forged close links between it, Harvard, Dunsink (Dublin) and Armagh. He was an excellent lecturer on popular astronomy, and founded the Armagh Planetarium in 1966.

Lockyer, Joseph Norman. 1836–1920. English astronomer, and an independent discoverer of the method of studying the solar chromosphere and prominences at time of non-eclipse. He was knighted in 1897. He founded the Norman Lockyer Observatory at Sidmouth in Devon, which still exists and is open to the public by arrangement.

Lohrmann, Wilhelm Gotthelf. 1796–1840. German land surveyor, who began an elaborate lunar map but was unable to complete it owing to ill-health. The map was completed 40 years later by Julius Schmidt.

Lomonosov, Mikhail. 1711–65. Russian astronomer; he was also termed 'the founder of Russian literature'. His father was a fisherman. In 1735 he went to the University of St Petersburg, and then to Marburg in Germany to study chemistry. On his return to Russia in 1741 he insulted some of his colleagues at the St Petersburg Academy and was imprisoned for several months, during which time he wrote two of his most famous poems. However, he later became Professor of Chemistry at St Petersburg, and in 1746 became a Secretary of State. He drew up the first accurate map of the Russian Empire, described a 'solar furnace', and investigated electrical phenomena. He also studied aurorae. In 1761 he observed the transit of Venus, and rightly concluded that Venus has a considerable atmosphere. His most important contribution was his championship of the Copernican theory and of Newton's theories, neither of which had really taken root in Russia before Lomonosov's work.

Lowell, Percival. 1855–1916. American astronomer, who founded the Lowell Observatory at Flagstaff, Arizona, in 1894. He paid great attention to Mars, and believed the 'canals' to be artificial waterways. His calculations led to the discovery of the planet Pluto, though the planet was not actually found until 1930 – by Clyde Tombaugh, at the Lowell Observatory. Lowell himself was a great astronomer who did much for science, and it is regrettable that he is today remembered mainly because of his erroneous theories about the Martian canals.

Lyot, Bernard. 1897–1953. Great French astronomer; Director of the Meudon Observatory. He made many advances in instrumental techniques, and invented the coronagraph, which enables the inner corona to be studied at times of non-eclipse.

He died suddenly while taking part in an eclipse expedition to Africa.

Maclear, Thomas. 1794–1879. Irish astronomer, who in 1833 succeeded Henderson as HM Astronomer at the Cape. He made an accurate measurement of an arc of meridian as well as verifying Henderson's parallax of α Centauri; he also studied comets and nebulæ. He was knighted in 1860.

Mädler, Johann Heinrich von. 1794–1874. German astronomer, who was the main observer in the great lunar map by himself and Beer, published in 1837–8 – a map which remained the standard for several decades. In 1840 he left his Berlin home to become Director of the Dorpat Observatory in Estonia. He erroneously believed that η Tauri (Alcyone) was the star lying at the centre of the Galaxy. He retired in 1865 and spent his last years in Hanover.

Maraldi, Giacomo Filippo. 1665–1729. Italian astronomer; nephew of G. D. Cassini. He was renowned for his observations of the planets, particularly Mars, and assisted his uncle at the Paris Observatory.

Maskelyne, Nevil. 1732–1811. English astronomer (Astronomer Royal 1765–1811). Educated at Cambridge; he then went to St Helena, at the suggestion of Bradley, to observe the transit of Venus, and decided to make a serious study of navigation. During his régime as Astronomer Royal he founded the *Nautical Almanac.*

Masursky, Harold. 1923–90. A leading planetary geologist, who was closely involved with the NASA missions, playing a key rôle in their planning. He spent his entire career with the US Geological Survey.

McVittie, George. 1904–88. British astronomer, born in Smyrna (his mother was Greek). He graduated from Edinburgh, and worked in England and America. He was a pioneer in studies of relativity, and made many important contributions to cosmology.

Méchain, Pierre François Andre. 1744–1805. French astronomer, who discovered 8 comets 1781–99.

Menzel, Donald H. 1901–76. American astronomer, celebrated for his research into problems of the Sun and planets as well as in stellar studies. He was also an excellent lecturer, and a skilled writer of popular books.

Messier, Charles. 1730–1817. French astronomer, interested mainly in comets. Though he discovered 13 comets, he is remembered chiefly because of his catalogue of star-clusters and nebulæ, published in 1781.

Michell, John. 1725–93. English clergyman, and an amateur astronomer who made the first suggestion that many double stars may be physically associated or binary systems.

Milne, Edward Arthur. 1896–1950. English astronomer, who graduated from Cambridge and then went successively to Manchester and Oxford. He made important contributions to astrophysics, and developed his theory of 'kinematic relativity', which was for a time regarded as an alternative to general Einsteinian relativity.

Minkowski, Rudolf. 1895–1976. German astronomer, who went to Mount Wilson in 1935 and remained there. He was one of the leading authorities on novæ and planetary nebulæ, and after the war became a pioneer in the new science of radio astronomy. His studies of rapidly-moving gases in radio galaxies led to the rejection of the 'colliding galaxies' theory.

Mitchell, Maria (1812–88). America's first woman astronomer. She was the first woman to be elected to the American Academy of Arts and Sciences; she also discovered a comet.

Montanari, Geminiano. 1633–87. Italian astronomer, who worked at Bologna and then at Padua. In 1669 he discovered the variability of Algol.

Nevill, Edmund Neison. 1851–1940. English astronomer, who published an important book and map concerning the Moon in 1876; he wrote under the name of Neison. He was Director of the Natal Observatory at Durban in South Africa 1882–1910, returning to England when the Observatory was closed.

Newcomb, Simon. 1835–1909. American astronomer, for some years head of the American Nautical Almanac office. His chief work was in mathematical astronomy, to which he made valuable contributions. He is also remembered as the man who proved to his own satisfaction that no heavier-than-air machine could ever fly!

Newton, Isaac. 1643–1727. Probably the greatest of all astronomers. To list all his contributions here would be pointless; suffice to say that his *Principia*, published in 1687, has been described as the greatest mental effort ever made by one man. In addition to his scientific work, he sat briefly in Parliament, and served as Master of the Mint. He was knighted in 1705, and on his death was buried in Westminster Abbey.

Olbers, Heinrich Wilhelm Matthias. 1758–1840. German doctor; also a skilled amateur astronomer; established his private observatory in Bremen. He discovered two of the first four asteroids (Pallas and Vesta) and rediscovered the first (Ceres); he carried out important work in connection with cometary orbits, and discovered a periodical comet which has a period of 69·5 years, and last returned in 1956. Olbers also wrote about his celebrated paradox: 'Why is it dark at night?'

Oort, Jan Hendrik. 1900–92. Dutch astronomer, who was the first (with B. Lindblad) to recognize the rotation of the Galaxy. He was a pioneer of radio astronomy and also of cometary studies. He was Director of the Leyden Observatory from 1945 to 1970, and President of the International Astronomical Union from 1958 to 1961.

Parmenides of Elea (second half of the 6th century BC) believed the stars to be of compressed fire, but agreed that the Earth was spherical, and in equilibrium because it was equidistant from all points on the sphere representing the universe.

Penston, Michael Victor. 1943–90. He was born in London, but spent much of his career at Cambridge. He made many contributions to astrophysics, and in 1983 was able to 'weigh' a black hole in the centre of the galaxy NGC 4151. He described himself as one of the LAGS – Lovers of Active Galaxies! Sadly, he died in 1990 after a long and brave fight against cancer.

Perrine, Charles Dillon. 1867–1951. American astronomer, who discovered two of Jupiter's satellites as well as nine comets. He worked at the Lick Observatory until 1909, when he became Director of the Cordoba Observatory in Argentina, where he constructed a 30 in reflector and made many observations of southern galaxies. He also planned a major star catalogue, but was politically unpopular, and after a narrow escape from assassination he retired (1936).

Peters, Christian Heinrich Friedrich. 1813–90. Danish astronomer, who emigrated to America in 1848. He discovered 48 asteroids.

Piazzi, Giuseppe. 1746–1826. Italian astronomer, who became Director of the Palermo Observatory in Sicily. During the compilation of a star catalogue he discovered the first asteroid, Ceres (on 1 Jan. 1801, the 1st day of the new century).

Pickering, Edward Charles. 1846–1919. American astronomer; for 43 years, from 1876, Director of the Harvard College Observatory. He concentrated upon photometry, variable stars, and above all stellar spectra; in the famous *Draper Catalogue*, the stars were classified according to their spectra. During his régime the Harvard Observatory was

modernized, and a southern outstation was set up at Arequipa in Peru.

Pickering, William Henry. 1858–1938. Brother of E. C. Pickering, who worked with him at Harvard and also served for a while as astronomer-in-charge of the Arequipa out-station. In 1898 he discovered Saturn's 9th satellite, Phœbe. He made extensive studies of the Moon and Mars, mainly from the Harvard station in Jamaica which was set up in 1900. Independently of Lowell, he calculated the position of the planet Pluto.

Plutarch. *c.* 46–120. Greek biographer, mentioned here because of his authorship of *De Facie in Orbe Lunæ* – On the Face in the Orb of the Moon – in which he claims that the Moon is a world of mountains and valleys.

Pond, John. 1767–1836. English astronomer (Astronomer Royal, 1811–35). Though an excellent and painstaking astronomer, Pond was handicapped by ill-health during the latter part of his régime at Greenwich, and was eventually asked to resign. He tried unsuccessfully to obtain star-distances by the parallax method.

Pons, Jean Louis. 1761–1831. French astronomer, whose first post at an observatory (Marseilles) was that of caretaker! He was self-taught and concentrated on hunting for comets; he found 36 in all, and ended his career as Director of the Museum Observatory in Florence.

Proctor, Richard Anthony. 1837–88. English astronomer, who was an excellent cosmologist but is best known for his many popular books. In 1881 he emigrated to America, and remained there for the rest of his life. Proctor paid considerable attention to the planets, and constructed a map of Mars.

Ptolemy (Claudius Ptolemæus). *c.* 120–180. The 'Prince of Astronomers', who lived and worked in Alexandria. Nothing is known about his life, but his great work has come down to us through its Arab translation (the *Almagest*). Ptolemy's star catalogue was based on that of Hipparchus but with many contributions of his own; he also brought the geocentric system to its highest state of perfection, so that it is always known as the Ptolemaic theory. He constructed a reasonable map of the Mediterranean world, and even showed Britain, though it is true that he joined Scotland on to England in a back-to-front position.

Purbach, Georg von. 1423–61. Austrian astronomer, who became a professor at Vienna in 1450. He founded a new school of astronomy, compiled tables, and began to write an *Epitome of Astronomy* based on Ptolemy's *Almagest*. After Purbach's premature death, the book was completed by his friend and pupil Regiomontanus.

Pythagoras. *c.* 572–500 BC. The great Greek geometer, mentioned here because he was one of the very first to maintain that the Earth is spherical rather than flat. He seems also to have studied the movements of the planets.

Ramsden, Jesse. 1735–1800. English maker of astronomical instruments. His meridian circles were the first to be lit through the hollow axis.

Rayet, Georges Antoine. 1839–1906. French astronomer; in 1867, with Wolf, drew attention to the Wolf-Rayet stars, which have bright lines in their spectra. He went from Paris to Bordeaux, and became Director of the Observatory there.

Redman, Richard Oliver. 1905–75. English astronomer, who graduated from Cambridge. He made extensive studies of the Sun, stellar velocities, galactic rotation and the photometry of galaxies. In 1937 he went to the Radcliffe Observatory, Pretoria, and designed the spectrograph for the 74 in reflector. In 1947 he returned to Cambridge as Professor of Astrophysics and Director of the Observatories. Many programmes were carried through, and Redman also devoted much time and energy in the planning and construction of the 153 in Anglo-Australian telescope at Siding Spring.

Regiomontanus. 1436–76. The Latinized name of Johann Müller, Purbach's pupil. He completed the *Epitome of Astronomy*, and at Nürnberg set up a printing press, publishing the first printed astronomical ephemerides. He died in Rome, where he had been invited to help in reforming the calendar.

Reinmuth, Karl. 1892–1979. He spent his entire career at the Königstuhl Observatory in Heidelberg, concentrating upon asteroids and comets. He discovered over 250 asteroids, among them the 1932 Earth-grazer Apollo.

Rhæticus, Georg Joachim. 1514–76. German astronomer, who became Professor of Astronomy, at Wittenberg in 1536. An early convert to the Copernican system, he visited Copernicus at Frombork, and persuaded him to send his great book for publication.

Riccioli, Giovanni Battista. 1598–1671. Italian Jesuit astronomer, who taught at Padua and Bologna. He was a pioneer telescopic observer, and drew up a lunar map, inaugurating the system of nomenclature which is still in use. Oddly enough, he never accepted the Copernican system!

Robinson, Romney. 1792–1882. Irish astronomer, who was Director of the Armagh Observatory from 1823 to his death. He published the Armagh catalogue of over 5000 stars, and made many other contributions; he also invented the cup anemometer. It is on record that when the railway company planned to build a line close to Armagh, Robinson managed to have it diverted, since he maintained that the trains would shake his telescopes!

Rømer, Ole. 1644–1710. Danish astronomer. In 1675 he used the eclipse times of Jupiter's satellites to make an accurate measurement of the velocity of light. In 1681 he became Director of the Copenhagen Observatory. Among his numerous inventions are the transit instrument and the meridian circle.

Rosse, 3rd Earl of. 1800–67. Irish amateur astronomer, who in 1845 completed the building of a 72 in reflector and erected it at his home at Birr Castle. The 72 in with its metal mirror, was much the largest telescope ever built up to that time. His greatest discovery was that many of the galaxies are spiral in form.

Rosse, 4th Earl of. 1840–1908. Continued his father's work, and was also the first man to measure the tiny quantity of heat coming from the Moon. The telescope is at present (1995) being restored.

Rowland, Henry Augustus. 1848–1901. American scientist; Professor of Physics in Baltimore from 1876. His great map of the solar spectrum was published in 1895–7; it showed 20 000 absorption lines.

Russell, Henry Norris. 1877–1957. American astronomer; Director, Princeton Observatory from 1908. He devoted much of his energy to studies of stellar constitution and evolution, and independently of Hertzsprung he discovered the giant and dwarf sub-divisions of stars of late spectral type. This led on to the H-R or Hertzsprung-Russell Diagram, in which luminosity (or the equivalent) is plotted against spectral type.

Rutherfurd, Lewis Morris. 1816–92. American barrister, who gave up his profession to devote himself to astronomy. A pioneer in astronomical photography, his pictures of the Moon were outstanding; his ruled solar gratings for solar spectra were the best of their time.

Ryle, Martin. 1918–84. British pioneer of radar and radio astronomy. He graduated from Oxford, and then went to Cambridge as Professor of Radio Astronomy. In 1972 he succeeded Woolley as Astronomer Royal. He developed the technique of aperture synthesis, and made many very important contributions.

Scheiner, Christoph. 1575–1650. German Jesuit,

who was for some time a professor of mathematics in Rome. He discovered sunspots independently of his contemporaries, and wrote a book, *Rosa Ursina*, which contains solar drawings and observations for the years 1611–25. He was unfriendly towards Galileo, and played a rather discreditable part in the events leading up to Galileo's trial and condemnation.

Schiaparelli, Giovanni Virginio. 1835–1910. Italian astronomer, who graduated from Turin and became Director of the Brera Observatory in Milan. He discovered the connection between meteors and comets, but his most famous work was in connection with the planets. It was he who first drew attention to the 'canal network' on Mars, in 1877.

Schlesinger, Frank. 1871–1943. American astronomer, *b.* New York. His main work was in con-nection with stellar parallaxes. He was Director successively of the Yale and Allegheny Observatories, and was responsible for the Yale 'southern station' in Johannesburg. His major works, *General Catalogue of Parallaxes* and its supplement, dealt with more than 2000 stars. He also pioneered the determination of star positions by using wide-angle cameras.

Schmidt, Julius (actually Johann Friedrich Julius). 1825–84. German astronomer, who became Director of the Athens Observatory in 1858 and spent most of his life in Greece. He concentrated upon lunar work, producing an elaborate map (based on Lohrmann's early work) and making great improvements in selenography. He drew attention to the alleged change in the lunar crater Linné, in 1866. He also discovered the outburst of the recurrent nova T Coronæ, in 1866.

Schönfeld, Eduard. 1828–91. German astronomer, who collaborated with Argelander in preparing the *Bonn Durchmusterung* and later extended it to the southern hemisphere.

Schröter, Johann Hieronymus. 1745–1816. Chief magistrate of Lilienthal, nr Bremen. He set up a private observatory, and made outstanding observations of the Moon and planets. Unfortunately many of his notebooks were lost in 1813, with the destruction of his observatory by the invading French troops.

Schwabe, Heinrich. 1789–1875. German apothecary, who became a noted amateur astronomer concentrating on the Sun. His great discovery was that of the 11-year sunspot cycle.

Schwarzschild, Karl. 1873–1916. German astronomer, who worked successively at Vienna, Göttingen and (as Observatory Director) Potsdam. His early work dealt with photometry, but he was also a pioneer of theoretical astrophysics. Military service during World War I broke his health and led to his premature death.

Secchi, Angelo. 1818–78. Italian Jesuit astronomer; one of the great pioneers of stellar spectroscopy, classifying the stars into four types (a system superseded later by that of Harvard). He was also an authority on solar work, and his planetary observations were equally outstanding.

Seyfert, Carl. 1911–60. American astronomer, who concentrated upon studies of galaxies. In 1942 he drew attention to those galaxies with very condensed nuclei, now always termed Seyfert galaxies.

Shajn, Grigorij Abramovich. 1892–1956. After serving in the Russian Army during World War I, he joined the staff of the Pulkovo Observatory, working upon meteoric astronomy. In 1924 he became Director of the Simeis Observatory, and began his work on stellar spectroscopy. During his latter years he was concerned mainly with the distribution of faint galactic nebulæ.

Shapley, Harlow. 1885–1972. Great American astronomer, who began his main work at Princeton under H. N. Russell. In 1914 he advanced the pulsation theory of Cepheid variables, and was soon able to use the variables in globular clusters to give the first accurate picture of the shape and size of the Galaxy. In 1921 he became Director of the Harvard College Observatory. In later years he concentrated upon studies of galaxies and upon the international aspect of astronomy. He was also an excellent lecturer, and a skilled writer of popular books.

Shklovskii, Iosif. 1916–85. He graduated from Moscow University, and became professor there, later heading the radio astronomy department at the Sternberg Institute. He specialized in extraterrestrial radio sources, and showed that the emission from the Crab Nebula is synchrotron radiation. With Carl Sagan he wrote about 'intelligent life in the universe'.

Smyth, William Henry. 1788–1865. English naval officer, rising to the rank of Admiral, who in 1830 established a private observatory at Bedford and made numerous observations. He is best remembered for his famous book, *Cycle of Celestial Objects.*

Smyth, Piazzi (actually Charles Piazzi). 1819–1900. Son of Admiral Smyth. Astronomer Royal for Scotland from 1844 until his death. He was a skilled astronomer who carried out much valuable work, including spectroscopic examinations of the Zodiacal Light, but also an eccentric who wrote a large and totally valueless volume about the significance of the Great Pyramid!

Sosigenes. Greek astronomer, who flourished about 46 BC. He was entrusted by Julius Cæsar with the reform of the calendar. Nothing is known about his life.

South, James. 1785–1867. English amateur astronomer, who founded a private observatory in Southwark and collaborated with John Herschel in studies of double stars. In 1822 he observed an occultation of a star by Mars, and the virtually instantaneous disappearance convinced him that the Martian atmosphere must be extremely tenuous.

Spörer, Friedrich Wilhelm Gustav. 1822–95. German astronomer, who joined the staff at Potsdam Observatory. He concentrated upon the Sun, and discovered the variation in latitude of spot zones over the course of a solar cycle (Spörer's Law).

Steavenson, William Herbert. 1894–1975. Steavenson was never a professional astronomer; he was a medical doctor whose practice was in Outer London. He was an expert observer, and one of the few amateurs to serve as President of the Royal Astronomical Society (1957–59). His knowledge of telescopes and optics was encyclopædic, and his advice was often sought by professional astronomers.

Strömgren, Bengt. 1908–87. Born in Sweden, but raised in Denmark. He became Director of the Royal Copenhagen Observatory in succession to his father Elias. He went to America to direct the Yerkes and McDonald Observatories during the 1950s, returning to Denmark in 1967. He was particularly known for his studies of H.II (ionized hydrogen) regions round hot stars often called Strömgren spheres.

Struve, Friedrich Georg Wilhelm. 1793–1864. German astronomer, *b.* Altona. He went to Dorpat in Estonia and in 1818 became Director of the Observatory. Using the 9 in Fraunhofer refractor – the first telescope to be clock-driven – he began his classic work on double stars. In 1839 he went to Pulkova, to become director of the new observatory set up by Tsar Nicholas. Here he continued his double-star work, and his *Mensuræ Micrometricæ* gives details of over 3000 pairs. Struve also measured the parallax of Vega; his value was announced in 1840.

Struve, Otto (Wilhelm). 1819–1905. Son of F. G. W. Struve; *b.* Dorpat. He became assistant to his father, accompanying him to Pulkova. He continued his father's work, and became a leading authority on double stars. He succeeded to the directorship of Pulkova Observatory in 1861, retiring in 1889 and returning to Germany.

Struve, Karl Hermann. 1854–1920. Son of Otto Struve; *b.* Pulkova, later becoming assistant to his father. His main work was in connection with planetary satellites. In 1895 he went to Königsberg, and in 1904 became Director of the Berlin Observatory, which was reorganized by him and transferred to Babelsberg during his term of office.

Struve, Gustav Wilhelm Ludwig. 1858–1920. Son of Otto Struve, and brother of Karl. He too was born at Pulkova and acted as assistant to his father. He went to Dorpat in 1886, and from 1894 was Director of the Kharkov Observatory. He was concerned mainly with statistical astronomy and with the motion of the Sun.

Struve, Otto. 1897–1963. Son of Gustav; often known as Otto Struve II. He was born in Kharkov, and fought during World War I; joined the White Army under Wrangel and Denikin, and after their defeat reached Constantinople, where he worked as a labourer. Finally he was offered a post at the Yerkes Observatory, where he arrived in 1921. He spent the rest of his life in America; in 1932 he became Director at Yerkes, after which he founded the McDonald Observatory in Texas and was its director from 1939–47, when he became Chairman of the Department of Astronomy at Chicago. In 1959 he began a new career as the first Director of the National Radio Astronomy Observatory, but ill-health forced his resignation in 1962. He was a brilliant astrophysicist, dealing mainly with spectroscopic binaries, stellar rotation and interstellar matter; he was also an author of popular books. He is (so far!) the last of the famous Struve astronomers. It is notable that all four were in succession awarded the Gold Medal of the Royal Astronomical Society – a sequence unique in astronomical history.

Swift, Lewis. 1820–1913. American astronomer who specialized in hunting for comets and nebulæ. He found 13 comets (including the Great Comet of 1862) and 900 nebulæ.

Tempel, Ernst Wilhelm. 1821–89. German astronomer, who became Director of the Arcetri Observatory. In 1859 he discovered the nebula in the Pleiades; he also discovered 6 asteroids and several comets, including the comet of 1865–6 which is associated with the Leonid meteors.

Thales. *c.* 624–547 BC. The first of the great Greek philosophers. He believed the Earth to be flat, and floating in an ocean, but he was a pioneer mathematician and observer, and successfully predicted the eclipse of 585 BC which stopped the war between the Lydians and the Medes.

Timocharis. Fl. *c.* 280 BC. Greek astronomer, who made some accurate measurements of star positions; one of these (of Spica) enabled Hipparchus, 150 years later, to demonstrate the precession of the equinoxes.

Turner, Herbert Hall. 1861–1930. English astronomer who played an important rôle in the organization and preparation of the International Astrographic Chart. In 1903 he discovered Nova Geminorum.

Tycho Brahe. 1546–1601. The great Danish observer – probably the best of pre-telescopic times. He studied the supernova of 1572 and from 1576 to 1596 worked at his observatory at Hven, an island in the Baltic, making amazingly accurate measurements of star positions and the movements of the planets, particularly Mars. His observatory – Uraniborg – became a scientific centre, but Tycho was haughty and tactless (during his student days he had part of his nose sliced off in a duel, and made himself a replacement out of gold, silver and wax!), and after quarrels with the Danish Court he left Hven and went to Prague as Imperial Mathematician to the Holy Roman Emperor, Rudolph II. Here he was joined by Kepler, who acted as his assistant. When Tycho died, Kepler came into possession of the Hven observations, and used them to prove that the Earth moves round the Sun – something which Tycho himself could never accept.

Ulugh Beigh (more properly Ulugbek). 1394–1449. Grandson of the Oriental conqueror Tamerlane. About 1420 he established an observatory at Samarkand, which became an astronomical centre; he compiled a star catalogue, and compiled tables of the Moon and planets. He was assassinated in 1449.

Van Maanen, Adriaan. 1884–1947. Dutch astronomer, who emigrated to America and joined the Mount Wilson staff in 1912. He specialized in stellar parallaxes and proper motions, and accomplished much valuable work, though his alleged detection of movements in the spiral arms of galaxies later proved to be erroneous. He also discovered the white dwarf still known as Van Maanen's Star.

Vehrenberg, Hans, 1910–91. German amateur astronomer, whose superb stellar photographs, contained in his *Atlas Stellarum,* are widely used.

Vogel, Hermann Carl. 1842–1907. German astronomer, born and educated in Leipzig. He went to Potsdam in 1874, and concentrated upon stellar spectroscopy, pioneering research into spectroscopic binaries. In 1883 he published the first catalogue of stellar spectra.

Walther, Bernard. 1430–1504. (Often spelled 'Walter'.) German amateur astronomer, who lived in Nürnberg; he financed Regiomontanus' equipment, and carried on the work when Regiomontanus died. He was a very accurate observer, whose measurements of star and planetary positions were of great value to later astronomers.

Wargentin, Pehr Vilhelm. 1717–83. Swedish astronomer, and Director of the Stockholm Observatory. His best work was in the preparation of accurate tables of Jupiter's satellites.

Webb, Thomas William, 1806–85. Vicar of Hardwicke in Herefordshire. He was an excellent observer, but is best remembered for his book *Celestial Objects for Common Telescopes.*

Wilkins, Hugh Percy. 1896–1960. Welsh amateur astronomer (by profession a Civil Servant) who concentrated upon lunar observation, and produced a 300 in map of the Moon. He was for many years Director of the Lunar Section of the British Astronomical Association.

Wolf, Maximilian Franz Joseph Cornelius. 1863–1932. (Better known as Max Wolf.) German astronomer, who was born and lived in Heidelberg. He studied comets, and discovered his periodical comet in 1884; he was the first to hunt for asteroids photographically. He also carried out research into dark nebulæ and discovered well over 1000.

Woolley, Richard van der Riet. 1906–. Pioneer astrophysicist, who graduated from Cambridge; he succeeded Spencer Jones as Astronomer Royal, and upon retirement returned to his native South Africa to become Director of the South African National Observatories.

Wright, Thomas. 1711–85. *b.* nr Durham. Trained as a clockmaker, though he afterwards taught mathematics. He is remembered for his book published in 1750, in which he suggested that the Galaxy is disk-shaped. He also believed Saturn's rings to be composed of small particles.

Xenophanes. *c.* 570–478 BC. Greek philosopher, *b.* Colophon. His astronomical theories sound strange today; an infinitely thick flat Earth, a new Sun each day, and celestial bodies which – apart from the Moon – were made of fire!

Zach, Franz Xavier von. 1754–1832. Hungarian baron, who became renowned as an amateur astronomer. He published tables of the Sun and Moon, and was one of the chief organizers of the 'Celestial Police' who banded together to hunt for the supposed planet between Mars and Jupiter. He became Director of the Seeberg Observatory at Gotha, and did much for international co-operation among astronomers.

Zwicky, Fritz. 1898–1974. Swiss astronomer; *b.* Bulgaria, but remained a Swiss citizen throughout his life. He graduated from Zürich, and in 1925 went to the California Institute of Technology, where he remained permanently, becoming Professor of Astrophysics from 1942 until his retirement in 1968. He became famous for his studies of galaxies and intergalactic matter; he predicted the existence of neutron stars (1934) and even black holes. He discovered many supernovæ in external galaxies, and masterminded a catalogue of compact galaxies. He was also active in the development of astronomical instrumentation, and was a pioneer worker with Schmidt telescopes. He received the Gold Medal of the Royal Astronomical Society in 1973.

GLOSSARY

Aberration of starlight. As light does not move infinitely fast, but at a rate of practically 300 000 km/s, and as the Earth is moving round the Sun at an average velocity of 28 km/s, the stars appear to be shifted slightly from their true positions. The best analogy is to picture a man walking along in a rainstorm, holding an umbrella. If he wants to keep himself dry, he will have to slant the umbrella forward; similarly, starlight seems to reach us 'from an angle'. Aberration may affect a star's position by up to 20·5 seconds of arc.

Absolute magnitude. The **apparent magnitude** that a star would have if it could be observed from a standard distance of 10 **parsecs** (32·6 light-years).

Achromatic object-glass. An **object-glass** which has been corrected so as to eliminate **chromatic aberration** or false colour as much as possible.

Aerolite. A **meteorite** whose main composition is stony.

Albedo. The reflecting power of a planet or other non-luminous body. The Moon is a poor reflector; its albedo is a mere 7% on average.

Altazimuth mounting for a telescope. A mounting on which the telescope may swing freely in any direction.

Altitude. The angular distance of a celestial body above the horizon.

Ångström unit. One hundred-millionth part of a centimetre.

Aphelion. The furthest distance of a planet or other body from the Sun in its orbit.

Apogee. The furthest point of the Moon from the Earth in its orbit.

Apparent magnitude. The apparent brightness of a celestial body. The lower the magnitude, the brighter the object: thus the sun is approximately –27, the Pole Star +2, and the faintest stars detectable by modern techniques around +29.

Areography. The official name for 'the geography of Mars'.

Asteroids. One of the names for the minor planet swarm.

Astrograph. An astronomical telescope designed specially for astronomical photography.

Astrolabe. An ancient instrument used to measure the altitudes of celestial bodies.

Astronomical unit. The mean distance between the Earth and the Sun. It is equal to 149 598 500 km.

Aurora. Auroræ are 'polar lights'; Aurora Borealis (northern) and Aurora Australis (southern). They occur in the Earth's upper atmosphere, and are caused by charged particles emitted by the Sun.

Azimuth. The bearing of an object in the sky, measured from north (0°) through east, south and west.

Bailly's beads. Brilliant points seen along the edge of the Moon just before and just after a total solar eclipse. They are caused by the sunlight shining through valleys at the Moon's limb.

Barycentre. The centre of gravity of the Earth-Moon system. Because the Earth is 81 times as massive as the Moon, the barycentre lies well inside the Earth's globe.

Binary star. A stellar system made up of two stars, genuinely associated, and moving round their common centre of gravity. The revolution periods range from millions of years for very widely-separated visual pairs down to less than half an hour for pairs in which the components are almost in contact with each other. With very close pairs, the components cannot be seen separately, but may be detected by spectroscopic methods.

Black hole. A region round a very small, very massive collapsed star from which not even light can escape.

BL Lacertæ objects. Variable objects which are powerful emitters of infra-red radiation, and are very luminous and remote. They are of the same nature as quasars.

Bode's law. A mathematical relationship linking the distances of the planets from the Sun. It may or may not be genuinely significant. Strictly speaking it should be called Titius' Law, since it was discovered by J. D. Titius some years before J. E. Bode popularized it in 1772.

Bolide. A brilliant exploding meteor.

Bolometer. An instrument used to measure small quantities of heat radiation.

Carbon stars. Red stars of spectral types R and N with unusually carbon-rich atmospheres.

Cassegrain reflector. A reflecting telescope in which the secondary mirror is convex; the light is passed back through a hole in the main mirror. Its main advantage is that it is more compact than the Newtonian reflector.

Celestial sphere. An imaginary sphere surrounding the Earth, whose centre is the same as that of the Earth's globe.

Cepheid. A short-period **variable star**, very regular in behaviour; the name comes from the prototype star, Delta Cephei. Cepheids are astronomically important because there is a definite law linking their variation periods with their real luminosities, so that their distances may be obtained by sheer observation.

Chromatic aberration. A defect in all lenses, due to the fact that light is a mixture of all wavelengths – and these wavelengths are refracted unequally, so that false colour is produced round a bright object such as a star. The fault may be reduced by making the lens a compound arrangement, using different kinds of glasses.

Chromosphere. That part of the Sun's atmosphere which lies above the bright surface or photosphere.

Circumpolar star. A star which never sets. For instance, Ursa Major (the Great Bear) is circumpolar as seen from England; Crux Australis (the Southern Cross) is circumpolar as seen from New Zealand.

Cluster variables. An obsolete name for the stars now known as RR Lyræ variables.

Cœlostat. An optical instrument making use of two mirrors, one of which is fixed, while the other is movable and is mounted parallel to the Earth's axis; as the Earth rotates, the light from the star (or other object being observed) is caught by the rotatable mirror and is reflected in a fixed direction on to the second mirror. The result is that the eyepiece of the instrument need not move at all.

Collapsar. The end product of a very massive star, which has collapsed and has surrounded itself with a **black hole**.

Colour index. The difference between a star's visual magnitude and its photographic magnitude. The redder the star, the greater the positive value of the colour index; bluish stars have negative colour indices. For stars of type Ao, colour index is zero.

Colures. Great circles on the celestial sphere.

Conjunction. (1) A planet is said to be in conjunction with a star, or with another planet, when the two bodies are apparently close together in the sky. (2) For the inferior planets, Mercury and Venus, inferior conjunction occurs when the planet is approximately between the Earth and the Sun; superior conjunction, when the planet is on the far side of the Sun and the three bodies are again lined up. Planets beyond the Earth's orbit can never come to inferior conjunction, for obvious reasons!

Corona. The outermost part of the Sun's atmosphere, made up of very tenuous gas. It is visible with the naked eye only during a total solar eclipse.

Coronagraph. A device used for studying the inner **corona** at times of non-eclipse.

Cosmic rays. High-velocity particles reaching the Earth from outer space. The heavier cosmic-ray particles are broken up when they enter the upper atmosphere.

Cosmogony. The study of the origin and evolution of the universe.

Cosmology. The study of the universe considered as a whole.

Counterglow. The English name for the sky-glow more generally called by its German name of the **Gegenschein**.

Culmination. The maximum altitude of a celestial body above the horizon.

Dawes' limit. The practical limit for the resolving power of a telescope; it is $4·56/d$, were d is the aperture of the telescope in inches.

Day, sidereal. The interval between successive meridian passages, or **culminations**, of the same star: 23 h 56 m 4 s·091.

Day, solar. The mean interval between successive meridian passages of the Sun: 24 h 3 m 56 s·555 of mean sidereal time. It is longer than the sidereal day because the Sun seems to move eastward against the stars at an average rate of approximately one degree per day.

Declination. The angular distance of a celestial body north or south of the celestial equator. It corresponds to latitude on the Earth.

Dewcap. An open tube fitted to the upper end of a refracting telescope. Its rôle is to prevent condensation upon the object-glass.

Dichotomy. The exact half-phase of the Moon or an **inferior planet**.

Diffraction grating. A device used for splitting up light; it consists of a polished metallic surface upon which thousands of parallel lines are ruled. It may be regarded as an alternative to the prism.

Direct motion. Movement of revolution or rotation in the same sense as that of the Earth.

Doppler effect. The apparent change in wavelength of the light from a luminous body which is in motion relative to the observer. With an approaching object, the wavelength is apparently shortened, and the spectral lines are shifted to the blue end of the spectral band; with a receding body there is a red shift, since the wavelength is apparently lengthened.

Double star. A star made up of two components – either genuinely associated (binary systems) or merely lined up by chance (optical pairs).

Driving clock. A mechanism for driving a telescope round at a rate which compensates for the axial rotation of the Earth, so that the object under observation remains fixed in the field of view.

Dwarf novæ. A term sometimes applied to the U Geminorum (or SS Cygni) variable stars.

Earthshine. The faint luminosity on the night side of the Moon, frequently seen when the Moon is in its crescent phase. It is due to light reflected on to the Moon from the Earth.

Eclipse, lunar. The passage of the Moon through the shadow cast by the Earth. Lunar eclipses may be either total or partial. At some eclipses, totality may last for approximately 1¾ hours, though most are shorter.

Eclipse, solar. The blotting-out of the Sun by the Moon, so that the Moon is then directly between the Earth and the Sun. Total eclipses can last for over 7 minutes under exceptionally favourable circumstances. In a partial eclipse, the Sun is incompletely covered. In an annular eclipse, exact alignment occurs when the Moon is in the far part of its orbit, and so appears smaller than the Sun; a ring of sunlight is left showing round the dark body of the Moon. Strictly speaking, a solar 'eclipse' is the **occultation** of the Sun by the Moon.

Eclipsing variable (or Eclipsing Binary). A **binary star** in which one component is regularly **occulted** by the other, so that the total light which we receive from the system is reduced. The prototype eclipsing variable is Algol (Beta Persei).

Ecliptic. The apparent yearly path of the Sun among the stars. It is more accurately defined as the projection of the Earth's orbit on to the celestial sphere.

Electron. Part of an atom; a fundamental particle carrying a negative electric charge.

Electron density. The number of 'free' (unattached) electrons in unit volume of space.

Elongation. The angular distance of a planet from the Sun, or of a satellite from its primary planet.

Ephemeris. A table showing the predicted positions of a celestial body such as a comet, asteroid or planet.

Epoch. A date chosen for reference purposes in quoting astronomical data.

Equator, celestial. The projection of the Earth's equator on to the **celestial sphere.**

Equatorial mounting for a telescope. A mounting in which the telescope is set up on an axis which is parallel with the axis of the Earth. This means that one movement only (east to west) will suffice to keep an object in the field of view.

Equinox. The equinoxes are the two points at which the **ecliptic** cuts the **celestial equator.** The vernal equinox or First Point of Aries now lies in the constellation of Pisces; the Sun crosses it about 21 March each year. The autumnal equinox is known as the First Point of Libra; the Sun reaches it about 22 September yearly.

Escape velocity. The minimum velocity which an object must have in order to escape from the surface of a planet, or other celestial body, without being given any extra impetus.

Evection. An inequality in the Moon's motion, due to slight changes in the shape of the lunar orbit.

Exosphere. The outermost part of the Earth's atmosphere.

Extinction. The apparent reduction in brightness of a star or planet when low down in the sky, so that more of its light is absorbed by the Earth's atmosphere. With a star 1° above the horizon, extinction amounts to 3 magnitudes.

Eyepiece (or Ocular). The lens, or combination of lenses, at the eye-end of a telescope. It is responsible for magnifying the image of the object under study. With a positive eyepiece (for instance, a Ramsden, Orthoscopic or Monocentric) the image plane lies between the eyepiece and the object-glass (or main mirror); with a negative eyepiece (such as a Huyghenian or Tolles) the image plane lies inside the eyepiece. A Barlow lens is concave, and is mounted in a short tube which may be placed between the eyepiece and the object-glass (or mirror). It increases the effective focal length of the telescope, thereby providing increased magnification.

Faculæ. Bright, temporary patches on the surface of the sun.

Filar micrometer. A device used for measuring very small angular distances as seen in the eyepiece of a telescope.

Finder. A small, wide-field telescope attached to a larger one, used for sighting purposes.

Fireball. A very brilliant **meteor.**

Flares, solar. Brilliant eruptions in the outer part of the Sun's atmosphere. Normally they can be detected only by spectroscopic means (or the equivalent), though a few have been seen in integrated light. They are made up of hydrogen, and emit charged particles which may later reach the Earth, producing magnetic storms and displays of aurorae. Flares are generally, though not always, associated with sunspot groups.

Flare stars. Faint Red Dwarf stars which show sudden, short-lived increases in brilliancy, due possibly to intense flares above their surfaces.

Flash spectrum. The sudden change-over from dark to bright lines in the Sun's spectrum, just before the onset of totality in a **solar eclipse.** The phenomenon is due to the fact that at this time the Moon has covered up the bright surface of the Sun, so that the chromosphere is shining 'on its own'.

Flocculi. Patches of the Sun's surface, observable with spectroscopic equipment. They are of two main kinds: bright (calcium) and dark (hydrogen).

Forbidden lines. Lines in the spectrum of a celestial body which do not appear under normal conditions, but may be seen in bodies where conditions are exceptional.

Fraunhofer lines. The dark absorption lines in the spectrum of the Sun.

Galaxies. Systems made up of stars, nebulæ, and interstellar matter. Many, though by no means all, are spiral in form.

Galaxy, the. The system of which our Sun is a member. It contains approximately 100 000 million stars, and is a rather loose spiral.

Gamma-rays. Radiation of extremely short wavelength.

Gauss. Unit of measurement of a magnetic field. (The Earth's field, at the surface, is on average about 0·3 to 0·6 gauss.)

Gegenschein. A faint sky-glow, opposite to the Sun and very difficult to observe. It is due to thinly-spread interplanetary material.

Geodesy. The study of the shape, size, mass and other characteristics of the Earth.

Gibbous phase. The phase of the Moon or planet when between half and full.

Globules. Small dark patches inside gaseous nebulæ. They may be embryo stars.

Gnomon. In a sundial, it is a pointer whose function is to cast the Sun's shadow on to the dial. The gnomon always points to the celestial pole.

Great circle. A circle on the surface of a sphere whose plane passes through the centre of that sphere.

Green Flash. Sudden, brief green light seen as the last segment of the Sun disappears below the horizon. It is purely an effect of the Earth's atmosphere. Venus has also been known to show a Green Flash.

Gregorian reflector. A telescope in which the secondary mirror is concave, and placed beyond the focus of the main mirror. The image obtained is erect. Few Gregorian telescopes are in use nowadays.

H.I and H.II regions. Clouds of hydrogen in the Galaxy. In H.I regions the hydrogen is neutral; in H.II regions the hydrogen is ionized, and the presence of hot stars will make the cloud shine as a nebula.

Halo, galactic. The spherical-shaped cloud of stars round the main part of the Galaxy.

Heliacal rising. The rising of a star or planet at the same time as the Sun, though the term is generally used to denote the time when the object is first detectable in the dawn sky.

Herschelian reflector. An obsolete type of telescope in which the main mirror is tilted, thus removing the need for a secondary mirror.

Hertzsprung-Russell diagram (usually known as the H-R Diagram). A diagram in which stars are plotted according to the spectral types and their **absolute magnitudes.**

Horizon. The great circle on the celestial sphere which is everywhere 90 degrees from the observer's zenith.

Hour angle (of a celestial object). The time which has elapsed since the object crossed the meridian. If RA = right ascension of the object and LST = the local sidereal time, then Hour Angle = LST − RA.

Hour circle. A great circle on the **celestial sphere**, passing through both celestial poles. The zero hour circle coincides with the observer's meridian.

Hubble's constant. The rate of increase in the recession of a galaxy with increased distance from the Earth.

Inferior planets. Mercury and Venus, whose distances from the Sun are less than that of the Earth.

Infra-red radiation. Radiation with wavelength longer than that of visible light (approximately 7500 Ångströms).

Interferometer, stellar. An instrument for measuring star diameters. The principle is based upon light-interference.

Ion. An atom which has lost or gained one or more of its planetary electrons, and so has respectively a positive or negative charge.

Ionosphere. The region of the Earth's atmosphere lying above the stratosphere.

Irradiation. The effect which makes very brilliant bodies appear larger than they really are.

Julian day. A count of the days, starting from 12 noon on 1 January 4713 BC. Thus 1 January 1977 was Julian Day 2 443 145. (The name 'Julian' has nothing to do with Julius Cæsar! The system was invented in 1582 by the mathematician Scaliger, who named it in honour of his father, Julius Scaliger.)

Kepler's laws of planetary motion. These were laid down by Johannes Kepler, from 1609 to 1618. They are: (1) The planets move in elliptical orbits, with the Sun occupying one focus. (2) The radius vector, or imaginary line joining the centre of the planet to the centre of the Sun, sweeps out equal areas in equal times. (3) With a planet, the square of the sidereal period is proportional to the cube of the mean distance from the Sun.

Kiloparsec. One thousand **parsecs** (3260 light-years).

Latitude, celestial. The angular distance of a celestial body from the nearest point on the ecliptic.

Libration. The apparent 'tilting' of the Moon as seen from Earth. There are three librations: latitudinal, longitudinal and diurnal. The overall effect is that at various times an observer on Earth can see a total of 59% of the total surface of the Moon, though, naturally, no more than 50% at any one moment!

Light-year. The distance travelled by light in one year: 9·4607 million million kilometres.

Local group. A group of more than two dozen galaxies, one member of which is our own **Galaxy.** The largest member of the Local Group is the Andromeda Spiral, M.31.

Longitude, celestial. The angular distance of a celestial body from the **vernal equinox**, measured in degrees eastward along the ecliptic.

Lunation. The interval between successive new moons: 29 d 12 h 44 m. (Also known as the Synodic Month.)

Magnetosphere. The region of the magnetic field of a planet or other body. In the Solar System, only the Earth, Jupiter, Mercury, Saturn, Uranus and Neptune are known to have detectable magnetospheres.

Main Sequence. A band along an **H-R Diagram**, including most normal stars except for the giants.

Maksutov telescope. An astronomical telescope involving both mirrors and lenses.

Mass. The quantity of matter that a body contains. It is not the same as 'weight'.

Mean sun. An imaginary sun travelling eastward along the celestial equator, at a speed equal to the average rate of the real Sun along the **ecliptic**.

Megaparsec. One million **parsecs**.

Meridian, celestial. The great circle on the **celestial sphere** which passes through the **zenith** and both celestial poles.

Meteor. A small particle, friable in nature and usually smaller than a sand grain, moving round the Sun, and visible only when it enters the upper atmosphere and is destroyed by friction. Meteors may be regarded as cometary débris.

Meteorite. A larger object, which may fall to the ground without being destroyed in the upper atmosphere. A meteorite is fundamentally different from a **meteor**. Meteorites are not associated with comets, but may be closely related to asteroids.

Micrometeorite. A very small particle of interplanetary material, too small to cause a luminous effect when it enters the Earth's upper atmosphere.

Micrometer. A measuring device, used together with a telescope to measure very small angular distances – such as the separations between the components of double stars.

Micron. One-thousandth of a millimetre. The usual symbol is µ.

Month. (1) Anomalistic: the interval between successive **perigee** passages of the Moon (27·55 days). (2) Sidereal: the revolution period of the Moon with reference to the stars (27·32 days). (3) Synodical: the interval between successive new moons (29·53 days). (4) Nodical or Draconitic: the interval between successive passages of the Moon through one of its nodes (27·21 days). (5) Tropical: the time taken for the Moon to return to the same celestial longitude (about 7 seconds shorter than the sidereal month).

Nadir. The point on the celestial sphere directly below the observer.

Nebula. A cloud of gas and dust in space. Galaxies were once known as 'spiral nebulæ' or 'extragalactic nebulæ'.

Neutrino. A fundamental particle with little or no mass and no electric charge.

Neutron. A fundamental particle with no electric charge, but a mass practically equal to that of a **proton**.

Neutron star. The remnant of a very massive star which has exploded as a **supernova**. Neutron stars send out rapidly-varying radio emissions, and are therefore called 'pulsars'. Only two (the Crab and Vela pulsars) have as yet been identified with optical objects.

Newtonian reflector. A reflecting telescope in which the light is collected by a main mirror, reflected on to a smaller flat mirror set at an angle of 45°, and thence to the side of the tube.

Nodes. The points at which the orbit of the Moon, a planet or a comet cuts the plane of the **ecliptic**; south to north (Ascending Node) or north to south (Descending Node).

Nova. A star which suddenly flares up to many times its normal brilliancy, remaining bright for a relatively short time before fading back to obscurity.

Nutation. A slow, slight 'nodding' of the Earth's axis, due to the gravitational pull of the Moon on the Earth's equatorial bulge.

Object-glass (or Objective). The main lens of a refracting telescope.

Objective prism. A small prism placed in front of the **object-glass** of a telescope. It produces small-scale spectra of the stars in the field of view.

Obliquity of the ecliptic. The angle between the **ecliptic** and the celestial equator: 23 deg 26 min 45 sec.

Occultation. The covering-up of one celestial body by another.

Opposition. The position of a planet when exactly opposite to the Sun in the sky; the Sun, the Earth and the planet are then approximately lined up.

Orbit. The path of a celestial object.

Orrery. A model showing the Sun and the planets, capable of being moved mechanically so that the planets move round the Sun at their correct relative speeds.

Parallax, trigonometrical. The apparent shift of an object when observed from two different directions.

Parsec. The distance at which a star would have a parallax of one second of arc: 3·26 **light-years**, 206 265 **astronomical units**, or 30·857 million million kilometres.

Penumbra. (1) The area of partial shadow to either side of the main cone of shadow cast by the Earth. (2) The lighter part of a sunspot.

Perigee. The position of the Moon in its orbit when closest to the Earth.

Perihelion. The position in orbit of a planet or other body when closest to the Sun.

Perturbations. The disturbances in the orbit of a celestial body produced by the gravitational effects of other bodies.

Phases. The apparent changes in shape of the Moon and the inferior planets from new to full. Mars may show a **gibbous phase**, but with the other planets there are no appreciable phases as seen from Earth.

Photoelectric cell. An electronic device; light falling on the cell produces an electric current, the strength of which depends upon the intensity of the light.

Photoelectric photometer. A **photoelectric cell** used together with a telescope for measuring the magnitudes of celestial bodies.

Photometer. An instrument used to measure the intensity of light from any particular source.

Photometry. The measurement of the intensity of light.

Photon. The smallest 'unit' of light.

Photosphere. The bright surface of the Sun.

Planetary nebula. A small, dense, hot star surrounded by a shell of gas. The name is ill-chosen, since planetary nebulæ are neither planets nor nebulæ!

Planetoid. An asteroid or minor planet.

Poles, celestial. The north and south points of the celestial sphere.

Populations, stellar. Two main types of star regions: I (in which the brightest stars are hot and white), and II (in which the brightest stars are old Red Giants).

Position angle. The apparent direction of one object with reference to another, measured from the north point of the main object through east, south and west.

Precession. The apparent slow movement of the celestial **poles**. This also means a shift of the celestial equator, and hence of the equinoxes; the vernal **equinox** moves by 50 sec of arc yearly, and has moved out of Aries into Pisces. Precession is due to the pull of the Moon and Sun on the Earth's equatorial bulge.

Prime meridian. The meridian on the Earth's surface which passes through the Airy Transit Circle at Greenwich Observatory. It is taken as longitude 0°.

Prominences. Masses of glowing gas rising from the surface of the Sun. They are made up chiefly of hydrogen.

Proper motion, stellar. The individual movement of a star on the celestial sphere.

Proton. A fundamental particle with a positive electric charge. The nucleus of the hydrogen atom is made up of a single proton.

Quadrant. An ancient astronomical instrument used for measuring the apparent positions of celestial bodies.

Quadrature. The position of the Moon or a planet when at right-angles to the Sun as seen from the Earth.

Quantum. The amount of energy possessed by one photon of light.

Quasar. A very remote, superluminous object. Quasars are now known to be the cores of very active galaxies, though the source of their energy is still a matter for debate.

Radial Velocity. The movement of a celestial body toward or away from the observer; positive if receding, negative if approaching.

Radiant. The point in the sky from which the meteors of any particular shower seem to radiate.

Regression of the nodes. The nodes of the Moon's orbit move slowly westward, making one complete revolution in 18·6 years. This regression is caused by the gravitational pull of the Sun.

Retardation. The difference in the time of moonrise between one night and the next.

Retrograde motion. Orbital or rotational movement in the sense opposite to that of the Earth's motion.

Reversing layer. The gaseous layer above the Sun's **photosphere**.

Right ascension. The angular distance of a celestial body from the vernal equinox, measured eastward. It is usually given in hours, minutes and seconds of time, so that the right ascension is the time-difference between the **culmination** of the vernal **equinox** and the culmination of the body.

Roche limit. The distance from the centre of a planet within which a second body would be broken up by the planet's gravitational pull. Note, however, that this would be the case only for a body which had no appreciable gravitational cohesion.

Saros. The period after which the Earth, Moon and Sun return to almost the same relative positions: 18 years 11·3 days. The saros may be used in eclipse prediction, since it is usual for an eclipse to be followed by a similar eclipse exactly one saros later.

Schmidt camera (or Schmidt telescope). An instrument which collects its light by means of a spherical mirror; a correcting plate is placed at the top of the tube. It is a purely photographic instrument.

Schwarzschild radius. The radius that a body must have if its **escape velocity** is to be equal to the velocity of light.

Scintillation. Twinkling of a star; it is due to the Earth's atmosphere. Planets may also show scintillation when low in the sky.

Secular acceleration of the Moon. The apparent speeding-up of the Moon in its orbit as measured over a long period of time, caused by the gradual slowing of the Earth's rotation (by 0·000 000 02 second per day).

Selenography. The study of the Moon's surface.

Sextant. An instrument used for measuring the altitude of a celestial object.

Seyfert galaxies. Galaxies with relatively small, bright nuclei and weak spiral arms. Some of them are strong radio emitters.

Sidereal period. The revolution period of a planet round the Sun, or of a satellite round its primary planet.

Sidereal time. The local time reckoned according to the apparent rotation of the **celestial sphere**. When the vernal **equinox** crosses the observer's **meridian**, the sidereal time is 0 hours.

Solar wind. A flow of atomic particles streaming out constantly from the Sun in all directions.

Solstices. The times when the Sun is at its maximum **declination** of approximately 23½ degrees; around 22 June (summer solstice, with the Sun in the northern hemisphere of the sky) and 22 December (winter solstice, Sun in the southern hemisphere).

Specific gravity. The density of any substance, taking that of water as 1. For instance, the Earth's specific gravity is 5·5, so that the Earth 'weighs' 5·5 times as much as an equal volume of water would do.

Spectroheliograph. An instrument used for photographing the Sun in the light of one particular wavelength only. The visual equivalent of the spectroheliograph is the spectrohelioscope.

Spectroscopic binary. A binary system whose components are too close together to be seen individually, but which can be studied by means of spectroscopic analysis.

Speculum. The main mirror of a reflecting telescope.

Spherical aberration. Blurring of a telescope image; it is due to the fact that the lens (or mirror) does not bring the light-rays falling on its edge and on its centre to exactly the same focal point.

Superior planets. All the planets lying beyond the orbit of the Earth in the Solar System (that is to say, all the principal planets apart from Mercury and Venus).

Supernova. A colossal stellar outburst, involving (1) the total destruction of the white dwarf member of a binary system, or (2) the collapse of a very massive star.

Synodic period. The interval between successive **oppositions** of a **superior planet.**

Syzygy. The position of the Moon in its orbit when new or full.

Tektites. Small, glassy objects found in a few localized parts of the Earth. Nobody is yet certain whether or not they come from the sky!

Terminator. The boundary between the day- and night-hemispheres of the Moon or a planet.

Thermocouple. An instrument used for measuring very small amounts of heat.

Transit. (1) The passage of a celestial body across the observer's meridian. (2) The projection of Mercury or Venus against the face of the Sun.

Transit instrument. A telescope mounted so that it can move only in **declination**; it is kept pointing to the meridian, and is used for timing the passages of stars across the meridian. Transit instruments were once the basis of all practical timekeeping. The Airy transit instrument at Greenwich is accepted as the zero for all longitudes on the Earth.

Troposphere. The lowest part of the Earth's atmosphere; its top lies at an average height of about 11 km. Above it lies the stratosphere; and above the stratosphere come the ionosphere and the exosphere.

Twilight. The state of illumination when the sun is below the horizon by less than 18 degrees.

Umbra. (1) The main cone of shadow cast by the Earth. (2) The darkest part of a sunspot.

Van Allen zones. Zones of charged particles around the Earth. There are two main zones; the outer (made up chiefly of **electrons**) and the inner (made up chiefly of **protons**).

Variable stars. Stars which change in brilliancy over short periods. They are of various types.

Variation. An inequality in the Moon's motion, due to the fact that the pull of the Sun on the Moon is not constant for all positions in the lunar orbit.

White dwarf. A very small, very dense star which has used up its nuclear energy, and is in a very late stage of its evolution.

Widmanstätten patterns. If an iron **meteorite** is cut, polished and then etched with acid, characteristic figures of the iron crystals appear; these are the Widmanstätten patterns.

Wolf-Rayet stars. Very hot, greenish-white stars which are surrounded by expanding gaseous envelopes. Their spectra show bright (emission) lines.

Year. (1) Sidereal: the period taken for the Earth to complete one journey round the Sun (365·26 days). (2) Tropical: the interval between successive passages of the Sun across the vernal equinox (365·24 days). (3) Anomalistic: the interval between successive perihelion passages of the Earth (365·26 days; slightly less then 5 minutes longer than the sidereal year, because the position of the perihelion point moves along the Earth's orbit by about 11 seconds of arc every year). (4) Calendar: the mean length of the year according to the Gregorian calendar (365·24 days, or 365 d 5 h 49 m 12 s).

Zenith. The observer's overhead point (altitude 90°).

Zenith distance. The angular distance of a celestial object from the **Zenith.**

Zodiac. A belt stretching round the sky, 8° to either side of the **ecliptic**, in which the Sun, Moon and principal planets are to be found at any time. (Pluto is the only planet which can leave the Zodiac, though many asteroids do so.)

Zodiacal Light. A cone of light rising from the horizon and stretching along the **ecliptic**; visible only when the Sun is a little way below the horizon. It is due to thinly spread interplanetary material near the main plane of the Solar System.

INDEX